INTRODUCTION TO RANDOM SIGNALS AND APPLIED KALMAN FILTERING

SECOND EDITION

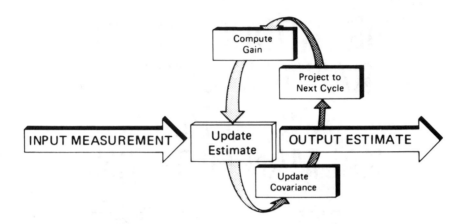

" . . . the Kalman filter represents the most widely applied and demonstrably useful result to emerge from the state variable approach of *modern control theory*."

Harold W. Sorenson
Kalman Filtering: Theory and Application, IEEE Press, 1985

INTRODUCTION TO RANDOM SIGNALS AND APPLIED KALMAN FILTERING

SECOND EDITION

ROBERT GROVER BROWN
Electrical Engineering Department
Iowa State University

PATRICK Y. C. HWANG
Rockwell International Corporation

JOHN WILEY & SONS, INC.
New York • Chichester • Brisbane • Toronto • Singapore

Acquisitions Editor	Steve Elliot
Production Manager	Linda Muriello
Designer	Laura Nicholls
Production Supervisor	Micheline Frederick
Manufacturing Manager	Lorraine Fumoso
Copy Editor	Deborah Herbert

Recognizing the importance of preserving what has been
written, it is a policy of John Wiley & Sons, Inc. to have books
of enduring value published in the United States printed on
acid-free paper, and we exert our best efforts to that end.

ISBN 0471-52573-1

Printed in the United States of America

Printed and bound by the Hamilton Printing Company.

10 9 8 7 6 5 4 3 2

Preface

This text is a second edition of *Introduction to Random Signal Analysis and Kalman Filtering*. It has a small, but important, change in title. We have added the word *Applied* to emphasize the application orientation of the book. The importance of Kalman filtering in engineering has certainly not diminished since the first edition was published in 1983. Applications of Kalman filtering, which began with navigation and trajectory determination, have now spread into many other diverse areas. This should come as no surprise, because Kalman filtering naturally fits the modern computer way of life.

We believe that the main addition to this edition is the inclusion of an educational Kalman-filter software package. This is supplied gratis on a diskette, and we consider it to be an integral part of the text. The programs on the diskette make it possible for the student (or independent-study reader) to work an important class of problems that cannot be done with paper-and-pencil methods. In this regard we have included a limited set of "starred" problems that require computer solutions. It is not that these problems are difficult. Quite to the contrary, the problems are conceptually easy, but they are labor intensive when done by hand. One can be spared this drudgery in this age of personal computers. Hence, readers should look for the starred problems. They illustrate conceptual ideas that can be explored through the aid of computers. The software package is described in detail in Appendix C.

In addition to the Kalman filtering software, this edition contains more examples of Kalman filter applications than were included in the first edition. Chapters 5, 6, and 9 have been revised extensively, and Chapter 10 is new. This chapter is devoted entirely to the application of Kalman filtering to the new navigation satellite system known as Global Positioning System (GPS).

The basic organization of the book is similar to the first edition. The first three chapters provide necessary background material in random signal analysis. The remaining chapters deal with minimum-mean-square-error filtering with emphasis on applications. The book contains enough material for a two-semester course for students who need an introduction to random signal theory as well as the material on optimal filtering. On the other hand, a one-semester course on Kalman filtering can easily be organized around Chapters 5 through 10, provided that the students have the necessary prerequisites in linear system theory and random signal analysis. Thus, the book can be used in a variety of ways. In addition to use in formal courses, it is also our hope that practicing engineers will find this book readable and useful for self-study.

We express our thanks to our many colleagues and students for their helpful suggestions in preparing this revised edition. Special thanks go to Janet Neumann for her comments on the new material in this edition. Also special thanks go to Dr. Larry Levy, Professor M. J. Caputi, and Dr. Dennis Duven for their careful review of the revised manuscript and software. We express our appreciation to Willard True for his suggestions on the Kalman filtering software package. Finally, we especially thank Lova (Mrs. Robert Grover Brown) for her patience and help in preparing the final manuscript.

Robert Grover Brown
Patrick Y. C. Hwang

Contents

1

Probability and Random Variables

1.1
RANDOM SIGNALS

Nearly everyone has some notion of random or noiselike signals. One has only to tune an ordinary AM radio away from a station, turn up the volume, and the result is static, or noise. If one were to look at a strip-chart recording of such a signal, it would appear to wander on aimlessly with no apparent order in its amplitude pattern, as shown in Fig. 1.1. Signals of this type cannot be described with explicit mathematical functions such as sine waves, step functions, and the like. Their description must be put in probabilistic terms. Early investigators recognized that random signals could be described loosely in terms of their spectral content, but a rigorous mathematical description of such signals was not formulated until the 1940s, most notably with the work of Wiener and Rice (1, 2).

Noise is usually unwanted. The additive noise in the radio signal disturbs our enjoyment of the music or interferes with our understanding of the spoken word; noise in an electronic navigation system induces position errors that can be disastrous in critical situations; noise in a digital data transmission system can cause bit errors with obvious undesirable consequences; and on and on. Any noise that corrupts the desired signal is bad; it is just a question of how bad! Even after the designer has done his best to eliminate all the obvious noise-producing mechanisms, there always seems to be some noise left over that must be suppressed with more subtle means, such as filtering. To do so effectively, one must understand noise in quantitative terms.

Probability plays a key role in the description of noiselike signals. Our treatment of this subject must necessarily be brief and directed toward the

1

Figure 1.1.   Typical noise waveform.

specific needs of subsequent chapters. The scope is thus limited in this regard. We make no apology for this, because many fine books have been written on probability in the broader sense. Our main objective here is the study of random signals and optimal filtering, and we wish to move on to this area as quickly as possible. First, though, we must at least review the bare essentials of probability with special emphasis on random variables.

1.2
INTUITIVE NOTION OF PROBABILITY

Most engineering and science students have had some acquaintance with the intuitive concepts of probability. Typically, with the intuitive approach we first consider all possible outcomes of a chance experiment as being equally likely, and then the probability of a particular event, say event A, is defined as

$$P(A) = \frac{\text{Possible outcomes favoring event } A}{\text{Total possible outcomes}} \qquad (1.2.1)$$

where we read $P(A)$ as "probability of event A." This concept is then expanded to include the relative-frequency-of-occurrence or statistical viewpoint of probability. With the relative-frequency concept, we imagine a large number of trials of some chance experiment and then define probability as the relative frequency of occurrence of the event in question. Considerations such as what is meant by "large" and the existence of limits are normally avoided in elementary treatments. This is for good reason. The idea of limit in a probabilistic sense is subtle.

Although the older intuitive notions of probability have limitations, they still play an important role in probability theory. The ratio-of-possible-events concept is a useful problem-solving tool in many instances. The relative-frequency concept is especially helpful in visualizing the statistical significance of the results of probability calculations. That is, it provides the necessary tie between the theory and the physical situation. Two examples that illustrate the usefulness of these intuitive notions of probability should now prove useful.

EXAMPLE 1.1

In straight poker, each player is dealt 5 cards face down from a deck of 52 playing cards. We pose two questions:

    (a) What is the probability of being dealt four of a kind, that is, four aces, four kings, and so forth?

    (b) What is the probability of being dealt a straight flush, that is, a continuous sequence of five cards in any suit?

Solution to Question (a)   This problem is relatively complicated if you think in terms of the sequence of chance events that can take place when the cards are dealt one at a time. Yet the problem is relatively easy when viewed in terms of the ratio of favorable to total number of outcomes. These are easily counted in this case. There are only 48 possible hands containing 4 aces; another 48 containing 4 kings; etc. Thus, there are $13 \cdot 48$ possible four-of-a-kind hands. The total number of possible poker hands of any kind is obtained from the combination formula for "52 things taken 5 at a time" (3). This is given by the binomial coefficient

$$\binom{52}{5} = \frac{52!}{5!(52-5)!} = \frac{52 \cdot 51 \cdot 50 \cdot 49 \cdot 48}{5 \cdot 4 \cdot 3 \cdot 2 \cdot 1} = 2{,}598{,}960 \qquad (1.2.2)$$

Therefore, the probability of being dealt four of a kind is

$$P(\text{Four of a kind}) = \frac{13 \cdot 48}{2{,}598{,}960} = \frac{624}{2{,}598{,}960} \approx .00024 \qquad (1.2.3)$$

Solution to Question (b)   Again, the direct itemization of favorable events is the simplest approach. The possible sequences in each of four suits are: $AKQJ10, KQJ109, \ldots, 5432A$. (*Note.* We allow the ace to be counted either high or low.) Thus, there are 10 possible straight flushes in each suit (including the royal flush of the suit) giving a total of 40 possible straight flushes. The probability of a straight flush is, then,

$$P(\text{Straight flush}) = \frac{40}{2{,}598{,}960} \approx .000015 \qquad (1.2.4)$$

We note in passing that in poker a straight flush wins over four of a kind; and, rightly so, since it is the rarer of the two hands. ∎

EXAMPLE 1.2

Craps is a popular gambling game played in casinos throughout the world (4). The player rolls two dice and plays against the house (i.e., the casino). If the first roll is 7 or 11, the player wins immediately; if it is 2, 3, or 12, the player loses immediately. If the first roll results in 4, 5, 6, 8, 9, or 10, the player continues to roll until either the same number appears, which constitutes a win, or a 7 appears, which results in the player losing. What is the player's probability of winning when throwing the dice?

This example was chosen to illustrate the shortcoming of the direct count-the-outcomes approach. In this case, one cannot enumerate all the possible outcomes. For example, if the player's first roll is a 4, the play continues until another outcome is reached. Presumably, the rolling could continue on ad infinitum without a 4 or 7 appearing, which is what is required to terminate the game. Thus, the direct enumeration approach fails in this situation. On the other hand, the relative-frequency-of-occurrence approach works quite well. Table 1.1 shows the relative frequency of occurrence of the various numbers on the first roll. The numbers in the column labeled "probability" were obtained by enumerating the 36 possible outcomes and allotting $\frac{1}{36}$ for each outcome that yields a sum corresponding to the number in the first column. For example, a 4 may be obtained with the combinations $(1,3)$, $(2,2)$, or $(1,3)$. For the cases where the game continues after the first throw, the subsequent probabilities were obtained simply by observing the *relative* frequency of occurrence of the numbers involved. For example, a 7 is twice as likely as a 4. Thus, the relative frequency of rolling a 7 before a 4 should be twice that of "4 before 7," and the respective probabilities are $\frac{2}{3}$ and $\frac{1}{3}$. The total probability of winning with a 4 on the first throw was reasoned as follows. A 4 appears on the first roll only $\frac{3}{36}$ of the time; and, of this fraction, only $\frac{1}{3}$ of the time will this result in an ultimate win. Thus, the relative frequency of winning via this route is the product of $\frac{3}{36} \cdot \frac{1}{3}$. Admittedly, this line of reasoning is quite intuitive, but that is the very nature of the relative-frequency-of-occurrence approach to probability.

For the benefit of those who like to gamble, it should be noted that craps is a very close game. The edge in favor of the house is only about $1\frac{1}{2}$ percent. (Also see Problem 1.8.)   ∎

1.3
AXIOMATIC PROBABILITY

It should be apparent that the intuitive concepts of probability have their limitations. The ratio-of-outcomes approach requires the equal-likelihood assumption for all outcomes. This may fit many situations, but often we wish to consider "unfair" chance situations as well as "fair" ones. Also, as demonstrated in Example 1.2, there are many problems for which all possible outcomes simply cannot be enumerated. The relative-frequency approach is intuitive by its very nature. Intuition should never be ignored; but, on the other hand, it can lead one astray in complex situations. For these reasons, the axiomatic formulation of probability theory is now almost universally favored among both applied and theoretical scholars in this area. As we would expect, axiomatic probability is compatible with the older, more heuristic probability theory.

Table 1.1 Probabilities in Craps

Number of First Throw	Probability	Result of First Throw	Subsequent Probabilities and Results	Relative Frequency of Winning with Various First Throws
2	$\frac{1}{36}$	Lose		0
3	$\frac{2}{36}$	Lose		0
4	$\frac{3}{36}$	Continue	$P(4 \text{ before } 7) = \frac{1}{3} \text{ (win)}$ $P(7 \text{ before } 4) = \frac{2}{3} \text{ (lose)}$	$\frac{3}{36} \cdot \frac{1}{3}$
5	$\frac{4}{36}$	Continue	$P(5 \text{ before } 7) = \frac{2}{5} \text{ (win)}$ $P(7 \text{ before } 5) = \frac{3}{5} \text{ (lose)}$	$\frac{4}{36} \cdot \frac{2}{5}$
6	$\frac{5}{36}$	Continue	$P(6 \text{ before } 7) = \frac{5}{11} \text{ (win)}$ $P(7 \text{ before } 6) = \frac{6}{11} \text{ (lose)}$	$\frac{5}{36} \cdot \frac{5}{11}$
7	$\frac{6}{36}$	Win		$\frac{6}{36}$
8	$\frac{5}{36}$	Continue	$P(8 \text{ before } 7) = \frac{5}{11} \text{ (win)}$ $P(7 \text{ before } 8) = \frac{6}{11} \text{ (lose)}$	$\frac{5}{36} \cdot \frac{5}{11}$
9	$\frac{4}{36}$	Continue	$P(9 \text{ before } 7) = \frac{2}{5} \text{ (win)}$ $P(7 \text{ before } 9) = \frac{3}{5} \text{ (lose)}$	$\frac{4}{36} \cdot \frac{2}{5}$
10	$\frac{3}{36}$	Continue	$P(10 \text{ before } 7) = \frac{1}{3} \text{ (win)}$ $P(7 \text{ before } 10) = \frac{2}{3} \text{ (lose)}$	$\frac{3}{36} \cdot \frac{1}{3}$
11	$\frac{2}{36}$	Win		$\frac{2}{36}$
12	$\frac{1}{36}$	Lose		0

Total probability of winning $= \frac{244}{495} \approx .4929$

Axiomatic probability begins with the concept of a *sample space*. We first imagine a conceptual chance experiment. The sample space is the set of all possible *outcomes* of this experiment. The individual outcomes are called *elements* or *points* in the sample space. We denote the sample space as S and its set of elements as $\{s_1, s_2, s_3, \ldots\}$. The number of points in the sample space may be finite, countably infinite, or simply infinite, depending on the experiment under consideration. A few examples of sample spaces should be helpful at this point.

EXAMPLE 1.3

The Experiment   Make a single draw from a deck of 52 playing cards. Since there are 52 possible outcomes, the sample space contains 52 discrete points. If we wished, we could enumerate them as Ace of Clubs, King of Clubs, Queen of Clubs, and so forth. Note that the points of the sample space in this case are "things," not numbers. ∎

EXAMPLE 1.4

The Experiment   Two fair dice are thrown and the number of dots on the top of each is observed. There are 36 discrete outcomes that can be enumerated as $(1,1),(1,2),(1,3),\ldots,(6,5),(6,6)$. The first number in parentheses identifies the number of dots on die 1 and the second is the number on die 2. Thus, 36 distinct 2-tuples describe the possible outcomes, and our sample space contains 36 points or elements. Note that the points in this sample space retain the identity of each individual die and the number of dots shown on its top face. ∎

EXAMPLE 1.5

The Experiment   Two fair dice are thrown and the sum of the number of dots is observed. In this experiment, we do not wish to retain the identity of the numbers on each die; only the sum is of interest. Therefore, it would be perfectly proper to say the possible outcomes of the experiment are $\{2,3,4,5,6,7,8,9,10,11,12\}$. Thus, the sample space would contain 11 discrete elements. From this and the preceding example, it can be seen that we have some discretion in how we define the sample space corresponding to a certain experiment. It depends to some extent on what we wish to observe. If certain details of the experiment are not of interest, they often may be suppressed with some resultant simplification. However, once we agree on what items are to be grouped together and called *outcomes*, the sample space must include all the defined outcomes; and, similarly, the result of an experiment must always yield one of the defined outcomes. ∎

EXAMPLE 1.6

The Experiment   A dart is thrown at a target and the location of the hit is observed. In this experiment we imagine that the random mechanisms affecting the throw are such that we get a continuous spread of data centered around the bull's-eye when the experiment is repeated over and over. In this case, even if we bound the hit locations within a certain region determined by reasonableness, we still cannot enumerate all possible hit locations. Thus, we have an infinite number of points in our sample space in this example. Even though we cannot enumerate the points one by one, they are, of course, identifiable in terms of either rectangular or polar coordinates. ∎

It should be noted that elements of a sample space must always be *mutually exclusive* or *disjoint*. On a given trial, the occurrence of one excludes the occurrence of another. There is no overlap of points in a sample space.

In axiomatic probability, the term event has special meaning and should not be used interchangeably with outcome. An *event* is a special subset of the sample space S. We usually wish to consider various events defined on a sample space, and they will be denoted with uppercase letters such as $A, B, C, \ldots$, or perhaps $A_1, A_2, \ldots$, etc. Also, we will have occasion to consider the set of operations of union, intersection, and complement of our defined events. Thus, we must be careful in our definition of events to make the set sufficiently complete such that these set operations also yield properly defined events. In discrete problems, this can always be done by defining the set of events under consideration to be all possible subsets of the sample space S. We will tacitly assume that the null set is a subset of every set, and that every set is a subset of itself.

One other comment about events is in order before proceeding to the basic axioms of probability. The event A is said to occur if *any* point in A occurs.

The three axioms of probability may now be stated. Let S be the sample space and A be any event defined on the sample space. The first two axioms are

$$\textit{Axiom 1}: \quad P(A) \geq 0 \tag{1.3.1}$$

$$\textit{Axiom 2}: \quad P(S) = 1 \tag{1.3.2}$$

Now, let $A_1, A_2, A_3, \ldots$ be mutually exclusive (disjoint) events defined on S. The sequence may be finite or countably infinite. The third axiom is then

$$\textit{Axiom 3}: \quad P(A_1 \cup A_2 \cup A_3 \cup \ldots)$$

$$= P(A_1) + P(A_2) + P(A_3) + \cdots \tag{1.3.3}$$

Axiom 1 simply says that the probability of an event cannot be negative. This certainly conforms to the relative-frequency-of-occurrence concept of probability. Axiom 2 says that the event S, which includes all possible outcomes, must have a probability of unity. It is sometimes called the certain event. The first two axioms are obviously necessary if axiomatic probability is to be compatible with the older relative-frequency probability theory. The third axiom is not quite so obvious, perhaps, and it simply must be assumed. In words, it says that when we have nonoverlapping (disjoint) events, the probability of the union of these events is the sum of the probabilities of the individual events. If this were not so, one could easily think of counterexamples that would not be compatible with the relative-frequency concept. This would be most undesirable.

We now recapitulate. There are three essential ingredients in the formal approach to probability. First, a sample space must be defined that includes all possible outcomes of our conceptual experiment. We have some discretion in what we call outcomes, but caution is in order here. The outcomes must be disjoint and all-inclusive such that $P(S) = 1$. Second, we must carefully define a set of events on the sample space, and the set must be closed such that the operations of union, intersection, and complement also yield events in the set. Finally, we must assign probabilities to all events in accordance with the basic axioms of probability. In physical problems, this assignment is chosen to be compatible with what we feel to be reasonable in terms of relative frequency of occurrence of the events. If the sample space S contains a finite number of elements, the probability assignment is usually made directly on the elements of S. They are, of course, elementary events themselves. This, along with Axiom 3, then indirectly assigns a probability to all other events defined on the sample space. However, if the sample space consists of an infinite "smear" of points, the probability assignment must be made on events and not on points in the sample space. This will be illustrated later in Example 1.8.

Once we have specified the sample space, the set of events, and the probabilities associated with the events, we have what is known as a *probability space*. This provides the theoretical structure for the formal solution of a wide variety of probability problems.

EXAMPLE 1.7

Consider a single throw of two dice, and let us say we are only interested in the sum of the dots that appear on the top faces. This chance situation fits many games that are played with dice. In this case, we will define our sample space to be

$$S = \{2, 3, 4, 5, 6, 7, 8, 9, 10, 11, 12\}$$

and it is seen to contain 11 discrete points. Next, we define the set of possible events to be all subsets of S, including the null set and S itself. Note that the elements of S are elementary events, and they are disjoint, as they should be. Also, $P(S) = 1$. Finally, we need to assign probabilities to the events. This could be done arbitrarily (within the constraints imposed by the axioms of probability), but in this case we want the results of our formal analysis to coincide with the relative-frequency approach. Therefore, we will assign probabilities to the elements in accordance with Table 1.2, which, in turn, indirectly specifies probabilities for all other events defined on S. We now have a properly defined probability space, and we can pose a variety of questions relative to the single throw of two dice.

Suppose we ask: What is the probability of throwing either a 7 or an 11? From Axiom 3, and noting that "7 or 11" is the equivalent of saying "7 $\cup$ 11," we have

$$P(7 \text{ or } 11) = \frac{6}{36} + \frac{2}{36} = \frac{2}{9} \tag{1.3.4}$$

Table 1.2 Probabilities for Two-Dice Example

Sum of Two Dice	Assigned Probability
2	$\frac{1}{36}$
3	$\frac{2}{36}$
4	$\frac{3}{36}$
5	$\frac{4}{36}$
6	$\frac{5}{36}$
7	$\frac{6}{36}$
8	$\frac{5}{36}$
9	$\frac{4}{36}$
10	$\frac{3}{36}$
11	$\frac{2}{36}$
12	$\frac{1}{36}$

Next, suppose we ask: What is the probability of not throwing 2, 3, or 12? This calls for the complement of event "2 or 3 or 12," which is the set $\{4, 5, 6, 7, 8, 9, 10, 11\}$. Recall that we say the event occurs if *any* element in the set occurs. Therefore, again using Axiom 3, we have

$$P(\text{Not throwing 2, 3, or 12}) = \frac{3 + 4 + 5 + 6 + 5 + 4 + 3 + 2}{36}$$

$$= \frac{8}{9} \tag{1.3.5}$$

Suppose we now pose the further question: What is the probability that two "4s" are thrown? In our definition of the sample space, we suppressed the identity of the individual dice, so this simply is not a proper question for the probability space, as defined. This example will be continued, but first we digress for a moment to consider intersection of events. ∎

In addition to the set operations of union and complementation, the operation of intersection is also useful in probability theory. The intersection of two events A and B is the event containing points that are common to both A and B. This is illustrated in Fig. 1.2 with what is sometimes called a Venn diagram. The points lying within the heavy contour comprise the union of A and B, denoted as $A \cup B$ or "A or B." The points within the shaded region are the event "A intersection B," which is denoted $A \cap B$, or sometimes just

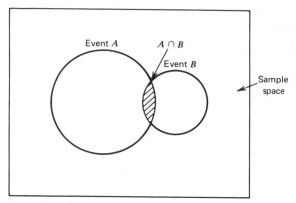

Figure 1.2. Venn diagram for two events A and B.

"A and B."* The following relationship should be apparent just from the geometry of the Venn diagram:

$$P(A \cup B) = P(A) + P(B) - P(A \cap B) \qquad (1.3.6)$$

The subtractive term in Eq. (1.3.6) is present because the probabilities in the overlapping region have been counted twice in the summation of $P(A)$ and $P(B)$.

The probability $P(A \cap B)$ is known as the *joint* probability of A and B and will be discussed further in Section 1.4. We digress for the moment, though, and look at two examples.

EXAMPLE 1.7 (continued)

We return to the two-dice example. Suppose we define event A as throwing a 4, 5, 6, or 7. Event B will be defined as throwing a 7, 8, 9, 10, or 11. We now pose two questions:

(a) What is the probability of event "A and B" (i.e., $A \cap B$)?
(b) What is the probability of event "A or B" (i.e., $A \cup B$)?

The answer to (a) is found by looking for the elements of the sample space that are common to both events A and B. Since the number 7 is the only common element,

$$P(A \cap B) = P(7) = \frac{1}{6}$$

---

* In many references, the notation for "A union B" is "$A + B$," and "A intersection B" is shortened to just "AB." We will be proceeding to the study of random variables shortly, and then the chance occurrences will be related to real numbers, not "things." Thus, in order to avoid confusion, we will stay with the more cumbersome notation of $\cup$ and $\cap$ for set operations, and reserve $X + Y$ and XY to mean the usual arithmetic operations on real variables.

The answer to (b) can be found either by itemizing the elements in $A \cup B$ or from Eq. (1.3.6). Using Eq. (1.3.6) leads to

$$P(A \cup B) = P(A) + P(B) - P(A \cap B)$$

$$= \frac{18}{36} + \frac{20}{36} - \frac{6}{36}$$

$$= \frac{8}{9}$$

This is easily verified using the direct itemization method. ∎

EXAMPLE 1.8

Let us reconsider the dart-throwing experiment. We might consider a simple game where we draw a circle of radius R on the wall. If the player's dart lands on or within the circle, the player wins; if the dart lands outside R, the player loses. We pose the simple question: What is the probability that the player wins on the single throw of a dart?

In this example, the sample space is all the points on the wall. (We assume that the player can at least hit the wall.) Thus, there are an infinite number of points in our sample space S.

In this game, we define only two events on the sample space—event W, player wins; and event L, player loses. This is a legitimate set of events because the set operations of union, intersection, and complement yield defined events of the set.

The assignment of probabilities in this case must be made directly on the two events rather than on the sample space, because the probability of hitting *exactly* any particular point on the wall is zero and this tells us nothing about events. To add a note of realism, we might speculate that we are devising a simple gambling game for the house, one that the patrons will enjoy and that will also be profitable to the house. Our observations indicate that the typical player throws about 10 percent of his or her darts within a circle of radius R_1. In addition, the hits are more or less uniformly spaced within the circle. It would then be appropriate to assign probabilities to the two events as

$$P(\text{Hit lies on or within } R_1) = .1$$

$$P(\text{Hit lies outside } R_1) \quad = .9$$

If the establishment were to offer 5 to 1 odds, this game might well please the players and at the same time produce revenue for the house.

Before leaving this example, it should be mentioned that the radius R may be treated as a parameter, and hence we have the structure for looking at a whole family of problems, not just a single one. To add variety to the game, the proprietor might occasionally wish to decrease the diameter of the circle and increase the odds. Clearly, if the hits within the circle are nearly uniformly distributed, reducing the area by a factor of 2 will reduce the relative frequency of occurrence by a similar factor. Yet the corresponding reduction in radius is only a factor of $\sqrt{2}$, which should make the game "look" attractive. The appropriate probability assignments to this case would be .05 for "win" and .95 for "lose." Presumably, the house could offer 10 to 1 odds and still net the same average return as received on the 5-to-1 game. ■

1.4
JOINT AND CONDITIONAL PROBABILITY

In complex situations it is often desirable to arrange the elements of the sample space in arrays. This provides an orderly way of grouping related points in the space and is especially useful when considering successive trials of similar experiments. Consider the following example.

EXAMPLE 1.9

The Experiment Draw a card from a deck of 52 playing cards. Then replace the card, reshuffle the deck, and draw a card a second time. This is known as sampling with replacement. The sample space for this experiment, when viewed as a whole, consists of all possible pairs of cards for the two draws. This amounts to $(52)^2$ or 2704 elements—clearly an unwieldy number of elements to keep track of without some systematic way of ordering things. Yet, in spite of its size, the sample space is quite manageable when viewed as an array as shown in Table 1.3. This is a two-dimensional array, but it should be obvious that the concept is easily extended to higher-order situations—conceptually at least. Had we specified three draws rather than two, we would have a three-dimensional array, and so forth. Thus the idea of an n-dimensional array associated with n trials of an experiment is an important concept. Note that summing out

Table 1.3 Joint Probabilities for Two Draws with Replacement

First Draw \ Second Draw	Ace of Spades	King of Spades	$\cdots$	Deuce of Clubs
Ace of spades	$\frac{1}{2704}$	$\frac{1}{2704}$		$\frac{1}{2704}$
King of spades	$\frac{1}{2704}$	$\frac{1}{2704}$		$\frac{1}{2704}$
$\vdots$	$\cdots$	$\cdots$		$\cdots$
(52 elements)		(Total of 2704 entries)		
Deuce of clubs	$\frac{1}{2704}$	$\frac{1}{2704}$		$\frac{1}{2704}$

Table 1.4 Array of Joint and Marginal Probabilities

A \\ B	Event B_1	Event B_2	$\cdots$	Event B_n	Marginal Probabilities
Event A_1	$P(A_1 \cap B_1)$	$P(A_1 \cap B_2)$	$\cdots$	$P(A_1 \cap B_n)$	$P(A_1)$
Event A_2	$P(A_2 \cap B_1)$	$P(A_2 \cap B_2)$	$\cdots$	$P(A_2 \cap B_n)$	$P(A_2)$
$\vdots$	$\vdots$	$\vdots$		$\vdots$	$\vdots$
Event A_m	$P(A_m \cap B_1)$	$P(A_m \cap B_2)$	$\cdots$	$P(A_m \cap B_n)$	$P(A_m)$
Marginal Probabilities	$P(B_1)$	$P(B_2)$	$\cdots$	$P(B_n)$	Sum = 1

the numbers along any particular row yields the probability of drawing that particular card on the first draw irrespective of the result of the second draw. This leads to the idea of marginal probability, which will now be considered in a more general setting. ∎

Let A and B loosely refer to "first" and "second" chance experiments. Time is not of the essence, so the experiments may be either successive or simultaneous as the situation dictates. Let $A_1, A_2, \ldots, A_m$ denote disjoint events associated with experiment A and, similarly, $B_1, B_2, \ldots, B_n$ denote disjoint events for experiment B. This leads to the joint-probability array shown in Table 1.4. Note first that n does not have to equal m, and therefore the array is not necessarily square. Also, the so-called *marginal probabilities* are shown in Table 1.4 to the right and bottom of the joint-probability array. These are the result of summing out rows or columns, as the case may be, and the term marginal arose because these probabilities are often written in the margins outside the $m \times n$ array. Clearly, summing out horizontally yields the probability of a particular event in the experiment A, irrespective of results of experiment B. Similarly, summing columns yields $P(B_1)$, $P(B_2)$, and so forth. Also, we tacitly assume that events $A_1, A_2, \ldots, A_m$ and $B_1, B_2, \ldots, B_n$ are all-inclusive as well as disjoint, so that the sum of the marginal probabilities, either vertically or horizontally, is unity.

Table 1.4 also contains information about the relative frequency of occurrence of various events in one set, given a particular event in the other set. For example, look at column 2, which lists $P(A_1 \cap B_2)$, $P(A_2 \cap B_2)$, ..., $P(A_m \cap B_2)$. Since no other entries in the table involve B_2, this list of numbers gives the relative distribution of events $A_1, A_2, \ldots, A_m$, given B_2 has occurred. However, the set of numbers appearing in column 2 is not a legitimate probability distribution because the sum is $P(B_2)$, not unity. So, imagine "renormalizing" all the entries in the column by dividing by $P(B_2)$. The new set of numbers is then $P(A_1 \cap B_2)/P(B_2)$, $P(A_2 \cap B_2)/P(B_2)$, ..., $P(A_m \cap B_2)/P(B_2)$, the sum is unity, and the relative distribution corresponds to the relative frequency of occurrence of $A_1, A_2, \ldots, A_m$, given B_2. This heuristic reasoning leads us to the formal definition of conditional probability. The *conditional probability* of A_i given B_j is defined as

$$P(A_i|B_j) = \frac{P(A_i \cap B_j)}{P(B_j)} \qquad (1.4.1)$$

Similarly, the conditional probability of B_j given A_i is

$$P(B_j|A_i) = \frac{P(B_j \cap A_i)}{P(A_i)} \qquad (1.4.2)$$

It is tacitly assumed in the above equations that $P(B_j)$ and $P(A_i)$ are not zero. Otherwise, conditional probability is not defined. It should also be emphasized that conditional probability is a *defined* concept and is not derived from other concepts. The discussion leading up to Eqs. (1.4.1) and (1.4.2) was presented to give an intuitive rationale for the definition and was not intended as a proof.

A useful relationship is obtained when Eqs. (1.4.1) and (1.4.2) are combined. Each equation may be solved for the probability of A_i intersection B_j and the results equated. This leads to *Bayes rule* (or *Bayes theorem*):

$$P(A_i|B_j) = \frac{P(B_j|A_i)P(A_i)}{P(B_j)} \qquad (1.4.3)$$

This relationship is useful in reversing the conditioning of events. Note that the joint probability array $P(A_i \cap B_j)$ contains all the necessary information for computing *all* marginal and conditional probabilities. Conversely, if you know $P(A_i|B_j)$ and $P(B_j)$ [or $P(B_j|A_i)$ and $P(A_i)$], there is sufficient information to find $P(A_i \cap B_j)$.

EXAMPLE 1.10

For variety, we now consider an urn problem. The urn contains two red and two black balls. Two balls are drawn sequentially from the urn without replacement.

(a) What is the array of joint probabilities for the first and second draws?
(b) What is the conditional probability that the second draw is red, given the first draw is red?

To obtain the joint probability table, we first define a sample space consisting of all possible outcomes, including the identity of the individual balls. The four balls will be referred to as Red 1, Red 2, Black 1, and Black 2. The joint probability array for the first and second draws is given in Table 1.5. Note the effect of the "without replacement" statement. It gives rise to zeros along the major diagonal, because drawing Red 1 on the first draw precludes Red 1

Table 1.5 Joint Probability Table for Four-Ball Urn Example

First Draw \ Second Draw	Red 1	Red 2	Black 1	Black 2
Red 1	0	$\frac{1}{12}$	$\frac{1}{12}$	$\frac{1}{12}$
Red 2	$\frac{1}{12}$	0	$\frac{1}{12}$	$\frac{1}{12}$
Black 1	$\frac{1}{12}$	$\frac{1}{12}$	0	$\frac{1}{12}$
Black 2	$\frac{1}{12}$	$\frac{1}{12}$	$\frac{1}{12}$	0

being drawn again on the second draw, and so forth. In effect, there are really only 12 nontrivial outcomes for this experiment, and we assume they are all equally likely.

In the original problem, there was no mention of retaining the individuality of the two red and two black balls—only "red" and "black" were specified. Therefore, we can consolidate outcomes in accordance with the partitioning shown by dashed lines in Table 1.5. This leads to the two-by-two array shown in Table 1.6. This, then, is the answer to part (a).

For the conditional probability we will use the basic definition given by Eq. (1.4.2). Writing this out explicitly for the conditional situation posed in question (b), we have

$$P(\text{Second draw red}|\text{First draw red})$$

$$= \frac{P(\text{First draw red and second draw red})}{P(\text{First draw red})} \tag{1.4.4}$$

The numerator of Eq. (1.4.4) is the upper-left entry in Table 1.6. The denominator is the marginal probability obtained by summing the elements of the first

Table 1.6 Joint Probability Table Reduced to Two-by-Two Array

First Draw \ Second Draw	Red	Black
Red	$\frac{1}{6}$	$\frac{1}{3}$
Black	$\frac{1}{3}$	$\frac{1}{6}$

row in Table 1.6. This yields

$$P(\text{Second draw red} | \text{First draw red}) = \frac{\frac{1}{6}}{\frac{1}{2}} = \frac{1}{3} \qquad (1.4.5)$$

This is the solution to part (b). Note this checks with the result one would obtain by considering the three balls that remain in the urn after a red one is withdrawn. ∎

1.5
INDEPENDENCE

In qualitative terms, two events are said to be independent if the occurrence of one does not affect the likelihood of the other. If we toss two coins simultaneously, we would not expect the outcome of one of the coins to affect the other. Similarly, if we draw a card from a deck of 52 playing cards, then replace it, reshuffle, and draw a second time, we would not expect the second outcome to be affected by the first. However, if we draw the second card without replacing the first, it is a much different matter. For example, the probability of drawing an ace on the second draw with replacement is $\frac{4}{52}$. However, if we draw an ace on the first draw and do not replace it, the probability of getting an ace on the second draw is only $\frac{3}{51}$. In the "without replacement" experiment, the outcome of the first draw certainly affects the chances on the second draw, so the two events are not independent.

Formally, events A and B are said to be *independent* if

$$P(A \cap B) = P(A)P(B) \qquad (1.5.1)$$

Also, it should be evident from Eq. (1.5.1) and the defining equations for conditional probability, Eqs. (1.4.1) and (1.4.2), that if A and B are independent

$$\left. \begin{array}{l} P(A|B) = P(A) \\ P(B|A) = P(B) \end{array} \right\} \quad \begin{array}{l} \text{For } A \text{ and } B \\ \text{independent only} \end{array} \qquad (1.5.2)$$

We might also note that the defining equation for independence, Eq. (1.5.1), usually provides the simplest test for independence. This is illustrated with two examples.

EXAMPLE 1.11

The joint probability array for the simultaneous toss of two coins is given in Table 1.7. The marginal probabilities are also shown in the "margins" with their significance stated in words in parentheses. Note that each of the four joint

Table 1.7 Joint and Marginal Probabilities for Toss of Two Coins

First Coin \ Second Coin	Heads	Tails		
Heads	$\frac{1}{4}$	$\frac{1}{4}$	$\frac{1}{2}$	(Prob. first coin is heads)
Tails	$\frac{1}{4}$	$\frac{1}{4}$	$\frac{1}{2}$	(Prob. first coin is tails)
	$\frac{1}{2}$ (Prob. second coin is heads)	$\frac{1}{2}$ (Prob. second coin is heads)		

probabilities of the array may be written as the product of their respective marginal probabilities. Thus, all events are independent in this case. ∎

EXAMPLE 1.12

Let us reconsider the urn experiment described in Example 1.10, Section 1.4. Recall that the two balls were withdrawn sequentially *without* replacement, and that this led to the joint probability array shown in Table 1.8. The marginal probabilities are also included, just as in the previous example. However, in this case none of the joint probabilities can be written as the product of the respective marginal probabilities. Thus, all event pairs are dependent. ∎

To recapitulate, we say events A and B are independent if their joint probability can be written as the product of the individual total probabilities, $P(A)$ and $P(B)$. Otherwise, they are said to be dependent.

Table 1.8 Joint and Marginal Probabilities for Urn Example

First Draw \ Second Draw	Red	Black	
Red	$\frac{1}{6}$	$\frac{1}{3}$	$\frac{1}{2}$
Black	$\frac{1}{3}$	$\frac{1}{6}$	$\frac{1}{2}$
	$\frac{1}{2}$	$\frac{1}{2}$	

1.6
RANDOM VARIABLES

In the study of noiselike signals, we are nearly always dealing with physical quantities such as voltage, torque, distance, and so forth, which can be measured in physical units. In these cases, the chance occurrences are related to real numbers, not just "things" like heads or tails, black balls or red balls, and the like. This brings us to the notion of a random variable. Let us say we have a conceptual experiment for which we have defined a suitable sample space, an appropriate set of events, and a probability assignment for the set of events. A *random variable* is simply a function that maps every point in the same space (things) on to the real line (numbers). A simple example of this mapping is shown in Fig. 1.3. Note that each face of the die is embossed with a pattern of dots, not a number. The assignment of numbers is our own doing and could be almost anything. In Fig. 1.3 we just happened to choose the most common numerical assignment, namely the sum of the number of dots, but this was not necessary.

Presumably, in our probability space, probabilities have been assigned to the events in the sample space in accordance with the basic axioms of probability. Associated with each event in the original sample space (things) there will be a corresponding event in the random-variable space (numbers). These will be called *equivalent events,* and it is only natural that we should assign probabilities to the random-variable events in the same manner as for the original sample-space events. Stated formally, we have

$$P(\text{equivalent event on the real line})$$

$$= P(\text{corresponding event in the original sample space}) \quad (1.6.1)$$

Two more examples will illustrate the concept of a random variable further.

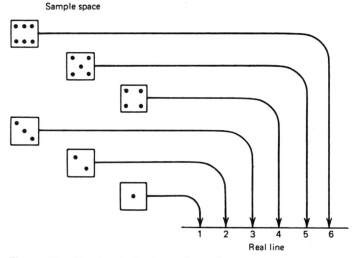

Figure 1.3.   Mapping for the throw of one die.

EXAMPLE 1.13

The mapping that defines a random variable must fit the chance situation at hand if it is to be useful. Sometimes this leads to unusual but perfectly legitimate functional relationships. For example, in the game of pitch, a portion of the scoring is done by summing the card values of the cards each player takes in tricks during the course of play. The card values, by arbitrary rules of the game, are as follows:

Card of Any Suit	Card Value
2 through 9	0
10	10
Jack	1
Queen	2
King	3
Ace	4

Thus, in exploring your chances relative to this aspect of the game, it would be appropriate to map the 52 points in the sample space (for a single card) into real numbers in accordance with the above table. This is also shown in Fig. 1.4. Note that multiple points in the sample space map into the *same* number on the real line. This is perfectly legitimate. The mapping must not be ambiguous in

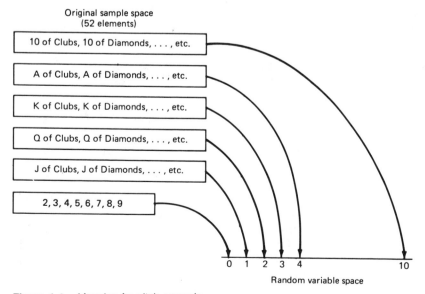

Figure 1.4. Mapping for pitch example.

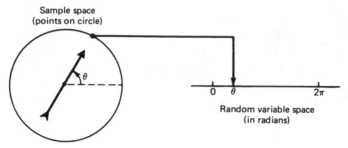

Figure 1.5. Mapping for spin-the-pointer example.

going from the sample space to the real line; however, it may be ambiguous going the other way. That is, the mapping need not be one-to-one. ∎

EXAMPLE 1.14 _____

In many games, the player spins a pointer that is mounted on a circular card of some sort and is free to spin about its center. This is depicted in Fig. 1.5 and the circular card is intentionally shown without any markings along its edge. Suppose we define the outcome of an experiment as the location on the periphery of the card at which the pointer stops. The sample space then consists of an infinite number of points along a circle. For analysis purposes, we might wish to identify each point in the sample space in terms of an angular coordinate measured in radians. The functional mapping that maps all points on a circle to corresponding points on the real line between 0 and 2π would then define an appropriate random variable. ∎

1.7
PROBABILITY DISTRIBUTION
AND DENSITY FUNCTIONS

When the sample space consists of a finite number of elements, the probability assignment can be made directly on the sample-space elements in accordance with what we feel to be reasonable in terms of likelihood of occurrence. This then defines probabilities for all events defined on the sample space. These probabilities, in turn, transfer directly to equivalent events in the random-variable space. The allowable realizations (i.e., real numbers) in the random-variable space are elementary equivalent events themselves, so the result is a probability associated with each allowable realization in the random-variable space. The sum of these probabilities must be unity, just as in the original sample space, but the distribution need not be the same. A continuation of Example 1.13, Section 1.6, will illustrate this.

EXAMPLE 1.15

The mapping from the sample space to the real line for the pitch card game was shown in Fig. 1.4. Let us assign equal probabilities for all elements in the original sample space. The probabilities for the allowable realizations in the random-variable space would then be:

Random Variable Realization	Probability
0	$\frac{32}{52}$
1	$\frac{4}{52}$
2	$\frac{4}{52}$
3	$\frac{4}{52}$
4	$\frac{4}{52}$
10	$\frac{4}{52}$

Note that the probabilities are not distributed uniformly in the random-variable space. The end result of the mapping is a set of real numbers representing the possible realizations of the random variable and a corresponding set of probabilities that sum to unity. Once this correspondence has been established, the original sample space is usually ignored. ∎

The random variable of Example 1.15 is an example of a *discrete* random variable in that its allowable realizations are discrete (i.e., countable) rather than continuous. The associated discrete set of probabilities is sometimes referred to as the *probability mass distribution* or simply *probability distribution*.

We also have occasion to work with continuous random variables. As a matter of fact, the usual electronic noise that is encountered in a wide variety of applications is of this type, that is, the voltage (or current) may assume a continuous range of values. The corresponding sample space then also contains an infinite number of points, so we cannot assign probabilities directly on the points of the sample space; this must be done on the defined events. We will continue the spin-the-pointer example of Section 1.6 (Example 1.14) to illustrate how this is done.

Let X denote a continuous random variable corresponding to the angular position of the pointer after it stops. Presumably, this could be any angle between 0 and 2π radians; therefore, the probability of *exactly* any particular position is zero. Thus, we assign a probability to the event that the pointer stops within a certain angular range, say between 0 and θ radians. If all positions are equally likely, it is reasonable to assign probabilities as follows:

$$P(X \leq \theta) = \begin{cases} 0, & \theta < 0 \\ \dfrac{1}{2\pi}\theta, & 0 \leq \theta \leq 2\pi \\ 1, & \theta > 2\pi \end{cases} \qquad (1.7.1)$$

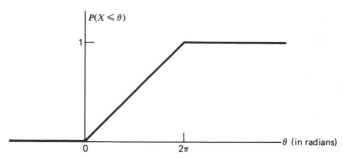

Figure 1.6.   Probability distribution function for pointer example.

Note that the probability assignment is a function of the parameter θ and the function is sketched in Fig. 1.6. The linear portion of the function between 0 and 2π is due to the "equally likely" assumption.

The function sketched in Fig. 1.6 is known as a *cumulative distribution function* (or just probability distribution function), and it simply describes the probability assignment as it reflects onto equivalent events in the random variable space. Specifically, the probability distribution function associated with the random variable X is defined as

$$F_X(\theta) = P(X \le \theta) \qquad (1.7.2)$$

where θ is a parameter representing a realization of X.

Just as in the discrete case, once this distribution is established in the random-variable space, the original sample space is usually ignored. It should be clear from the definition that a probability distribution function always has the following properties:

1. $F_X(\theta) \to 0, \qquad$ as $\theta \to -\infty$.
2. $F_X(\theta) \to 1, \qquad$ as $\theta \to \infty$.
3. $F_X(\theta)$ is a nondecreasing function of θ.

The information contained in the distribution function (e.g., Fig. 1.6) may also be presented in derivative form. Specifically, let $f_X(\theta)$ be defined as

$$f_X(\theta) = \frac{d}{d\theta} F_X(\theta) \qquad (1.7.3)$$

The function $f_X(\theta)$ is known as the *probability density function* associated with the random variable X. The density function for the pointer example is shown in Fig. 1.7. From properties 1, 2, and 3 just cited for the distribution function, it

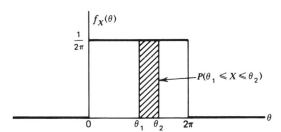

Figure 1.7. Probability density function for pointer example.

should be apparent that the density function has the following properties:

1. $f_X(\theta)$ is a nonnegative function.
2. $\int_{-\infty}^{\infty} f_X(\theta)\, d\theta = 1$.

It should also be apparent from elementary calculus that the shaded area shown in Fig. 1.7 represents the probability that X lies between θ_1 and θ_2. If θ_1 and θ_2 are separated by an infinitesimal amount, $\Delta\theta$, the area is approximately $f_X(\theta_1)\,\Delta\theta$, and thus we have the term probability *density*.

The probability density and distribution functions are alternative ways of describing the relative distribution of the random variable. Both functions are useful in random-variable analysis, and you should always keep in mind the derivative/integral relationship between the two. As a matter of notation, we will normally use an uppercase symbol for the distribution function and the corresponding lowercase symbol for the density function. The subscript in each case indicates the random variable being considered. The argument of the function is a dummy variable and may be almost anything.

1.8
EXPECTATION, AVERAGES,
AND CHARACTERISTIC FUNCTION

The idea of averaging is so commonplace that it may not seem worthy of elaboration. Yet there are subtleties, especially as averaging relates to probability. Thus we need to formalize the notion of average.

Perhaps the first thing to note is that we always average over numbers and not "things." There is no such thing as the average of apples and oranges. When we compute a student's average grades, we do not average over A, B, C, and so on; instead, we average over numerical equivalents that have been arbitrarily assigned to each grade. Also, the quantities being averaged may or may not be governed by chance. In either case, random or deterministic, the average is just the sum of the numbers divided by the number of quantities being averaged. In the random case, the *sample average* or *sample mean* of a random variable X is defined as

$$\bar{X} = \frac{X_1 + X_2 + \cdots + X_N}{N} \tag{1.8.1}$$

where the bar over X indicates average, and $X_1, X_2, \ldots,$ are sample realizations obtained from repeated trials of the chance situation under consideration. We will use the terms *average* and *mean* interchangeably, and the adjective *sample* serves as a reminder that we are averaging over a finite number of trials as in Eq. (1.8.1).

In the study of random variables we also like to consider the conceptual average that would occur for an infinite number of trials. This idea is basic to the relative-frequency concept of probability. This hypothetical average is called *expected value* and is aptly named; it simply refers to what one would "expect" in the typical statistical situation. Beginning with discrete probability, imagine a random variable whose n possible realizations are $x_1, x_2, \ldots, x_n$. The corresponding probabilities are $p_1, p_2, \ldots, p_n$. If we have N trials, where N is large, we would expect approximately $p_1 N$ x_1's, $p_2 N$ x_2's, etc. Thus the sample average would be

$$\bar{X} \approx \frac{(p_1 N) x_1 + (p_2 N) x_2 + \cdots + (p_n N) x_n}{N} \tag{1.8.2}$$

This suggests the following definition for expected value for the discrete probability case:

$$\text{Expected value of } X = E(X) = \sum_{i=1}^{n} p_i x_i \tag{1.8.3}$$

where n is the number of allowable values of the random variable X.

Similarly, for a continuous random variable X, we have

$$\text{Expected value of } X = E(X) = \int_{-\infty}^{\infty} x f_X(x)\, dx \tag{1.8.4}$$

It should be mentioned that Eqs. (1.8.3) and (1.8.4) are definitions, and the arguments leading up to these definitions were presented to give a sensible rationale for the definitions, and not as a proof. We can use these same arguments for defining the expectation of a function of X, as well as for X.

Thus we have
Discrete case:

$$E(g(X)) = \sum_i^n p_i g(x_i) \qquad (1.8.5)$$

Continuous case:

$$E(g(X)) = \int_{-\infty}^{\infty} g(x) f_X(x)\, dx \qquad (1.8.6)$$

As an example of the use of Eq. (1.8.6), let the function $g(X)$ be X^k. Equation (1.8.6) [or its discrete counterpart Eq. (1.8.5)] then provides an expression for the kth *moment* of X, that is,

$$E(X^k) = \int_{-\infty}^{\infty} x^k f_X(x)\, dx \qquad (1.8.7)$$

The second moment of X is of special interest, and it is given by

$$E(X^2) = \int_{-\infty}^{\infty} x^2 f_X(x)\, dx \qquad (1.8.8)$$

The first moment is, of course, just the expectation of X, which is also known as the *mean* or *average value* of X. Note that when the term *sample* is omitted, we tacitly assume that we are referring to the hypothetical infinite-sample average.

We also have occasion to look at the second moment of X "about the mean." This quantity is called the *variance* of X and is defined as

$$\text{Variance of } X = E\big[(X - E(X))^2\big] \qquad (1.8.9)$$

In a qualitative sense, the variance of X is a measure of the dispersion of X about its mean. Of course, if the mean is zero, the variance is identical to the second moment.

The expression for variance given by Eq. (1.8.9) can be reduced to a more convenient computational form by expanding the quantity within the brackets and then noting that the expectation of the sum is the sum of the expectations. This leads to

$$\text{Var } X = E\big[X^2 - 2X \cdot E(X) + (E(X))^2\big]$$

$$= E(X^2) - (E(X))^2 \qquad (1.8.10)$$

The square root of the variance is also of interest, and it has been given the name *standard deviation*, that is,

$$\text{Standard deviation of } X = \sqrt{\text{variance of } X} \qquad (1.8.11)$$

EXAMPLE 1.16

Let X be uniformly distributed in the interval $(0, 2\pi)$. This leads to the probability density function (see Example 1.14).

$$f_X(x) = \begin{cases} \dfrac{1}{2\pi}, & 0 \le x < 2\pi \\ 0, & \text{elsewhere} \end{cases}$$

Find the mean, variance, and standard deviation of X.

The mean is just the expectation of X and is given by Eq. (1.8.4).

$$\text{Mean of } X = E(X) = \int_0^{2\pi} x \cdot \frac{1}{2\pi} \, dx$$

$$= \frac{1}{2\pi} \cdot \frac{x^2}{2}\Big|_0^{2\pi} = \pi \qquad (1.8.12)$$

Now that we have computed the mean, we are in a position to find the variance from Eq. (1.8.10).

$$\text{Var } X = \int_0^{2\pi} x^2 \frac{1}{2\pi} \, dx - \pi^2$$

$$= \frac{4}{3}\pi^2 - \pi^2$$

$$= \frac{1}{3}\pi^2 \qquad (1.8.13)$$

The standard deviation is now just the square root of the variance:

$$\text{Standard deviation of } X = \sqrt{\text{Var } X}$$

$$= \sqrt{\frac{1}{3}\pi^2} = \frac{1}{\sqrt{3}}\pi \qquad (1.8.14)$$

■

The *characteristic function* associated with the random variable X is defined as

$$\psi_X(\omega) = \int_{-\infty}^{\infty} f_X(x)e^{j\omega x} \, dx \qquad (1.8.15)$$

It can be seen that $\psi_X(\omega)$ is just the Fourier transform of the probability density function with a reversal of sign on ω. Thus the theorems (and tables) of Fourier transform theory can be used to advantage in evaluating characteristic functions and their inverses.

The characteristic function is especially useful in evaluating the moments of X. This can be demonstrated as follows. The moments of X may be written as

$$E(X) = \int_{-\infty}^{\infty} x f_X(x) \, dx \qquad (1.8.16)$$

$$E(X^2) = \int_{-\infty}^{\infty} x^2 f_X(x) \, dx$$

$$\vdots$$

$$\text{etc.} \qquad (1.8.17)$$

Now consider the derivatives of $\psi_X(\omega)$ evaluated at $\omega = 0$.

$$\left[\frac{d\psi_X}{d\omega}\right]_{\omega=0} = \left[\int_{-\infty}^{\infty} jx f_X(x)e^{j\omega x} \, dx\right]_{\omega=0} = \int_{-\infty}^{\infty} jx f_X(x) \, dx \qquad (1.8.18)$$

$$\left[\frac{d^2\psi_X}{d\omega^2}\right]_{\omega=0} = \left[\int_{-\infty}^{\infty} (jx)^2 f_X(x)e^{j\omega x} \, dx\right]_{\omega=0} = \int_{-\infty}^{\infty} j^2 x^2 f_X(x) \, dx$$

$$\vdots$$

$$\text{etc.} \qquad (1.8.19)$$

It can be seen that

$$E(X) = \frac{1}{j}\left[\frac{d\psi_X}{d\omega}\right]_{\omega=0} \qquad (1.8.20)$$

$$E(X^2) = \frac{1}{j^2}\left[\frac{d^2\psi_X}{d\omega^2}\right]_{\omega=0}$$

$$\vdots$$

$$\text{etc.} \qquad (1.8.21)$$

Thus, with the help of a table of Fourier transforms, you can often evaluate the moments without performing the integrations indicated in their definitions. [See Problems 1.24(b) and 2.31 for applications of the characteristic function.]

1.9
NORMAL OR GAUSSIAN RANDOM VARIABLES

The random variable X is called *normal* or *Gaussian* if its probability density function is

$$f_X(x) = \frac{1}{\sqrt{2\pi}\,\sigma} \exp\left[-\frac{1}{2\sigma^2}(x - m_X)^2 \right] \tag{1.9.1}$$

Note that this density function contains two parameters m_X and σ^2. These are the random variable's mean and variance. That is, for the f_X specified by Eq. (1.9.1),

$$\int_{-\infty}^{\infty} x f_X(x)\, dx = m_X \tag{1.9.2}$$

and

$$\int_{-\infty}^{\infty} (x - m_X)^2 f_X(x)\, dx = \sigma^2 \tag{1.9.3}$$

Note that the normal density function is completely specified by assigning numerical values to the mean and variance. Thus a shorthand notation has come into common usage to designate a normal random variable. When we write

$$X \sim N\left(m_X, \sigma^2\right) \tag{1.9.4}$$

we mean X is normal with its mean given by the first argument in parentheses and its variance by the second argument. Also, as a matter of terminology, the terms normal and Gaussian are used interchangeably in describing normal random variables, and we will make no distinction between the two.

The normal density and distribution functions are sketched in Fig. 1.8a and 1.8b. Note that the density function is symmetric and peaks at its mean. Qualitatively, then, the mean is seen to be the most likely value, with values on either side of the mean gradually becoming less and less likely as the distance from the mean becomes larger. Since many natural random phenomena seem to exhibit this central-tendency property, at least approximately, the normal distribution is encountered frequently in applied probability. Recall that the variance

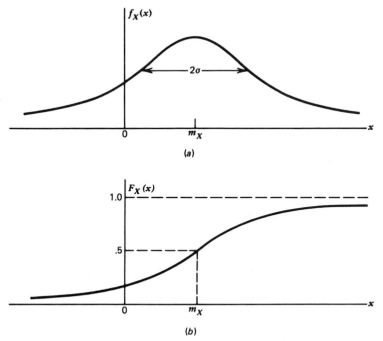

Figure 1.8. Normal density and distribution functions. (a) Normal density function. (b) Normal distribution function.

is a measure of dispersion about the mean. Thus small σ corresponds to a sharp-peaked density curve, whereas large σ will yield a curve with a flat peak.

The normal distribution function is, of course, the integral of the density function:

$$F_X(x) = \int_{-\infty}^{x} f_X(u)\, du = \int_{-\infty}^{x} \frac{1}{\sqrt{2\pi}\,\sigma} \exp\left[-\frac{1}{2\sigma^2}(u - m_X)^2\right] du \quad (1.9.5)$$

Unfortunately, this integral cannot be represented in closed form in terms of elementary functions. Thus, its value must be tabulated as a function of the upper limit. A brief tabulation for zero mean and unity variance is given in Table 1.9. A quick glance at the table will show that the distribution function is very close to unity for values of the argument greater than 4.0. In our table, which was taken from Feller (5), the tabulation quits at about this point, and this is adequate for most purposes. However, in some instances, the difference between F_X and unity, be it ever so small, is very much of interest and we need to be able to compute it. An approximate formula for doing so for zero mean and unity variance is given by

$$1 - F_X(x) \approx \frac{1}{\sqrt{2\pi}\,x} e^{-\frac{1}{2}x^2} \quad (1.9.6)$$

Table 1.9 The Normal Density and Distribution Functions for Zero Mean and Unity Variance

$$f_X(x) = \frac{1}{\sqrt{2\pi}}e^{-\frac{1}{2}x^2}, \qquad F_X(x) = \int_{-\infty}^{x}\frac{1}{\sqrt{2\pi}}e^{-\frac{1}{2}u^2}\,du$$

x	$f_X(x)$	$F_X(x)$	x	$f_X(x)$	$F_X(x)$
.0	.398 942	.500 000	2.3	.028 327	.989 276
.1	.396 952	.539 828	2.4	.022 395	.991 802
.2	.391 043	.579 260	2.5	.017 528	.993 790
.3	.381 388	.617 911	2.6	.013 583	.995 339
.4	.368 270	.655 422	2.7	.010 421	.996 533
.5	.352 065	.691 462	2.8	.007 915	.997 445
.6	.333 225	.725 747	2.9	.005 953	.998 134
.7	.312 254	.758 036	3.0	.004 432	.998 650
.8	.289 692	.788 145	3.1	.003 267	.999 032
.9	.266 085	.815 940	3.2	.002 384	.999 313
1.0	.241 971	.841 345	3.3	.001 723	.999 517
1.1	.217 852	.864 334	3.4	.001 232	.999 663
1.2	.194 186	.884 930	3.5	.000 873	.999 767
1.3	.171 369	.903 200	3.6	.000 612	.999 841
1.4	.149 727	.919 243	3.7	.000 425	.999 892
1.5	.129 518	.933 193	3.8	.000 292	.999 928
1.6	.110 921	.945 201	3.9	.000 199	.999 952
1.7	.094 049	.955 435	4.0	.000 134	.999 968
1.8	.078 950	.964 070	4.1	.000 089	.999 979
1.9	.065 616	.971 283	4.2	.000 059	.999 987
2.0	.053 991	.977 250	4.3	.000 039	.999 991
2.1	.043 984	.982 136	4.4	.000 025	.999 995
2.2	.035 475	.986 097	4.5	.000 016	.999 997

Feller (5) may be consulted for a justification of this formula. The accuracy of Eq. (1.9.6) is within only a few percent for x equal to 4. It improves rapidly, though, as x increases.

It is worth mentioning that tables of normal distribution are presented in different ways in various references. For example, some tables present the one-sided area under the normal density curve from 0 to x, rather than from $-\infty$ to x as we have done in Table 1.9. Other tables do something similar by tabulating a function known as the error function (6), which is normalized differently than the usual distribution function. Thus a word of warning is in order. Be wary in the use of unfamiliar tables! An example illustrating the use of our brief tabulation in Table 1.9 is now in order.

EXAMPLE 1.17

Let us say that the random variable X is normal with a mean of 1 and a variance of 4. In our shortened notation we could then write $X \sim N\,(1, 4)$. Suppose we wish to find:

    (a) The value of the density function at its peak.
    (b) The probability that $X \geq 2$.
    (c) The probability that $0 \leq X \leq 2$.

We must first remember that the tabulated quantities are normalized for unity variance and zero mean. Thus the $f_X(x)$ of the tables is

$$f_X(x) = \frac{1}{\sqrt{2\pi}} e^{-\frac{1}{2}x^2}$$

In part (a) we are asked to find the peak value of

$$\frac{1}{\sqrt{2\pi} \cdot 2} \exp\left[-\frac{1}{2 \cdot 4}(x - 1)^2\right]$$

Clearly, this occurs where $x = 1$ and is $1/2\sqrt{2\pi} \approx .199$. This could be obtained from Table 1.9 by noting the value of $f_X(x)$ for $x = 0$ and dividing by 2. The division is necessary because σ (2 in this case) appears in the denominator of the normal density function. Of course, with modern pocket calculators, one does not need tables to evaluate the density function. Not so with the distribution function though. Most calculators are not programmed with an algorithm for this function, so tables (or perhaps appropriate computer software) are still necessary.

The areas involved in the solution of parts (b) and (c) are shown in Fig. 1.9. Formally, we wish to evaluate the following two integrals:

$$P(0 \leq X \leq 2) = \int_0^2 \frac{1}{\sqrt{2\pi} \cdot 2} \exp\left[-\frac{1}{2 \cdot 4}(u - 1)^2\right] du \qquad (1.9.7)$$

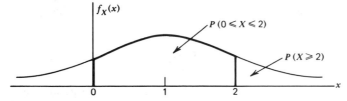

Figure 1.9. Key areas for Example 1.17.

and

$$P(X \geq 2) = \int_2^\infty \frac{1}{\sqrt{2\pi} \cdot 2} \exp\left[-\frac{1}{2 \cdot 4}(u-1)^2 \right] du$$

$$= 1 - \int_{-\infty}^2 \frac{1}{\sqrt{2\pi} \cdot 2} \exp\left[-\frac{1}{2 \cdot 4}(u-1)^2 \right] du \qquad (1.9.8)$$

First we make the obvious change of variable:

$$v = \frac{u-1}{2}$$

Then $du = 2\, dv$, and the integrals become

$$P(0 \leq X \leq 2) = \int_{-\frac{1}{2}}^{\frac{1}{2}} \frac{1}{\sqrt{2\pi}} e^{-\frac{1}{2}v^2} dv \qquad (1.9.9)$$

and

$$P(X \geq 2) = 1 - \int_{-\infty}^{\frac{1}{2}} \frac{1}{\sqrt{2\pi}} e^{-\frac{1}{2}v^2} dv \qquad (1.9.10)$$

The tables can be used directly for the second term of Eq. (1.9.10), and therefore the answer for part (b) is

$$P(X \geq 2) = 1 - .691462 = .308538 \qquad (1.9.11)$$

By taking advantage of symmetry, Eq. (1.9.9) can be written in the form

$$P(0 \leq X \leq 2) = 2\int_0^{\frac{1}{2}} \frac{1}{\sqrt{2\pi}} e^{-\frac{1}{2}v^2} dv$$

$$= 2\left[\int_{-\infty}^{\frac{1}{2}} \frac{1}{\sqrt{2\pi}} e^{-\frac{1}{2}v^2} dv - .5 \right] \qquad (1.9.12)$$

The tables may now be used to evaluate Eq. (1.9.12), and this gives the solution to part (c):

$$P(0 \leq X \leq 2) = 2[.691462 - .5] = .382924 \qquad (1.9.13)$$

We can, of course, be less formal and take a more geometric approach in the solution of parts (b) and (c). This involves translating and rescaling the horizontal axis in accordance with the mean and variance of the random variable under consideration. The areas under the curve relative to the new axis then have a direct relationship to the quantities tabulated in Table 1.9. This is relatively obvious and will not be pursued further. ■

1.10
IMPULSIVE PROBABILITY DENSITY FUNCTIONS

In the case of the normal distribution, and many others, the probability associated with the random variable X is smoothly distributed over the real line from $-\infty$ to ∞. The corresponding probability density function is then continuous, and the probability that any particular value of X, say x_0, is realized is zero. This situation is common in physical problems, but we also have occasion to consider cases where the random variable has a mixture of discrete and smooth distribution. Rectification or any sort of hard limiting of noise leads to this situation, and an example will illustrate how this affects the probability density and distribution functions.

EXAMPLE 1.18

Consider a simple half-wave rectifier driven by noise as shown in Fig. 1.10. For our purpose here, it will suffice to assume that the amplitude of the noise is normally distributed with zero mean. That is, if we were to sample the input at any particular time t_1, the resultant sample is a random variable, say X, whose distribution is $N(0, \sigma_X^2)$. The corresponding output sample is, of course, a different random variable; it will be denoted as Y.

Because of our assumption of normality, X will have probability density and distribution functions as shown in Fig. 1.8, but with $m_X = 0$. The sample space in this case may be thought of as all points on the real line, and the function defining the random variable X is just a one-to-one mapping of the sample space into X. Not so with Y though; all positive points in the sample space map one-to-one, but the negative points all map into zero because of the diode! This means that a total probability of $\frac{1}{2}$ in the sample space gets squeezed into a single point, zero, in the Y space. The effect of this on the density and distribution functions for Y is shown in Fig. 1.11a and 1.11b. Note that in order to have the area under the density function be $\frac{1}{2}$ at $y = 0$, we must

Figure 1.10.   Half-wave rectifier driven by noise.

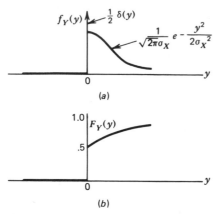

Figure 1.11. Output density and distribution functions for Example 1.18. (*a*) Probability density function for Y. (*b*) Probability distribution function for Y.

have a Dirac delta (impulse) function at the origin. This, in turn, gives rise to a jump or discontinuity in the corresponding distribution function. It should be mentioned that at the point of discontinuity, the value of the distribution function is $\frac{1}{2}$. That is, the distribution function is continuous from the right and not from the left. This is due to the "*equal to* or less than . . ." statement in the definition of the probability distribution function (see Section 1.8). ∎

1.11
MULTIPLE RANDOM VARIABLES

In subsequent chapters, we will frequently have occasion to deal with multiple random variables and their mutual relationships. Multiple random variables are often referred to as *multivariate* random variables, with *bivariate* being the special case of two variables. The various probabilistic relationships will be illustrated here for the bivariate case. The extension to three or more random variables is straightforward, so it will not be discussed specifically.

Let us first consider two *discrete* random variables X and Y. By discrete we mean that X and Y may take on discrete values x_i and y_j, where i and j are certain allowed integers. We will define the *joint probability distribution* (or *joint probability mass distribution*) as

$$p_{XY}(x_i, y_j) = P(X = x_i \text{ and } Y = y_j) \qquad (1.11.1)$$

Note that this is not a cumulative distribution and, as an extra reminder, we will denote the discrete distribution with p rather than F. Just as in the case of joint events A and B (Section 1.4), the joint distribution of X and Y can be thought of as a two-dimensional array of probabilities, with each element in the array representing the probability of occurrence of a particular combination of X and

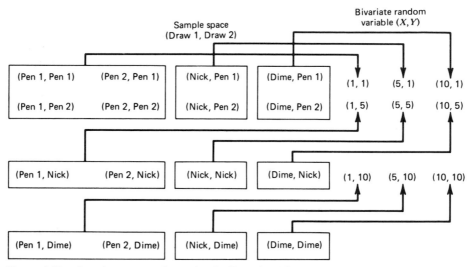

Figure 1.12. Sample space and mapping for Example 1.19.

Y. The sum of the numbers in the array obviously must be unity. Also, summing horizontally or vertically yields the marginal probabilities, just as in the "events" case. An example will illustrate these concepts.

EXAMPLE 1.19

A sack contains 2 pennies, 1 nickel, and 1 dime. A coin is withdrawn at random and then replaced; then a second coin is withdrawn. The face value of the first draw (i.e., possible values are 1, 5, or 10) will be called random variable X, and the value of the second coin will be Y. We will assume that the outcome of the first draw does not affect the second in any way. The sample space in this case consists of 16 elements, and it, along with mapping into the bivariate random variable (X, Y), is shown in Fig. 1.12. The two pennies are distinguished as Pen 1 and Pen 2. Note that the 16 elements of the sample space map into 9 2-tuples in the random-variable space. We have shown the 2-tuples in Fig. 1.12 as two numbers separated by commas. We could, of course, have shown them as two-element column vectors; it's purely a matter of notation. If we assume that all elementary events in the sample space are equally likely, each should be assigned a probability of $\frac{1}{16}$. The joint probability distribution for X and Y is then as shown in Table 1.10. ∎

Just as in Section 1.4, we can write certain relationships among the joint, marginal, and conditional probabilities for random variables X and Y.

Marginal (unconditional) probability:

$$p_X(x_i) = \sum_j p_{XY}(x_i, y_j) \qquad (1.11.2)$$

$$p_Y(y_j) = \sum_i p_{XY}(x_i, y_j) \qquad (1.11.3)$$

Table 1.10 Joint and Marginal Probabilities for X and Y (X Is First Draw and Y Is Second Draw)

X \\ Y	1	5	10	Marginal Probability $p(x_i)$
1	$\frac{1}{4}$	$\frac{1}{8}$	$\frac{1}{8}$	$\frac{1}{2}$
5	$\frac{1}{8}$	$\frac{1}{16}$	$\frac{1}{16}$	$\frac{1}{4}$
10	$\frac{1}{8}$	$\frac{1}{16}$	$\frac{1}{16}$	$\frac{1}{4}$
Marginal probability $p(y_j)$	$\frac{1}{2}$	$\frac{1}{4}$	$\frac{1}{4}$	Sum is unity when summed either horizontally or vertically

Conditional probability:

$$p_{X|Y} = \frac{p_{XY}}{p_Y} \tag{1.11.4}$$

$$p_{Y|X} = \frac{p_{XY}}{p_X} \tag{1.11.5}$$

Bayes rule:

$$p_{X|Y} = \frac{p_{Y|X} p_X}{p_Y} \tag{1.11.6}$$

The arguments in Eqs. (1.11.4) to (1.11.6) were omitted because the permissible x_i and y_j are obvious.

The discrete random variables X and Y are defined to be *statistically independent* if

$$p_{XY}(x_i, y_j) = p_X(x_i) p_Y(y_j) \tag{1.11.7}$$

for all allowable x_i and y_j. As an example, the random variables in Example 1.19 are statistically independent because all nine probabilities in Table 1.10 satisfy the product relationship given by Eq. (1.11.7).

Let us now turn our attention to continuous random variables. Just as in the single-variable case, the relative distribution in the multivariate case must be described in terms of either a cumulative distribution function or a density function. Let X and Y be continuous random variables. The *joint cumulative distribution function* is defined* as

$$F_{XY}(x, y) = P(X \le x \text{ and } Y \le y) \tag{1.11.8}$$

---

* The term "distribution" is a bit overworked in probability theory. The cumulative distribution of continuous random variables is much different than the probability distribution associated with discrete random variables. Thus we will append the word "cumulative" here to avoid confusion.

Clearly, F_{XY} has the following properties:

(a) $F_{XY}(-\infty, -\infty) = 0$ (1.11.9)

(b) $F_{XY}(\infty, \infty) = 1$ (1.11.10)

(c) F_{XY} is nondecreasing in x and y (1.11.11)

The joint density function of continuous random variables X and Y is given by

$$f_{XY}(x, y) = \frac{\partial^2 F_{XY}(x, y)}{\partial x \, \partial y} \qquad (1.11.12)$$

Note that there is an integral/derivative relationship between the cumulative distribution and density functions, just as in the single-variate case. Thus, to find the probability that a joint realization of X and Y will lie within a certain region R in the xy plane, we use the double integral formula

$$P(X \text{ and } Y \text{ lie within } R) = \iint_R f_{XY}(x, y)\,dx\,dy \qquad (1.11.13)$$

This is shown geometrically in Fig. 1.13. Of course, if the region R is a differential rectangle, as shown in Fig. 1.14, the probability of X and Y lying within the rectangle is

$$P(x_0 \le X \le x_0 + dx \text{ and } y_0 \le Y \le y_0 + dy) = f_{XY}(x_0, y_0)\,dx\,dy \qquad (1.11.14)$$

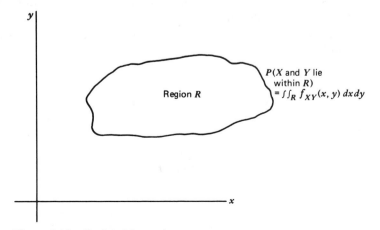

Figure 1.13. Region R in xy plane.

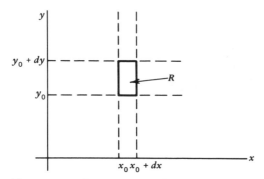

Figure 1.14. Differential area R in xy plane.

Thus it should be apparent that f_{XY} has the usual meaning of density in a two-dimensional sense.

The *marginal* or *unconditional densities* are obtained in a manner similar to the discrete case, except that summation is replaced with integration. Thus we have

$$f_X(x) = \int_{-\infty}^{\infty} f_{XY}(x, y)\, dy \qquad (1.11.15)$$

and

$$f_Y(y) = \int_{-\infty}^{\infty} f_{XY}(x, y)\, dx \qquad (1.11.16)$$

Equations (1.11.4) and (1.11.5) for discrete conditional probabilities can be applied to density functions for differential regions. This results in similar equations for conditional densities. The differential regions shown in Fig. 1.14 will be used in the following development. From the basic definition of conditional probability, we have

$$P(X \text{ is in strip } dx | Y \text{ is in strip } dy) = \frac{f_{XY}(x_0, y_0)\, dx\, dy}{f_Y(y_0)\, dy} \qquad (1.11.17)$$

Canceling the dy's and noting the statement "Y is in strip dy" is approximately the same as saying "Y is equal to y_0" leads to

$$P(x_0 \leq X \leq x_0 + dx | Y = y_0) = \left[\frac{f_{XY}(x_0, y_0)}{f_Y(y_0)} \right] dx \qquad (1.11.18)$$

Note that the quantity in brackets in Eq. (1.11.18) has all the earmarks of a density function, and its probabilistic interpretation is on the left side of the equation. Thus we define conditional density as

$$f_{X|Y}(x) = \frac{f_{XY}(x, y)}{f_Y(y)} \qquad (1.11.19)$$

The functional dependence on the second argument y has been intentionally omitted in $f_{X|Y}$. This is to emphasize that $f_{X|Y}$ is a density function on x, not y [see Eq. (1.11.18)]. Of course, y does appear in $f_{X|Y}$ as a parameter, which may be thought of as a given, deterministic quantity. It is not the primary argument of $f_{X|Y}$ though.

The analogous definition of $f_{Y|X}$ is

$$f_{Y|X}(y) = \frac{f_{XY}(x, y)}{f_X(x)} \qquad (1.11.20)$$

Once Eqs. (1.11.19) and (1.11.20) have been established, Bayes rule follows directly.

Bayes rule:

$$f_{X|Y}(x) = \frac{f_{Y|X}(y)f_X(x)}{f_Y(y)} \qquad (1.11.21)$$

Statistical independence for continuous random variables X and Y is defined in the same manner as for discrete variables; that is, X and Y are statistically independent if

$$f_{XY}(x, y) = f_X(x)f_Y(y) \qquad (1.11.22)$$

An example is now in order.

EXAMPLE 1.20

Consider a dart-throwing game in which the target is a conventional xy coordinate system. The player aims each throw at the origin according to his or her best ability. Since there are many vagaries affecting each throw, we can expect a scatter in the hit pattern. Also, after some practice, the scatter should be unbiased, left-to-right and up-and-down. Let the coordinate of a hit be a bivariate random variable (X, Y). In this example we would not expect the x coordinate to affect y in any way; therefore statistical independence of X and Y is a reasonable assumption. Also, because of the central tendency in X and Y, the assumption of normal distribution would appear to be reasonable. Thus we

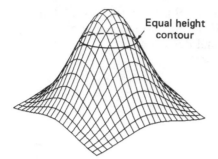

Figure 1.15. Bivariate normal density function.

assume the following unconditional densities:

$$f_X(x) = \frac{1}{\sqrt{2\pi}\,\sigma} e^{-x^2/2\sigma^2} \tag{1.11.23}$$

$$f_Y(y) = \frac{1}{\sqrt{2\pi}\,\sigma} e^{-y^2/2\sigma^2} \tag{1.11.24}$$

The joint density is then given by

$$f_{XY}(x, y) = \frac{1}{\sqrt{2\pi}\,\sigma} e^{-x^2/2\sigma^2} \cdot \frac{1}{\sqrt{2\pi}\,\sigma} e^{-y^2/2\sigma^2} = \frac{1}{2\pi\sigma^2} e^{-(x^2+y^2)/2\sigma^2} \tag{1.11.25}$$

Equation (1.11.25) is the special case of a bivariate normal density function in which X and Y are independent, unbiased, and have equal variances. This is an important density function and it is sketched in Fig. 1.15. It is often described as a smooth "hill-shaped" function and, for the special case here, the hill is symmetric in every respect. Thus the equal-height contour shown in Fig. 1.15 is an exact circle. ∎

The joint density function for a discrete bivariate random variable (X, Y) can be represented with impulsive-type functions, just as in the single-variate case. However, in the bivariate case, remember that volume (not area) represents probability. Thus volume-type impulse functions are appropriate. When X and Y are independent, the appropriate two-dimensional impulse function is obtained as a simple product of Dirac delta functions. For example, a unit volume impulse at the origin is given by $\delta(x)\delta(y)$. However, the situation is

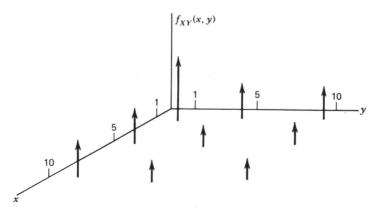

$f_{XY}(x, y)$

Figure 1.16. Impulsive joint probability density function for coin-drawing example, Example 1.19.

more complicated when X and Y are not independent, and we simply have to imagine appropriate volume-type impulsive functions. We will consider examples of this in subsequent chapters. The impulsive density function representation for the coin-drawing example, Example 1.19, is shown in Fig. 1.16.

1.12
CORRELATION, COVARIANCE, AND ORTHOGONALITY

The expectation of the product of two random variables X and Y is of special interest. In general, it is given by

$$E(XY) = \int_{-\infty}^{\infty} \int_{-\infty}^{\infty} xy f_{XY}(x, y) \, dx \, dy \qquad (1.12.1)$$

There is a special simplification of Eq. (1.12.1) that occurs when X and Y are independent. In this case, f_{XY} may be factored (see Eq. 1.11.22). Equation (1.12.1) then reduces to

$$E(XY) = \int_{-\infty}^{\infty} x f_X(x) \, dx \int_{-\infty}^{\infty} y f_Y(y) \, dy = E(X)E(Y) \qquad (1.12.2)$$

If X and Y possess the property of Eq. (1.12.2), that is, the expectation of the product is the product of the individual expectations, they are said to be *uncorrelated*. Obviously, if X and Y are independent, they are also uncorrelated. However, the converse is not true, except in a few special cases (see Section 1.16).

As a matter of terminology, if

$$E(XY) = 0 \qquad (1.12.3)$$

X and Y are said to be *orthogonal*.

The *covariance* of X and Y is also of special interest, and it is defined as

$$\text{Cov of } X \text{ and } Y = E[(X - m_X)(Y - m_Y)] \qquad (1.12.4)$$

With the definition of Eq. (1.12.4) we can now define the *correlation coefficient* for two random variables as

$$\text{Correlation coefficient} = \rho = \frac{\text{Cov of } X \text{ and } Y}{\sqrt{\text{Var } X}\,\sqrt{\text{Var } Y}}$$

$$= \frac{E[(X - m_X)(Y - m_Y)]}{\sqrt{\text{Var } X}\,\sqrt{\text{Var } Y}} \qquad (1.12.5)$$

The correlation coefficient is a normalized measure of the degree of correlation between two random variables, and the normalization is such that ρ always lies within the range $-1 \le \rho \le 1$. This will be demonstrated (not proved) by looking at three special cases:

1. $Y = X$ (maximum positive correlation):
   When $Y = X$, Eq. (1.12.5) becomes

$$\rho = \frac{E[(X - m_X)(X - m_X)]}{\sqrt{E(X - m_X)^2}\,\sqrt{E(X - m_X)^2}} = 1$$

2. $Y = -X$ (maximum negative correlation):
   When $Y = -X$, Eq. (1.12.5) becomes

$$\rho = \frac{E[(X - m_X)(-X + m_X)]}{\sqrt{E(X - m_X)^2}\,\sqrt{E(-X + m_X)^2}} = -1$$

3. X and Y uncorrelated, that is, $E(XY) = E(X)E(Y)$:
   Expanding the numerator of Eq. (1.12.5) yields

$$E(XY - m_X Y - m_Y X + m_X m_Y)$$

$$= E(XY) - m_X E(Y) - m_Y E(X) + m_X m_Y$$

$$= m_X m_Y - m_X m_Y - m_Y m_X + m_X m_Y = 0$$

Thus $\rho = 0$.

We have now examined the extremes of positive and negative correlation and zero correlation; there can be all shades of gray in between. [For further details, see Papoulis (7)].

1.13
SUM OF INDEPENDENT RANDOM VARIABLES AND TENDENCY TOWARD NORMAL DISTRIBUTION

Since we frequently need to consider additive combinations of independent random variables, this will now be examined in some detail. Let X and Y be independent random variables with probability density functions $f_X(x)$ and $f_Y(y)$. Define another random variable Z as the sum of X and Y:

$$Z = X + Y \qquad (1.13.1)$$

Given the density functions of X and Y, we wish to find the corresponding density of Z.

Let z be a particular realization of the random variable Z, and think of z as being fixed. Now consider all possible realizations of X and Y that yield z. Clearly, they satisfy the equation

$$x + y = z \qquad (1.13.2)$$

and the locus of points in the x, y plane is just a straight line, as shown in Fig. 1.17.

Next, consider an incremental perturbation of z to $z + dz$, and again consider the locus of realizations of X and Y that will yield $z + dz$. This locus

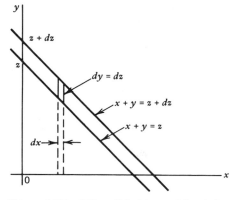

Figure 1.17. Differential strip used for deriving $f_z(z)$.

is also a straight line, and it is shown as the upper line in Fig. 1.17. It should be apparent that all x and y lying within the differential strip between the two lines map into points between z and $z + dz$ in the z space. Therefore,

$$P(z \leq Z \leq z + dz) = P(x \text{ and } y \text{ lie in differential strip})$$

$$= \iint_{\substack{\text{Diff.} \\ \text{strip}}} f_X(x)f_Y(y)\,dx\,dy \qquad (1.13.3)$$

But within the differential strip, y is constrained to x according to

$$y = z - x \qquad (1.13.4)$$

Also, since the strip is only of differential width, the double integral of Eq. (1.13.3) reduces to a single integral. Choosing x as the variable of integration and noting that $dy = dz$ leads to

$$P(z \leq Z \leq z + dz) = \left[\int_{-\infty}^{\infty} f_X(x)f_Y(z - x)\,dx \right] dz \qquad (1.13.5)$$

It is now apparent from Eq. (1.13.5) that the quantity within the brackets is the desired probability density function for Z. Thus

$$f_Z(z) = \int_{-\infty}^{\infty} f_X(x)f_Y(z - x)\,dx \qquad (1.13.6)$$

It is of interest to note that the integral on the right side of Eq. (1.13.6) is a convolution integral. Thus, from Fourier transform theory, we can then write

$$\mathscr{F}[f_Z] = \mathscr{F}[f_X] \cdot \mathscr{F}[f_Y] \qquad (1.13.7)$$

where $\mathscr{F}[\cdot]$ denotes "Fourier transform of $[\cdot]$." We now have two ways of evaluating the density of Z: (1) We can evaluate the convolution integral directly, or (2) we can first transform f_X and f_Y, then form the product of the transforms, and finally invert the product to get f_Z. Examples that illustrate each of these methods follow.

EXAMPLE 1.21

Let X and Y be independent random variables with identical rectangular density functions as shown in Fig. 1.18a. We wish to find the density function for their sum, which we will call Z.

Note first that the density shown in Fig. 1.18a has even symmetry. Thus $f_Y(z - x) = f_Y(x - z)$. The convolution integral expression of Eq. (1.13.6) is

(a)

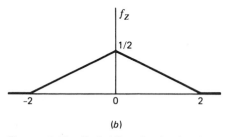

(b)

Figure 1.18. Probability density functions for X and Y and their sum. (a) Density function for both X and Y. (b) Density function for Z, where $Z = X + Y$.

then the integral of a rectangular pulse multiplied by a similar pulse shifted to the right amount z. When $z > 2$ or $z < 2$ there is no overlap in the pulses so their product is zero. When $-2 \le z \le 0$ there is a nontrivial overlap that increases linearly beginning at $z = -2$ and reaching a maximum at $z = 0$. The convolution integral then increases accordingly as shown in Fig. 1.18b. A similar argument may be used to show that $f_Z(z)$ decreases linearly in the interval where $0 \le z \le 2$. This leads to the triangular density function of Fig. 1.18b.

We can go one step further now and look at the density corresponding to the sum of three random variables. Let W be defined as

$$W = X + Y + V \qquad (1.13.8)$$

where X, Y, and V are mutually independent and have identical rectangular densities as shown in Fig. 1.18a. We have already worked out the density for $X + Y$, so the density of W is the convolution of the two functions shown in Fig. 1.18a and 1.18b. We will leave the details of this as an exercise, and the result is shown in Fig. 1.19. Each of the segments labeled 1, 2, and 3 is an arc of a parabola. Notice the smooth central tendency. With a little imagination one can see a similarity between this and a zero-mean normal density curve. If we were to go another step and convolve a rectangular density with that of Fig. 1.19, we would get the density for the sum of four independent random variables. The resulting function would consist of connected segments of cubic functions extending from -4 to $+4$. Its appearance, though not shown, would resemble

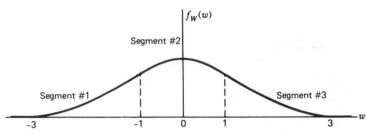

Figure 1.19. Probability density for the sum of three independent random variables with identical rectangular density functions.

the normal curve even more than that of Fig. 1.19. And on and on—each additional convolution results in a curve that resembles the normal curve more closely than the preceding one.

This simple example is intended to demonstrate (not prove) that a superposition of independent random variables always tends toward normality, regardless of the distribution of the individual random variables contributing to the sum. This is known as the *central limit theorem* of statistics. It is a most remarkable theorem, and its validity is subject to only modest restrictions (3). In engineering applications the noise we must deal with is frequently due to a superposition of many small contributions. When this is so, we have good reason to make the assumption of normality. The central limit theorem says to do just that. Thus we have here one of the reasons for our seemingly exaggerated interest in normal random variables—they are a common occurrence in nature. ∎

EXAMPLE 1.22

Let X and Y be independent normal random variables with zero means and variances σ_X^2 and σ_Y^2. We wish to find the probability density function for the sum of X and Y, which will again be denoted as Z. For variety, we illustrate the Fourier transform approach. The explicit expressions for f_X and f_Y are

$$f_X(t) = \frac{1}{\sqrt{2\pi}\,\sigma_X}e^{-t^2/2\sigma_X^2} \tag{1.13.9}$$

$$f_Y(t) = \frac{1}{\sqrt{2\pi}\,\sigma_Y}e^{-t^2/2\sigma_Y^2} \tag{1.13.10}$$

Note that we have used t as the dummy argument of the functions. It is of no consequence because it is integrated out in the transformation to the ω domain. Using Fourier transform tables, we find the transforms of f_X and f_Y to be

$$\mathcal{F}[f_X] = e^{-\sigma_x^2\omega^2/2} \tag{1.13.11}$$

$$\mathcal{F}[f_Y] = e^{-\sigma_y^2\omega^2/2} \tag{1.13.12}$$

Forming their product yields

$$\mathscr{F}[f_X]\mathscr{F}[f_Y] = e^{-(\sigma_X^2+\sigma_Y^2)\omega^2/2}$$ (1.13.13)

Then the inverse gives the desired f_Z:

$$f_Z(z) = \mathscr{F}^{-1}\left[e^{-(\sigma_X^2+\sigma_Y^2)\omega^2/2}\right]$$

$$= \frac{1}{\sqrt{2\pi(\sigma_X^2 + \sigma_Y^2)}}e^{-z^2/2(\sigma_X^2+\sigma_Y^2)}$$ (1.13.14)

Note that the density function for Z is also normal in form, and its variance is given by

$$\sigma_Z^2 = \sigma_X^2 + \sigma_Y^2$$ (1.13.15)

The summation of any number of random variables can always be thought of as a sequence of summing operations on two variables; therefore, it should be clear that summing any number of independent normal random variables leads to a normal random variable. This rather remarkable result can be generalized further to include the case of dependent normal random variables, which we will discuss later. ∎

1.14 TRANSFORMATION OF RANDOM VARIABLES

A mathematical transformation that takes one set of variables (say, inputs) into another set (say, outputs) is a common situation in systems analysis. Let us begin with a simple single-input, single-output situation where the input–output relationship is governed by the algebraic equation

$$y = g(x)$$ (1.14.1)

Here we are interested in random inputs, so think of x as a realization of the input random variable X, and y as the corresponding realization of the output Y. Assume we know the probability density function for X, and would like to find the corresponding density for Y. It is tempting to simply replace x in $f_X(x)$ with its equivalent in terms of y and pass it off at that. However, it is not quite that simple, as will be seen presently.

First, let us assume that the transformation $g(x)$ is one-to-one for all permissible x. By this we mean that the functional relationship given by Eq.

(1.14.1) can be reversed, and x can be written uniquely as a function of y. Let the "reverse" relationship be

$$x = h(y) \tag{1.14.2}$$

The probabilities that X and Y lie within corresponding differential regions must be equal. That is,

$$P(X \text{ is between } x \text{ and } x + dx) = P(Y \text{ is between } y \text{ and } y + dy) \tag{1.14.3}$$

or

$$\int_x^{x+dx} f_X(u) \, du = \begin{cases} \int_y^{y+dy} f_Y(u) \, du, & \text{for } dy \text{ positive} \\ -\int_y^{y+dy} f_Y(u) \, du, & \text{for } dy \text{ negative} \end{cases} \tag{1.14.4}$$

One of the subtleties of this problem should now be apparent from Eq. (1.14.4). If positive dx yields negative dy (i.e., a negative derivative), the integral of f_Y must be taken from $y + dy$ to y in order to yield a positive probability. This is the equivalent of interchanging the limits and reversing the sign as shown in Eq. (1.14.4).

The differential equivalent of Eq. (1.14.4) is

$$f_X(x) \, dx = f_Y(y)|dy| \tag{1.14.5}$$

where we have tacitly assumed dx to be positive. Also, x is constrained to be $h(y)$. Thus we have

$$f_Y(y) = \left| \frac{dx}{dy} \right| f_X(h(y)) \tag{1.14.6}$$

or, equivalently,

$$f_Y(y) = |h'(y)| f_X(h(y)) \tag{1.14.7}$$

where $h'(y)$ indicates the derivative of h with respect to y. Two examples will now be presented.

EXAMPLE 1.23

Find the appropriate output density functions for the case where the input X is $N(0, \sigma_X^2)$ and the transformation is

(a) $y = Kx$ (K is a given constant) \hfill (1.14.8)

(b) $y = x^3$ \hfill (1.14.9)

(c) $y = x^2$ \hfill (1.14.10)

We begin with the scale-factor transformation indicated by Eq. (1.14.8). We first solve for x in terms of y and then form the derivative. Thus,

$$x = \frac{1}{K}y \qquad (1.14.11)$$

$$\left|\frac{dx}{dy}\right| = \left|\frac{1}{K}\right| \qquad (1.14.12)$$

We can now obtain the equation for f_Y from Eq. (1.14.6). The result is

$$f_Y(y) = \frac{1}{|K|} \cdot \frac{1}{\sqrt{2\pi}\,\sigma_X} \exp\left[-\frac{\left(\dfrac{y}{K}\right)^2}{2\sigma_X^2}\right] \qquad (1.14.13)$$

Or rewriting Eq. (1.14.13) in standard normal form yields

$$f_Y(y) = \frac{1}{\sqrt{2\pi(K\sigma_X)^2}} \exp\left[-\frac{y^2}{2(K\sigma_X)^2}\right] \qquad (1.14.14)$$

It can now be seen that transforming a zero-mean normal random variable with a simple scale factor yields another normal random variable with a corresponding scale change in its standard deviation. It is important to note that normality is preserved in a linear transformation.

Next, consider part (b). This transformation is also one-to-one, so solving for x yields

$$x = \sqrt[3]{y} \qquad (1.14.15)$$

In Eq. (1.14.15) we take the positive real root for $y > 0$ and the negative real root for $y < 0$. The derivative of x is then

$$\frac{dx}{dy} = \frac{1}{3}y^{-2/3} \qquad (1.14.16)$$

The quantity $y^{2/3}$ can be written as $(y^{1/3})^2$, so $y^{2/3}$ is always positive, provided $y^{1/3}$ is real. This is consistent with the geometric interpretation of $y = x^3$, which always has a nonnegative slope. The density function for Y is then

$$f_Y(y) = \frac{1}{3y^{2/3}} \cdot \frac{1}{\sqrt{2\pi}\,\sigma_X}e^{-(y^{1/3})^2/2\sigma_X^2} \qquad (1.14.17)$$

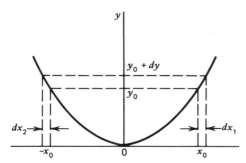

Figure 1.20.   Sketch of $y = x^2$.

Since this does not reduce to normal form, we have here an example of a nonlinear transformation that converts a normal random variable to non-Gaussian form.

The transformation $y = x^2$ for part (c) is sketched in Fig. 1.20. It is obvious from the sketch that two values of x yield the same y. Thus the function is not one-to-one, which violates one of the assumptions in deriving Eq. (1.14.6). The problem is solvable, though; we simply must go back to fundamentals and derive a new $f_Y(y)$ relationship. Note that we can consider $y = x^2$ in terms of two branches. These will be defined as

$$x = \begin{cases} \sqrt{y}, & x \geq 0 \ (\text{Branch 1}) \\ -\sqrt{y}, & x < 0 \ (\text{Branch 2}) \end{cases} \qquad (1.14.18)$$

Now think of perturbing y a positive differential amount dy as shown in Fig. 1.20. This results in two corresponding differential regions on the x axis. Thus we have the following relationship for probabilities:

$$P(y_0 \leq Y \leq y_0 + dy) = P(x_0 \leq X \leq X_0 + dx_1)$$

$$+ P(-x_0 + dx_2 \leq X \leq -x_0) \quad (1.14.19)$$

Note that dx_1 (Branch 1) is positive and dx_2 (Branch 2) is negative. Next, Eq. (1.14.19) can be rewritten in terms of integrals as

$$\int_{y_0}^{y_0 + dy} f_Y(u) \, du = \int_{x_0}^{x_0 + dx_1} f_X(v) \, dv + \int_{-x_0 + dx_2}^{-x_0} f_X(v) \, dv \quad (1.14.20)$$

We now note that

$$dx_1 = \frac{1}{2\sqrt{y_0}}\, dy$$

$$dx_2 = -\frac{1}{2\sqrt{y_0}}\, dy \qquad (1.14.21)$$

Substituting Eq. (1.14.21) into Eq. (1.14.20) and letting the range of integration be incremental yields

$$f_Y(y_0)\, dy = f_X\left(\sqrt{y_0}\right) \cdot \frac{1}{2\sqrt{y_0}}\, dy + f_X\left(\sqrt{y_0}\right) \cdot \frac{1}{2\sqrt{y_0}}\, dy, \qquad y_0 \geq 0$$

$$(1.14.22)$$

Now, since y_0 can be any y greater than zero, Eq. (1.14.22) reduces to

$$f_Y(y) = \frac{1}{\sqrt{y}} f_X\left(\sqrt{y}\right), \qquad y \geq 0 \qquad (1.14.23)$$

From the equation $y = x^2$, we see that no real values of x map into negative y. Therefore, $f_Y(y) = 0$ for negative y. The final f_Y is then

$$f_Y(y) = \begin{cases} \dfrac{1}{\sqrt{y}} f_X\left(\sqrt{y}\right), & y \geq 0 \\[2mm] 0, & y < 0 \end{cases} \qquad (1.14.24)$$

We can now apply Eq. (1.14.24) to the case where X is normal with zero mean. This yields

$$f_Y(y) = \begin{cases} \dfrac{1}{\sqrt{2\pi}\,\sigma_X \cdot \sqrt{y}} e^{-y/2\sigma_X^2}, & y \geq 0 \\[2mm] 0, & y < 0 \end{cases} \qquad (1.14.25)$$

Note that Y is not a normal random variable. Rather, it is a special case of what is known as a chi-square random variable. (See Problem 1.20 for a more general discussion of the chi-square distribution.) ■

We also have occasion to deal with transformations of multiple random variables. Consider a bivariate transformation example. Since the extension to higher-order transformations is fairly obvious, only the bivariate case will be discussed in detail.

EXAMPLE 1.24

In the target-shooting example (Example 1.20), the hit location was described in terms of rectangular coordinates. This led to the joint probability density function

$$f_{XY}(x, y) = \frac{1}{2\pi\sigma^2} e^{-(x^2+y^2)/2\sigma^2} \tag{1.14.26}$$

This is a special case of the bivariate normal density function where the two variates are independent and have zero means and equal variances. Suppose we wish to find the corresponding density in terms of polar coordinates r and θ. Formally, then, we wish to define new random variables R and Θ such that pairwise realizations (x, y) transform to (r, θ) in accordance with

$$r = \sqrt{x^2 + y^2}, \quad r \geq 0$$

$$\theta = \tan^{-1}\frac{y}{x}, \quad 0 \leq \theta < 2\pi \tag{1.14.27}$$

Or, equivalently,

$$x = r \cos \theta$$

$$y = r \sin \theta \tag{1.14.28}$$

We wish to find $f_{R\Theta}(r, \theta)$ and the unconditional density functions $f_R(r)$ and $f_\Theta(\theta)$.

The probability that a hit will lie within an area bounded by a closed contour C is given by

$$P(\text{Hit lies within } C) = \iint\limits_{\substack{\text{Area} \\ \text{enclosed} \\ \text{by } C}} f_{XY}(x, y)\, dx\, dy \tag{1.14.29}$$

We know from multivariable calculus that the double integral in Eq. (1.14.29) can also be evaluated in the r, θ coordinate frame as

$$\iint\limits_{\substack{\text{Region} \\ \text{enclosed} \\ \text{by } C}} f_{XY}(x, y)\, dx\, dy = \iint\limits_{\substack{\text{Region} \\ \text{enclosed} \\ \text{by } C'}} f_{XY}(x(r,\theta), y(r,\theta)) \left| J\left(\frac{x, y}{r, \theta}\right) \right| dr\, d\theta$$

$$\tag{1.14.30}$$

Figure 1.21. Corresponding differential regions for transformation of Example 1.24.

where C' is the contour in the r, θ plane corresponding to C in the x, y plane. That is, points within C map into points within C'. (Note that it is immaterial as to how we draw the "picture" in the r, θ coordinate frame. For example, if we choose to think of r and θ as just another set of Cartesian coordinates for plotting purposes, C' becomes a distortion of C.) The J quantity in Eq. (1.14.30) is the Jacobian of the transformation, defined as

$$J\left(\frac{x, y}{r, \theta}\right) = \mathrm{Det} \begin{bmatrix} \dfrac{\partial x}{\partial r} & \dfrac{\partial y}{\partial r} \\ \dfrac{\partial x}{\partial \theta} & \dfrac{\partial y}{\partial \theta} \end{bmatrix} \qquad (1.14.31)$$

The vertical bars around J in Eq. (1.14.30) indicate absolute magnitude. We can argue now that if Eq. (1.14.30) is true for regions in general, it must also be true for differential regions. Let the differential region in the r, θ domain be bounded by r and $r + dr$ in one direction and by θ and $\theta + d\theta$ in the other. If it helps, think of plotting r and θ as Cartesian coordinates. The differential region in the r, θ domain is then rectangular, and the corresponding one in the x, y domain is a curvilinear differential rectangle (see Fig. 1.21). Now, by the very definition of joint density, the quantity multiplying $dr \, d\theta$ in Eq. (1.14.30) is seen to be the desired density function. That is,

$$f_{R\Theta}(r, \theta) = f_{XY}[x(r, \theta), y(r, \theta)]\left|J\left(\frac{x, y}{r, \theta}\right)\right| \qquad (1.14.32)$$

In the transformation of this example,

$$J\left(\frac{x, y}{r, \theta}\right) = \mathrm{Det} \begin{bmatrix} \cos \theta & \sin \theta \\ -r \sin \theta & r \cos \theta \end{bmatrix} = r \qquad (1.14.33)$$

Since the radial coordinate r is always positive, $|J| = r$. We can now substitute

Eqs. (1.14.26) and (1.14.33) into (1.14.32) and obtain

$$f_{R\Theta}(r, \theta) = r\frac{1}{2\pi\sigma^2} \exp\left[-\frac{(r\cos\theta)^2 + (r\sin\theta)^2}{2\sigma^2}\right]$$

$$= \frac{r}{2\pi\sigma^2} e^{-r^2/2\sigma^2} \tag{1.14.34}$$

Note that the density function of this example has no explicit dependence on θ. In other words, all angles between 0 and 2π are equally likely, which is what we would expect in the target-throwing experiment.

We get the unconditional density functions by integrating $f_{R\Theta}$ with respect to the appropriate variables. That is,

$$f_R(r) = \int_0^{2\pi} f_{R\Theta}(r, \theta)\, d\theta = \frac{r}{2\pi\sigma^2} e^{-r^2/2\sigma^2} \int_0^{2\pi} d\theta$$

$$= \frac{r}{\sigma^2} e^{-r^2/2\sigma^2} \tag{1.14.35}$$

and

$$f_\Theta(\theta) = \int_0^\infty f_{R\Theta}(r, \theta)\, dr = \begin{cases} \dfrac{1}{2\pi}, & 0 \le \theta < 2\pi \\ 0, & \text{otherwise} \end{cases} \tag{1.14.36}$$

The single-variate density function given by Eq. (1.14.35) is called the *Rayleigh* density function. It is of considerable importance in applied probability, and it is sketched in Fig. 1.22. It is easily verified that the mode (peak value) of the Rayleigh density is equal to standard deviation of the x and y normal random variables from which it was derived. Thus we see that similar independent, zero-mean normal densities in the x, y domain correspond to Rayleigh and uniform densities in the r, θ domain. ∎

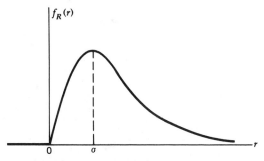

Figure 1.22.   Rayleigh density function.

1.15
MULTIVARIATE NORMAL DENSITY FUNCTION

In Sections 1.9 and 1.11 examples of the single and bivariate normal density functions were presented. We work so much with normal random variables that we need to elaborate further and develop a general form for the n-dimensional normal density function. One can write out the explicit equations for the single and bivariate cases without an undue amount of "clutter" in the equations. However, beyond this, matrix notation is a virtual necessity. Otherwise, the explicit expressions are completely unwieldy.

Consider a set of n random variables $X_1, X_2, \ldots, X_n$ (also called variates). We define a vector random variable $\mathbf{X}$ as*

$$\mathbf{X} = \begin{bmatrix} X_1 \\ X_2 \\ \vdots \\ X_n \end{bmatrix} \tag{1.15.1}$$

In general, the components of $\mathbf{X}$ may be correlated and have nonzero means. We denote the respective means as $m_1, m_2, \ldots, m_n$, and thus we define a mean vector $\mathbf{m}$ as

$$\mathbf{m} = \begin{bmatrix} m_1 \\ m_2 \\ \vdots \\ m_n \end{bmatrix} \tag{1.15.2}$$

Also, if $x_1, x_2, \ldots, x_n$ is a set of realizations of $X_1, X_2, \ldots, X_n$, we can define a vector realization of $\mathbf{X}$ as

$$\mathbf{x} = \begin{bmatrix} x_1 \\ x_2 \\ \vdots \\ x_n \end{bmatrix} \tag{1.15.3}$$

We next define a matrix that describes the variances and correlation structure of

---

* Note that uppercase $\mathbf{X}$ denotes a column vector in this section. This is a departure from the usual notation of matrix theory, but it is necessitated by a desire to be consistent with the previous uppercase notation for random variables. The reader will simply have to remember this minor deviation in matrix notation. It appears in this section and also occasionally in Chapter 2.

the n variates. The *covariance matrix* for $\mathbf{X}$ is defined as

$$
\mathbf{C} = \begin{bmatrix} E\big[(X_1 - m_1)^2\big] & E[(X_1 - m_1)(X_2 - m_2)] & \cdots \\ E[(X_2 - m_2)(X_1 - m_1)] & & \\ \vdots & & \ddots \\ & & E\big[(X_n - m_n)^2\big] \end{bmatrix}
$$

(1.15.4)

The terms along the major diagonal of $\mathbf{C}$ are seen to be the variances of the variates, and the off-diagonal terms are the covariances.

The random variables $X_1, X_2, \ldots, X_n$ are said to be *jointly normal* or *jointly Gaussian* if their joint probability density function is given by

$$
f_X(\mathbf{x}) = \frac{1}{(2\pi)^{n/2}|\mathbf{C}|^{1/2}} \exp\left\{ -\frac{1}{2}\big[(\mathbf{x} - \mathbf{m})^T \mathbf{C}^{-1}(\mathbf{x} - \mathbf{m})\big]\right\} \quad (1.15.5)
$$

where $\mathbf{x}$, $\mathbf{m}$, and $\mathbf{C}$ are defined by Eqs. (1.15.2) to (1.15.4) and $|\mathbf{C}|$ is the determinant of $\mathbf{C}$. "Super -1" and "super T" denote matrix inverse and transpose, respectively. Note that the defining function for f_X is scalar and is a function of $x_1, x_2, \ldots, x_n$ when written out explicitly. We have shortened the indicated functional dependence to $\mathbf{x}$ just for compactness in notation. Also note that $\mathbf{C}^{-1}$ must exist in order for f_X to be properly defined by Eq. (1.15.5). Thus $\mathbf{C}$ must be nonsingular. More will be said of this later.

Clearly, Eq. (1.15.5) reduces to the standard normal form for the single variate case. For the bivariate case, we may write out f_X explicitly in terms of x_1 and x_2 without undue difficulty. Proceeding to do this, we have

$$
\mathbf{X} = \begin{bmatrix} X_1 \\ X_2 \end{bmatrix}, \qquad \mathbf{x} = \begin{bmatrix} x_1 \\ x_2 \end{bmatrix}, \qquad \mathbf{m} = \begin{bmatrix} m_1 \\ m_2 \end{bmatrix} \qquad (1.15.6)
$$

and

$$
\mathbf{C} = \begin{bmatrix} E\big[(X_1 - m_1)^2\big] & E[(X_1 - m_1)(X_2 - m_2)] \\ E[(X_1 - m_1)(X_2 - m_2)] & E\big[(X_2 - m_2)^2\big] \end{bmatrix}
$$

$$
= \begin{bmatrix} \sigma_1^2 & \rho\sigma_1\sigma_2 \\ \rho\sigma_1\sigma_2 & \sigma_2^2 \end{bmatrix} \qquad (1.15.7)
$$

The second form for $\mathbf{C}$ in Eq. (1.15.7) follows directly from the definitions of variance and correlation coefficient. The determinant of $\mathbf{C}$ and its inverse are given by

$$|\mathbf{C}| = \begin{vmatrix} \sigma_1^2 & \rho\sigma_1\sigma_2 \\ \rho\sigma_1\sigma_2 & \sigma_2^2 \end{vmatrix} = (1 - \rho^2)\sigma_1^2\sigma_2^2 \qquad (1.15.8)$$

$$\mathbf{C}^{-1} = \begin{bmatrix} \dfrac{\sigma_2^2}{|\mathbf{C}|} & -\dfrac{\rho\sigma_1\sigma_2}{|\mathbf{C}|} \\ -\dfrac{\rho\sigma_1\sigma_2}{|\mathbf{C}|} & \dfrac{\sigma_1^2}{|\mathbf{C}|} \end{bmatrix} = \begin{bmatrix} \dfrac{1}{(1 - \rho^2)\sigma_1^2} & \dfrac{-\rho}{(1 - \rho^2)\sigma_1\sigma_2} \\ \dfrac{-\rho}{(1 - \rho^2)\sigma_1\sigma_2} & \dfrac{1}{(1 - \rho^2)\sigma_2^2} \end{bmatrix}$$

$$(1.15.9)$$

Substituting Eqs. (1.15.8) and (1.15.9) into Eq. (1.15.5) then yields the desired density function in terms of x_1 and x_2.

$$f_{X_1X_2}(x_1, x_2) =$$

$$\frac{1}{2\pi\sigma_1\sigma_2\sqrt{1 - \rho^2}} \exp\left\{ -\frac{1}{2(1 - \rho^2)} \left[\frac{(x_1 - m_1)^2}{\sigma_1^2} \right.\right.$$

$$\left.\left. -\frac{2\rho(x_1 - m_1)(x_2 - m_2)}{\sigma_1\sigma_2} + \frac{(x_2 - m_2)^2}{\sigma_2^2} \right]\right\} \qquad (1.15.10)$$

It should be clear in Eq. (1.15.10) that $f_{X_1X_2}(x_1, x_2)$ is intended to mean the same as $f_X(\mathbf{x})$ in vector notation.

As mentioned previously, the third- and higher-order densities are very cumbersome to write out explicitly; therefore, we will examine the bivariate density in some detail in order to gain insight into the general multivariate normal density function. A sketch of $f_{X_1X_2}(x_1, x_2)$ is shown in Fig. 1.23. The bivariate normal density is seen to be a smooth hill-shaped surface over the

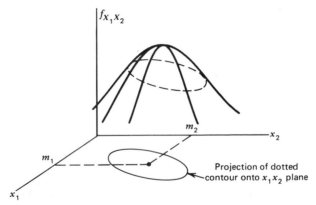

Figure 1.23. Bivariate normal density function.

x_1, x_2 plane with the peak of the hill occurring directly above the point (m_1, m_2) in the x_1, x_2 plane. Equal-height contours on the $f_{X_1X_2}$ surface project into ellipses in the x_1, x_2 plane, and the one shown in Fig. 1.23 is typical for a positive correlation coefficient ρ. Points on the ellipse may be thought of as equally likely combinations of x_1 and x_2. If $\rho = 0$, we have the case where X_1 and X_2 are uncorrelated, and the ellipses have their semimajor and semiminor axes parallel to the x_1 and x_2 axes. If we specialize further and let $\sigma_1 = \sigma_2$ (and $\rho = 0$), the ellipses degenerate to circles. In the other extreme, as $|\rho|$ approaches unity, the ellipses become more and more eccentric.

The uncorrelated case where $\rho = 0$ is of special interest, and in this case $f_{X_1X_2}$ reduces to the form given in Eq. (1.15.11).

For uncorrelated X_1 and X_2:

$$f_{X_1X_2}(x_1, x_2) = \frac{1}{2\pi\sigma_1\sigma_2} \exp\left\{-\frac{1}{2}\left[\frac{(x_1 - m_1)^2}{\sigma_1^2} + \frac{(x_2 - m_2)^2}{\sigma_2^2}\right]\right\}$$

$$= \frac{1}{\sqrt{2\pi}\,\sigma_1} e^{-(x_1-m_1)^2/2\sigma_1^2} \cdot \frac{1}{\sqrt{2\pi}\,\sigma_2} e^{-(x_2-m_2)^2/2\sigma_2^2} \quad (1.15.11)$$

The joint density function is seen to factor into the product of $f_{X_1}(x_1)$ and $f_{X_2}(x_2)$. Thus, *two normal random variables that are uncorrelated are also statistically independent*. It is easily verified from Eq. (1.15.5) that this is also true for any number of uncorrelated normal random variables. This is exceptional, because in general zero correlation does not imply statistical independence. It does, however, in the Gaussian case.

Now try to visualize the three-variate normal density function. The locus of constant $f_{X_1X_2X_3}(x_1, x_2, x_3)$ will be a closed elliptically shaped surface with three axes of symmetry. These aces will be aligned with the x_1, x_2, x_3 axes for the case of zero correlation among the three variates. If, in addition to zero correlation, the variates have equal variances, the surface becomes spherical.

If we try to extend the geometric picture beyond the three-variate case, we run out of Euclidean dimensions. However, conceptually, we can still envision equal-likelihood surfaces in hyperspace, but there is no way of sketching a picture of such surfaces. Some general properties of multivariate normal random variables will be explored further in the next section.

1.16
LINEAR TRANSFORMATION
AND GENERAL PROPERTIES
OF NORMAL RANDOM VARIABLES

The general linear transformation of one set of normal random variables to another is of special interest in noise analysis. This will now be examined in detail.

We have just seen that the density function for jointly normal random variables $X_1, X_2, \ldots, X_n$ can be written in matrix form as

$$f_X(\mathbf{x}) = \frac{1}{(2\pi)^{n/2}|\mathbf{C}_X|^{1/2}} \exp\left\{-\frac{1}{2}\left[(\mathbf{x} - \mathbf{m}_X)^T \mathbf{C}_X^{-1}(\mathbf{x} - \mathbf{m}_X)\right]\right\} \quad (1.16.1)$$

We have added the subscript X to $\mathbf{m}$ and $\mathbf{C}$ to indicate that these are the mean and covariance matrices associated with the $\mathbf{X}$ random variable. We now define a new set of random variables $Y_1, Y_2, \ldots, Y_n$ that are linearly related to $X_1, X_2, \ldots, X_n$ via the equation

$$\mathbf{y} = \mathbf{A}\mathbf{x} + \mathbf{b} \quad (1.16.2)$$

where lowercase $\mathbf{x}$ and $\mathbf{y}$ indicate realizations of $\mathbf{X}$ and $\mathbf{Y}$, $\mathbf{b}$ is a constant vector, and $\mathbf{A}$ is a square matrix that will be assumed to be nonsingular (i.e., invertible). We wish to find the density function for $\mathbf{Y}$, and we can use the methods of Section 1.14 to do so. In particular, the transformation is one-to-one; therefore, a generalized version of Eq. (1.14.32) may be used.

$$f_Y(\mathbf{y}) = f_X(\mathbf{x}(\mathbf{y}))\left|J\left(\frac{\mathbf{x}}{\mathbf{y}}\right)\right| \quad (1.16.3)$$

We must first solve Eq. (1.16.2) for $\mathbf{x}$ in terms of $\mathbf{y}$. The result is

$$\mathbf{x} = \mathbf{A}^{-1}\mathbf{y} - \mathbf{A}^{-1}\mathbf{b} \quad (1.16.4)$$

Let the individual terms of $\mathbf{A}^{-1}$ be denoted as

$$\mathbf{A}^{-1} = \begin{bmatrix} d_{11} & d_{12} & \cdots & d_{1n} \\ d_{21} & d_{22} & \cdots & \\ \vdots & \vdots & \cdots & \\ d_{n1} & \cdot & \cdots & d_{nn} \end{bmatrix} \quad (1.16.5)$$

The scalar equations represented by Eq. (1.16.4) are then

$$x_1 = (d_{11}y_1 + d_{12}y_2 + \cdots) - (d_{11}b_1 + d_{12}b_2 + \cdots)$$

$$x_2 = (d_{21}y_1 + d_{22}y_2 + \cdots) - (d_{21}b_1 + d_{22}b_2 + \cdots)$$

$$x_3 = \cdots \text{etc.} \quad (1.16.6)$$

The Jacobian of the transformation is then

$$J\left(\frac{\mathbf{x}}{\mathbf{y}}\right) = J\left(\frac{x_1, x_2, \dots}{y_1, y_2, \dots}\right) = \mathrm{Det}\begin{bmatrix} \dfrac{\partial x_1}{\partial y_1} & \dfrac{\partial x_2}{\partial y_1} & \cdots \\ \dfrac{\partial x_1}{\partial y_2} & \dfrac{\partial x_2}{\partial y_2} & \cdots \\ \cdots\cdots\cdots\cdots\cdots \end{bmatrix}$$

$$= \mathrm{Det}\begin{bmatrix} d_{11} & d_{21} & \cdots \\ d_{12} & d_{22} & \cdots \\ \cdots\cdots\cdots\cdots \end{bmatrix} = \mathrm{Det}\,(\mathbf{A}^{-1})^T = \mathrm{Det}\,(\mathbf{A}^{-1}) \quad (1.16.7)$$

We can now substitute Eqs. (1.16.4) and (1.16.7) into Eq. (1.16.3). The result is

$$f_Y(\mathbf{y}) = \frac{\left|\mathrm{Det}\,(\mathbf{A}^{-1})\right|}{(2\pi)^{n/2}|\mathbf{C}_X|^{1/2}}$$

$$\times \exp\left\{-\frac{1}{2}\left[\left(\mathbf{A}^{-1}\mathbf{y} - \mathbf{A}^{-1}\mathbf{b} - \mathbf{m}_X\right)^T \mathbf{C}_X^{-1}\left(\mathbf{A}^{-1}\mathbf{y} - \mathbf{A}^{-1}\mathbf{b} - \mathbf{m}_X\right)\right]\right\}$$

$$(1.16.8)$$

We find the mean of **Y** by taking the expectation of both sides of the linear transformation

$$\mathbf{Y} = \mathbf{AX} + \mathbf{b}$$

Thus

$$\mathbf{m}_Y = \mathbf{Am}_X + \mathbf{b} \qquad (1.16.9)$$

The exponent in Eq. (1.16.8) may now be written as

$$-\frac{1}{2}\left[\left(\mathbf{A}^{-1}\mathbf{y} - \mathbf{A}^{-1}\mathbf{b} - \mathbf{A}^{-1}\mathbf{Am}_X\right)^T \mathbf{C}_X^{-1}\left(\mathbf{A}^{-1}\mathbf{y} - \mathbf{A}^{-1}\mathbf{b} - \mathbf{A}^{-1}\mathbf{Am}_X\right)\right]$$

$$= -\frac{1}{2}\left[(\mathbf{y} - \mathbf{m}_Y)^T (\mathbf{A}^{-1})^T \mathbf{C}_X^{-1}\mathbf{A}^{-1}(\mathbf{y} - \mathbf{m}_Y)\right]$$

$$= -\frac{1}{2}\left[(\mathbf{y} - \mathbf{m}_Y)^T \left(\mathbf{A}\mathbf{C}_X\mathbf{A}^T\right)^{-1}(\mathbf{y} - \mathbf{m}_Y)\right] \qquad (1.16.10)$$

The last step in Eq. (1.16.10) is accomplished by using the reversal rule for the inverse of triple products and noting that the order of the transpose and inverse operations may be interchanged. Also note that

$$\left|\mathrm{Det}\,(\mathbf{A}^{-1})\right| = \frac{1}{|\mathrm{Det}\,\mathbf{A}|} = \frac{1}{|\mathrm{Det}\,\mathbf{A}|^{\frac{1}{2}} \cdot |\mathrm{Det}\,\mathbf{A}^{T}|^{\frac{1}{2}}} \qquad (1.16.11)$$

Substitution of the forms given in Eqs. (1.16.10) and (1.16.11) into Eq. (1.16.8) yields for f_Y

$$f_Y(\mathbf{y}) = \frac{1}{(2\pi)^{n/2}|\mathbf{A}\mathbf{C}_X\mathbf{A}^T|^{1/2}}$$

$$\times \exp\left\{-\frac{1}{2}\left[(\mathbf{y} - \mathbf{m}_Y)^T(\mathbf{A}\mathbf{C}_X\mathbf{A}^T)^{-1}(\mathbf{y} - \mathbf{m}_Y)\right]\right\} \quad (1.16.12)$$

It is apparent now that f_Y is also normal in form with the mean and covariance matrix given by

$$\mathbf{m}_Y = \mathbf{A}\mathbf{m}_X + \mathbf{b} \qquad (1.16.13)$$

and

$$\mathbf{C}_Y = \mathbf{A}\mathbf{C}_X\mathbf{A}^T \qquad (1.16.14)$$

Thus we see that *normality is preserved in a linear transformation*. All that is changed is the mean and the covariance matrix; the *form* of the density function remains unchanged.

There are, of course, an infinite number of linear transformations one can make on a set of normal random variables. The particular transformation, say $\mathbf{S}$, that produces a new covariance matrix $\mathbf{S}\mathbf{C}_X\mathbf{S}^T$ that is diagonal is of special interest. This transformation yields a new set of normal random variables that are uncorrelated, and thus they are also statistically independent. In a given problem, we may not choose to actually make this change of variables, but it is important just to know that the variables can be decoupled and under what circumstances this can be done. It works out that a diagonalizing transformation will always exist if $\mathbf{C}_X$ is positive definite.* In the case of a covariance matrix, this is the equivalent of saying that all the correlation coefficients are less than unity in magnitude. This will be demonstrated for the bivariate case, and the extension to higher-order cases is fairly obvious.

A symmetric matrix $\mathbf{C}$ is said to be positive definite if the scalar $\mathbf{x}^T\mathbf{C}\mathbf{x}$ is positive for all nontrivial $\mathbf{x}$, that is, $\mathbf{x} \neq 0$. Writing out $\mathbf{x}^T\mathbf{C}\mathbf{x}$ explicitly for the

---

* The diagonalizing transformation is sometimes called the *similarity* transformation. In the situation at hand, $\mathbf{C}_X$ is a symmetric, positive-definite matrix, and there will exist a set of linearly independent, orthogonal eigenvectors for such a matrix (8). The transforming matrix $\mathbf{S}$ has as its columns the normalized eigenvectors of $\mathbf{C}_X$. In this very special case, $\mathbf{S}^{-1} = \mathbf{S}^T$.

2×2 case yields

$$[x_1 x_2] \begin{bmatrix} c_{11} & c_{12} \\ c_{12} & c_{22} \end{bmatrix} \begin{bmatrix} x_1 \\ x_2 \end{bmatrix} = c_{11} x_1^2 + 2 c_{12} x_1 x_2 + c_{22} x_2^2 \qquad (1.16.15)$$

But if $\mathbf{C}$ is a covariance matrix,

$$c_{11} = \sigma_1^2, \qquad c_{12} = \rho \sigma_1 \sigma_2, \qquad c_{22} = \sigma_2^2 \qquad (1.16.16)$$

Therefore, $\mathbf{x}^T \mathbf{C} \mathbf{x}$ is

$$\mathbf{x}^T \mathbf{C} \mathbf{x} = (\sigma_1 x_1)^2 + 2\rho(\sigma_1 x_1)(\sigma_2 x_2) + (\sigma_2 x_2)^2 \qquad (1.16.17)$$

Equation (1.16.17) now has a simple geometric interpretation. Assume $|\rho| < 1$; ρ can then be related to the negative cosine of some angle θ, where $0 < \theta < \pi$. Equation (1.16.17) will then be recognized as the equation for the square of the "opposite side" of a general triangle; and this, of course, must be positive. Thus, a 2×2 covariance matrix is positive definite, provided $|\rho| < 1$.

It is appropriate now to summarize some of the important properties of multivariate normal random variables:

1. The probability density function describing a vector random variable $\mathbf{X}$ is completely defined by specifying the mean and covariance matrix of $\mathbf{X}$.

2. The covariance matrix of $\mathbf{X}$ is positive definite if the magnitudes of all correlation coefficients are less than unity.

3. If normal random variables are uncorrelated, they are also statistically independent.

4. A linear transformation of normal random variables leads to another set of normal random variables. A decoupling (decorrelating) transformation will always exist if the original covariance matrix is positive definite.

5. If the joint density function for n random variables is normal in form, all marginal and conditional densities associated with the n variates will also be normal in form. (This was not shown; see Problem 1.39.)

1.17
LIMITS, CONVERGENCE, AND
UNBIASED ESTIMATORS

No discussion of probability could be complete without at least some mention of limits and convergence. To put this in perspective, we first review the usual deterministic concept of convergence. As an example, recall that the Maclaurin series for e^x is

$$e^x = 1 + x + \frac{x^2}{2!} + \frac{x^3}{3!} + \cdots \qquad (1.17.1)$$

This series converges uniformly to e^x for all real x in any finite interval. By convergence we mean that if a given accuracy figure is specified, we can find an appropriate number of terms such that the specified accuracy is met by a truncated version of the series. In particular, note that once we have determined how many terms are needed in the truncated series, this same number is good for all x within the interval, and there is nothing "chancy" about it. In contrast, we will see presently that such "100 percent sure" statements cannot be made in probabilistic situations. A look at the sample mean of n random variables will serve to illustrate this.

Let $X_1, X_2, \ldots, X_n$ be independent random variables with identical probability density functions $f_X(x)$. In terms of an experiment, these may be thought of as ordered samples of the random variable X. Next, consider a sequence of random variables defined as follows:

$$Y_1 = X_1$$

$$Y_2 = \frac{X_1 + X_2}{2}$$

$$Y_3 = \frac{X_1 + X_2 + X_3}{3}$$

$$\vdots$$

$$Y_n = \frac{X_1 + X_2 + \cdots X_n}{n} \qquad (1.17.2)$$

The random variable Y_n is, of course, just the sample mean of the random variable X. We certainly expect Y_n to get closer to $E(X)$ as n becomes large. But closer in what sense? This is the crucial question. It should be clear that any particular experiment could produce an "unusual" event in which the sample mean would differ from $E(X)$ considerably. On the other hand, quite by chance, a similar experiment might yield a sample mean that was quite close to $E(X)$. Thus, in this probabilistic situation, we cannot expect to find a fixed number of samples n that will meet a specified accuracy figure for all experiments. No matter how large we make n, there is always some nonzero probability that the very unusual thing will happen, and a particular experiment will yield a sample mean that is outside the specified accuracy. Thus, we can only hope for convergence in some sort of average sense and not in an absolute (100 percent sure) sense.

Let us now be more specific in this example, and let X (and thus $X_1, X_2, \ldots, X_n$) be normal with mean m_X and variance σ_X^2. From Section 1.16 we also know that the sample mean Y_n is a normal random variable. Since a normal random variable is characterized by its mean and variance, we now examine these parameters for Y_n. The expectation of a sum of elements is the sum of the expectations of the elements. Thus

$$E(Y_n) = E\left(\frac{X_1 + X_2 + \cdots X_n}{n}\right)$$

$$= \frac{1}{n}[E(X_1) + E(X_2) + \cdots]$$

$$= \frac{1}{n}[nE(X)] = m_X \tag{1.17.3}$$

The sample mean is, of course, an estimate of the true mean of X, and we see from Eq. (1.17.3) that it at least yields $E(X)$ "on the average." Estimators that have this property are said to be unbiased. That is, an estimator is said to be *unbiased* if

$$E(\text{Estimate of } X) = E(X) \tag{1.17.4}$$

Consider next the variance of Y_n. Using Eq. (1.17.3) and recalling that the expectation of the sum is the sum of the expectations,

$$\text{Var } Y_n = E[Y_n - E(Y_n)]^2$$

$$= E(Y_n^2 - 2Y_n m_X + m_X^2)$$

$$= E(Y_n^2) - m_X^2 \tag{1.17.5}$$

The sample mean Y_n may now be replaced with $(1/n)(X_1 + X_2 \cdots + X_n)$; and, after squaring and some algebraic simplification, Eq. (1.17.5) reduces to

$$\text{Var } Y_n = \frac{1}{n}\text{Var } X$$

$$= \frac{\sigma_X^2}{n} \tag{1.17.6}$$

Thus we see that the variance of the sample mean decreases with increasing n and eventually goes to zero as $n \to \infty$.

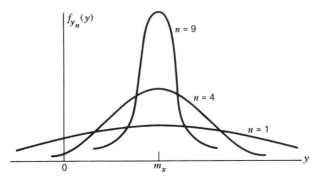

Figure 1.24. Probability density functions illustrating convergence of the sample mean.

The probability density functions associated with the sample mean are shown in Fig. 1.24 for three values of n. It should be clear from the figure that convergence of some sort takes place as $n \to \infty$. However, no matter how large we make n, there will still remain a nonzero probability that Y_n will fall outside some specified accuracy interval. Thus, we have convergence in only a statistical sense and not in an absolute deterministic sense.

There are a number of types of statistical convergence that have been defined and are in common usage (9, 10). We look briefly at two of these. Consider a sequence of random variables $Y_1, Y_2, \ldots, Y_n$. The sequence Y_n is said to *converge in the mean* (or mean square) to Y if

$$\lim_{n \to \infty} E\left[(Y_n - Y)^2\right] = 0 \qquad (1.17.7)$$

Convergence in the mean is sometimes abbreviated as

$$\text{l.i.m.} \, Y_n = Y \qquad (1.17.8)$$

where l.i.m. denotes "limit in the mean."

The sequence Y_n *converges in probability* to Y if

$$\lim_{n \to \infty} P\left(|Y_n - Y| \geq \varepsilon\right) = 0 \qquad (1.17.9)$$

where ε is an arbitrarily small positive number.

It should be clear from Eqs. (1.17.3) and (1.17.6) that the sample mean converges "in the mean" to m_X. It also converges in probability because the area under the "tails" of the probability density outside a specified interval about m_X goes to zero as $n \to \infty$. Roughly speaking, convergence in the mean indicates that the dispersion (variance) about the limiting value shrinks to zero in the limit. Similarly, convergence in probability means than an arbitrarily small accuracy criterion is met with a probability of one as $n \rightarrow \infty$. Davenport and Root (9) point out that convergence in the mean is a more severe requirement than convergence in probability. Thus, if a sequence converges in the mean, we are also assured that it will converge in probability. The converse is not true though, because convergence in probability is a "looser" sort of criterion than convergence in the mean.

PROBLEMS

1.1. In straight poker, five cards are dealt to each player from a deck of ordinary playing cards. What is the probability that a player will be dealt a flush (i.e., five cards all of one suit)?

1.2. In the game of blackjack, the player is initially dealt two cards from a deck of ordinary playing cards. Without going into all the details of the game, it will suffice to say here that the best possible hand one could receive on the initial deal is a combination of an ace of any suit and any face card or 10. What is the probability that the player will be dealt this combination?

1.3. An urn contains two white, one black, and four red balls. Three balls are drawn out simultaneously.
   (a) How many elements are in the sample space for this experiment?
   (b) What is the probability that the draw will produce one white and two red balls?

1.4. Five dice are rolled simultaneously.
   (a) Describe the sample space for this experiment.
   (b) What is the probability that exactly one 6 will be rolled?
   (c) What is the probability that at least one 6 will be rolled?

1.5. Consider a horse race with five entries. The approximate win odds posted by the track just prior to the race are:

Horse Number	Approximate Odds
1	2 to 1
2	3 to 1
3	5 to 1
4	5 to 1
5	11 to 1

Track "odds" simply refer to additional amount returned to the bettor in event of a win. For example, if the bettor places a bet on horse 1 and that horse wins with 2 to 1 odds, the track will return 1 (the original wager) plus 2 (for winning), or 3 dollars for each dollar wagered.

Assume the race is a chance event and let the sample space consist of five elements: Horse 1 wins, Horse 2 wins, etc. Ignore the small percentage kept by the track (the odds are approximate anyway), and make a reasonable probability assignment to each element in the sample space for this situation.

1.6. Sketch a Venn diagram for three events A, B, and C. Use the diagram to justify the following relationship:

$$P(A \cup B \cup C) = P(A) + P(B) + P(C) - P(A \cap B)$$
$$- P(B \cap C) - P(A \cap C) + P(A \cap B \cap C) \quad \text{(P1.6)}$$

1.7. Contract bridge is played with an ordinary deck of 52 playing cards. There are four players with players on opposite sides of the table being partners. One player acts as dealer and deals each player 13 cards. A bidding sequence then takes place, and this establishes the trump suit and names the player who is to attempt to take a certain number of tricks. This player is called the declarer. The play begins with the player to the declarer's left leading a card. The declarer's partner's hand is then laid out on the table face up for everyone to see. This then enables the declarer to see a total of 27 cards as the play begins. Knowledge of these cards will, of course, affect his or her strategy in the subsequent play.

Suppose the declarer sees 11 of the 13 trump cards as the play begins. We will assume that the opening lead was not a trump, which leaves 2 trumps outstanding in the opponents' hands. The disposition of these is, of course, unknown to the declarer. There are, however, a limited number of possibilities:

(a) Both trumps lie with the opponent to the left and none to the right.

(b) Both trumps are to the right and none to the left.

(c) The two trumps are split, one in each opponent's hand.

Compute the probabilities for each of the (a), (b), and (c) possibilities. (*Hint:* Rather than look at all possible combinations, look at numbers of combinations for 25 specific cards held by the opponents just after the opening lead. Two of these will, of course, be specific trump cards. The resulting probability will be the same regardless of the particular choice of specific cards.)

1.8. In the game of craps the casino also has ways of making money other than betting against the player rolling the dice. The other participants standing around the table may place side bets with the casino while waiting their turn with the dice. One such side bet at some tables is "Don't Pass—Bar 12." The participant placing such a side bet wins if the player rolling the dice loses, except for the case of a 12 on the first roll of the dice, and in that case the result is a standoff. In other words, the side bettor is betting against the roller, except that the rules are changed slightly in the side-bet wager to make it such that 12 on the first roll does not win for either the casino or the side bettor. Presumably, this tips the odds in favor of the casino for the side bet.

(a) Find the probability of winning when placing a side bet "Don't Pass—Bar 12."

(b) It is often said that the casinos make a higher percentage (on the average) on side bets than they do on the regular bet with the player rolling the dice. Compare the results of this problem with those of Example 1.2, relative to the average percentage "take" for the casino. [It works out that this side bet is fairly even. Others are not so favorable. See Goren (4).]

1.9. Assume equal likelihood for the birth of boys and girls. What is the probability that a four-child family chosen at random will have two boys and two girls, irrespective of the order of birth? (Note that the answer is not $\frac{1}{2}$ as might be suspected at first glance.)

1.10. Consider a sequence of random binary digits, zeros and ones. Each digit may be thought of as an independent sample from a sample space containing two elements, each having a probability of $\frac{1}{2}$. For a six-digit sequence, what is the probability of having:

(a) Exactly 3 zeros and 3 ones arranged in any order?
(b) Exactly 4 zeros and 2 ones arranged in any order?
(c) Exactly 5 zeros and 1 one arranged in any order?
(d) Exactly 6 zeros?

1.11. A certain binary message is n bits in length. If the probability of making an error in the transmission of a single bit is p, and if the error probability does not depend on the outcome of any previous transmissions, show that the probability of occurrence of exactly k bit errors in a message is

$$P(k \text{ errors}) = \binom{n}{k} p^k (1-p)^{n-k} \tag{P1.11}$$

The quantity $\binom{n}{k}$ denotes the number of combinations of n things taken k at a time. (This is a generalization of Problems 1.9 and 1.10.)

1.12. Video poker has recently become a popular game in casinos in the United States (11). The player plays against the machine in much the same way as with slot machines, except that the machine displays cards on a video screen instead of the familiar bells, bars, etc., on spinning wheels. When a coin is put into the machine, it immediately displays five cards on the screen. After this initial five-card deal, the player is allowed to discard one to five cards at his or her discretion and obtain replacement cards (i.e., this is the "draw"). The object, of course, is to try to improve the poker hand with the draw.

(a) Suppose the player is dealt the 3, 7, 8, 10 of hearts and the queen of spades on the initial deal. The player then decides to keep the four hearts and discard the queen of spades in hopes of getting another heart on the draw, and thus obtain a flush (five cards, all of the same suit). The typical video poker machine pays out five coins for a flush. Assume that this is the payout and that the machine is statistically fair. What is the expected (i.e., average) return for this draw situation? (Note that an average return of 1.0 is the break-even return.)

(b) Consider another deal where the initial deal is 3 of spades, 3 of hearts, 10 of spades, 10 of clubs, and 5 of clubs. Suppose the player decides to discard the 5 of clubs and draw one card in hope of improving the two-pair hand to a full house (i.e., any three-of-a-kind and a pair combination). The typical payouts for two-pair and full-house hands are 2 and 8, respectively. What is the expected return for this draw situation?

(c) Parts (a) and (b) of this problem can obviously be worked intuitively using the relative frequency of occurrence concept. Put your solutions in a more formal setting for each case by defining a suitable sample space, an associated random variable, and an appropriate expectation formula. (For this part, the conditioning may be assumed to be implied and thus suppressed in the sample space.)

1.13. The random variable X may take on all values between 0 and 2, with all values within this range being equally likely.
(a) Sketch the probability density function for X.
(b) Sketch the cumulative probability distribution function for X.
(c) Calculate $E(X)$, $E(X^2)$, and Var X.

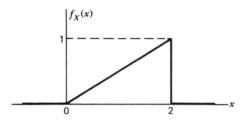

Problem 1.14

1.14. A random variable X has a probability density function as shown.
(a) Sketch the cumulative distribution function for X.
(b) What is the variance of X?

1.15. Resistors are a common component in nearly all electronic equipment. Suppose a resistor company manufactures a large lot of resistors whose nominal value is to be 100 Ω. Because of manufacturing tolerances this results in resistance values spread between 90 and 110 Ω (± 10 percent of nominal). For the sake of simplicity let us assume the spread is uniform between the two extremes. Let us further speculate that the manufacturer sorts out all the ± 5 percent tolerance resistors for sale at a premium price; the remaining resistors are then classified as "10 percent resistors" and are to be sold at the regular price. Let the resistance value of such a "10 percent resistor" be a random variable X.
(a) What is the probability density function for X?
(b) What is the probability that X will lie within 8 percent of the nominal value?

1.16. Suppose the resistance values of Problem 1.15 were normally distributed with a mean of 100 Ω and a standard deviation of 5 Ω. What is the probability that a sample resistance out of the batch will lie between 90 and 110 Ω? (*Note:* Negative resistance is not physically possible, so a normal distribution can only be an approximation in this situation.)

1.17. A random variable X whose probability density function is given by

$$f_X(x) = \begin{cases} \alpha\, e^{-\alpha x}, & x \geq 0 \\ 0, & x < 0 \end{cases}$$

is said to have an exponential probability density function. This density function is sometimes used to describe the failure of equipment components (12, 13). That is, the probability that a particular component will fail within time T is

$$P(\text{failure}) = \int_0^T f_X(x)\, dx \qquad (P1.17)$$

Note that α is a parameter that may be adjusted to fit the situation at hand.
  (a) Find α for an electronic component whose average lifetime is 10,000 hours. ("Average" is used synonymously with "expectation" here.)
  (b) Suppose we wish the probability of failure for the component of part (a) to be less than .01, that is, we wish the reliability to be .99. For what time span might we expect a reliability of .99?

1.18. Consider a sack containing several identical coins whose sides are labeled $+1$ and -1. A certain number of coins are withdrawn and tossed simultaneously. The algebraic sum of the numbers resulting from the toss is a discrete random variable. Sketch the probability density function associated with the random variable for the following situations:
  (a) One coin is tossed.
  (b) Two coins are tossed.
  (c) Five coins are tossed.
  (d) Ten coins are tossed.
The density functions in this case consist of impulses. In the sketches, represent impulses with "arrows" whose lengths are proportional to the magnitudes of the impulses. (This is the discrete analog of Example 1.21. Note the tendency toward central distribution.)

1.19. Two players are matching pennies. Each player begins with a stack of pennies and they uncover their pennies one at a time. By mutual agreement, pennies that match are taken by one player and those that do not match are taken by the other. When either player's original stack is exhausted, the players restack their available pennies and repeat the matching process. This is continued until one or the other player has won all the pennies.

Suppose player A has two pennies and player B has one. Sketch the probability density function for the number of trials required to end the game. The game ends when either player runs out of pennies. (Just as in Problem 1.18, use arrows to represent impulses in the sketch.)

1.20. Consider a random variable that is defined to be the sum of the squares of n independent normal random variables, all of which are $N(0, 1)$. The parameter n is any positive integer. Such a sum-of-the-squares random variable is called a chi-square random variable with n degrees of freedom. The probability density function associated with a chi-square random variable X is

$$f_X(x) = \begin{cases} \dfrac{x^{(n/2-1)}e^{-x/2}}{2^{n/2}\Gamma\left(\dfrac{n}{2}\right)}, & x > 0 \\[2mm] 0, & x \le 0 \end{cases}$$

where Γ indicates the gamma function (3, 10). It is not difficult to show that the mean and variance of X are given by

$$E(X) = n$$

$$\text{Var } X = 2n$$

(This is easily derived by noting that the defining integral expressions for the first and second moments of X are in the exact form of a single-sided Laplace transform with $s = \frac{1}{2}$. See Appendix A for a table of Laplace transforms and note that $n! = \Gamma(n + 1)$.)

    (a) Now, suppose we form the sum of the squares of M similar normal random variables that are $N(0, \sigma^2)$ rather than $N(0, 1)$. What would be the mean and variance for such an unnormalized chi-square-like random variable?

    (b) Make rough sketches of both the chi-square density and cumulative distribution functions for $n = 1, 2,$ and 3. (If you wish to be precise in your sketches, refer to Shanmugan and Breipohl (10) for numerical values of the chi-square distribution function.)

1.21. Let the sum of the dots resulting from the throw of two dice be a discrete random variable X. The probabilities associated with their permissible values $2, 3, \ldots, 12$ are easily found by itemizing all possible results of the throw (or from Table 1.2). Find $E(X)$ and Var X.

1.22. Discrete random variables X and Y may each take on integer values 1, 3, and 5, and the joint probability of X and Y is given in the table below.

X \ Y	1	3	5
1	$\frac{1}{18}$	$\frac{1}{18}$	$\frac{1}{18}$
3	$\frac{1}{18}$	$\frac{1}{18}$	$\frac{1}{6}$
5	$\frac{1}{18}$	$\frac{1}{6}$	$\frac{1}{3}$

    (a) Are random variables X and Y independent?

    (b) Find the unconditional probability $P(Y = 5)$.

    (c) What is the conditional probability $P(Y = 5 | X = 3)$?

1.23. The diagram shown below gives the error characteristics of a hypothetical binary transmission system. The numbers shown next to the arrows are the conditional probabilities of Y given X. The unconditional probabilities for X are shown to the left of the figure. Find:

    (a) The conditional probabilities $P(X = 0 | Y = 1)$ and $P(X = 0 | Y = 0)$.

    (b) The unconditional probabilities $P(Y = 0)$ and $P(Y = 1)$.

    (c) The joint probability array for $P(X, Y)$.

Total Probability $P(X)$	Transmitter X	Receiver Y

$P(X = 0) = .75$

$P(X = 1) = .25$

Problem 1.23

1.24. The Poisson distribution is defined as

$$P(k) = \frac{\lambda^k}{k!}e^{-\lambda} \qquad \text{(P1.24)}$$

where λ is a positive real parameter and $k = 0, 1, 2, \ldots$. This discrete distribution appears in a wide variety of applications in which events occur at random with time (6, 12). For example, a particular telephone call handled by an office might occur with equal likelihood at any time during the working day. Let the average number of calls handled per unit time be α. Then the probability of exactly k calls occurring in a time interval T is

$$P(k) = \frac{(\alpha T)^k}{k!}e^{-\alpha T}$$

Obviously, in this instance αT is the λ parameter of the Poisson distribution.
   (a) Let αT be 2 and sketch the corresponding probability density as a function of k for $k = 0, 1, 2, \ldots, 10$. (As before, use arrows for impulses.)
   (b) Find the mean for the Poisson distribution. (*Hint:* The characteristic function discussed in Section 1.8 is useful here. *Answer:* λ.)

1.25. A large power system has an average of three outages per year somewhere in the system. If the outages are assumed to occur randomly, what is the probability of going without an outage for the next year? (See Problem 1.24 for a discussion of the Poisson distribution.)

1.26. Records show that a certain city has had an average of 1 earthquake per decade when averaged over the past 200 years. Assume that earthquakes occur at random and find the probability that at least one earthquake will occur within:
   (a) The next decade.
   (b) A person's normal life span (about 70 years).
(See Problem 1.24 for a discussion of the Poisson distribution.)

1.27. The Rayleigh probability density function is defined as

$$f_R(r) = \frac{r}{\sigma^2} e^{-r^2/2\sigma^2} \tag{P1.27}$$

where σ^2 is a parameter of the distribution (see Example 1.24).
   (a) Find the mean and variance of a Rayleigh distributed random variable R.
   (b) Find the mode of R (i.e., the most likely value of R).

1.28. Three similar "unfair" coins are tossed simultaneously. The coins are unfair in that $P(\text{Heads}) = .6$ and $P(\text{Tails}) = .4$. Let the discrete random variable X be defined to be the number of heads that result from the toss of the three coins. Find the discrete probability distribution associated with the random variable X.

1.29. The target shooting example of Section 1.14 led to the Rayleigh density function specified by Eq. (1.14.35) (also repeated in Problem 1.27).
   (a) Show that the probability that a hit will lie within a specified distance R_0 from the origin is given by

$$P(\text{Hit lies within } R_0) = 1 - e^{-R_0^2/2\sigma^2} \tag{P1.29}$$

   (b) The value of R_0 in Eq. (P1.29) that yields a probability of .5 is known as the circular probable error (or circular error probable, CEP). Find the CEP in terms of σ.

1.30. Find the mean and variance of the output of a half-wave rectifier driven by Gaussian noise. (The probability density function for the output is given in Example 1.18.)

1.31. Consider a random variable X with an exponential probability density function

$$f_X(x) = \begin{cases} e^{-x}, & x \geq 0 \\ 0, & x < 0 \end{cases}$$

Find:
   (a) $P(X \geq 2)$
   (b) $P(1 \leq X \leq 2)$
   (c) $E(X)$, $E(X^2)$, and Var X.

1.32. Random variables X and Y have a joint probability density function defined as follows:

$$f_{XY}(x, y) = \begin{cases} .25, & -1 \leq x \leq 1 \quad \text{and} \quad -1 \leq y \leq 1 \\ 0, & \text{otherwise} \end{cases}$$

Are random variables X and Y statistically independent? (*Hint:* Integrate with respect to appropriate dummy variables to obtain $f_X(x)$ and $f_Y(y)$. Then check to see if the product of f_X and f_Y is equal to f_{XY}.)

1.33.   Random variables X and Y have a joint probability density function

$$f_{XY}(x, y) = \begin{cases} e^{-(x+y)}, & x \geq 0 \quad \text{and} \quad y \geq 0 \\ 0, & \text{otherwise} \end{cases}$$

Find:
   (a) $P(X \leq \frac{1}{2})$
   (b) $P[(X + Y) \leq 1]$
   (c) $P[(X \text{ or } Y) \geq 1]$
   (d) $P[(X \text{ and } Y) \geq 1]$

1.34.   Are the random variables of Problem 1.33 statistically independent?

1.35.   Random variables X and Y are statistically independent and their respective probability density functions are

$$f_X(x) = \tfrac{1}{2}e^{-|x|}$$

$$f_Y(y) = e^{-2|y|}$$

Find the probability density function associated with $X + Y$. (*Hint:* Fourier transforms are helpful here.)

1.36.   Random variable X has a probability density function

$$f_X(x) = \begin{cases} \tfrac{1}{2}, & -1 \leq x \leq 1 \\ 0, & \text{otherwise} \end{cases}$$

Random variable Y is related to X through the equation

$$y = x^3 + 1$$

What is the probability density function for Y?

1.37. X and Y are independent, zero-mean random variables with variances σ_X^2 and σ_Y^2. Another set of random variables U and V are related to X and Y through the equations

$$u = 2x + y$$

$$v = x - y$$

Find the correlation coefficient of U and V. Let $\sigma_X^2 = \sigma_Y^2$.

1.38. The vector Gaussian random variable

$$\mathbf{X} = \begin{bmatrix} X_1 \\ X_2 \end{bmatrix}$$

is completely described by its mean and covariance matrix. In this case they are

$$\mathbf{m}_X = \begin{bmatrix} 1 \\ 2 \end{bmatrix}$$

$$\mathbf{C}_X = \begin{bmatrix} 4 & 1 \\ 1 & 1 \end{bmatrix}$$

Now consider another vector random variable **Y** that is related to **X** by the equation

$$\mathbf{y} = \mathbf{Ax} + \mathbf{b}$$

where

$$\mathbf{A} = \begin{bmatrix} 2 & 1 \\ 1 & -1 \end{bmatrix}, \quad \mathbf{b} = \begin{bmatrix} 1 \\ 1 \end{bmatrix}$$

Find the mean and covariance matrix for **Y**.

1.39. The general bivariate normal density function is given by Eq. (1.16.1). It should be apparent that the unconditional density functions $f_{X_1}(x_1)$ and $f_{X_2}(x_2)$ are also normal in form. (A linear transformation yielding uncorrelated random variables may be made, and then the normal form is obvious.) Show that the conditional density functions $f_{X_1|X_2}(x_1)$ and $f_{X_2|X_1}(x_2)$ are also normal in form. (*Hint:* Simply noting the quadratic form in the exponential is sufficient justification of normality.)

1.40. A pair of random variables, X and Y, have a joint probability density function

$$f_{XY}(x, y) = \begin{cases} 1, & 0 \le y \le 2x \quad \text{and} \quad 0 \le x \le 1 \\ 0, & \text{elsewhere} \end{cases}$$

Find:

$$E(X|Y = .5)$$

[*Hint:* Use Eq. (1.11.19) to find $f_{X|Y}(x)$ for $y = .5$, and then integrate $x f_{X|Y}(x)$ to find $E(X|Y = .5)$.]

1.41. Two continuous random variables X and Y have a joint probability density function that is uniform inside the unit circle and zero outside, that is,

$$f_{XY}(x, y) = \begin{cases} 1/\pi, & (x^2 + y^2) \le 1 \\ 0, & (x^2 + y^2) > 1 \end{cases}$$

(a) Find the unconditional probability density function for the random variable Y and sketch the probability density as a function of Y.

(b) Are the random variables X and Y statistically independent?

1.42. A normal random variable X is described by $N(0, 4)$. Similarly, Y is normal and is described by $N(1, 9)$. X and Y are independent. Another random variable Z is defined by the additive combination

$$Z = X + 2Y$$

Write the explicit expression for the probability density function for Z. Use appropriate numerical values for all parameters.

1.43. Real-life problems that involve random processes are often quite formidable when approached analytically. Thus, in many cases we must resort to empirical methods. In the empirical approach, a hypothetical experiment is repeated over and over again under the same probabilistic conditions, and the results are then analyzed statistically. Such methods are often referred to as *Monte Carlo simulation* (for the famous casino in Monaco). In performing Monte Carlo experiments, there is a need for generating a large number of samples, chosen at random, from a population with a specified distribution. More often than not, a normal distribution is desired. It works out that it is relatively easy with a computer to generate independent random samples with a uniform distribution between 0 and 1. (These are actually pseudorandom rather than truly random.) The problem then is to work out an algorithm that will convert samples with a uniform distribution to a corresponding set with a normal distribution.

One method of obtaining an approximate normal distribution is to sum random variables with a uniform distribution and take advantage of the central tendency property that was discussed in Example 1.21, Section 1.13. One such algorithm is as follows: Let $X_1, X_2, \ldots, X_{12}$ be independent random variables with a uniform distribution between 0 and 1. Next, define a new random variable S to be

$$S = -6 + \sum_{i=1}^{12} X_i$$

Show that S has zero mean and unity variance. Then sketch a rough plot of the probability density for S, and show on your sketch where the density function truncates. Note that even though the density function for S may approximate the normal density quite well in the "fat" region, the "tails" of the two distributions are quite different. In one case certain very large numbers can *never* occur, whereas in the other they occur only *rarely*. This can be an important difference in some applications.

REFERENCES CITED IN CHAPTER 1

1. N. Wiener, *Extrapolation, Interpolation, and Smoothing of Stationary Time Series*, Cambridge, Mass.: MIT Press; and New York: Wiley; 1949.
2. S. O. Rice, "Mathematical Analysis of Noise," *Bell System Tech. J.*, *23*, 282–332 (1944); *24*, 46–256 (1945).
3. A. M. Mood, F. A. Graybill, and D. C. Boes, *Introduction to the Theory of Statistics*, 3rd ed., New York: McGraw-Hill, 1974.
4. C. H. Goren, *Go With the Odds*, New York: Macmillan, 1969.
5. W. Feller, *An Introduction to Probability Theory and Its Applications*, Vol. 1, 2nd ed., New York: Wiley, 1957.
6. P. Beckmann, *Probability in Communication Engineering*, New York: Harcourt, Brace, and World, 1967.
7. A. Papoulis, *Probability, Random Variables, and Stochastic Processes*, 2nd ed., New York: McGraw-Hill, 1984.
8. S. I. Grossman, *Elementary Linear Algebra*, Florence, KY: Wadsworth, Inc., 1980.
9. W. B. Davenport, Jr., and W. L. Root, *An Introduction to the Theory of Random Signals and Noise*, New York: McGraw-Hill, 1958.
10. K. S. Shanmugan and A. M. Breipohl, *Random Signals: Detection, Estimation, and Data Analysis*, New York: Wiley, 1988.
11. D. Gerhardt and T. Korfman, *Video Poker, Playing to Win*, Las Vegas, NV: Gaming Books International, Inc. 1987.
12. I. F. Blake, *An Introduction to Applied Probability*, New York: Wiley, 1979.
13. H. J. Larson and B. O. Shubert, *Probabilistic Models in Engineering Sciences*, Vol. 1, New York: Wiley, 1979.

Additional References on Probability

14. P. Z. Peebles, Jr., *Probability, Random Variables, and Random Signal Principles*, 2nd ed., New York: McGraw-Hill, 1987.
15. G. R. Cooper and C. D. McGillem, *Probabilistic Methods of Signal and System Analysis*, 2nd ed., New York: Holt, Rinehart, and Winston, 1986.
16. A. M. Breipohl, *Probabilistic Systems Analysis*, New York: Wiley, 1970.
17. J. L. Melsa and A. P. Sage, *An Introduction to Probability and Stochastic Processes*, Englewood Cliffs, N.J.: Prentice-Hall, 1973.

2

Mathematical Description
of Random Signals

The concept of frequency spectrum is familiar from elementary physics, and so it might seem appropriate to begin our discussion of noiselike signals with their spectral description. This approach, while intuitively appealing, leads to all sorts of difficulties. The only really careful way to describe noise is to begin with a probabilistic description and then proceed to derive the associated spectral characteristics from the probabilistic model. We now proceed toward this end.

2.1
CONCEPT OF A RANDOM PROCESS

We should begin by distinguishing between deterministic and random signals. Usually, the signals being considered here will represent some physical quantity such as voltage, current, distance, temperature, and so forth. Thus, they are real variables. Also, time will usually be the independent variable, although this does not necessarily need to be the case. A signal is said to be *deterministic* if it is exactly predictable for the time span of interest. Examples would be

(a) $x(t) = 10 \sin 2\pi t$ (Sine wave)

(b) $x(t) = \begin{cases} 1, & t \geq 0 \\ 0, & t < 0 \end{cases}$ (Unit step)

(c) $x(t) = \begin{cases} 1 - e^{-t}, & t \geq 0 \\ 0, & t < 0 \end{cases}$ (Exponential response)

Notice that there is nothing "chancy" about any of these signals. They are described by functions in the usual mathematical sense; that is, specify a numerical value of t and the corresponding value of x is determined. We are usually able to write the functional relationship between x and t explicitly.

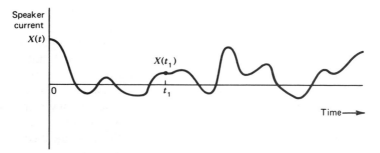

Figure 2.1. Typical radio noise signal.

However, this is not really necessary. All that is needed is to know conceptually that a functional relationship exists.

In contrast with a deterministic signal, a *random signal* always has some element of chance associated with it. Thus it is not predictable in a deterministic sense. Examples of random signals are:

(d) $X(t) = 10 \sin(2\pi t + \theta)$, where θ is a random variable uniformly distributed between 0 and 2π.

(e) $X(t) = A \sin(2\pi t + \theta)$, where θ and A are independent random variables with known distributions.

(f) $X(t) = $ A noiselike signal with no particular deterministic structure— one that just wanders on aimlessly ad infinitum.

Since all of these signals have some element of chance associated with them, they are random signals. Signals such as (d), (e), and (f) are formally known as *random* or *stochastic processes*, and we will use the terms *random* and *stochastic* interchangeably throughout the remainder of the book.*

Let us now consider the description of signal (f) in more detail. It might be the common audible radio noise that was mentioned in Chapter 1. If we looked at an oscillographic recording of the radio speaker current, it might appear as shown in Fig. 2.1. We might expect such a signal to have some kind of spectral description, because the signal is audible to the human ear. Yet the precise mathematical description of such a signal is remarkably elusive, and it eluded investigators prior to the 1940s (3, 4).

---

*We need to recognize a notational problem here. Denoting the random process as $X(t)$ implies that there is a functional relationship between X and t. This, of course, is not the case because $X(t)$ is governed by chance. For this reason, some authors (1, 2) prefer to use a subscript notation, that is, X_t rather than $X(t)$, to denote a random time signal. X_t then "looks" like a random variable with time as a parameter, which is precisely what it is. This notation, however, is not without its own problems. Suffice it to say, in most engineering literature, time random processes are denoted with "parentheses t" rather than "subscript t." We will do likewise, and the reader will simply have to remember that $X(t)$ does not function in the usual mathematical sense when $X(t)$ is a random process.

Imagine sampling the noise shown in Fig. 2.1 at a particular point in time, say t_1. The numerical value obtained would be governed largely by chance, which suggests it might be considered to be a random variable. However, with random variables we must be able to visualize a conceptual statistical experiment in which samples of the random variable are obtained under identical chance circumstances. It would not be proper in this case to sample X by taking successive time samples of the same signal, because, if they were taken very close together, there would be a close statistical connection among nearby samples. Therefore, the conceptual experiment in this case must consist of many "identical" radios, all playing simultaneously, all being tuned away from regular stations in different portions of the broadcast band, and all having their volumes turned up to the same sound level. This then leads to the notion of an ensemble of similar noiselike signals as shown in Fig. 2.2.

It can be seen then that a random process is a set of random variables that unfold with time in accordance with some conceptual chance experiment. Each of the noiselike time signals so generated is called a *sample realization* of the process. Samples of the individual signals at a particular time t_1 would then be sample realizations of the random variable $X(t_1)$. Four of these are illustrated in Fig. 2.2 as $X_A(t_1)$, $X_B(t_1)$, $X_C(t_1)$, and $X_D(t_1)$. If we were to sample at a different time, say t_2, we would obtain samples of a different random variable $X(t_2)$, and so forth. Thus, in this example, an infinite set of random variables is generated by the random process $X(t)$.

The radio experiment just described is an example of a *continuous-time random process* in that time evolves in a continuous manner. In this example, the probability density function describing the amplitude variation also happens to be continuous. However, random processes may also be discrete in either time or amplitude, as will be seen in the following two examples.

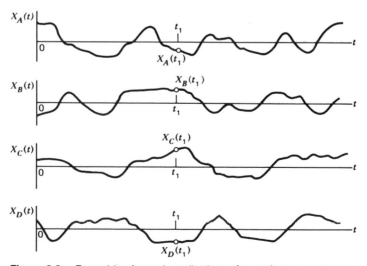

Figure 2.2.   Ensemble of sample realizations of a random process.

EXAMPLE 2.1

Consider a card player with a deck of standard playing cards numbered from 1 (ace) through 13 (king). The deck is shuffled and the player picks a card at random and observes its number. The card is then replaced, the deck reshuffled, and another card observed. This process is then repeated at unit intervals of time and continued on ad infinitum. The random process so generated would be discrete in both time and "amplitude," provided we say that the observed number applies only at the precise instant of time it is observed.

The preceding description would, of course, generate only one sample realization of the process. In order to obtain an ensemble of sample signals, we need to imagine an ensemble of card players, each having similar decks of cards and each generating a different (but statistically similar) sample realization of the process. ∎

EXAMPLE 2.2

Imagine a sack containing a large quantity of sample numbers taken from a zero-mean, unity-variance normal distribution. An observer reaches into the sack at unit intervals of time and observes a number with each trial. In order to avoid exact repetition, he does not replace the numbers during the experiment. This process would be discrete in time, as before, but continuous in amplitude. Also, the conceptual experiment leading to an ensemble of sample realizations of the process would involve many observers, each with a separate sack of random numbers. ∎

We will concentrate mostly on continuous-time processes in this chapter. However, a brief discussion of discrete-time Markov and white processes is given in Problems 2.34*, 2.35*, and 2.36* at the end of the chapter. Scalar discrete-time processes are then considered further in Chapter 3 with the discussion of the ARMA model in Section 3.9. Finally, vector discrete-time models are introduced in Chapter 5 and then used in all of the subsequent chapters.

2.2
PROBABILISTIC DESCRIPTION
OF A RANDOM PROCESS

As mentioned previously, one can usually write out the functional form for a deterministic signal explicitly; for example, $s(t) = 10 \sin 2\pi t$, or $s(t) = t^2$, and so on. No such deterministic description is possible for random signals because the numerical value of the signal at any particular time is governed by chance. Thus we should expect our description of noiselike signals to be somewhat vaguer than that for deterministic signals. One way to specify a random process is to describe in detail the conceptual chance experiment giving rise to the

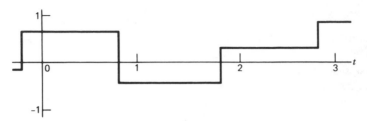

Figure 2.3. Sample signal for Example 2.3.

process. Examples 2.1 and 2.2 illustrate this way of describing a random process. The following two examples will illustrate this further.

EXAMPLE 2.3

Consider a time signal (e.g., a voltage) that is generated according to the following rules: (a) The waveform is generated with a sample-and-hold arrangement where the "hold" interval is 1 sec; (b) the successive amplitudes are independent samples taken from a set of random numbers with uniform distribution from -1 to $+1$; and (c) the first switching time after $t = 0$ is a random variable with uniform distribution from 0 to 1. (This is equivalent to saying the time origin is chosen at random.) A typical sample realization of this process is shown in Fig. 2.3. Note that the process mean is zero and its mean-square value works out to be one-third. [This is obtained from item (b) of the description.] ■

EXAMPLE 2.4

Consider another time function generated with a sample-and-hold arrangement with these properties: (a) The "hold" interval is 0.2 sec, (b) the successive amplitudes are independent samples obtained from a zero-mean normal distribution with a variance of one-third, and (c) the switching points occur at multiples of .2 units of time; that is, the time origin is not chosen at random in this case. A sketch of a typical waveform for this process is shown in Fig. 2.4. ■

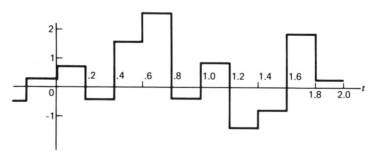

Figure 2.4. Typical waveform for Example 2.4.

Figure 2.5. Sample signal of a typical noise process.

Now, from Examples 2.3 and 2.4 it should be apparent that if we simply say, "Noiselike waveform with zero mean and mean-square value of one-third," we really are not being very definite. Both processes of Examples 2.3 and 2.4 would satisfy these criteria, but yet they are quite different. Obviously, more information than just mean and variance is needed to completely describe a random process. We will now explore the "description" problem in more detail.

A more typical "noiselike" signal is shown in Fig. 2.5. The times indicated, $t_1, t_2, \ldots, t_k$, have been arranged in ascending order, and the corresponding sample values $X_1, X_2, \ldots, X_k$, are of course, random variables. Note that we have abbreviated the notation and have let $X(t_1) = X_1, X(t_2) = X_2, \ldots$, and so on. Obviously the first-order probability density functions $f_{X_1}(x), f_{X_2}(x)$ $, \ldots, f_{X_k}(x)$ are important in describing the process because they tell use something about the process amplitude distribution. In Example 2.3, $f_{X_1}(x), f_{X_2}(x), \ldots, f_{X_k}(x)$, are all identical density functions and are given by [using $f_{X_1}(x)$ as an example]

$$f_{X_1}(x) = \begin{cases} \frac{1}{2}, & -1 \le x \le 1 \\ 0, & |x| > 1 \end{cases}$$

The density functions are not always identical for $X_1, X_2, \ldots, X_k$; they just happened to be in this simple example. In Example 2.4, the density functions describing the amplitude distribution of the $X_1, X_2, \ldots, X_k$ random variables are again all the same, but in this case they are normal in form with a variance of one-third. Note that the first-order densities tell us something about the relative distribution of the process amplitude as well as its mean and mean-square value.

It should be clear that the joint densities relating any pair of random variables, for example, $f_{X_1 X_2}(x_1, x_2), f_{X_1 X_3}(x_1, x_3)$, and so forth, are also important in our process description. It is these density functions that tell us something about how rapidly the signal changes with time, and these will eventually tell us something about the signal's spectral content. Continuing on, the third, fourth, and subsequent higher-order density functions provide even more detailed information about the process in probabilistic terms. However, this leads to a formidable description of the process, to say the least, because a k-variate density function is required where k can be any positive integer. Obviously, we will not usually be able to specify this kth order density function explicitly. Rather, this usually must be done more subtly by providing, with a word description or otherwise, enough information about the process to enable one to write out any desired higher-order density function; but the actual "writing it out" is usually not done.

Recall from probability theory that two random variables X and Y are said to be statistically independent if their joint density function can be written in product form

$$f_{XY}(x, y) = f_X(x)f_Y(y) \tag{2.2.1}$$

Similarly, random processes $X(t)$ and $Y(t)$ are statistically independent if the joint density for any combination of random variables of the two processes can be written in product form, that is, $X(t)$ and $Y(t)$ are independent if

$$f_{X_1 X_2 \ldots Y_1 Y_2 \ldots} = f_{X_1 X_2 \ldots} f_{Y_1 Y_2 \ldots} \tag{2.2.2}$$

In Eq. (2.2.2) we are using the shortened notation $X_1 = X(t_1)$, $X_2 = X(t_2), \ldots$, and $Y_1 = Y_1(t_1')$, $Y_2 = Y_2(t_2'), \ldots$, where the sample times do not have to be the same for the two processes.

In summary, the test for completeness of the process description is this: Is enough information given to enable one, conceptually at least, to write out the kth order probability density function for any k? If so, the description is as complete as can be expected; if not, it is incomplete to some extent, and radically different processes may fit the same incomplete description.

2.3
GAUSSIAN RANDOM PROCESS

There is one special situation where an explicit probability density description of the random process is both feasible and appropriate. This case is the *Gaussian* or *normal* random process. It is defined as one in which *all the density functions describing the process are normal in form*. Note that it is not sufficient that just the "amplitude" of the process be normally distributed; all higher-order density functions must also be normal! As an example, the process defined in Example 2.4 has a normal first-order density function, but closer scrutiny will reveal that its second-order density function is not normal in form. Thus the process is not a Gaussian process.

The multivariate normal density function was discussed in Section 1.15. It was pointed out there that matrix notation makes it possible to write out all k-variate density functions in the same compact matrix form, regardless of the size of k. All we have to do is specify the vector random-variable mean and covariance matrix, and the density function is specified. In the case of a Gaussian random process the "variates" are the random variables $X(t_1), X(t_2), \ldots, X(t_k)$, where the points in time may be chosen arbitrarily. Thus enough information must be supplied to specify the mean and covariance matrix regardless of the choice of $t_1, t_2, \ldots, t_k$. Examples showing how to do this will be deferred for the moment, because it is expedient first to introduce the basic ideas of stationarity and correlation functions.

2.4
STATIONARITY, ERGODICITY, AND CLASSIFICATION OF PROCESSES

A random process is said to be *time stationary* or simply *stationary* if the density functions describing the process are invariant under a translation of time. That is, if we consider a set of random variables $X_1 = X(t_1), X_2 = X(t_2), \ldots, X_k = X(t_k)$, and also a translated set $X_1' = X(t_1 + \tau)$, $X_2' = X(t_2 + \tau), \ldots, X_k' = X(t_k + \tau)$, the density functions $f_{X_1}, f_{X_1 X_2}, \ldots, f_{X_1 X_2 \ldots X_k}$ describing the first set would be identical in form to those describing the translated set. Note that this applies to all the higher-order density functions. The adjective *strict* is also used occasionally with this type of stationarity to distinguish it from wide-sense stationarity, which is a less restrictive form of stationarity. This will be discussed later in Section 2.5 on correlation functions.

A random process is said to be *ergodic* if time averaging is equivalent to ensemble averaging. In a qualitative sense this implies that a single sample time signal of the process contains all possible statistical variations of the process. Thus, no additional information is to be gained by observing an ensemble of sample signals over the information obtained from a one-sample signal, for example, one long strip-chart recording. An example will illustrate this concept.

EXAMPLE 2.5

Consider a somewhat trivial process defined to be a constant with time, the constant being a random variable with zero-mean normal distribution. An ensemble of sample realizations for this process is shown in Fig. 2.6. A common physical situation where this kind of process model would be appropriate is random instrument bias. In many applications, some small residual random bias will remain in spite of all attempts to eliminate it, and the bias will be different for each instrument in the batch. In Fig. 2.6 we see that time samples collected from a single sample signal, say the first one, will all have the same value a_0. The average of these is, of course, just a_0. On the other hand, if we were to collect samples in an ensemble sense, the values $a_0, a_1, a_2, \ldots, a_n$, would be obtained. These would have a normal distribution with zero mean. Obviously time and ensemble sampling do not lead to the same result in this case, so the process is not ergodic. It is, however, a stationary process because the "statistics" of the process do not change with time. ∎

In the case of physical noise processes, one can rarely justify strict stationarity or ergodicity in a formal sense. Thus, we often lean on heuristic knowledge of the processes involved and simply make assumptions accordingly.

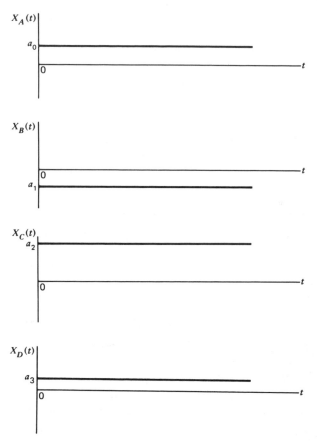

Figure 2.6. Ensemble of random constants.

Random processes are sometimes classified according to two categories, *deterministic* and *nondeterministic*. As might be expected, a deterministic random process resembles a deterministic nonrandom signal in that it has some special deterministic structure. Specifically, if the process description is such that knowledge of a sample signal's past enables exact prediction of its future, it is classified as a deterministic random process. Examples are:

1. $X(t) = a$; a is normal, $N(m, \sigma^2)$.
2. $X(t) = A \sin \omega t$; A is Rayleigh distributed.
3. $X(t) = A \sin(\omega t + \theta)$; A and θ are independent, and Rayleigh and uniformly distributed, respectively.

In each case, if one were to specify a particular sample signal prior to some time, say t_1, the sample realizations for that particular signal would be indirectly specified, and the signal's future values would be exactly predictable.

Random processes that are not deterministic are classified as nondeterministic. These processes have no special functional structure that enables their exact prediction by specification of certain key parameters or their past history. Typical "noise" is a good example of a nondeterministic random process. It wanders on aimlessly, as determined by chance, and has no particular deterministic structure.

2.5
AUTOCORRELATION FUNCTION

The autocorrelation function for a random process $X(t)$ is defined as*

$$R_X(t_1, t_2) = E[X(t_1)X(t_2)] \qquad (2.5.1)$$

where t_1 and t_2 are arbitrary sampling times. Clearly, it tells how well the process is correlated with itself at two different times. If the processes is stationary, its probability density functions are invariant with time, and the autocorrelation function depends only on the time difference $t_2 - t_1$. Thus R_X reduces to a function of just the time difference variable τ, that is,

$$R_X(\tau) = E[X(t)X(t + \tau)] \qquad \text{(Stationary case)} \qquad (2.5.2)$$

where t_1 is now denoted as just t and t_2 is $(t + \tau)$. Stationarity assures us that the expectation is not dependent on t.

Note that the autocorrelation function is the ensemble average (i.e., expectation) of the product of $X(t_1)$ and $X(t_2)$; therefore, it can formally be written as

$$R_X(t_1, t_2) = E[X_1 X_2] = \int_{-\infty}^{\infty} \int_{-\infty}^{\infty} x_1 x_2 f_{X_1 X_2}(x_1, x_2) \, dx_1 \, dx_2 \quad (2.5.3)$$

where we are using the shortened notation $X_1 = X(t_1)$ and $X_2 = X(t_2)$. However, Eq. (2.5.3) is often not the simplest way of determining R_X because the joint density function $f_{X_1 X_2}(x_1, x_2)$ must be known explicitly in order to evaluate the integral. If the ergodic hypothesis applies, it is often easier to compute R_X

---

* In describing the correlation properties of random processes, some authors prefer to work with the *autocovariance function* rather than the autocorrelation function as defined by Eq. (2.5.1). The autocovariance function is defined as

$$\text{Autocovariance function} = E\{[X(t_1) - m_x(t_1)][X(t_2) - m_x(t_2)]\}$$

The two functions are obviously related. In one case the mean is included in the product (autocorrelation), and in the other the mean is subtracted out (autocovariance). That is the essential difference. The two functions are, of course, identical for zero-mean processes. The autocorrelation function is probably the more common of the two in engineering literature, so it will be used throughout this text.

as a time average rather than an ensemble average. An example will illustrate this.

EXAMPLE 2.6

Consider the same process defined in Example 2.3. A typical sample signal for this process is shown in Fig. 2.7 along with the same signal shifted in time an amount τ. Now, the process under consideration in this case is ergodic, so we should be able to interchange time and ensemble averages. Thus, the autocorrelation function can be written as

$$R_X(\tau) = \text{time average of } X_A(t) \cdot X_A(t + \tau)$$

$$= \lim_{T \to \infty} \frac{1}{T} \int_0^T X_A(t) X_A(t + \tau) \, dt \qquad (2.5.4)$$

It is obvious that when $\tau = 0$, the integral of Eq. (2.5.4) is just the mean square value of $X_A(t)$, which is $\frac{1}{3}$ in this case. On the other hand, when τ is unity or larger, there is no overlap of the correlated portions of $X_A(t)$ and $X_A(t + \tau)$, and thus the average of the product is zero. Now, as the shift τ is reduced from 1 to 0, the overlap of correlated portions increases linearly until the maximum overlap occurs at $\tau = 0$. This then leads to the autocorrelation function shown in Fig. 2.8. Note that for stationary ergodic processes, the direction of time shift τ is immaterial, and hence the autocorrelation function is symmetric about the

Figure 2.7. Random waveform for Example 2.6.

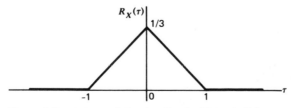

Figure 2.8. Autocorrelation function for Example 2.6.

origin. Also, note that we arrived at $R_X(\tau)$ without formally finding the joint density function $f_{X_1 X_2}(x_1, x_2)$. ∎

Sometimes, the random process under consideration is not ergodic, and it is necessary to distinguish between the usual autocorrelation function (ensemble average) and the time-average version. Thus, we define the *time autocorrelation function* as

$$\mathscr{R}_{X_A}(\tau) = \lim_{T \to \infty} \frac{1}{T} \int_0^T X_A(t) X_A(t + \tau) \, dt \tag{2.5.4a}$$

where $X_A(t)$ denotes a sample realization of the $X(t)$ process. There is the tacit assumption that the limit indicated in Eq. (2.5.4a) exists. Also note that script $\mathscr{R}$ rather than italic R is used as a remainder that this is a time average rather than an ensemble average.

EXAMPLE 2.7

To illustrate the difference between the usual autocorrelation function and the time autocorrelation function, consider the deterministic random process

$$X(t) = A \sin \omega t \tag{2.5.5}$$

where A is a normal random variable with zero mean and variance σ^2, and ω is a known constant. Suppose we obtain a single sample of A and its numerical value is A_1. The corresponding sample of $X(t)$ would then be

$$X_A(t) = A_1 \sin \omega t \tag{2.5.6}$$

According to Eq. (2.5.4), its time autocorrelation function would then be

$$\mathscr{R}_{X_A}(\tau) = \lim_{T \to \infty} \frac{1}{T} \int_0^T A_1 \sin \omega t \cdot A_1 \sin \omega(t + \tau) \, dt$$

$$= \frac{A_1^2}{2} \cos \omega \tau \tag{2.5.7}$$

On the other hand, the usual autocorrelation function is calculated as an ensemble average, that is, from Eq. (2.5.1). In this case it is

$$R_X(t_1, t_2) = E[X(t_1) X(t_2)]$$

$$= E[A \sin \omega t_1 \cdot A \sin \omega t_2]$$

$$= \sigma^2 \sin \omega t_1 \sin \omega t_2 \tag{2.5.8}$$

Note that this expression is quite different from that obtained for $\mathcal{R}_{X_A}(\tau)$. Clearly, time averaging does not yield the same result as ensemble averaging, so the process is not ergodic. Furthermore, the autocorrelation function given by Eq. (2.5.8) does not reduce to simply a function of $t_2 - t_1$. Therefore, the process is not stationary. ∎

General Properties of Autocorrelation Functions

There are some general properties that are common to all autocorrelation functions for stationary processes. These will now be enumerated with a brief comment about each:

1. $R_X(0)$ is the mean-square value of the process $X(t)$. This is self-evident from Eq. (2.5.2).

2. $R_X(\tau)$ is an even function of τ. This results from the stationarity assumption. [In the nonstationary case there is symmetry with respect to the two arguments t_1 and t_2. In Eq. (2.5.1) it certainly makes no difference in which order we multiply $X(t_1)$ and $X(t_2)$. Thus, $R_X(t_1, t_2) = R_X(t_2, t_1)$].

3. $|R_X(\tau)| \leq R_X(0)$ for all τ. We have assumed $X(t)$ is stationary and thus the mean-square values of $X(t)$ and $X(t + \tau)$ must be the same. Also the magnitude of the correlation coefficient relating two random variables is never greater than unity. Thus $R_X(\tau)$ can never be greater in magnitude than $R_X(0)$.

4. If $X(t)$ contains a periodic component, $R_X(t)$ will also contain a periodic component with the same period. This can be verified by writing $X(t)$ as the sum of the nonperiodic and periodic components and then applying the definition given by Eq. (2.5.2). It is of interest to note that if the process is ergodic as well as stationary and if the periodic component is sinusoidal, then $R_X(\tau)$ will contain no information about the phase of the sinusoidal component. The harmonic component always appears in the autocorrelation function as a cosine function, irrespective of its phase.

5. If $X(t)$ does not contain any periodic components, $R_X(\tau)$ tends to zero as $\tau \to \infty$. This is just a mathematical way of saying that $X(t + \tau)$ becomes completely uncorrelated with $X(t)$ for large τ if there are no hidden periodicities in the process. Note that a constant is a special case of a periodic function. Thus $R_X(\infty) = 0$ implies zero mean for the process.

6. The Fourier transform of $R_X(\tau)$ is real, symmetric, and nonnegative. The real, symmetric property follows directly from the even property of $R_X(\tau)$. The nonnegative property is not obvious at this point. It will be justified later in Section 2.7, which deals with the spectral density function for the process.

It was mentioned previously that strict stationarity is a severe requirement, because it requires that all the higher-order probability density functions be invariant under a time translation. This is often difficult to verify. Thus, a less demanding form of stationarity is often used, or assumed. A random process is said to be *covariance stationary* or *wide-sense stationary* if $E[X(t_1)]$ is independent of t_1 and $E[X(t_1)X(t_2)]$ is dependent only on the time difference $t_2 - t_1$. Obviously, if the second-order density $f_{X_1X_2}(x_1, x_2)$ is independent of the time origin, the process is covariance stationary.

Further examples of autocorrelation functions will be given as this chapter progresses. We will see that the autocorrelation function is an important descriptor of a random process and one that is relatively easy to obtain because it depends on only the second-order probability density for the process.

2.6
CROSSCORRELATION FUNCTION

The crosscorrelation function between the processes $X(t)$ and $Y(t)$ is defined as

$$R_{XY}(t_1, t_2) = E[X(t_1)Y(t_2)] \qquad (2.6.1)$$

Again, if the processes are stationary, only the time *difference* between sample points is relevant, so the crosscorrelation function reduces to

$$R_{XY}(\tau) = E[X(t)Y(t + \tau)] \qquad \text{(Stationary case)} \qquad (2.6.2)$$

Just as the autocorrelation function tells us something about how a process is correlated with itself, the crosscorrelation function provides information about the mutual correlation between the two processes.

Notice that it is important to order the subscripts properly in writing $R_{XY}(\tau)$. A skew-symmetric relation exists for stationary processes as follows. By definition,

$$R_{XY}(\tau) = E[X(t)Y(t + \tau)] \qquad (2.6.3)$$

$$R_{YX}(\tau) = E[Y(y)X(t + \tau)] \qquad (2.6.4)$$

The expectation in Eq. (2.6.4) is invariant under a translation of $-\tau$. Thus, $R_{YX}(\tau)$ is also given by

$$R_{YX}(\tau) = E[Y(t - \tau)X(t)] \qquad (2.6.5)$$

Now, comparing Eqs. (2.6.3) and (2.6.5), we see that

$$R_{XY}(\tau) = R_{YX}(-\tau) \tag{2.6.6}$$

Thus, interchanging the order of the subscripts of the crosscorrelation function has the effect of changing the sign of the argument.

EXAMPLE 2.8

Let $X(t)$ be the same random process of Example 2.6 and illustrated in Fig. 2.7. Let $Y(t)$ be the same signal as $X(t)$, but delayed one-half unit of time. The crosscorrelation $R_{XY}(\tau)$ would then be shown in Fig. 2.9. Note that $R_{XY}(\tau)$ is not an even function of τ; nor does its maximum occur at $\tau = 0$. Thus, the crosscorrelation function lacks the symmetry possessed by the autocorrelation function. ∎

We frequently need to consider additive combinations of random processes. For example, let the process $Z(t)$ be the sum of stationary processes $X(t)$ and $Y(t)$:

$$Z(t) = X(t) + Y(t) \tag{2.6.7}$$

The autocorrelation function of the summed process is then

$$R_Z(\tau) = E\{[X(t) + Y(t)][X(t + \tau) + Y(t + \tau)]\}$$

$$= E[X(t)X(t + \tau)] + E[Y(t)X(t + \tau)]$$

$$+ E[X(t)Y(t + \tau)] + E[Y(t)Y(t + \tau)]$$

$$= R_X(\tau) + R_{YX}(\tau) + R_{XY}(\tau) + R_Y(\tau) \tag{2.6.8}$$

Now, if X and Y are zero-mean uncorrelated processes, the middle terms of Eq. (2.6.8) are zero, and we have

$$R_Z(\tau) = R_X(\tau) + R_Y(\tau) \qquad \text{(for zero crosscorrelation)} \tag{2.6.9}$$

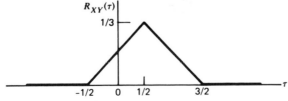

Figure 2.9.   Crosscorrelation function for Example 2.8.

This can obviously be extended to the sum of more than two processes. Equation (2.6.9) is a much-used relationship, and it should always be remembered that it applies only when the processes being summed have zero crosscorrelation.

2.7
POWER SPECTRAL DENSITY FUNCTION

It was mentioned in Section 2.6 that the autocorrelation function is an important descriptor of a random process. Qualitatively, if the autocorrelation function decreases rapidly with τ, the process changes rapidly with time; conversely, a slowly changing process will have an autocorrelation function that decreases slowly with τ. Thus, we would suspect that this important descriptor contains information about the frequency content of the process; and this is in fact the case. For stationary processes, there is an important relation known as the *Wiener–Khinchine relation*:

$$S_X(j\omega) = \mathscr{F}[R_X(\tau)] = \int_{-\infty}^{\infty} R_X(\tau)e^{-j\omega\tau}\,d\tau \qquad (2.7.1)$$

where $\mathscr{F}[\cdot]$ indicates Fourier transform and ω has the usual meaning of (2π) (frequency in hertz). S_X is called the *power spectral density function* or simply the *spectral density function* of the process.

The adjectives *power* and *spectral* come from the relationship of $S_X(j\omega)$ to the usual spectrum concept for a deterministic signal. However, some care is required in making this connection. If the process $X(t)$ is time stationary, it wanders on ad infinitum and is not absolutely integrable. Thus the defining integral for the Fourier transform does not converge. When considering the Fourier transform of the process, we are forced to consider a truncated version of it, say $X_T(t)$, which is truncated to zero outside a span of time T. The Fourier transform of a sample realization of the truncated process will then exist.

Let $\mathscr{F}\{X_T\}$ denote the Fourier transform of $X_T(t)$, where it is understood that for any given ensemble of samples of $X_T(t)$ there will be corresponding ensemble of $\mathscr{F}\{X_T(t)\}$. That is, $\mathscr{F}\{X_T(t)\}$ has stochastic attributes just as does $X_T(t)$. Now look at the following expectation:

$$E\left[\frac{1}{T}|\mathscr{F}\{X_T(t)\}|^2\right]$$

For any particular sample realization of $X_T(t)$, the quantity inside the brackets is known as the *periodogram* for that particular signal. It will now be shown that averaging over an ensemble of periodograms for large T yields the power spectral density function.

The expectation of the periodogram of a signal spanning the time interval $[0, T]$ can be manipulated as follows:

$$E\left[\frac{1}{T}|\mathscr{F}\{X_T(t)\}|^2\right]$$

$$= E\left[\frac{1}{T}\int_0^T X(t)e^{-j\omega t}\,dt\int_0^T X(s)e^{j\omega s}\,ds\right]$$

$$= \frac{1}{T}\int_0^T\int_0^T E[X(t)X(s)]e^{-j\omega(t-s)}\,dt\,ds \qquad (2.7.2)$$

Note that we were able to drop the subscript T on $X(t)$ because of the restricted range of integration. If we now assume $X(t)$ is stationary, $E[X(t)X(s)]$ becomes $R_X(t-s)$ and Eq. (2.7.2) becomes

$$E\left[\frac{1}{T}|\mathscr{F}\{X_T(t)\}|^2\right] = \frac{1}{T}\int_0^T\int_0^T R_X(t-s)e^{-j\omega(t-s)}\,ds\,dt \qquad (2.7.3)$$

The appearance of $t-s$ in two places in Eq. (2.7.3) suggests a change of variables. Let

$$\tau = t - s \qquad (2.7.4)$$

Equation (2.7.3) then becomes

$$\frac{1}{T}\int_0^T\int_0^T R_X(t-s)e^{-j\omega(t-s)}\,ds\,dt = -\frac{1}{T}\int_0^T\int_t^{t-T} R_X(\tau)e^{-j\omega\tau}\,d\tau\,dt \qquad (2.7.5)$$

The new region of integration in the τt plane is shown in Fig. 2.10.

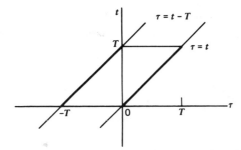

Figure 2.10. Region of integration in the τt plane.

Next we interchange the order of integration and integrate over the two triangular regions separately. This leads to

$$E\left[\frac{1}{T}|\mathscr{F}\{X_T(t)\}|^2\right]$$

$$= \frac{1}{T}\int_{-T}^0\int_0^{\tau+T} R_X(\tau)e^{-j\omega\tau}\,dt\,d\tau + \frac{1}{T}\int_0^T\int_\tau^T R_X(\tau)e^{-j\omega\tau}\,dt\,d\tau \quad (2.7.6)$$

We now integrate with respect to t with the result

$$E\left[\frac{1}{T}|\mathscr{F}\{X_T(t)\}|^2\right]$$

$$= \frac{1}{T}\int_{-T}^0 (\tau+T)R_X(\tau)e^{-j\omega\tau}\,d\tau + \frac{1}{T}\int_0^T (T-\tau)R_X(\tau)e^{-j\omega\tau}\,d\tau \quad (2.7.7)$$

Finally, Eq. (2.7.7) may be written in more compact form as

$$E\left[\frac{1}{T}|\mathscr{F}\{X_T(t)\}|^2\right] = \int_{-T}^T \left(1 - \frac{|\tau|}{T}\right)R_X(\tau)e^{-j\omega\tau}\,d\tau \quad (2.7.8)$$

The factor $1 - |\tau|/T$ that multiplies $R_X(\tau)$ may be thought of as a triangular weighting factor that approaches unity as T becomes large; at least this is true if $R_X(\tau)$ approaches zero as τ becomes large, which it will do if $X(t)$ contains no periodic components. Thus, as T becomes large, we have the following relationship:

$$E\left[\frac{1}{T}|\mathscr{F}\{X_T(t)\}|^2\right] \Rightarrow \int_{-\infty}^\infty R_X(\tau)e^{-j\omega\tau}\,d\tau, \qquad \text{as } T \to \infty \quad (2.7.9)$$

Or, in other words,

$$\text{Average periodogram for large } T \Rightarrow \text{power spectral density} \quad (2.7.10)$$

Equation (2.7.9) is a most important relationship, because it is this that ties the spectral function $S_X(j\omega)$ to "spectrum" as thought of in the usual deterministic sense. Remember that the spectral density function, as formally defined by Eq. (2.7.1), is a probabilistic concept. On the other hand, the periodogram is a spectral concept in the usual sense of being related to the Fourier transform of a time signal. The relationship given by Eq. (2.7.9) then provides the tie between the probabilistic and spectral descriptions of the process, and it is this equation that suggests the name for $S_X(j\omega)$, power *spectral* density function. More will be said of this in Section 2.14, which deals with the determination of the spectral function from experimental data.

Because of the special attributes of the autocorrelation function $R_X(\tau)$, its Fourier transform $S_X(j\omega)$ always works out to be a real, nonnegative, symmetric function of ω. This should be apparent from the left side of Eq. (2.7.9), and will be illustrated in Example 2.9.

EXAMPLE 2.9

Consider a random process $X(t)$ whose autocorrelation function is given by

$$R_X(\tau) = \sigma^2 e^{-\beta|\tau|} \qquad (2.7.11)$$

where σ^2 and β are known constants. The spectral density function for the $X(t)$ process is

$$S_X(j\omega) = \mathscr{F}[R_X(\tau)] = \frac{\sigma^2}{j\omega + \beta} + \frac{\sigma^2}{-j\omega + \beta} = \frac{2\sigma^2\beta}{\omega^2 + \beta^2} \quad (2.7.12)$$

Both R_X and S_X are sketched in Fig. 2.11. ∎

Occasionally it is convenient to write the spectral density function in terms of the complex frequency variable s rather than ω. This is done by simply replacing $j\omega$ with s; or, equivalently, replacing ω^2 with $-s^2$. For Example 2.9, the spectral density function in terms of s is then

$$S_X(s) = \frac{2\sigma^2\beta}{\omega^2 + \beta^2}\bigg|_{\omega^2 = -s^2} = \frac{2\sigma^2\beta}{-s^2 + \beta^2} \qquad (2.7.13)$$

It should be clear now why we chose to include the "j" with ω in $S_X(j\omega)$, even though $S_X(j\omega)$ always works out to be a real function of ω. By writing the argument of $S_X(j\omega)$ as $j\omega$, rather than just ω, we can use the same symbol for

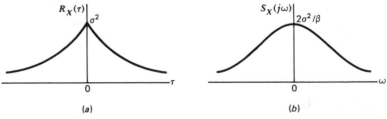

Figure 2.11. Autocorrelation and spectral density functions for Example 2.9. (a) Autocorrelation function. (b) Spectral function.

spectral function in either the complex or real frequency domain. That is,

$$S_X(s) = S_X(j\omega) \tag{2.7.14}$$

is correct notation in the usual mathematical sense.

From Fourier transform theory, we know that the inverse transform of the spectral function should yield the autocorrelation function, that is,

$$\mathscr{F}^{-1}[S_X(j\omega)] = \frac{1}{2\pi} \int_{-\infty}^{\infty} S_X(j\omega) e^{j\omega\tau} \, d\omega = R_X(\tau) \tag{2.7.15}$$

If we let $\tau = 0$ in Eq. (2.7.15), we get

$$R_X(0) = E[X^2(t)] = \frac{1}{2\pi} \int_{-\infty}^{\infty} S_X(j\omega) \, d\omega \tag{2.7.16}$$

Equation (2.7.16) provides a convenient means of computing the mean square value of a stationary process, given its spectral function. As mentioned before, it is sometimes convenient to use the complex frequency variable s rather than $j\omega$. If this is done, Eq. (2.7.16) becomes

$$E[X^2(t)] = \frac{1}{2\pi j} \int_{-j\infty}^{j\infty} S_X(s) \, ds \tag{2.7.17}$$

Equation (2.7.16) suggest that we can consider the signal power as being distributed in frequency in accordance with $S_X(j\omega)$; thus the terms *power* and *density* in power spectral density function. Using this concept, we can obtain the power in a finite band by integrating over the appropriate range of frequencies, that is,

$$\begin{bmatrix} \text{``Power'' in} \\ \text{range } \omega_1 \le \omega \le \omega_2 \end{bmatrix} = \frac{1}{2\pi} \int_{-\omega_2}^{-\omega_1} S_X(j\omega) \, d\omega + \frac{1}{2\pi} \int_{\omega_1}^{\omega_2} S_X(j\omega) \, d\omega \tag{2.7.18}$$

An example will now be given to illustrate the use of Eqs. (2.7.16) and (2.7.17).

EXAMPLE 2.10

Consider the spectral function of Example 2.9:

$$S_X(j\omega) = \frac{2\sigma^2 \beta}{\omega^2 + \beta^2} \tag{2.7.19}$$

Application of Eq. (2.7.16) should yield the mean square value σ^2. This can be verified using conventional integral tables.

$$E(X^2) = \frac{1}{2\pi}\int_{-\infty}^{\infty}\frac{2\sigma^2\beta}{\omega^2+\beta^2}\,d\omega = \frac{\sigma^2\beta}{\pi}\left[\frac{1}{\beta}\tan^{-1}\frac{\omega}{\beta}\right]_{-\infty}^{\infty} = \sigma^2 \quad (2.7.20)$$

Or equivalently, in terms of s:

$$E(X^2) = \frac{1}{2\pi j}\int_{-j\infty}^{j\infty}\frac{2\sigma^2\beta}{-s^2+\beta^2}\,ds = \sigma^2 \qquad (2.7.21)$$

More will be said about evaluating integrals of the type in Eq. (2.7.21) later, in Chapter 3. ∎

In summary, we see that the autocorrelation function and spectral density function are Fourier transform pairs. Thus, both contain the same basic information about the process, but in different forms. Since we can easily transform back and forth between the time and frequency domains, the manner in which the information is presented is purely a matter of convenience for the problem at hand.

2.8
CROSS SPECTRAL DENSITY FUNCTION

Cross spectral density functions for stationary processes $X(t)$ and $Y(t)$ are defined as

$$S_{XY}(j\omega) = \mathscr{F}[R_{XY}(\tau)] = \int_{-\infty}^{\infty}R_{XY}(\tau)e^{-j\omega\tau}\,d\tau \qquad (2.8.1)$$

$$S_{YX}(j\omega) = \mathscr{F}[R_{YX}(\tau)] = \int_{-\infty}^{\infty}R_{YX}(\tau)e^{-j\omega\tau}\,d\tau \qquad (2.8.2)$$

The crosscorrelation functions $R_{XY}(\tau)$ and $R_{YX}(\tau)$ are not necessarily even functions of τ, and thus the corresponding cross spectral densities are usually not real functions of ω. It was noted in Section 2.6 that $R_{XY}(\tau) = R_{YX}(-\tau)$. Thus S_{XY} and S_{YX} are complex conjugates of each other:

$$S_{XY}(j\omega) = S_{YX}^*(j\omega) \qquad (2.8.3)$$

and the sum of S_{XY} and S_{YX} is real.

Another function that is closely related to the cross spectral density is the *coherence function*. It is defined as

$$\gamma_{XY}^2 = \frac{|S_{XY}(j\omega)|^2}{S_X(j\omega)S_Y(j\omega)} \tag{2.8.4}$$

The coherence function can be seen to be normalized, and it is sort of a "correlation coefficient in the frequency domain." To see the normalization, let $X(t) = Y(t)$ (maximum correlation) and then

$$\gamma_{XX}^2 = \frac{|S_{XX}(j\omega)|^2}{S_X(j\omega)S_X(j\omega)} = 1 \tag{2.8.5}$$

On the other extreme, if $X(t)$ and $Y(t)$ have zero crosscorrelation, $S_{XY}(j\omega) = 0$ and $\gamma_{XY}^2 = 0$. Both the cross spectral density and coherence functions are useful in analysis of experimental data, because modern computer technology has made it possible to transform time data to the frequency domain with ease. (See Section 2.15 and references 5 and 6 for more on the analysis of experimental data. Also, see Problems 3.23 and 3.25.)

If $Z(t)$ is the sum of zero-mean processes $X(t)$ and $Y(t)$, the spectral density of $Z(t)$ is given by

$$S_Z(j\omega) = \mathscr{F}\left[R_{X+Y}(\tau)\right] \tag{2.8.6}$$

Referring to Eq. (2.6.8), we then have

$$S_Z(j\omega) = S_X(j\omega) + S_{YX}(j\omega) + S_{XY}(j\omega) + S_Y(j\omega) \tag{2.8.7}$$

Just as in the case of the autocorrelation function, the two middle terms in Eq. (2.8.7) are zero if the X and Y processes have zero crosscorrelation. So, for this special case,

$$S_{X+Y}(j\omega) = S_X(j\omega) + S_Y(j\omega), \quad \text{(for zero crosscorrelation)} \tag{2.8.8}$$

2.9
WHITE NOISE

White noise is defined to be a stationary random process having a constant spectral density function. The term "white" is a carryover from optics, where white light is light containing all visible frequencies. Denoting the white-noise spectral amplitude as A, we then have

$$S_{wn}(j\omega) = A \qquad (2.9.1)$$

The corresponding autocorrelation function for white noise is then

$$R_{wn}(\tau) = A\delta(\tau) \qquad (2.9.2)$$

In analysis, one frequently makes simplifying assumptions in order to make the problem mathematically tractable. White noise is a good example of this. However, by assuming the spectral amplitude of white noise to be constant for all frequencies (for the sake of mathematical simplicity), we find ourselves in the awkward situation of having defined a process with infinite variance. Qualitatively, white noise is sometimes characterized as noise that is jumping around infinitely far, infinitely fast! This is obviously physical nonsense but it is a useful abstraction. The saving feature is that all physical systems are bandlimited to some extent, and a bandlimited system driven by white noise yields a process that has finite variance; that is, the end result makes sense. We will elaborate on this further in Chapter 3.

Bandlimited white noise is a random process whose spectral amplitude is constant over a finite range of frequencies, and zero outside that range. If the bandwidth includes the origin (sometimes called baseband), we then have

$$S_{bwn}(j\omega) = \begin{cases} A, & |\omega| \le 2\pi W \\ 0, & |\omega| > 2\pi W \end{cases} \qquad (2.9.3)$$

where W is the physical bandwidth in hertz. The corresponding autocorrelation function is

$$R_{bwn}(\tau) = 2WA\frac{\sin(2\pi W\tau)}{2\pi W\tau} \qquad (2.9.4)$$

Both the autocorrelation and spectral density functions for baseband bandlimited white noise are sketched in Fig. 2.12. It is of interest to note that the autocorrelation function for baseband bandlimited white noise is zero for $\tau = 1/2W, 2/2W, 3/2W$, etc. From this we see that if the process is sampled at a rate of $2W$ samples/second (sometimes called the Nyquist rate), the resulting set of random variables are uncorrelated. Since this usually simplifies the analysis, the white bandlimited assumption is frequently made in bandlimited situations.

The frequency band for bandlimited white noise is sometimes offset from the origin and centered about some center frequency W_0. It is easily verified

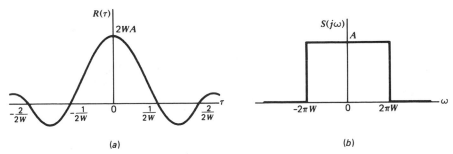

Figure 2.12. Baseband bandlimited white noise. (a) Autocorrelation function. (b) Spectral density function.

that the autocorrelation/spectral-function pair for this situation is

$$S(j\omega) = \begin{cases} A, & 2\pi W_1 \leq |\omega| \leq 2\pi W_2 \\ 0, & |\omega| < 2\pi W_1 \quad \text{and} \quad |\omega| > 2\pi W_2 \end{cases} \qquad (2.9.5)$$

$$R(\tau) = A\left[2W_2 \frac{\sin 2\pi W_2 \tau}{2\pi W_2 \tau} - 2W_1 \frac{\sin 2\pi W_1 \tau}{2\pi W_1 \tau}\right]$$

$$= A2\,\Delta W \frac{\sin \pi \Delta W \tau}{\pi \Delta W \tau} \cos 2\pi W_0 \tau \qquad (2.9.6)$$

where

$$\Delta W = W_2 - W_1 \ \text{Hz}$$

$$W_0 = \frac{W_1 + W_2}{2} \ \text{Hz}$$

These functions are sketched in Fig. 2.13.

It is worth noting the bandlimited white noise has a finite mean-square value, and thus it is physically plausible, whereas pure white noise is not. However, the mathematical forms for the autocorrelation and spectral functions in the bandlimited case are more complicated than for pure white noise.

Before leaving the subject of white noise, it is worth mentioning that the analogous discrete-time process is referred to as a white sequence. A *white sequence* is defined simply as a sequence of zero-mean, uncorrelated random variables. That is, all members of the sequence have zero means and are mutually uncorrelated with all other members of the sequence. If the random variables are also normal, then the sequence is a *Gaussian* white sequence. (See Problem 2.36* for more on the properties of a Gaussian white sequence.)

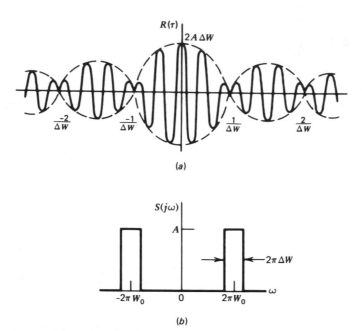

Figure 2.13.   Bandlimited white noise with center frequency W_0. (*a*) Autocorrelation function. (*b*) Spectral density.

2.10
GAUSS–MARKOV PROCESS

A stationary Gaussian process $X(t)$ that has an exponential autocorrelation is called a *Gauss–Markov* process. The autocorrelation and spectral functions for this process are then of the form

$$R_X(\tau) = \sigma^2 e^{-\beta|\tau|} \tag{2.10.1}$$

$$S_X(j\omega) = \frac{2\sigma^2\beta}{\omega^2 + \beta^2} \quad \left[\text{or } S_X(s) = \frac{2\sigma^2\beta}{-s^2 + \beta^2}\right] \tag{2.10.2}$$

These functions are sketched in Fig. 2.14. The mean-square value and time constant for the process are given by the σ^2 and $1/\beta$ parameters, respectively. The process is nondeterministic, so a typical sample time function would show no deterministic structure and would look like typical "noise." The exponential autocorrelation function indicates that sample values of the process gradually become less and less correlated as the time separation between samples increases. The autocorrelation function approaches zero as $\tau \to \infty$, and thus the mean value of the process must be zero. The reference to Markov in the name

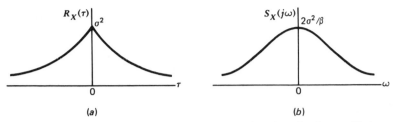

Figure 2.14. Autocorrelation and spectral density functions for Gauss – Markov process. (*a*) Autocorrelation function. (*b*) Spectral density.

of this process is not obvious at this point, but it will be after the discussion on optimal prediction in Chapter 4.

The Gauss–Markov process is an important process in applied work because (1) it seems to fit a large number of physical processes with reasonable accuracy, and (2) it has a relatively simple mathematical description. As in the case of all stationary Gaussian processes, *specification of the process autocorrelation function completely defines the process*. This means that any desired higher-order probability density function for the process may be written out explicitly, given the autocorrelation function. An example will illustrate this.

EXAMPLE 2.11

Let us say that a Gauss–Markov process $X(t)$ has autocorrelation function

$$R_X(\tau) = 100e^{-2|\tau|} \tag{2.10.3}$$

We wish to write out the third-order probability density function

$$f_{X_1 X_2 X_3}(x_1, x_2, x_3) \qquad \text{where } X_1 = X(0),\ X_2 = X(.5),\ \text{and } X_3 = X(1)$$

First we note that the process mean is zero. The covariance matrix in this case is a 3×3 matrix and is obtained as follows:

$$
\mathbf{C}_X =
\begin{bmatrix}
E(X_1^2) & E(X_1 X_2) & E(X_1 X_3) \\
E(X_2 X_1) & E(X_2^2) & E(X_2 X_3) \\
E(X_3 X_1) & E(X_3 X_2) & E(X_3^2)
\end{bmatrix}
=
\begin{bmatrix}
R_X(0) & R_X(.5) & R_X(1) \\
R_X(.5) & R_X(0) & R_X(.5) \\
R_X(1) & R_X(.5) & R_X(0)
\end{bmatrix}
$$

$$
=
\begin{bmatrix}
100 & 100e^{-1} & 100e^{-2} \\
100e^{-1} & 100 & 100e^{-1} \\
100e^{-2} & 100e^{-1} & 100
\end{bmatrix}
\tag{2.10.4}
$$

Now that $\mathbf{C}_X$ has been written out explicitly, we can use the general normal form given by Eq. (1.15.5). The desired density function is then

$$f_X(\mathbf{x}) = \frac{1}{(2\pi)^{3/2}|\mathbf{C}_X|^{1/2}} e^{-\frac{1}{2}[\mathbf{x}^T\mathbf{C}_X^{-1}\mathbf{x}]} \tag{2.10.5}$$

where $\mathbf{x}$ is the 3-tuple

$$\mathbf{x} = \begin{bmatrix} x_1 \\ x_2 \\ x_3 \end{bmatrix} \tag{2.10.6}$$

and $\mathbf{C}_x$ is given by Eq. (2.10.4). ∎

2.11
RANDOM TELEGRAPH WAVE

Consider a binary voltage waveform that is generated according to the following rules:

1. The voltage is either $+1$ or -1 V.
2. The state at $t = 0$ may be either $+1$ or -1 V with equal likelihood.
3. As time progresses, the voltage switches from one state to the other at random. Specifically, the probability of k switches in a time interval T is governed by the Poisson distribution

$$P(k) = \frac{(aT)^k}{k!} e^{-aT} \tag{2.11.1}$$

where a is the average number of switches per unit time.

This random process is called the random telegraph wave, and a sample waveform is shown in Fig. 2.15. It is worth noting that the likelihood of

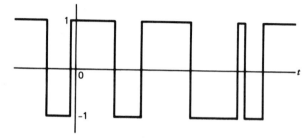

Figure 2.15. Random telegraph wave.

Figure 2.16. Gauss–Markov signal with about the same time constant as for the random telegraph wave of Figure 2.15.

switching at any point in time does not depend on the particular state or the length of time the system has been in that state.

A rigorous derivation of the autocorrelation function for the random telegraph wave can be found in a number of references (7, 8). For our purposes here, we can use the following heuristic argument. Consider the product of successive time samples $X(t_1)$ and $X(t_2)$; the result must be either of two possibilities, $+1$ or -1. If the time interval $t_2 - t_1$ is small, $X(t_1)$ and $X(t_2)$ are highly correlated, so that $X_1(t_1)X(t_2)$ is nearly unity. Then, as the spacing between samples is increased, their correlation gradually decreases and approaches zero as the spacing between samples goes to infinity. This leads to an exponential autocorrelation function of the form

$$R_X(\tau) = e^{-2a|\tau|} \tag{2.11.2}$$

where a is the average number of switches per unit time.

It should be apparent that the random telegraph wave is not a Gaussian process—far from it! Yet the Gauss–Markov process described in Section 2.10 has an autocorrelation function identical in form to that of the random telegraph wave. For the purposes of comparison, a typical Gauss–Markov signal with about the same time constant as the random telegraph wave of Fig. 2.15 is shown in Fig. 2.16. The difference is striking. This is a vivid example of two processes that have radically different random structures, but still have the same autocorrelation functions and, of course, identical spectral characteristics. The moral is this: The autocorrelation function and/or spectral characteristics do not tell the whole story; all the probability density functions must be specified in order for the process to be completely described. In the Gauss–Markov case, this was done by specifying "Gaussian process" in addition to the autocorrelation function. In the random telegraph wave case, the higher-order densities were indirectly specified by describing a conceptual chance experiment creating the waveform. Obviously the probability density functions are radically different for the two processes in this case.

2.12
NARROWBAND GAUSSIAN PROCESS

In both control and communication systems, we frequently encounter situations where a very narrowband system is excited by wideband Gaussian noise. A high-Q tuned circuit and/or a lightly damped mass-spring arrangement are examples of narrowband systems. The resulting output is a noise process with essentially all its spectral content concentrated in a narrow frequency range. If one were to observe the output of such a system, the time function would appear to be nearly sinusoidal, especially if just a few cycles of the output signal were observed. However, if one were to carefully examine a long record of the signal, it would be seen that the quasi-sinusoid is slowly varying in both amplitude and phase. Such a signal is called narrowband noise and, if it is the result of passing wideband Gaussian noise through a linear narrowband system, then it is also Gaussian. We are assured of this because any linear operation on a set of normal variates results in another set of normal variates. The quasi-sinusoidal character depends only on the narrowband property, and the exact spectral shape within the band is immaterial.

The mathematical description of narrowband Gaussian noise follows. We first write the narrowband signal as

$$S(t) = X(t) \cos \omega_c t - Y(t) \sin \omega_c t \qquad (2.12.1)$$

where $X(t)$ and $Y(t)$ are independent Gaussian random processes with similar narrowband spectral functions that are centered about zero frequency. The frequency ω_c is usually called the carrier frequency, and the effect of multiplying $X(t)$ and $Y(t)$ by $\cos \omega_c t$ and $\sin \omega_c t$ is to translate the baseband spectrum up to a similar spectrum centered about ω_c (see Problem 2.33). The independent $X(t)$ and $Y(t)$ processes are frequently called the in-phase and quadrature components of $S(t)$. Now, think of time t as a particular time, and think of $X(t)$ and $Y(t)$ as the corresponding random variables. Then make the usual rectangular to polar transformation via the equations

$$X = R \cos \Theta$$
$$Y = R \sin \Theta \qquad (2.12.2)$$

or, equivalently,

$$R = \sqrt{X^2 + Y^2}$$

$$\Theta = \tan^{-1} \frac{Y}{X} \qquad (2.12.3)$$

By substituting Eq. (2.12.2) into Eq. (2.12.1), we can now write $S(t)$ in the form

$$S(t) = R(t) \cos \Theta(t) \cos \omega_c t - R(t) \sin \Theta(t) \sin \omega_c t$$

$$= R(t) \cos \left[\omega_c t + \Theta(t) \right] \qquad (2.12.4)$$

It is from Eq. (2.12.4) that we get the physical interpretation of "slowly varying envelope (amplitude) and phase."

Before proceeding, a word or two about the probability densities for X, Y, R, and Θ is in order. If X and Y are independent normal random variables with the same variance σ^2, their individual and joint densities are

$$f_X(x) = \frac{1}{\sqrt{2\pi}\,\sigma}e^{-x^2/2\sigma^2} \tag{2.12.5}$$

$$f_Y(y) = \frac{1}{\sqrt{2\pi}\,\sigma}e^{-y^2/2\sigma^2} \tag{2.12.6}$$

and

$$f_{XY}(x, y) = \frac{1}{2\pi\sigma^2}e^{-(x^2+y^2)/2\sigma^2} \tag{2.12.7}$$

The corresponding densities for R and Θ are Rayleigh and uniform (see Example 1.24). The mathematical forms are

$$f_R(r) = \frac{r}{\sigma^2}e^{-r^2/2\sigma^2}, \qquad r \geq 0 \qquad \text{(Rayleigh)} \tag{2.12.8}$$

$$f_\Theta(\theta) = \begin{cases} \dfrac{1}{2\pi}, & 0 \leq \theta < 2\pi \\ 0, & \text{otherwise} \end{cases} \qquad \text{(Uniform)} \tag{2.12.9}$$

Also, the joint density function for R and Θ is

$$f_{R\Theta}(r, \theta) = \frac{r}{2\pi\sigma^2}e^{-r^2/2\sigma^2}, \qquad r \geq 0 \quad \text{and} \quad 0 \leq \theta < 2\pi \tag{2.12.10}$$

It is of interest to note here that if we consider simultaneous time samples of envelope and phase, the resulting random variables are statistically independent. However, the *processes* $R(t)$ and $\Theta(t)$ are not statistically independent (7). This is due to the fact that the joint probability density associated with adjacent samples cannot be written in product form, that is,

$$f_{R_1R_2\Theta_1\Theta_2}(r_1, r_2, \theta_1, \theta_2) \neq f_{R_1R_2}(r_1, r_2)f_{\Theta_1\Theta_2}(\theta_1, \theta_2) \tag{2.12.11}$$

We have assumed that $S(t)$ is a Gaussian process, and from Eq. (2.12.1) we see that

$$\operatorname{Var} S = \tfrac{1}{2}(\operatorname{Var} X) + \tfrac{1}{2}(\operatorname{Var} Y) = \sigma^2 \tag{2.12.12}$$

Thus

$$f_S(s) = \frac{1}{\sqrt{2\pi}\,\sigma} e^{-s^2/2\sigma^2} \qquad (2.12.13)$$

The higher-order density functions for S will, of course, depend on the specific shape of the spectral density for the process.

2.13
WIENER OR
BROWNIAN-MOTION PROCESS

Suppose we start at the origin and take n steps forward or backward at random, with equal likelihood of stepping in either direction. We pose two questions: After taking n steps, (1) what is the average distance traveled, and (2) what is the variance of the distance? This is the classical random-walk problem of statistics. The averages considered here must be taken in an ensemble sense; for example, think of running simultaneous experiments and then averaging the results for a given number of steps. It should be apparent that the average distance traveled is zero, provided we say "forward" is positive and "backward" is negative. However, the square of the distance is always positive (or zero), so its average for a large number of trials will not be zero. It is shown in elementary statistics that the variance after n unit steps is just n, or the standard deviation is $\sqrt{n}$ (see Problem 2.21). Note that this increases without bound as n increases, and thus the process is nonstationary.

The continuous analog of random-walk is the output of an integrator driven with white noise. This is shown in block-diagram form in Fig. 2.17a. Here we consider the input switch as closing at $t = 0$ and the initial integrator output as being zero. An ensemble of typical output time signals is shown in Fig. 2.17b. The system response $X(t)$ is given by

$$X(t) = \int_0^t F(u)\,du \qquad (2.13.1)$$

Clearly, the average of the output is

$$E[X(t)] = E\left[\int_0^t F(u)\,du\right] = \int_0^t E[F(u)]\,du = 0 \qquad (2.13.2)$$

Also, the mean-square-value (variance) is

$$E[X^2(t)] = E\left[\int_0^t F(u)\,du \int_0^t F(v)\,dv\right] = \int_0^t \int_0^t E[F(u)F(v)]\,du\,dv \qquad (2.13.3)$$

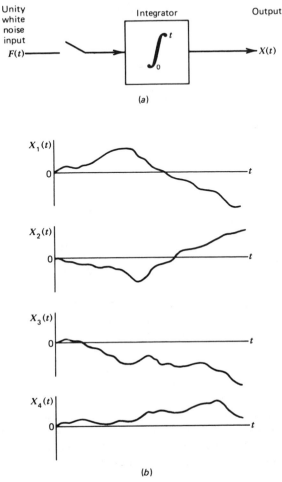

Figure 2.17. Continuous analog of random walk. (a) Block diagram. (b) Ensemble of output signals.

But $E[F(u)F(v)]$ is just the autocorrelation function $R_F(u - v)$, which in this case is a Dirac delta function. Thus

$$E[X^2(t)] = \int_0^t \int_0^t \delta(u - v)\, du\, dv = \int_0^t dv = t \qquad (2.13.4)$$

Thus, $E[X^2(t)]$ increases linearly with time and the rms value increases in accordance with $\sqrt{t}$ (for unity white noise input).

Now, add the further requirement that the input be *Gaussian* white noise. The output will then be a Gaussian process because integration is a linear operation on the input. The resulting continuous random-walk process is known as the *Wiener* or *Brownian-motion* process. The process is nonstationary, it is Gaussian, and its mean, mean-square value, and autocorrelation function are given by

$$E[X(t)] = 0 \tag{2.13.5}$$

$$E[X^2(t)] = t \tag{2.13.6}$$

$$R_X(t_1, t_2) = E[X(t_1)X(t_2)] = E\left[\int_0^{t_1} F(u)\, du \cdot \int_0^{t_2} F(v)\, dv\right]$$

$$= \int_0^{t_2}\int_0^{t_1} E[F(u)F(v)]\, du\, dv = \int_0^{t_2}\int_0^{t_1} \delta(u - v)\, du\, dv$$

Evaluation of the double integral yields

$$R_X(t_1, t_2) = \begin{cases} t_2, & t_1 \geq t_2 \\ t_1, & t_1 < t_2 \end{cases} \tag{2.13.7}$$

Since the process is nonstationary, the autocorrelation function is a general function of the two arguments t_1 and t_2. With a little imagination, Eq. (2.13.7) can be seen to describe two faces of a pyramid with the sloping ridge of the pyramid running along the line $t_1 = t_2$.

It was mentioned before that there are difficulties in defining directly what is meant by Gaussian white noise. This is because of the "infinite variance" problem. The Wiener process is well-behaved, though. Thus we can reverse the argument given here and begin by arbitrarily defining it as a Gaussian process with an autocorrelation function given by Eq. (2.13.7). This completely specifies the process. We can now describe Gaussian white noise in terms of its integral. That is, Gaussian white noise is that hypothetical process which, when integrated, yields a Wiener process.

2.14
PSEUDORANDOM SIGNALS

As the name implies, pseudorandom signals have the appearance of being random, but are not truly random. In order for a signal to be truly random, there must be some uncertainty about it that is governed by chance. Pseudorandom signals do not have this "chance" property. Two examples of pseudorandom signals will now be presented.

EXAMPLE 2.12

Consider a sample realization of finite length T of a Gauss–Markov process. Let the time length T of the sample be large relative to the time constant of the process. After the sample is taken, of course, the time function in all its intimate

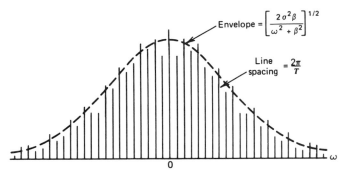

Figure 2.18. Spectrum of noise sample looped back on itself.

detail is known to the observer. After the fact, nothing remains to chance insofar as the observer in concerned. Next, imagine folding this record back on itself into a single loop (it might be on magnetic tape), and then imagine playing the loop continuously. It should be clear that the resulting signal would be periodic and completely known (determined), at least to the original observer. Yet to a second observer casually looking at a small portion of the loop, the signal would *appear* to be just random noise. It should be mentioned that this is not a completely hypothetical situation; experimental spectral analysis was frequently implemented in just this way in times prior to the modern on-line digital methods.

The "looped" signal that goes on ad infinitum is periodic, so it would have line type rather than continuous spectral characteristics. See Fig. 2.18. ■

Line-type spectra are characteristic of all pseudorandom signals. The line spacing may be extremely small, as is the case for very large T, but it is there, nevertheless. Note that the envelope of the lines would approximate the square root of the spectral density of the process from which the sample was taken, provided the record length T is large.* In this case the usual laboratory analog spectrum analyzer would not be able to resolve individual lines, and it would indicate a smooth spectrum proportional to the average spectrum. This would be just fine and what was desired if the original experimental problem was to determine the spectral characteristics of random process from a single long sample of the process. The point of all this is that the typical analog spectrum analyzer could not distinguish between pseudorandom noise and true random noise if the line spacing for the pseudorandom noise is very small.

EXAMPLE 2.13

Binary sequences generated by shift registers with feedback have periodic properties and have found extensive application in ranging and communication systems (9, 10, 11, 12). We will use the simple 5-bit shift register shown in Fig.

---

* Strictly speaking, the *average* envelope would approximate the square root of the spectral density (see Section 2.15).

Figure 2.19. Binary shift register with feedback.

2.19 to demonstrate how a pseudorandom binary signal can be generated. In this system the bits are shifted to the right with each clock pulse, and the input on the left is determined by the feedback arrangement. For the initial condition shown, it can be verified that the output sequence is

$$\underbrace{1111100011011101010000100101100}_{\text{31 bits}} \underbrace{1111100}_{\text{same 31 bits}} \underbrace{\ldots\ldots}_{\text{etc.}}$$

Note that the sequence repeats itself after 31 bits. This periodic property is characteristic of a shift register with feedback. The maximum length of the sequence (before repetition) is given by $(2^n - 1)$ where n is the register length (9). The 5-bit example used here is then a maximum-length sequence. These sequences are especially interesting because of their pseudorandom appearance. Note that there are nearly the same number of zeros and ones in the sequence (16 ones and 15 zeros), and that they appear to occur more or less at random. If we consider a longer shift register, say one with 10 bits, its maximum-length sequence would be 1023 bits. It would have 512 ones and 511 zeros; and, again, these would appear to be distributed at random. A casual look at any interior string of bits would not reveal anything systematic. Yet the string of bits so generated is entirely deterministic. Once the feedback arrangement and initial condition are specified, the output is determined forever after. Thus, the sequence is pseudorandom, not random. ∎

When converted to voltage waveforms, maximum-length sequences also have special autocorrelation properties. Returning to the 31-bit example, let

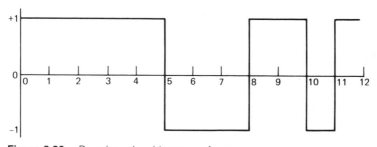

Figure 2.20. Pseudorandom binary waveform.

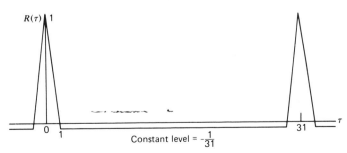

Figure 2.21. Time autocorrelation function for waveform of Figure 2.20.

Figure 2.22. Spectral density for pseudorandom binary waveform.

binary one be 1 V and binary zero be -1 V, and let the voltage level be held constant during the clock-pulse interval. The resulting time waveform is shown in Fig. 2.20, and its time autocorrelation function is shown in Fig. 2.21. Note that the unique distribution of zeros and ones for this sequence is such that the autocorrelation function is a small constant value after a shift of one unit of time (i.e., one bit). This is typical of all maximum-length sequences. When the sequence length is long, the correlation after a shift of one unit is quite small. This has obvious advantages in correlation detection schemes, and such schemes have been used extensively in electronic ranging applications (10, 11).

The spectral density function for the waveform of Fig. 2.20 is shown in Fig. 2.22. As with all pseudorandom signals, the spectrum is line-type rather than continuous (9, 12).

2.15
DETERMINATION OF AUTOCORRELATION AND SPECTRAL DENSITY FUNCTIONS FROM EXPERIMENTAL DATA

The determination of spectral characteristics of a random process from experimental data is a common engineering problem. All of the optimization techniques presented in the following chapters depend on prior knowledge of the spectral density of the processes involved. Thus, the designer needs this information and it usually must come from experimental evidence. Spectral determination is a relatively complicated problem with many pitfalls, and one should approach it with a good deal of caution. It is closely related to the larger problem of digital data processing, because the amount of data needed is usually large, and processing it either manually or in analog form is often not feasible. We first consider the span of observation time of the experimental data, which is a fundamental limitation, irrespective of the means of processing the data.

The time span of the data to be analyzed must, of course, be finite; and, as a practical matter, we prefer not to analyze any more data than is necessary to achieve reasonable results. Remember that since this is a matter of statistical inference, there will always remain some statistical uncertainty in the result. One way to specify the accuracy of the experimentally determined spectrum or autocorrelation function is to say that its variance must be less than a specified value. General accuracy bounds applicable to all processes are not available but there is one special case, the Gaussian process, that is amenable to analysis. We will not give the proof here, but it is shown in a number of references (5, 13) that the variance of an experimentally determined autocorrelation function satisfies the inequality

$$\text{Var } V_X(\tau) \leq \frac{4}{T} \int_0^\infty R_X^2(\tau) \, d\tau \qquad (2.15.1)$$

where it is assumed that a single sample realization of the process is being analyzed, and

$T =$ time length of the experimental record

$R_X(\tau) =$ autocorrelation function of the Gaussian process under consideration

$V_X(\tau) =$ time average of $X_T(t)X_T(t + \tau)$ where $X_T(t)$ is the finite-length sample of $X(t)$ (that is, $V_X(\tau)$ is the experimentally determined autocorrelation function based on a finite record length)

It should be mentioned that in determining the time average of $X_T(t)X_T(t + \tau)$, we cannot use the whole span of time T, because $X_T(t)$ must be shifted an amount of τ with respect to itself before multiplication. The true extension of $X_T(t)$ beyond the experimental data span is unknown; therefore, we simply omit the nonoverlapped portion in the integration:

$$V_X(\tau) = \left[\text{time ave. of } X_T(t)X_T(t + \tau)\right] = \frac{1}{T - \tau} \int_0^{T-\tau} X_T(t)X_T(t + \tau) \, dt$$

$$(2.15.2)$$

It will be assumed from this point on that the range of τ being considered is much less than the total data span T, that is, $\tau \ll T$.

We first note that $V_X(\tau)$ is the result of analyzing a single time signal; therefore, $V_X(\tau)$ is itself just a sample function from an ensemble of functions. It is hoped that $V_X(\tau)$ as determined by Eq. (2.15.2) will yield a good estimate of $R_X(\tau)$ and, in order to do so, it should be an unbiased estimator. This can be verified by computing its expectation:

$$E[V_X(\tau)] = E\left[\frac{1}{T-\tau}\int_0^{T-\tau} X_T(t)X_T(t+\tau)\,dt\right]$$

$$= \frac{1}{T-\tau}\int_0^{T-\tau} E[X_T(t)X_T(t+\tau)]\,dt$$

$$= \frac{1}{T-\tau}\int_0^{T-\tau} R_X(\tau)\,dt = R_X(\tau) \tag{2.15.3}$$

Thus, $V_X(\tau)$ is an unbiased estimator of $R_X(\tau)$. Also, it can be seen from the equation for Var $V_X(\tau)$, Eq. (2.15.1), that if the integral of R_X^2 converges (e.g., R_X decreases exponentially with τ), then the variance of $V_X(\tau)$ approaches zero as T becomes large. Thus $V_X(\tau)$ would appear to be a well-behaved estimator of $R_X(\tau)$, that is, $V_X(\tau)$ converges in the mean to $R_X(\tau)$. We will now pursue the estimation accuracy problem further.

Equation (2.15.1) is of little value if the process autocorrelation function is not known. So, at this point, we assume that $X(t)$ is a Gauss–Markov process with an autocorrelation function

$$R_X(\tau) = \sigma^2 e^{-\beta|\tau|} \tag{2.15.4}$$

The σ^2 and β parameters may be difficult to determine in a real-life problem, but we can get at least a rough estimate of the amount of experimental data needed for a given required accuracy. Substituting the assumed Markov autocorrelation function into Eq. (2.15.1) then yields

$$\text{Var}[V_X(\tau)] \le \frac{2\sigma^4}{\beta T} \tag{2.15.5}$$

We now look at an example illustrating the use of Eq. (2.15.5).

EXAMPLE 2.14

Let us say that the process being investigated is thought to be a Gauss–Markov process with an estimated time constant $(1/\beta)$ of 1 sec. Let us also say that we wish to determine its autocorrelation function within an accuracy of 10 percent, and we want to know the length of experimental data needed. By "accuracy" we mean that the experimentally determined $V(\tau)$ should have a standard deviation less than .1 of the σ^2 of the process, at least for a reasonably small range of τ near zero. Therefore, the ratio of $\text{Var}[V(\tau)]$ to $(\sigma^2)^2$ must be less than .01. Using Eq. (2.15.5), we can write

$$\frac{\text{Var}\left[V(\tau)\right]}{\sigma^4} \leq \frac{2}{\beta T}$$

Setting the quantity on the left side equal to .01 and using the equality condition yields

$$T = \frac{1}{(.1)^2} \cdot \frac{2}{\beta} = 200 \text{ sec} \qquad (2.15.6)$$

A sketch indicating a typical sample experimental autocorrelation function is shown in Fig. 2.23. Note that 10 percent accuracy is really not an especially demanding requirement, but yet the data required is 200 times the time constant of the process. To put this in more graphic terms, if the process under investigation were random gyro drift with an estimated time constant of 10 hours, 2000 hours of continuous data would be needed to achieve 10 percent accuracy. This could very well be in the same range as the mean time to failure for the gyro. Were we to be more demanding and ask for 1 percent accuracy, about 23 years of data would be required! It can be seen that accurate

Figure 2.23. Experimental and true autocorrelation functions for Example 2.14.

determination of the autocorrelation function is not a trivial problem in some applications. ■

The main point to be learned from this example is that reliable determination of the autocorrelation function takes considerably more experimental data than one might expect intuitively. The spectral density function is just the Fourier transform of the autocorrelation function, so we might expect a similar accuracy problem in its experimental determination.

As just mentioned, the spectral density function for a given sample signal may be estimated by taking the Fourier transform of the experimentally determined autocorrelation function. This, of course, involves a numerical procedure of some sort because the data describing $V_X(\tau)$ will be in numerical form. The spectral function may also be estimated directly from the periodogram of the sample signal. Recall from Section 2.7 that the average periodogram (the square of the magnitude of the Fourier transform of X_T) is proportional to the spectral density function (for large T). Unfortunately, since we do not usually have the luxury of having a large ensemble of periodograms to average, there are pitfalls in this approach, just as there are in going the autocorrelation route. Nevertheless, modern digital processing methods using Fast Fourier Transform (FFT) techniques have popularized the periodogram approach. Thus it is important to understand its limitations (5, 6).

First, there is the truncation problem. When the time record being analyzed is finite in length, we usually assume that the signal will "jump" abruptly to zero outside the valid data interval. This causes frequency spreading and gives rise to high-frequency components that are not truly representative of the process under consideration, which is assumed to ramble on indefinitely in a continuous manner. An extreme case of this would occur if we were to chop up one long record into many very short records and then average the periodograms of the short records. The individual periodograms, with their predominance of high-frequency components due to the truncation, would not be at all representative of the spectral content of the original signal; nor would their average! Thus the first rule is that we must have a long time record relative to the typical time variations in the signal. This is true regardless of the method used in analyzing the data. There is, however, a statistical convergence problem that arises as the record length becomes large, and this will now be examined.

In Section 2.7 it was shown that the expectation of the periodogram approaches the spectral density of the process for large T. This is certainly desirable, because we want the periodogram to be an unbiased estimate of the spectral density. It is also of interest to look at the behavior of the variance of the periodogram as T becomes large. Let us denote the periodogram of $X_T(\tau)$ as $M(\omega, T)$, that is,

$$M(\omega, T) = \frac{1}{T}\left|\mathscr{F}\{X_T(t)\}\right|^2 \qquad (2.15.7)$$

Note that the periodogram is a function of the record length T as well as ω.

The variance of $M(\omega, T)$ is

$$\text{Var } M = E(M^2) - [E(M)]^2 \qquad (2.15.8)$$

Since we have already found $E(M)$ as given by Eqs. (2.7.8) and (2.7.9), we now need to find $E(M)^2$. Squaring Eq. (2.15.7) leads to

$$E(M^2) = \frac{1}{T^2} E\left[\int_0^T\int_0^T\int_0^T\int_0^T X(t)X(s)X(u)X(v)e^{-j\omega(t-s+u-v)}\,dt\,ds\,du\,dv\right]$$

$$(2.15.9)$$

It can be shown that if $X(t)$ is a Gaussian process,*

$$E[X(t)X(s)X(u)X(v)] = R_X(t-s)R_X(u-v)$$

$$+ R_X(t-u)R_X(s-v)$$

$$+ R_X(t-v)R_X(s-u) \quad (2.15.10)$$

Thus, moving the expectation operator inside the integration in Eq. (2.15.9) and using Eq. (2.15.10) leads to

$$E(M^2) = \frac{1}{T^2}\int_0^T\int_0^T\int_0^T\int_0^T [R_X(t-s)R_X(u-v) + R_X(t-u)R_X(s-v)$$

$$+ R_X(t-v)R_X(s-u)]e^{-j\omega(t-s+u-v)}\,dt\,ds\,du\,dv$$

$$= \frac{1}{T^2}\int_0^T\int_0^T R_X(t-s)e^{-j\omega(t-s)}\,dt\,ds\int_0^T\int_0^T R_X(u-v)e^{-j\omega(u-v)}\,du\,dv$$

$$+ \frac{1}{T^2}\int_0^T\int_0^T R_X(t-v)e^{-j\omega(t-v)}\,dt\,dv\int_0^T\int_0^T R_X(s-u)e^{-j\omega(s-u)}\,ds\,du$$

$$+ \frac{1}{T^2}\left|\int_0^T\int_0^T R_X(t-u)e^{-j\omega(t+u)}\,dt\,du\right|^2 \qquad (2.15.11)$$

Next, substituting Eq. (2.7.3) into (2.15.11) leads to

$$E(M^2) = 2[E(M)]^2 + \frac{1}{T^2}\left|\int_0^T\int_0^T R_X(t-u)e^{-j\omega(t+u)}\,dt\,du\right|^2 \quad (2.15.12)$$

*See Problem 2.31.

Therefore

$$\text{Var } M = E(M^2) - [E(M)]^2$$

$$= [E(M)]^2 + \frac{1}{T^2} \left| \int_0^T \int_0^T R_X(t-u) e^{-j\omega(t+u)} \, dt \, du \right|^2 \quad (2.15.13)$$

The second term of Eq. (2.15.13) is nonnegative, so it should be clear that

$$\text{Var } M \geq [E(M)]^2 \qquad (2.15.14)$$

But $E(M)$ approaches the spectral function as $T \rightarrow \infty$. Thus, the variance of the periodogram does not go to zero as $T \rightarrow \infty$ (except possibly at those exceptional points where the spectral function is zero). In other words, the periodogram does not converge in the mean as $T \rightarrow \infty$! This is most disturbing, especially in view of the popularity of the periodogram method of spectral determination. The dilemma is summarized in Fig. 2.24. Increasing T will not help reduce the ripples in the individual periodogram. It simply makes M "jump around" faster with ω. This does help, though, with the subsequent averaging that must accompany the spectral analysis. Recall that it is the *average* periodogram that is the measure of the spectral density function. Averaging may not be essential in the analysis of deterministic signals, but it is for random signals. Averaging in both frequency and time is easily accomplished in analog spectrum analyzers by appropriate adjustment of the width of the scanning window and the sweep speed. In digital analyzers, similar averaging over a band of discrete frequencies can be implemented in software. Also, further averaging in time may be accomplished by averaging successive periodograms before displaying the spectrum graphically. In either event, analog or digital, some form of averaging is essential when analyzing noise.

Our treatment of the general problem of autocorrelation and spectral determination from experimental data must be brief. However, the message here should be clear. Treat this problem with respect. It is fraught with subtleties and pitfalls. Engineering literature abounds with reports of shoddy

Figure 2.24. Typical periodogram for long record length.

spectral analysis methods and the attendant questionable results. Know your digital signal processing methods and recognize the limitations of the results.

We will pursue the subject of digital spectral analysis further in Section 2.17. But first we digress to present Shannon's sampling theorems, which play an important role in digital signal processing.

2.16
SAMPLING THEOREM

Consider a time function $g(t)$ that is bandlimited, that is,

$$\mathscr{F}[g(t)] = G(\omega) = \begin{cases} \text{Nontrivial}, & |\omega| \leq 2\pi W \\ 0, & |\omega| > 2\pi W \end{cases} \qquad (2.16.1)$$

Under the conditions of Eq. (2.16.1), the time function can be written in the form

$$g(t) = \sum_{n=-\infty}^{\infty} g\left(\frac{n}{2W}\right) \frac{\sin(2\pi Wt - n\pi)}{2\pi Wt - n\pi} \qquad (2.16.2)$$

This remarkable theorem is due to C. E. Shannon (14, 15), and it has special significance when dealing with bandlimited noise.* The theorem says that if one were to specify an infinite sequence of sample values $\ldots, g_1, g_2, g_3, \ldots$, uniformly spaced $1/2W$ sec apart as shown in Fig. 2.25, then there would be one and only one bandlimited function that would go through all the sample values. In other words, specifying the signal sample values and requiring $g(t)$ to be bandlimited indirectly specifies the signal between the sample points as well. The sampling rate of $2W$ Hz is known as the *Nyquist rate*. This represents the minimum sampling rate needed to preserve all the information content in the continuous signal. If we sample $g(t)$ at less than the Nyquist rate, some information will be lost, and the original signal cannot be exactly reconstructed on the basis of the sequence of samples. Sampling at a rate higher than the Nyquist rate is not necessary, but it does no harm because this simply extends the allowable range of signal frequencies beyond W Hz. Certainly, a signal lying within the bandwidth W also lies within a bandwidth greater than W.

In describing a stationary random process that is bandlimited, it can be seen that we need to consider only the statistical properties of samples taken at the Nyquist rate of $2W$ Hz. This simplifies the process description considerably. If we add the further requirement that the process is Gaussian and white within

---

* The basic concept of sampling at twice the highest signal frequency is usually attributed to H. Nyquist (16). However, the explicit form of the sampling theorem given by Eq. (2.16.2) and its associated signal bandwidth restriction was first introduced into communication theory by C. E. Shannon. You may wish to refer to Shannon (15) or Black (17) for further comments on the history of sampling theory.

Figure 2.25. Samples of bandlimited signal $g(t)$.

the bandwidth W, then the joint probability density for the samples may be written as a simple product of single-variate normal density functions. This simplification is frequently used in noise analysis in order to make the problem mathematically tractable.

Since there is symmetry in the direct and inverse Fourier transforms, we would expect there to be a corresponding sampling theorem in the frequency domain. It may be stated as follows. Consider the time function $g(t)$ to be time limited, that is, nontrivial over a span of time T and zero outside this interval; then its Fourier transform $G(\omega)$ may be written as

$$G(\omega) = \sum_{n=-\infty}^{\infty} G\left(\frac{2\pi n}{T}\right) \frac{\sin\left(\frac{\omega T}{2} - n\pi\right)}{\frac{\omega T}{2} - n\pi} \tag{2.16.3}$$

All of the previous comments relative to time domain sampling have their corresponding frequency-domain counterparts.

Frequently, it is useful to consider time functions that are limited in both time and frequency. Strictly speaking, this is not possible, but it is a useful approximation. This being the case, the time function can be uniquely represented by $2WT$ samples. These may be specified either in the time or frequency domain.

Sampling theorems have also been worked out for the nonbaseband case (18, 19). These are somewhat more involved than the baseband theorems and will not be given here.

2.17
DISCRETE FOURIER TRANSFORM
AND FAST FOURIER TRANSFORM

The subject of digital signal processing has received considerable attention in the past few decades, and it is only natural that this would occur concurrently with the advancement of computer technology. Whole books have been devoted to the subject (20, 21), so that we cannot expect to do the matter justice in one short section. We can, however, present a brief overview in order to place digital spectral analysis in proper perspective. We will then proceed on to the main subject of this book, namely, filter analysis.

Modern computer technology has made it possible to perform an efficient discrete version of the Fourier transform. Thus, nearly all spectral analysis is now done using the direct periodogram approach rather than the more round-about approach via the autocorrelation function. The sampling theorems presented in Section 2.16 dictate some constraints in the choice of sampling rates and the total amount of data analyzed in any one batch. Since these constraints play an important role in digital signal processing, we will examine their consequences in some detail.

In spectral analysis, we usually have at least a rough idea as to the bandwidth of the signal to be analyzed; therefore, let us say this is approximately 0 to W Hz. The sampling theorem, Eq. (2.16.2), says that the sampling rate should be $2W$ samples/sec or, equivalently, the sample spacing should be $1/2W$ sec. Let us further say that we wish to analyze N samples in one batch where N is yet to be determined. The total time span of the samples would then be

$$\text{Total time span of data} = T = \frac{N}{2W} \tag{2.17.1}$$

The frequency-domain sampling theorem, Eq. (2.16.3), states that our truncated time signal could be represented in the spectral domain with discrete samples spaced $1/T$ or $2W/N$ apart. That is, we have N samples uniformly spaced in the time domain and $N/2$ corresponding samples spaced uniformly from 0 to W Hz in the frequency domain. Since each spectral sample is a complex number, the number of degrees of freedom is N in either the time or frequency domain. This one-to-one correspondence of scalar elements in the two domains suggests a transform-pair relationship, and this will now be formalized.

As a matter of notation let the truncated time signal be $g(t)$ and its Fourier transform be

$$G(j\omega) = \int_0^T g(t)e^{-j\omega t}\,dt \tag{2.17.2}$$

It is tacitly assumed here that $g(t)$ is real. Consider next a discrete approximation for $G(j\omega)$ as follows:

$$G(jn2\pi\,\Delta f) \approx \sum_{k=0}^{N-1} g_k e^{-jn2\pi\,\Delta fk\,\Delta T}\,\Delta T \tag{2.17.3}$$

where
g_k = sequence of N time samples of $g(t)$, $\quad k = 0,1,\dots,N-1$
$\Delta T = \dfrac{1}{2W}$ = sample spacing in time domain
$\Delta f = \dfrac{2W}{N}$ = sample spacing in frequency domain (in hertz)

We note now that

$$\Delta f \Delta T = \frac{1}{N} \tag{2.17.4}$$

Thus, Eq. (2.17.3) can be rewritten in the form

$$G(jn2\pi \Delta f)\, \Delta f \approx \frac{1}{N}\sum_{k=0}^{N-1} g_k \exp\left(-j\frac{2\pi nk}{N}\right) \tag{2.17.5}$$

Now, given the time sequence $g_0, g_1, \ldots, g_{N-1}$, think of the right side of Eq. (2.17.5) as defining another sequence $\mathscr{G}_0, \mathscr{G}_1, \ldots, \mathscr{G}_{N-1}$ as n is indexed from 0 to $N-1$. That is, the $\mathscr{G}_n$ sequence is defined as

$$\mathscr{G}_n = \frac{1}{N}\sum_{k=0}^{N-1} g_k \exp\left(-j\frac{2\pi nk}{N}\right), \qquad n = 0,1,\ldots,N-1 \tag{2.17.6}$$

Note that we do not claim the $\mathscr{G}_n$'s to be exact samples of $G(j\omega)$, but, hopefully, they will be reasonable approximations thereof.

The $\mathscr{G}_n$ sequence as defined by Eq. (2.17.6) exhibits certain symmetry that is worth noting. First, if we extend the index n beyond $N-1$, we simply get a periodic extension of the sequence:

$$\mathscr{G}_N = \mathscr{G}_0$$

$$\mathscr{G}_{N+1} = \mathscr{G}_1$$

$$\cdot$$
$$\cdot$$
$$\cdot$$

$$\text{etc.} \tag{2.17.7}$$

Also, there is symmetry about the midpoint of the sequence in that

$$\mathscr{G}_{N-1} = \mathscr{G}_1^*$$

$$\mathscr{G}_{N-2} = \mathscr{G}_2^*$$

$$\cdot$$
$$\cdot$$
$$\cdot$$

$$\text{etc.} \tag{2.17.8}$$

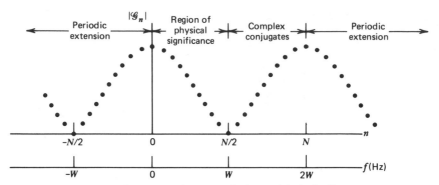

Figure 2.26.   Discrete Fourier transform magnitudes and their significance.

In other words, half of the defined $\mathscr{G}_n$'s are complex conjugates of the other half and are thus redundant (see Problem 2.32). This is as expected; there can only be N degrees of freedom in the frequency domain, just as in the time domain. Figure 2.26 summarizes the symmetry properties of the $\mathscr{G}$ sequence.

Once the $\mathscr{G}_n$ sequence is defined as per Eq. (2.17.6), it can be shown that an exact inverse relationship exists as follows (20, 21):

$$g_k = \sum_{n=0}^{N-1} \mathscr{G}_n \exp\left(j\frac{2\pi nk}{N}\right), \qquad k = 0, 1, \ldots, N-1 \qquad (2.17.9)$$

The $\mathscr{G}_n$ and g_k sequences then form a transform pair in the usual sense; that is, given one, we can find the other, and vice versa.* As might be expected, the $\mathscr{G}_n$ sequence is called the *discrete Fourier transform* (DFT) of g_k and, of course, g_k is the discrete inverse transform of $\mathscr{G}_n$. It should be emphasized that there is no approximation involved in the discrete transform pair relationship. The approximation comes in the *interpretation* of $\mathscr{G}_n$ as samples of the continuous signal spectrum. This is a relatively complicated matter and the references cited (5, 20, 21) may be consulted for further details. It suffices here to say that this has been studied extensively and, with appropriate cautions and proper weighting of the time data, the discrete Fourier transform can provide meaningful results.

We might also note that all the preceding remarks about digital signal processing apply to deterministic as well as random signals. In analyzing deterministic signals, we usually compute the magnitude of the discrete Fourier transform and call this the signal spectrum. In the random signal case, usually the square of the magnitudes of the $\mathscr{G}_n$ terms are formed, and this sequence (i.e., $|\mathscr{G}_0|^2, |\mathscr{G}_1|^2, \ldots$) becomes an approximation of the periodogram of the signal being analyzed. The periodogram is, in turn, statistically related to the power spectral density function, which is usually the desired end result of the analysis. (See Problem 2.30 for comments about the crossperiodogram.)

---

*Some authors prefer to associate the $1/N$ factor of Eq. (2.17.6) with the inverse relationship rather than the direct transform. Since it is a simple proportionality factor, this is perfectly proper.

Digital implementation of the discrete Fourier transform is not a trivial matter. High resolution and reliable results in the frequency domain are obtained by making N large. However, if one programs the transform literally as given by Eq. (2.17.6), the number of multiplications required is of the order N^2. This can easily get out of hand, especially in "on-line" applications. Fortunately, fast, efficient algorithms have been developed for which the number of required multiplications is of the order $N \log_2 N$ rather than N^2 (20, 21). The computational saving is spectacular for large N. For example, let N be $2^{10} = 1024$, which is a modest number of time samples for many applications. Then N^2 would be about 10^6, whereas $N \log_2 N$ is only about 10^4. This represents a saving of about a factor of 100 and reflects directly into the time required for the transformation.

All of the fast discrete Fourier transform algorithms require that the number of samples be an integer power 2 (which usually presents no particular problem), and they all go under the generic name of *Fast Fourier Transform* (FFT). The FFT cannot perform any wondrous, magical tricks on the basic data (as some seem to believe); it is simply an efficient means of implementing the DFT. Thus, all of the cautions that apply to the DFT also apply to the FFT. Because of its efficiency, though, the FFT is used almost universally in on-line spectral analysis applications. Occasionally, though, in off-line applications where speed is of little concern, the straightforward programming of the DFT, as given by Eq. (2.17.6), is advantageous. For example, if only a limited amount of data is available and it is desirable to achieve as fine a resolution as possible in the frequency domain, the straightforward DFT might as well be preferred, because N is not restricted to integer powers of 2 as it is with the FFT.

Before leaving the subject of the discrete Fourier transform, it is worth mentioning that the DFT is easily generalized to the case where the time-domain samples are complex, rather than real as assumed here. The direct and inverse transform Eqs. (2.17.6) and (2.17.9) still apply in the complex case. If one begins with N complex samples in the time domain, then there will be a corresponding set of N complex samples in the frequency domain. Also, if no special symmetry exists among the time-domain samples, then there will be no special symmetry or "folding over" in the frequency domain. The periodic extension stated in Eq. (2.17.7) still applies in the complex case, though.

PROBLEMS

2.1. Noise measurements at the output of a certain amplifier (with its input shorted) indicate that the rms output voltage due to internal noise is $100 \, \mu\text{V}$. If we assume that the frequency spectrum of the noise is flat from 0 to 10 MHz and zero above 10 MHz, find:

  (a) The spectral density function for the noise.
  (b) The autocorrelation function for the noise.
Give proper units for both the spectral density and autocorrelation functions.

2.2. A sketch of a sample realization of a random process $X(t)$ is shown in the figure. The pulse amplitudes a_i are independent samples of a normal random

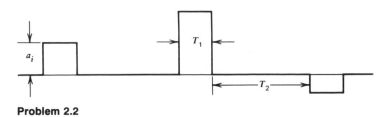

Problem 2.2

variable with zero mean and variance σ^2. The time origin is completely random. Find the autocorrelation function for the process.

2.3. The waveform shown is an example of a digital-coded waveform. The signal is equally likely to be zero or one in the intervals $(t_0, t_0 + 1), (t_0 + 2, t_0 + 3)$, etc., and it is always zero in the "in between" intervals $(t_0 + 1, t_0 + 2), (t_0 + 3, t_0 + 4)$, etc. The switching time t_0 is random and uniformly distributed between zero and two. The presence or absence of a pulse in the "pulse possible" intervals is the code for a binary digit. There is no statistical correlation among any of the bits of the message. Find the autocorrelation and spectral density functions for this process.

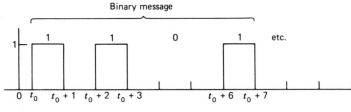

Problem 2.3

2.4. Find the autocorrelation function corresponding to the spectral density function

$$S(j\omega) = \delta(\omega) + \tfrac{1}{2}\delta(\omega - \omega_0) + \tfrac{1}{2}\delta(\omega + \omega_0) + e^{-2|\omega|}$$

2.5. A stationary Gaussian random process $X(t)$ has an autocorrelation function of the form

$$R_X(\tau) = 4e^{-|\tau|}$$

What fraction of the time will $|X(t)|$ exceed four units?

2.6. A random process $X(t)$ is generated according to the following rules:
    (a) The waveform is generated with a sample-and-hold arrangement where the "hold" interval is 1 sec.

(b) The successive amplitudes for each 1-sec interval are independent samples taken from a zero-mean normal distribution with a variance of σ^2.
(c) The first switching time is a random variable with uniform distribution from 0 to 1 (i.e., the time origin is random).

Let X_1 denote $X(t_1)$ and X_2 denote $X(t_1 + \tau)$.
(a) Find the joint probability density function $f_{X_1 X_2}(x_1, x_2)$.
(b) Is this process a Gaussian process?

2.7.   Find the autocorrelation function for the process described in Problem 2.6 using the expectation formula

$$R_X(\tau) = E[X_1 X_2] = \int_{-\infty}^{\infty} \int_{-\infty}^{\infty} x_1 x_2 f_{X_1 X_2}(x_1, x_2) \, dx_1 \, dx_2$$

2.8.   It is suggested that a certain real process has an autocorrelation function as shown in the figure. Is this possible? Justify your answer. (*Hint:* Calculate the spectral density function and see if it is plausible.)

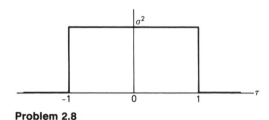

Problem 2.8

2.9.   Consider the random process $X(t) = 2 \sin \omega t$ where ω is a random variable with uniform distribution between $\omega = 2$ and $\omega = 6$. Is the process (a) stationary, (b) ergodic, and (c) deterministic or nondeterministic?

2.10.   The input to an ideal rectifier (unity forward gain, zero reverse gain) is a stationary Gaussian process.
(a) Is the output stationary?
(b) Is the output a Gaussian process?
Give a brief justification for both answers.

2.11.   A random process $X(t)$ has sample realizations of the form

$$X(t) = at + Y$$

where a is a known constant and Y is a random variable whose distribution is $N(0, \sigma^2)$. Is the process (a) stationary and (b) ergodic? Justify your answers.

2.12.   What is the autocorrelation function for $X(t)$ of Problem 2.11?

2.13. A sample realization of a random process $X(t)$ is shown in the figure. The time t_0 when the transition from the -1 state to the $+1$ state takes place is a random variable that is uniformly distributed between 0 and 2 units.
 (a) Is the process stationary?
 (b) Is the process deterministic or nondeterministic?
 (c) Find the autocorrelation function and spectral density function for the process.

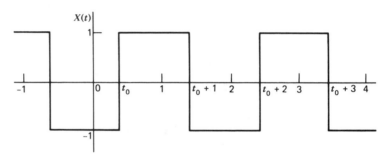

Problem 2.13

2.14. A common autocorrelation function encountered in physical problems is

$$R(\tau) = \sigma^2 e^{-\beta|\tau|} \cos \omega_0 \tau$$

 (a) Find the corresponding spectral density function.
 (b) $R(\tau)$ will be recognized as a damped cosine function. Sketch both the autocorrelation and spectral density functions for the lightly damped case.

2.15. Show that a Gauss–Markov process described by the autocorrelation function

$$R(\tau) = \sigma^2 e^{-\beta|\tau|}$$

becomes Gaussian white noise if we let $\beta \to \infty$ and $\sigma^2 \to \infty$ in such a way that the area under the autocorrelation-function curve remains constant in the limiting process.

2.16. A stationary random process $X(t)$ has a spectral density function of the form

$$S_X(\omega) = \frac{6\omega^2 + 12}{(\omega^2 + 4)(\omega^2 + 1)}$$

What is the mean-square value of $X(t)$? (*Hint:* $S_X(\omega)$ may be resolved into a sum of two terms of the form: $[A/(\omega^2 + 4)] + [B/(\omega^2 + 1)]$. Each term may then be integrated using standard integral tables.)

2.17. The stationary process $X(t)$ has an autocorrelation function of the form

$$R_X(\tau) = \sigma^2 e^{-\beta|\tau|}$$

Another process $Y(t)$ is related to $X(t)$ by the deterministic equation

$$Y(t) = aX(t) + b$$

where a and b are known constants.

    (a) What is the autocorrelation function for $Y(t)$?

    (b) What is the crosscorrelation function $R_{XY}(\tau)$?

2.18. The crosscorrelation function $R_{XY}(\tau)$ for Example 2.8 is sketched below. What is the corresponding cross spectral density function?

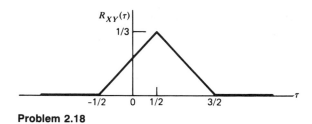

Problem 2.18

2.19. The random telegraph wave is described in Section 2.11. Let this process be the input to an ideal rectifier (unity forward gain, zero reverse gain).

    (a) What is the autocorrelation function of the output?

    (b) What is the crosscorrelation function $R_{XY}(\tau)$ (X is input, Y is output)?

2.20. Two deterministic random processes are defined by

$$X(t) = A \sin(\omega t + \theta)$$
$$Y(t) = B \sin(\omega t + \theta)$$

where θ is a random variable with uniform distribution between 0 and 2π, and ω is a known constant. The A and B coefficients are both normal random variables $N(0, \sigma^2)$ and are correlated with a correlation coefficient ρ. What is the crosscorrelation function $R_{XY}(\tau)$? (Assume A and B are independent of θ.)

2.21. The discrete random walk process is discussed in Section 2.13. Assume each step is of length l and that the steps are independent and equally likely to be positive or negative. Show that the variance of the total distance D traveled in N steps is given by

$$\text{Var } D = l^2 N$$

[*Hint:* First write D as the sum $l_1 + l_2 + \cdots l_N$ and note that $l_1, l_2, \ldots, l_N$ are independent random variables. Then form $E(D)$ and $E(D^2)$ and compute Var D as $E(D^2) - [E(D)]^2$.]

2.22. Let the process $Z(t)$ be the product of two independent stationary processes $X(t)$ and $Y(t)$. Show that the spectral density function for $Z(t)$ is given by (in the s domain)

$$S_Z(s) = \frac{1}{2\pi j} \int_{-j\infty}^{j\infty} S_X(w) S_Y(s - w) \, dw$$

[*Hint:* First show that $R_Z(\tau) = R_X(\tau) R_Y(\tau)$.]

2.23. The spectral density function for the stationary process $X(t)$ is

$$S_X(j\omega) = \frac{1}{\left(1 + \omega^2\right)^2}$$

Find the autocorrelation function for $X(t)$.

2.24. A stationary process $X(t)$ is Gaussian and has an autocorrelation function of the form

$$R_X(\tau) = 4e^{-|\tau|}$$

Let the random variable X_1 denote $X(t_1)$ and X_2 denote $X(t_1 + 1)$. Write the expression for the joint probability density function $f_{X_1 X_2}(x_1, x_2)$.

2.25. A stationary Gaussian process $X(t)$ has a power spectral density function

$$S_X(j\omega) = \frac{2}{\omega^4 + 1}$$

Find $E(X)$ and $E(X^2)$.

2.26. A typical sample function of a stationary Gauss–Markov process is shown in the sketch. The process has a mean-square value of 9 units, and the random variables X_1 and X_2 indicated on the waveform have a correlation coefficient of 0.5. Write the expression for the autocorrelation function of $X(t)$.

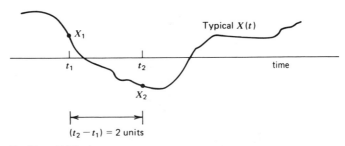

Problem 2.26

2.27. The Global Positioning System (GPS) is a satellite navigation system that is scheduled to be fully operational in the early 1990s (10, 11). The signals transmitted by the satellites provide timing information that enables the receiver to determine its position and velocity in an earth-fixed reference frame. One of the signals provided by each of the system satellites is of the form

$$s(t) = C(t) \sin 2\pi f_c t$$

where

$$f_c = \text{carrier frequency} = 1575.42 \text{ MHz}$$

and $C(t)$ is a pseudorandom binary ($+1$ or -1) code signal similar to the one described in Example 2.13. The code repeats itself after 1023 bits, and the repetition period is exactly 1 ms. Thus, the "code rate" is 1.023 MHz. (*Note:* There are exactly 1540 cycles of the carrier per bit of the pseudorandom code.)

(a) Assume that the autocorrelation function of the binary code can be approximated as being triangular just as in Example 2.13, except that the repetition period is 1023 bits rather than 31. Sketch the spectrum of $C(t)$. Note on your sketch the Δf spacing between spectral lines.

(b) Sketch the spectrum of the radiofrequency signal $s(t)$. Note on your sketch the bandwidth of the main lobe of the spectrum.

(Actually, the transmittal signal has an additional low-frequency modulation signal that contains the navigational message. This has the effect of spreading the spectral lines slightly. This may be ignored in this problem.)

2.28. We wish to determine the autocorrelation function a random signal empirically from a single time record. Let us say we have good reason to believe the process is ergodic and at least approximately Gaussian and, furthermore, that the autocorrelation function of the process decays exponentially with a time constant no greater than 10 sec. Estimate the record length needed to achieve 5 percent accuracy in the determination of the autocorrelation function. (By 5 percent accuracy, assume we mean that for any τ, the standard deviation of the experimentally determined autocorrelation function will not be more than 5 percent of the maximum value of the true autocorrelation function.)

2.29. In Problem 2.28 the signal is not known to be truly bandlimited, but it is reasonable to assume that essentially all the signal energy will lie between 0 and .1 Hz. Let us assume .1 Hz to be the signal bandwidth, and let us say we wish to sample the signal at the Nyquist rate.

(a) How many discrete samples would be required to describe the record length of Problem 2.28?

(b) Suppose that we wish to compute the discrete Fourier transform of the finite-time signal using the Fast Fourier Transform. This requires that the number of samples N be an integer power of 2. What should N be in this case?

(c) The value of N in part (b) should work out to be greater than that found in part (a). In order to achieve the appropriate N for the FFT algorithm, would we be better off (in terms of accuracy) to increase the sampling rate for the length of time computed in Problem 2.28, or should we keep the sampling rate at .2 Hz and increase the time length of the record accordingly? Presumably, the computational effort would be the same either way.

2.30. Calculating either the autocorrelation or crosscorrelation function can be an onerous task, especially if there is a large amount of data to be processed. It is sometimes easier to compute the appropriate spectral function first and then inverse transform it, rather than do the computation as an averaging process directly in the time domain. This may seem to be a "roundabout" approach at first glance, but the ease of using the FFT often makes this approach attractive.

Suppose we have time records of length T of two stationary random processes $x(t)$ and $y(t)$, and suppose we are interested in obtaining the crosscorrelation $R_{xy}(\tau)$ of these two processes. Show that a crossperiodogram-type function can be formed just as was done for the usual power spectral density, and that it is given by

$$\text{Crossperiodogram} = M_{xy}(\omega, T) \triangleq \frac{1}{T} X_T(-j\omega) Y_T(j\omega)$$

where X_T and Y_T are the Fourier transforms of the truncated $x_T(t)$ and $y_T(t)$ functions. Note that it is relatively easy to form the indicated product function in the complex domain and then do another FFT to get the crosscorrelation function in the time domain. (It should be mentioned that all of the same statistical convergence problems mentioned in Section 2.15 relative to the periodogram also apply to the crossperiodogram.)

2.31. Let X_1, X_2, X_3, X_4 be zero-mean Gaussian random variables. Show that

$$E(X_1 X_2 X_3 X_4) = E(X_1 X_2) E(X_3 X_4) + E(X_1 X_3) E(X_2 X_4)$$

$$+ E(X_1 X_4) E(X_2 X_3) \tag{P2.31.1}$$

(*Hint:* The characteristic function was discussed briefly in Section 1.8.) The multivariate version of the characteristic function is useful here. Let $\psi(\omega_1, \omega_2, \ldots, \omega_n)$ be the multidimensional Fourier transform of $f_{X_1 X_2 \cdots X_n}(x_1, x_2, \ldots, x_n)$ (but with the signs reversed on $\omega_1, \omega_2, \ldots, \omega_n$). Then it can be readily verified that

$$(-j)^n \frac{\partial^n \psi(\omega_1, \omega_2, \ldots, \omega_n)}{\partial \omega_1 \partial \omega_2, \ldots, \partial \omega_n} \Bigg|_{\substack{\omega_1 = 0 \\ \omega_2 = 0 \\ \text{etc.}}}$$

$$= \int_{-\infty}^{\infty} \int_{-\infty}^{\infty} \cdots \int_{-\infty}^{\infty} x_1, x_2, \ldots,$$

$$x_n f_{X_1 X_2, \ldots, X_n}(x_1, x_2, \ldots, x_n) \, dx_1, dx_2, \ldots, dx_n$$

$$= E(X_1, X_2, \ldots, X_n) \tag{P2.31.2}$$

The characteristic function for a zero-mean, vector Gaussian random variable $\mathbf{X}$ is

$$\psi(\boldsymbol{\omega}) = e^{-\frac{1}{2}\boldsymbol{\omega}^T \mathbf{C}_X \boldsymbol{\omega}} \tag{P2.31.3}$$

where $\mathbf{C}_X$ is the covariance matrix for $\mathbf{X}$. This, along with Eq. (P2.31.2), may now be used to justify the original statement given by Eq. (P2.31.1).

2.32. It was mentioned in Section 2.17 that half of the discrete Fourier transform elements are complex conjugates of the other half. This statement deserves closer scrutiny because N may be either odd or even insofar as the basic discrete transform is concerned (and not its efficient FFT implementation).
   (a) For the case where N is even, show that $\mathscr{G}_0$ and $\mathscr{G}_{N/2}$ are both real, and thus the total number of nonredundant scalar elements in the frequency domain is N, just as in the time domain.
   (b) For the case where N is odd, show that only $\mathscr{G}_0$ is constrained to be real, and thus the count of nonredundant scalar elements is N, just as in the previous case.

2.33. The accompanying figure shows a means of generating narrowband noise from two independent baseband noise sources. (See Section 2.12.) The bandwidth of the resulting narrowband noise is controlled by the cutoff frequency of the low-pass filters, which are assumed to have identical characteristics. Assume that $F_1(t)$ and $F_2(t)$ are independent white Gaussian noise processes with similar spectral amplitudes. The resulting noise processes after low-pass filtering will then have identical autocorrelation functions that will be denoted $R_X(\tau)$.
   (a) Show that the narrowband noise $S(t)$ is a stationary Gaussian random process whose autocorrelation function is

$$R_S(\tau) = R_X(\tau) \cos \omega_c \tau$$

   (b) Also show that both the in-phase and quadrature channels are needed to produce stationary narrowband noise. (That is, if either of the $\sin \omega_c t$ or $\cos \omega_c t$ multiplying operations is omitted, the resultant output will not be strictly stationary.)

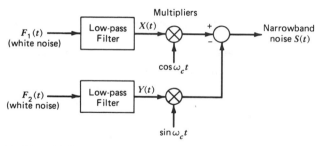

Problem 2.33

2.34.* A sequence of discrete samples of a Gauss–Markov process can be generated from the following difference equation (see Examples 5.3 and 5.6):

$$x_{k+1} = e^{-\beta \Delta t} x_k + w_k, \qquad k = 0, 1, 2, \ldots$$

$$w_k = \text{white sequence} \sim N\left[0, \sigma^2 (1 - e^{-2\beta \Delta t})\right]$$

$$\sigma^2 = \text{variance of the Markov process}$$

$$\beta = \text{reciprocal time constant of the Markov process}$$

$$\Delta t = \text{time interval between samples}$$

If x_0 is chosen at random from a population $N(0, \sigma^2)$, the sequence so generated will be a sample realization of stationary Gauss–Markov process.

(a) Generate 512 samples of a Gauss–Markov process with $\sigma^2 = 1$, $\beta = 1$, and $\Delta t = .04$ sec. Store the samples in a file.

(b) Using only the first 128 samples of the process, work out an experimental autocorrelation function [i.e., $V(\tau)$] for $\tau = 0, .04, .08, \ldots, 2.0$. View a plot of the resulting $V(\tau)$.

(c) Repeat part (b) using the first 256 samples in the determination of $V(\tau)$.

(d) Finally, repeat part (b) using all 512 samples. Note that as the time span of the data increases, the $V(\tau)$ becomes smoother and a better approximation to the true $R_x(\tau)$.

(e) The process generated in part (a) is, of course, just one sample realization of the Markov process. One must expect the unusual situation to happen occasionally as well as the usual in any such statistical experiment. So, if the results of parts (b), (c), and (d) are not satisfying, try generating another sample realization of the process and repeating the experiment. This is easily done by choosing a different randomizing seed in the process generation.

2.35.* A recursion equation for generating a Gauss–Markov process was given in Problem 2.34*. We wish to use the same process here, but the sampling rate will be changed to make it more suitable for a demonstration of experimental determination of the process spectral function. Therefore, the parameters will be the same as in Problem 2.34*, except that we will let $\Delta t = .5$ sec in this problem.

(a) Generate 256 samples of the specified Gauss–Markov process, and store the samples in a file.

(b) Using only the first 64 samples of the process, work out the discrete periodogram for the process, i.e., the square of the magnitude of the DFT. View a plot of the data points of the periodogram. Note that the data do not follow a nice smooth bell-shaped curve for a Markov spectral function.

(c) Repeat part (b) using the first 128 samples. Note that the resulting frequency span of the periodogram is the same as in part (b) when considered in terms of real frequency in Hz. The spacing between spectral lines is halved, though.

(d) Repeat part (b) using all 256 samples. Again, note that the spacing of the spectral lines is half that of the previous 128-point periodogram. Also note that increasing the data span in time does not mitigate the noisiness of the data in the frequency domain. It does, however, increase the density of the spectral lines, which, in turn, makes it possible to do some averaging in the frequency domain.

(e) The 256-point periodogram obtained in part (d) is a "noisy" approximation of the power spectral density function (within a scale factor). Now form the inverse DFT of the 256-point periodogram and view a plot of the discrete data. This, of course, should be an approximation to the true autocorrelation function for the process. Note that the approximation is reasonably good for small τ, and that the plot is much less noisy than the corresponding plot in the frequency domain. In effect, a degree of smoothing (averaging) is automatically achieved in the process of transforming the numerical data from the frequency domain to the time domain. This is as it should be. We did not see extreme noisiness in determining the autocorrelation function directly from the time data in Problem 2.34*; nor should we when going the roundabout way via the periodogram and its inverse.

2.36.* The recursion equation given in Problem 2.34* for generating a Gauss–Markov process can also be used to generate a stationary white sequence. This is done by letting $e^{-\beta \Delta t}$ (i.e., the transition matrix) be zero and initializing x_0 to be a zero-mean Gaussian random variable. The step size Δt may be thought of as unity for convenience. If we also let Q_k (i.e., the variance of w_k) be unity, we generate a unity-variance white sequence.

(a) Generate a 256-sample unity-variance white sequence using the recursion formula given in Problem 2.34*. Store the resulting data in a file.

(b) Work out the experimental autocorrelation function for this sequence for $\tau = 0, 1, 2, \ldots, 50$. The result should approximate a unit pulse at the origin.

(c) Compute the 256-point discrete Fourier transform (DFT) for the sequence of part (a), and view the plot of the magnitude of the DFT. The spectral lines will vary wildly with frequency, but the overall trend should be flat with frequency.

REFERENCES CITED IN CHAPTER 2

1. E. Wong, *Stochastic Processes in Information and Dynamical Systems*, New York: McGraw-Hill, 1971.
2. A. H. Jazwinski, *Stochastic Processes and Filtering Theory*, New York: Academic Press, 1970.
3. N. Wiener, *Extrapolation, Interpolation, and Smoothing of Stationary Time Series*, Cambridge, Mass.: MIT Press, and New York: Wiley; 1949.
4. S. O. Rice, "Mathematical Analysis of Noise," *Bell System Tech. J.*, *23*, 282–332 (1944); *24*, 46–256 (1945).

5. J. S. Bendat and A. G. Piersol, *Random Data: Analysis and Measurement Procedures*, New York: Wiley-Interscience, 1971.
6. K. S. Shanmugan and A. M. Breipohl, *Random Signals: Detection, Estimation, and Data Analysis*, New York: Wiley, 1988.
7. W. B. Davenport, Jr., and W. L. Root, *An Introduction to the Theory of Random Signals and Noise*, New York: McGraw-Hill, 1958.
8. A. Papoulis, *Probability, Random Variables, and Stochastic Processes*, 2nd ed., New York: McGraw-Hill, 1984.
9. R. C. Dixon, *Spread Spectrum Systems*, New York: Wiley, 1976.
10. R. P. Denaro, "Navstar: The All-Purpose Satellite," *IEEE Spectrum, 18:5*, 35–40 (May 1981).
11. B. W. Parkinson and S. W. Gilbert, "NAVSTAR: Global Positioning System—Ten Years Later," *Proc. of the IEEE, 71:10*, 1177–1186, (October 1983).
12. R. E. Ziemer and R. L. Peterson, *Digital Communications and Spread Spectrum Systems*, New York: Macmillan, 1985.
13. J. H. Laning, Jr., and R. H. Battin, *Random Processes in Automatic Control*, New York: McGraw-Hill, 1956.
14. C. E. Shannon, "The Mathematical Theory of Communication," *Bell System Tech. J.*, July and October 1948. (Later reprinted in book form by the University of Illinois Press, 1949.)
15. C. E. Shannon, "Communication in the Presence of Noise," *Proc. of the Inst. of Radio Engr., 37:1*, 10–21, (January 1949).
16. H. Nyquist, "Certain Topics in Telegraph Transmission Theory," *Trans. of the Am. Inst. of Elect. Engr., 47*, 617–644, (April 1928).
17. H. S. Black, *Modulation Theory*, New York: D. Van Nostrand Co., 1950.
18. S. Goldman, *Information Theory*, Englewood Cliffs, N.J.: Prentice-Hall, 1953.
19. K. S. Shanmugam, *Digital and Analog Communication Systems*, New York: Wiley, 1979.
20. A. V. Oppenheim and R. W. Schafer, *Digital Signal Processing*, Englewood Cliffs, N.J.: Prentice-Hall, 1975.
21. D. Childers and A. Durling, *Digital Filtering and Signal Processing*, St. Paul: West Publishing Co., 1975.
22. P. Z. Peebles, Jr., *Probability, Random Variables, and Random Signal Principles*, 2nd ed., New York: McGraw-Hill, 1987.
23. H. J. Larson and B. O. Shubert, *Probabilistic Models in Engineering Sciences (Vols. 1 and 2)*, New York: Wiley, 1979.
24. G. R. Cooper and C. D. McGillem, *Probabilistic Methods of Signal and System Analysis*, 2nd ed., New York: Holt, Rinehart, and Winston, 1986.
25. J. L. Melsa and A. P. Sage, *An Introduction to Probability and Stochastic Processes*, Englewood Cliffs, N.J.: Prentice-Hall, 1973.
26. A. M. Breipohl, *Probabilistic Systems Analysis*, New York: Wiley, 1970.
27. J. V. Candy, *Signal Processing: The Model-Based Approach*, New York: McGraw-Hill, 1986.

3

Response of Linear Systems to Random Inputs

The central problem of linear systems analysis is: Given the input, what is the output? In the deterministic case, we usually seek an explicit expression for the response or output. In the random-input problem no such explicit expression is possible, except for the special case where the input is a so-called deterministic random process (and not always in this case). Usually, in random-input problems, we must settle for a considerably less-complete description of the output than we get for corresponding deterministic problems. In the case of random processes the most convenient descriptors to work with are autocorrelation function, spectral density function, and mean-square value. We now examine the input–output relationships of linear systems in these terms.

3.1
INTRODUCTION:
THE ANALYSIS PROBLEM

In any system satisfying a set of linear differential equations, the solution may be written as a superposition of an initial-condition part and another part due to the driving or forcing functions. Both the initial conditions and forcing functions may be random; and, if so, the resultant response is a random process. We direct our attention here to such situations, and it will be assumed that the reader has at least an elementary acquaintance with deterministic methods of linear system analysis (1, 2, 3).

With reference to Fig. 3.1, the analysis problem may be simply stated: Given the initial conditions and the input and the system's dynamical character-

$f(t)$ = Input
$x(t)$ = Output
$G(s)$ = Transfer function
= $X(s)/F(s)$

Figure 3.1. Block Diagram for elementary analysis problem.

istics [i.e., $G(s)$ in Fig. 3.1], what is the output? Of course, in the stochastic problem, the input and output will have to be described in probabilistic terms.

We need to digress here for a moment and discuss a notational matter. In Chapters 1 and 2 we were careful to use uppercase symbols to denote random variables and lowercase symbols for the corresponding arguments of their probability density functions. This is the custom in most current books on probability. There is, however, a long tradition in engineering books on automatic control and linear systems analysis of using lowercase for time functions and uppercase for the corresponding Laplace or Fourier transforms. Hence, we are confronted with notational conflict. We will resolve this in favor of the traditional linear analysis notation, and from this point on we will use lowercase symbols for time signals—either random or deterministic—and uppercase for their transforms. This seems to be the lesser of the two evils. The reader will simply have to interpret the meaning of symbols such as $x(t)$, $f(t)$, and the like, within the context of the subject matter under discussion. This usually presents no problem. For example, with reference to Fig. 3.1, $g(t)$ would mean inverse transform of $G(s)$, and it clearly is a deterministic time function. On the other hand, the input and output, $f(x)$ and $x(t)$, will usually be random processes in the subsequent material.

Generally, analysis problems can be divided into two categories:

1. Stationary (steady-state) analysis. Here the input is assumed to be time stationary, and the system is assumed to have fixed parameters with a stable transfer function. This leads to a stationary output, provided the input has been present for a long period of time relative to the system time constants.

2. Nonstationary (transient) analysis. Here we usually consider the driving function as being applied at $t = 0$, and the system may be initially at rest or have nontrivial initial conditions. The response in this case is usually nonstationary. We note that analysis of unstable systems falls into this category, because no steady-state (stationary) condition will exist.

The similarity between these two categories and the corresponding ones in deterministic analysis should be apparent. Just as in circuit analysis, we would expect the transient solution to lead to the steady-state response as $t \rightarrow \infty$. However, if we are only interested in the stationary result, this is getting at the solution the "hard way." Much simpler methods are available for the stationary solution, and these will now be considered.

3.2
STATIONARY (STEADY-STATE) ANALYSIS

We assume in Fig. 3.1 that $G(s)$ represents a stable, fixed-parameter system and that the input is covariance (wide-sense) stationary with a known spectral function. In deterministic analysis, we know that if the input is Fourier transformable, the input spectrum is simply modified by $G(j\omega)$ in going through the filter. In the random process case, one interpretation of the spectral function is that it is proportional to the magnitude of the *square* of the Fourier transform. Thus the equation relating the input and output spectral functions is*

$$S_x(s) = G(s)G(-s)S_f(s) \qquad (3.2.1)$$

Note that Eq. (3.2.1) is written in the s domain where the imaginary axis has the meaning of real angular frequency ω. If you prefer to write Eq. (3.2.1) in terms of ω, just replace s with $j\omega$. Equation (3.2.1) then becomes

$$S_x(j\omega) = G(j\omega)G(-j\omega)S_f(j\omega)$$

$$= |G(j\omega)|^2 S_f(j\omega) \qquad (3.2.2)$$

Because of the special properties of spectral functions, both sides of Eq. (3.2.2) work out to be real functions of ω. Also note that the autocorrelation function of the output can be obtained as the inverse Fourier transform of $S_x(s)$. Two examples will now illustrate the use of Eq. (3.2.1).

EXAMPLE 3.1
Consider a first-order low-pass filter with unity white noise as the input. With reference to Fig. 3.1, then

$$S_f(s) = 1$$

$$G(s) = \frac{1}{1 + Ts}$$

where T is the time constant of the filter. The output spectral function is then

$$S_x(s) = \frac{1}{1 + Ts} \cdot \frac{1}{1 + T(-s)} \cdot 1$$

$$= \frac{(1/T)^2}{-s^2 + (1/T)^2}$$

---

* See Problem 3.1 for a formal justification of Eq. (3.2.1).

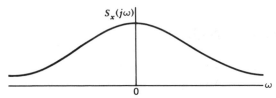

Figure 3.2.   Spectral function for low-pass filter output with white noise input.

Or, in terms of real frequency ω,

$$S_x(j\omega) = \frac{(1/T)^2}{\omega^2 + (1/T)^2}$$

This is sketched as a function of ω in Fig. 3.2. As would be expected, most of the spectral content is concentrated at low frequencies and then it gradually diminishes as $\omega \to \infty$.

It is also of interest to compute the mean-square value of the output. It is given by Eq. (2.7.17).

$$E(x^2) = \frac{1}{2\pi j} \int_{-j\infty}^{j\infty} \frac{1}{1 + Ts} \cdot \frac{1}{1 + T(-s)} \, ds \qquad (3.2.3)$$

The integral of Eq. (3.2.3) is easily evaluated in this case by substituting $j\omega$ for s and using a standard table of integrals. This leads to

$$E(x^2) = \frac{1}{2T}$$

The "standard" table-of-integrals approach is of limited value, though, as will be seen in the next example.   ■

EXAMPLE 3.2

Consider the input process to have an exponential autocorrelation function and the filter to be the same as in the previous example. Then

$$R_f(\tau) = \sigma^2 e^{-\beta|\tau|}$$

$$G(s) = \frac{1}{1 + Ts}$$

First we transform R_f to obtain the input spectral function.

$$\mathscr{F}\left[R_f(\tau)\right] = \frac{2\sigma^2\beta}{-s^2 + \beta^2}$$

The output spectral function is then

$$S_x(s) = \frac{2\sigma^2\beta}{-s^2 + \beta^2} \cdot \frac{(1/T)}{[s + (1/T)]} \cdot \frac{(1/T)}{[-s + (1/T)]} \qquad (3.2.4)$$

Now, if we wish to find $E(x^2)$ in this case, it will involve integrating a function that is fourth order in the denominator, and most tables of integrals will be of no help. We note, though, that the input spectral function can be factored and the terms of Eq. (3.2.4) can be rearranged as follows:

$$S_x(s) = \left[\frac{\sqrt{2\sigma^2\beta}}{(s + \beta)} \cdot \frac{(1/T)}{[s + (1/T)]}\right]\left[\frac{\sqrt{2\sigma^2\beta}}{(-s + \beta)} \cdot \frac{(1/T)}{[-s + (1/T)]}\right] \qquad (3.2.5)$$

The first term has all its poles and zeros in the left half-plane, and the second term has mirror-image poles and zeros in the right half-plane. This regrouping of terms is known as *spectral factorization* and can always be done if the spectral function is rational in form (that is, if it can be written as a ratio of polynomials in even powers of s).

Since special tables of integrals have been worked out for integrating complex functions of the type given by Eq. (3.2.5), we defer evaluating $E(x^2)$ until these have been presented in the next section. ∎

3.3
INTEGRAL TABLES FOR COMPUTING MEAN-SQUARE VALUE

In linear analysis problems, the spectral function can often be written as a ratio of polynomials in s^2. If this is the case, spectral factorization can be used to write the function in the form

$$S_x(s) = \frac{c(s)}{d(s)} \cdot \frac{c(-s)}{d(-s)} \qquad (3.3.1)$$

where $c(s)/d(s)$ has all its poles and zeros in the left half-plane and $c(-s)/d(-s)$ has mirror-image poles and zeros in the right half-plane. No roots of $d(s)$ are permitted on the imaginary axis. The mean-square value of x can now be

Table 3.1 Table of Integrals

$$I_n = \frac{1}{2\pi j} \int_{-j\infty}^{j\infty} \frac{c(s)c(-s)}{d(s)d(-s)} \, ds \tag{3.3.3}$$

$$c(s) = c_{n-1}s^{n-1} + c_{n-2}s^{n-2} + \cdots + c_0$$

$$d(s) = d_n s^n + d_{n-1}s^{n-1} + \cdots + d_0$$

$$I_1 = \frac{c_0^2}{2d_0 d_1}$$

$$I_2 = \frac{c_1^2 d_0 + c_0^2 d_2}{2d_0 d_1 d_2}$$

$$I_3 = \frac{c_2^2 d_0 d_1 + \left(c_1^2 - 2c_0 c_2\right) d_0 d_3 + c_0^2 d_2 d_3}{2d_0 d_3 (d_1 d_2 - d_0 d_3)}$$

$$I_4 = \frac{c_3^2 \left(-d_0^2 d_3 + d_0 d_1 d_2\right) + \left(c_2^2 - 2c_1 c_3\right) d_0 d_1 d_4 + \left(c_1^2 - 2c_0 c_2\right) d_0 d_3 d_4 + c_0^2 \left(-d_1 d_4^2 + d_2 d_3 d_4\right)}{2d_0 d_4 \left(-d_0 d_3^2 - d_1^2 d_4 + d_1 d_2 d_3\right)}$$

written as

$$E(x^2) = \frac{1}{2\pi j} \int_{-j\infty}^{j\infty} \frac{c(s)c(-s)}{d(s)d(-s)} \, ds \tag{3.3.2}$$

R. S. Phillips (4) was the first to prepare a table of integrals for definite integrals of the type given by Eq. (3.3.2). His table has since been repeated in many texts with a variety of minor modifications (5, 6, 7). An abbreviated table in terms of the complex s domain follows. An example will now illustrate the use of Table 3.1.

EXAMPLE 3.3

The solution in Example 3.2 was brought to the point where the spectral function had been written in the form

$$S_x(s) = \left[\frac{\sqrt{2\sigma^2 \beta} \cdot 1/T}{(s + \beta)(s + 1/T)} \right]\left[\frac{\sqrt{2\sigma^2 \beta} \cdot 1/T}{(-s + \beta)(-s + 1/T)} \right] \tag{3.3.4}$$

Clearly, S_x has been factored properly with its poles separated into left and right half-plane parts. The mean-square value of x is given by

$$E(x^2) = \frac{1}{2\pi j} \int_{-j\infty}^{j\infty} S_x(s) \, ds \tag{3.3.5}$$

Comparing the form of $S_x(s)$ in Eq. (3.3.4) with the standard form given in Eq.

(3.3.3), we see that

$$c(s) = \frac{\sqrt{2\sigma^2\beta}}{T}$$

$$d(s) = s^2 + (\beta + 1/T)s + \beta/T$$

Thus we can use the I_2 integral of Table 3.1. The coefficients for this case are

$$c_1 = 0 \qquad\qquad d_2 = 1$$

$$c_0 = \frac{\sqrt{2\sigma^2\beta}}{T} \qquad d_1 = (\beta + 1/T)$$

$$d_0 = \beta/T$$

and $E(x^2)$ is then

$$E(x^2) = \frac{c_0^2}{2d_0d_1} = \frac{2\sigma^2\beta/T^2}{2(\beta/T)(\beta + 1/T)} = \frac{\sigma^2}{1 + \beta T} \qquad\blacksquare$$

3.4
PURE WHITE NOISE AND
BANDLIMITED SYSTEMS

We are now in a position to demonstrate the validity of using the pure white-noise model in certain problems, even though white noise has infinite variance. This will be done by posing two hypothetical mean-square analysis problems:

1. Consider a simple first-order low-pass filter with bandlimited white noise as the input. Specifically, with reference to Fig. 3.1, let

$$S_f(j\omega) = \begin{cases} A, & |\omega| \le \omega_c \\ 0, & |\omega| > \omega_c \end{cases} \qquad (3.4.1)$$

$$G(s) = \frac{1}{1 + Ts} \qquad (3.4.2)$$

2. Consider the same low-pass filter as in problem 1, but with pure white noise as the input:

$$S_f(j\omega) = A, \qquad \text{for all } \omega \tag{3.4.3}$$

$$G(s) = \frac{1}{1 + Ts} \tag{3.4.4}$$

Certainly problem 1 is physically plausible because bandlimited white noise has finite variance. Conversely, problem 2 is not because the input has infinite variance. The preceding theory enables us to evaluate the mean-square value of the output for both problems. As a matter of convenience, we do this in the real frequency domain rather than the complex s domain.

Problem 1:

$$S_x(j\omega) = \begin{cases} \dfrac{A}{1 + (T\omega)^2}, & |\omega| \leq \omega_c \\ 0, & |\omega| > \omega_c \end{cases} \tag{3.4.5}$$

$$E(x^2) = \frac{1}{2\pi} \int_{-\omega_c}^{\omega_c} \frac{A}{1 + (T\omega)^2} \, d\omega = \frac{A}{\pi T} \tan^{-1}(\omega_c T) \tag{3.4.6}$$

Problem 2:

$$S_x(j\omega) = \frac{A}{1 + (T\omega)^2}, \qquad \text{for all } \omega \tag{3.4.7}$$

$$E(x^2) = \frac{1}{2\pi} \int_{-\infty}^{\infty} \frac{A}{1 + (T\omega)^2} \, d\omega$$

$$= \frac{A}{\pi T} \tan^{-1}(\infty) = \frac{A}{2T} \tag{3.4.8}$$

Now, we see by comparing the results given by Eqs. (3.4.6) and (3.4.8) that the difference is just that between $\tan^{-1}(\omega_c T)$ and $\tan^{-1}(\infty)$. The bandwidth of the input is ω_c and the filter bandwidth is $1/T$. Thus, if their ratio is large, $\tan^{-1}(\omega_c T) \approx \tan^{-1}(\infty)$. For a ratio of 100:1, the error is less than 1 percent. Thus, if the input spectrum is flat considerably out beyond the point where the system response is decreasing at 20 db/decade (or faster), there is relatively little error introduced by assuming that the input is flat out to infinity. The resulting simplification in the analysis is significant.

3.5
NOISE EQUIVALENT BANDWIDTH

In filter theory, it is sometimes convenient to think of an idealized filter whose frequency response is unity over a prescribed bandwidth B (in hertz) and zero outside this band. This response is depicted in Fig. 3.3a. If this ideal filter is driven by white noise with amplitude A, its mean-square response is

$$E(x^2)(\text{ideal}) = \frac{1}{2\pi} \int_{-2\pi B}^{2\pi B} A \, d\omega = 2AB \qquad (3.5.1)$$

Next, consider an actual filter $G(s)$ whose gain has been normalized to yield a peak response of unity. An example is shown in Fig. 3.3b. The mean-square response of the actual filter to white noise of amplitude A is given by

$$E(x^2)(\text{actual}) = \frac{1}{2\pi j} \int_{-j\infty}^{j\infty} A G(s) G(-s) \, ds \qquad (3.5.2)$$

Now, if we wish to find the idealized filter that will yield this same response, we simply equate $E(x^2)$(ideal) and $E(x^2)$(actual) and solve for the bandwidth that gives equality. The resultant bandwidth B is known as the *noise equivalent*

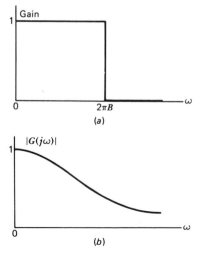

Figure 3.3. Ideal and actual filter responses. (a) Ideal. (b) Actual.

bandwidth. It may, of course, be written explicitly as

$$B(\text{in hertz}) = \frac{1}{2}\left[\frac{1}{2\pi j}\int_{-j\infty}^{j\infty} G(s)G(-s)\,ds\right] \qquad (3.5.3)$$

EXAMPLE 3.4

Suppose we wish to find the noise equivalent bandwidth of the second-order low-pass filter

$$G(s) = \frac{1}{(1 + Ts)^2}$$

Since the peak response of $G(s)$ occurs at zero frequency and is unity, the gain scale factor is set properly. We must next evaluate the integral in brackets in Eq. (3.5.3). Clearly, $G(s)$ is second-order in the denominator, and therefore we use I_2 of the integral tables given in Section 3.3. The coefficients in this case are

$$c_1 = 0 \qquad d_2 = T^2$$
$$c_0 = 1 \qquad d_1 = 2T$$
$$d_0 = 1$$

and thus I_2 is

$$I_2 = \frac{c_0^2}{2d_0 d_1} = \frac{1}{4T}$$

The filter's noise equivalent noise bandwidth is then

$$B = \frac{1}{8T}\ \text{Hz}$$

This says, in effect, that an idealized filter with a bandwidth of $1/8T$ Hz would pass the same amount of noise as the actual second-order filter.   ∎

3.6
SHAPING FILTER

With reference to Fig. 3.4, we have seen that the output spectral function can be written as

$$S_x(s) = 1 \cdot G(s)G(-s) \qquad (3.6.1)$$

Figure 3.4. Shaping filter.

If $G(s)$ is minimum phase and rational in form,* Eq. (3.6.1) immediately provides a factored form for $S_x(s)$ with poles and zeros automatically separated into left and right half-plane parts.

Clearly, we can reverse the analysis problem and pose the question: What minimum-phase transfer function will shape unity white noise into a given spectral function $S_x(s)$? The answer should be apparent. If we can use spectral factorization on $S_x(s)$, the part with poles and zeros in the left half-plane provides the appropriate shaping filter. This is a useful concept, both as a mathematical artifice and also as a physical means of obtaining a noise source with desired spectral characteristics from a wideband source.

EXAMPLE 3.5

Suppose we wish to find the shaping filter that will shape unity white noise into noise with a spectral function

$$S_x(j\omega) = \frac{\omega^2 + 1}{\omega^4 + 64} \tag{3.6.2}$$

First, we write S_x in the s domain as

$$S_x(s) = \frac{-s^2 + 1}{s^4 + 64} \tag{3.6.3}$$

Next, we find the poles and zeros of S_x.

$$\text{Zeros} = \pm 1$$

$$\text{Poles} = -2 \pm j2, \quad 2 \pm j2$$

Finally, we group together left and right half-plane parts. $S_x(s)$ can then be written as

$$S_x(s) = \frac{s + 1}{s^2 + 4s + 8} \cdot \frac{-s + 1}{s^2 - 4s + 8} \tag{3.6.4}$$

---

* This condition requires $G(s)$ to have a finite number of poles and zeros, all of which must be in the left half-plane.

The desired shaping filter is then

$$G(s) = \frac{s + 1}{s^2 + 4s + 8} \qquad (3.6.5)$$

■

3.7
NONSTATIONARY
(TRANSIENT) ANALYSIS —
INITIAL CONDITION RESPONSE

As mentioned previously, the response of a linear system may always be considered as a superposition of an initial-condition part and a driven part. The response due to the initial conditions is often ignored in tutorial discussions, but it should not be because there are many applications where the initial conditions are properly modeled as random variables. If this is the case, one simply solves the problem using standard deterministic methods leaving the initial conditions in the solution in general terms. An example will illustrate the procedure.

EXAMPLE 3.6

Suppose the circuit shown in Fig. 3.5 has been in operation for a long time with the switch open, and then the switch is closed at a random time, which we denote as $t = 0$. We are asked to describe the voltage across the capacitor after $t = 0$.

We first look at the steady-state condition just prior to closing the switch. The transfer function relating the capacitor voltage to the white noise source is given by

$$G(s) = \frac{1}{1 + 2s} \qquad (3.7.1)$$

$$R = 1 \ M\Omega$$
$$C = 1 \ \mu F$$

Figure 3.5. Circuit for Example 3.6.

Therefore, using the methods of Sections 3.2 and 3.3, the mean-square value of the capacitor voltage v_c is found to be

$$E(v_c^2) = \frac{1}{2\pi j} \int_{-j\infty}^{j\infty} \frac{1}{1+2s} \cdot \frac{1}{1-2s} \, ds = \frac{1}{4} \qquad (3.7.2)$$

We have now established that the initial condition is a normal random variable with zero mean and a variance of $\frac{1}{4}$.

After the switch is closed the system differential equation is

$$C\dot{v}_c + \frac{1}{R}v_c = 0 \qquad (3.7.3)$$

Taking the Laplace transform of both sides yields

$$C\left[sV_c(s) - v_c(0)\right] + \frac{1}{R}V_c(s) = 0$$

or

$$V_c(s) = \frac{v_c(0)}{s + (1/RC)} \qquad (3.7.4)$$

The explicit expression for the time-domain waveform is now obtained by taking the inverse transform of Eq. (3.7.4). It is

$$v_c(t) = v_c(0)e^{-t/RC} \qquad (3.7.5)$$

where $v_c(0)$ is a random variable characterized by $N(0, \frac{1}{4})$. Note that the solution of the initial-condition problem leads to a deterministic random process; that is, the process has deterministic structure and any particular realization of the process is exactly predictable once the initial condition is known.

The mean-square value of a random process is usually of prime interest. In this case it is easily computed as

$$E(v_c^2) = E\left[v_c(0)e^{-t/RC}\right]^2$$

$$= e^{-2t/RC} \cdot E\left[v_c^2(0)\right]$$

$$= \tfrac{1}{4}e^{-2t/RC} \qquad (3.7.6)$$

It is, of course, a function of time. ∎

The extension of the procedure of Example 3.6 to more complicated situations is fairly obvious, so this will not be pursued further.

3.8
NONSTATIONARY (TRANSIENT) ANALYSIS — FORCED RESPONSE

The block diagram of Fig. 3.1 is repeated as Fig. 3.6 with the addition of a switch in the input. Imagine the system to be initially at rest, and then close the switch at $t = 0$. A transient response takes place in the stochastic problem just as in the corresponding deterministic problem. If the input $f(t)$ is a nondeterministic random process, we would expect the response also to be nondeterministic, and its autocorrelation function may be computed in terms of the input autocorrelation function. This is done as follows.

The system response can be written as a convolution integral

$$x(t) = \int_0^t g(u) f(t - u) \, du \qquad (3.8.1)$$

where $g(u)$ is the inverse Laplace transform of $G(s)$ and is usually referred to as the system weighting function. To find the autocorrelation function we simply evaluate $E[x(t_1)x(t_2)]$.

$$R_x(t_1, t_2) = E[x(t_1)x(t_2)]$$

$$= E\left[\int_0^{t_1} g(u) f(t_1 - u) \, du \int_0^{t_2} g(v) f(t_2 - v) \, dv\right]$$

$$= \int_0^{t_2}\int_0^{t_1} g(u) g(v) E[f(t_1 - u) f(t_2 - v)] \, du \, dv \qquad (3.8.2)$$

Now, if $f(t)$ is stationary, Eq. (3.8.2) can be written as

$$R_x(t_1, t_2) = \int_0^{t_2}\int_0^{t_1} g(u) g(v) R_f(u - v + t_2 - t_1) \, du \, dv \qquad (3.8.3)$$

$f(t) =$ Input
$x(t) =$ Output
$G(s) =$ Transfer function
$\quad = X(s)/F(s)$

Figure 3.6. Block diagram for nonstationary analysis problem.

and we now have an expression for the output autocorrelation function in terms of the input autocorrelation function and system weighting function.

Equation (3.8.3) is difficult to evaluate except for relatively simple systems. Thus, we are often willing to settle for less information about the response and just compute its mean-square value. This is done by letting $t_2 = t_1 = t$ in Eq. (3.8.3) with the result

$$E[x^2(t)] = \int_0^t \int_0^t g(u)g(v)R_f(u - v)\, du\, dv \qquad (3.8.4)$$

Three examples will now illustrate the use of Eqs. (3.8.3) and (3.8.4).

EXAMPLE 3.7

Let $G(s)$ be a first-order low-pass filter, and let $f(t)$ be white noise with amplitude A. Then

$$G(s) = \frac{1}{1 + Ts}$$

$$S_f(\omega) = A$$

Taking inverse transforms gives

$$g(u) = \frac{1}{T}e^{-u/T}$$

$$R_f(\tau) = A\delta(\tau)$$

Next, substituting in Eq. (3.8.4) yields

$$E[x^2(t)] = \int_0^t \int_0^t \frac{A}{T^2}e^{-u/T}e^{-v/T}\delta(u - v)\, du\, dv$$

$$= \frac{A}{T^2}\int_0^t e^{-2v/T}\, dv$$

$$= \frac{A}{2T}[1 - e^{-2t/T}] \qquad (3.8.5)$$

Note that as $t \to \infty$, the mean-square value approaches $A/2T$, which is the same result obtained in Section 3.2 using spectral analysis methods. ■

EXAMPLE 3.8

Let $G(s)$ be an integrator with zero initial conditions, and let $f(t)$ be a Gauss–Markov process with variance σ^2 and time constant $1/\beta$. We desire the mean-square value of the output x. The transfer function and input autocorrelation function are

$$G(s) = \frac{1}{s} \quad \text{or} \quad g(u) = 1$$

and

$$R_f(\tau) = \sigma^2 e^{-\beta|\tau|}$$

Next, we use Eq. (3.8.4) to obtain $E[x^2(t)]$.

$$E[x^2(t)] = \int_0^t \int_0^t 1 \cdot 1 \cdot \sigma^2 e^{-\beta|u-v|} \, du \, dv \tag{3.8.6}$$

Some care is required in evaluating Eq. (3.8.6) because one functional expression for $e^{-\beta|u-v|}$ applies for $u > v$, and a different one applies for $u < v$. This is shown in Fig. 3.7. Recognizing that the region of integration must be split into two parts, we have

$$E[x^2(t)] = \int_0^t \int_0^v \sigma^2 e^{-\beta(v-u)} \, du \, dv + \int_0^t \int_v^t \sigma^2 e^{-\beta(u-v)} \, du \, dv \tag{3.8.7}$$

Since there is symmetry in the two integrals of Eq. (3.8.7), we can simply

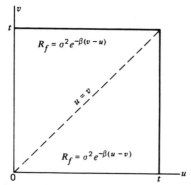

Figure 3.7. Regions of integration for Example 3.8.

evaluate the first one and multiply by 2. The mean-square value of x is then

$$E[x^2(t)] = 2\int_0^t \int_0^v \sigma^2 e^{-\beta(v-u)}\, du\, dv = 2\int_0^t \sigma^2 e^{-\beta v}\int_0^v e^{\beta u}\, du\, dv$$

$$= \frac{2\sigma^2}{\beta}\int_0^t e^{-\beta v}(e^{\beta v} - 1)\, dv$$

$$= \frac{2\sigma^2}{\beta^2}\left[\beta t - (1 - e^{-\beta t})\right] \tag{3.8.8}$$

Note that $E[x^2(t)]$ increases without bound as $t \to \infty$. This might be expected because an integrator is an unstable system. ∎

EXAMPLE 3.9

As our final example, we find the autocorrelation function of the output of a simple integrator driven by unity-amplitude Gaussian white noise. The transfer function and input autocorrelation function are

$$G(s) = \frac{1}{s} \quad \text{or} \quad g(u) = 1$$

$$R_f(\tau) = \delta(\tau)$$

We obtain $R_x(t_1, t_2)$ from Eq. (3.8.3):

$$R_x(t_1, t_2) = \int_0^{t_2}\int_0^{t_1} 1 \cdot 1 \cdot \delta(u - v + t_2 - t_1)\, du\, dv \tag{3.8.9}$$

The region of integration for this double integral is shown in Fig. 3.8 for $t_2 > t_1$. The argument of the Dirac delta function in Eq. (3.8.9) is zero along the dashed line in the figure. It can be seen that it is convenient to integrate first with respect to v if $t_2 > t_1$ as shown in the figure. Considering this case first (i.e., $t_2 > t_1$), we have

$$R_x(t_1, t_2) = \int_0^{t_1}\int_0^{t_2}\delta(u - v + t_2 - t_1)\, dv\, du = \int_0^{t_1} 1 \cdot du = t_1$$

Similarly, when $t_2 < t_1$,

$$R_x(t_1, t_2) = t_2$$

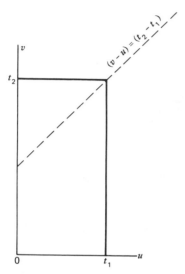

Figure 3.8. Region of integration for Example 3.9.

The final result is then

$$R_x(t_1, t_2) = \begin{cases} t_1, & t_2 \geq t_1 \\ t_2, & t_2 < t_1 \end{cases} \tag{3.8.10}$$

Note that this is the same result obtained for the Wiener process in Chapter 2. ∎

In concluding this section, we might comment that if the transient response includes both forced and initial-condition components, the total response is just the superposition of the two. The mean-square value must be evaluated with care, though, because the total mean-square value is the sum of the two only when the crosscorrelation is zero. If the crosscorrelation between the two responses is not zero, it must be properly accounted for in computing the mean-square value.

3.9
DISCRETE-TIME PROCESS MODELS AND ANALYSIS

Our emphasis thus far has been on continuous-time random signals and the associated response of linear systems to such signals. There is a discrete-time counterpart to all of this, and we shall now take a brief look at discrete-time random processes. In order to keep the discussion as brief as possible, we will confine our attention to single-input, single-output constant-parameter systems. Then later, in Chapter 5, we will consider multiple input–output systems.

It was mentioned in Section 3.6 that a continuous-time process with a rational spectral density can be thought of as the result of passing white noise through a linear filter. The filter, in turn, specifies an input–output differential equation relationship between the response and the input white noise. This equation has the general form

$$\left(D^n + a_{n-1} D^{n-1} + a_{n-2} D^{n-2} + \cdots + a_0 \right) x(t)$$

$$= \left(b_m D^m + b_{m-1} D^{m-1} + \cdots + b_0 \right) u(t), \qquad m = n - 1 \quad (3.9.1)$$

where D is derivative operator and $u(t)$ is unity white noise. (The scale factor on the input is absorbed in the b coefficients.) The system transfer function $G(s)$ is then

$$G(s) = \frac{b_m s^m + b_{m-1} s^{m-1} + \cdots + b_0}{s^n + a_{n-1} s^{n-1} + a_{n-2} s^{n-2} + \cdots + a_0} \qquad (3.9.2)$$

It should be apparent that the continuous-time process $x(t)$ generated by Eq. (3.9.1) can be either stationary or nonstationary, depending on the stability characteristics of $G(s)$ and the initial conditions. [We assume here that the $x(t)$ process is initiated at $t = 0$.]

In the discrete-time world, the input–output relationship corresponding to Eq. (3.9.1) is a difference equation, and it has the general form

$$y(k + n) + \alpha_{n-1} y(k + n - 1) + \alpha_{n-2} y(k + n - 2) + \cdots + \alpha_0 y(k)$$

$$= \beta_m w(k + m) + \beta_{m-1} w(k + m - 1) + \cdots + \beta_0 w(k), \qquad m = n - 1$$

$$(3.9.3)$$

The corresponding quantities in the continuous and discrete models can be summarized as follows:

Continuous-time	Discrete-time
Continuous time t	Integer index k, $k = 0, 1, 2, \ldots$
Response $x(t)$	Response $y(k)$
nth derivative	n units of advance
Input continuous unity white noise $u(t)$	Input discrete white sequence $w(k)$ with unity variance

Note that we have intentionally used different symbols in the continuous and discrete models in order to emphasize that there need be no direct connection between the two models. The difference equation, Eq. (3.9.3), is called the ARMA model for the $y(k)$ process; the left side of the equation is the AR part of the model (for autoregressive), and the right side is the MA part (for moving average).* Just as in the continuous problem, we can think of generating a sample response as the result of inputting a particular $w(k)$ sequence (chosen by chance, of course) and then solving for the resulting $y(k)$ sequence. We cannot write out an explicit solution, but we can conceptually think of the ARMA model as shaping the input white sequence into a corresponding colored sequence. Also, just as in the continuous case we can generate either stationary or nonstationary processes, depending on the stability characteristics of the ARMA difference equation. Z-transforms play the same role in difference equations that Laplace transforms do in differential equations, so they can be used in examining the stability of discrete-time systems. Two examples will illustrate this.

EXAMPLE 3.10

Consider the simple first-order ARMA model

$$y(k + 1) - y(k) = w(k), \qquad k = 0, 1, 2, \ldots \qquad (3.9.4)$$

Note that the MA part of the model is trivial in this case because there is only the $w(k)$ term. We inquire as to the stability of the $y(k)$ process. All we need to do is look at the transfer function of the system in the z-domain. Toward this end we take the z-transform of both sides of Eq. (3.9.4) and form the output-to-input ratio.

$$zY(z) - Y(z) = W(z) \qquad (3.9.5)$$

$$\frac{Y(z)}{W(z)} = \frac{1}{z - 1} \qquad (3.9.6)$$

Immediately, we see that we have a pole on the unit circle at $z = 1$, so the

---

* We are being somewhat restrictive in our ARMA model in that k is allowed to increment only in a positive sense beginning at zero, and that the order of the MA part of the model is less than the order of the AR part. The resulting transfer function in the z-domain is then strictly proper. Also, the transform algebra used in the subsequent analysis can be single-sided, in contrast to two-sided, because k is nonnegative. All of this is in anticipation of our particular use of the ARMA model in Chapter 5. See references 8 and 9 for more on the ARMA model.

system is unstable. In this situation we would expect the output $y(k)$ to be nonstationary. This example, of course, is the sampled form of a Wiener process, provided the initial $y(0)$ is zero and $w(k)$ is a white Gaussian sequence. We know that its variance increases linearly with the number of steps from the origin. ■

EXAMPLE 3.11

We will now look at a slightly more complicated example where stability is not obvious at first glance. Let the ARMA model be

$$y(k + 2) - y(k + 1) + .5y(k) = .5w(k + 1) + .25w(k), \qquad k = 0, 1, 2, \dots$$

$$(3.9.7)$$

Again, we inquire about the stability of $y(k)$; and, just as in the previous example, we can form the system transfer function in the z-domain by taking the z-transform of both sides of Eq. (3.9.7).

$$z^2 Y(z) - zY(z) + .5Y(z) = .5zW(z) + .25W(z)$$

Or

$$\frac{Y(z)}{W(z)} = \frac{.5z + .25}{z^2 - z + .5} \tag{3.9.8}$$

Solving for the roots of the denominator tells us that we have a pair of complex poles at $-.5 \pm j.5$. They are within the unit circle (and the zero at $-.5$ does not affect the stability), so we would expect the output to reach a stationary condition after the transient dies out. Furthermore, the pole locations also provide information about how fast this should occur. We will not pursue this further here (see Example 5.4). ■

We could go on and on exploiting the analogy between continuous- and discrete-time models, but much of this is better done in a state-space setting with vector models. Thus, further discussion of this will be deferred until Chapter 5. One further comment is in order here before closing, though. We have referred to our discrete models here as discrete-*time* models, but time may not actually be important in many physical applications. For example, if one looks at a sequence of rolls of a pair of dice, it is only the sum of the dots that matters in many games, and the result of such an experiment is simply a sequence of discrete events. There is nothing "in between," so to speak, and time is not of the essence! On the other hand, there are many situations where the discrete-time sequence arises as a result of sampling a continuous-time process, and the sampling interval and other time-related matters are quite important. Both of these physical situations fit into the mathematical framework of what we have called discrete-time systems, and we append the word *time* just as a reminder that the sequences being considered are discrete in their argument and not in the random variable space.

3.10
SUMMARY

The mean-square value and spectral density function (or autocorrelation function) are the output process descriptors of prime interest. This is partially a matter of mathematical convenience, because they are usually the only descriptors that can be readily computed. If only the stationary solution is desired, the output spectral density is readily computed from Eq. (3.2.1) or (3.2.2). The mean-square value may then be obtained from the integral of the spectral function. Integral tables are available to assist in this task, provided the transfer and input spectral functions are both in rational form.

In transient problems, the portion of the system response due to random initial conditions is computed using standard deterministic methods. The resulting response is always a deterministic random process. The autocorrelation function of the driven response may be computed from Eq. (3.8.3). In many cases the computation is quite involved; therefore, only the mean-square value is computed. This computation is relatively simple and is given by Eq. (3.8.4). The total response in the transient problem is, of course, the superposition of both the initial-condition and driven parts.

Linear multiple-input, multiple-output problems were not discussed in this chapter. Complicated problems of this type are best handled by state-variable methods, and discussion of such systems will be deferred to Chapter 5. Simple multiple-input, multiple-output problems are sometimes manageable using scalar methods, and in these cases one simply uses superposition in computing the various responses. One must, of course, be careful in computing spectral functions and mean-square values to account properly for any nontrivial cross-correlations that may exist.

One further comment about process models for physical systems is in order. The models, be they continuous or discrete, must come from somewhere, of course. Most often they are derived from experimental data, but occasionally they come from purely theoretical considerations (perhaps backed up by experimental evidence). Analysis of experimental data is a separate subject in its own right, and much has been written about it (8, 9, 10). Usually, the end result of the model formulation is either a spectral description of the process (or equivalently, its correlation structure) or a discrete time-series model, that is, an ARMA model. Often the experimentally derived models are themselves uncertain because of limited data or other experimental vagaries. Be that as it may, reasonable process models must be assumed before the engineer/analyst can go forward and proceed with the analysis and system design. We are primarily concerned here with the second half of the problem, that is, the analysis and design part. Thus, we will assume in subsequent chapters that somehow or other reasonable process models have been provided, and we simply pick up the problem at that point and proceed to look at analysis and design methods that are based on minimizing the mean-square error.

PROBLEMS

3.1. In section 3.1 the equation relating the input and output spectral densities $[S_x(j\omega) = |G(j\omega)|^2 S_f(j\omega)]$ was justified with heuristic arguments. This can be formalized by proceeding through the following steps:

(a) Write the output $x(t)$ as a convolution integral using Fourier rather than Laplace transforms.

(b) Do likewise for the shifted output $x(t + \tau)$.

(c) Multiply the expressions for $x(t)$ and $x(t + \tau)$ and (symbolically) form the expectation of the product.

(d) Now note that the Fourier transform of the autocorrelation function is the spectral density; transform both sides, interchange the order of integration and the desired result is apparent.

Proceed through the steps just described and formally justify the $S_x(j\omega) = |G(j\omega)|^2 S_f(j\omega)$.

3.2. Find the steady-state mean-square value of the output for the following filters. The input is white noise with a spectral density amplitude A.

(a) $G(s) = \dfrac{Ts}{(1 + Ts)^2}$

(b) $G(s) = \dfrac{\omega_0^2}{s^2 + 2\zeta\omega_0 s + \omega_0^2}$

(c) $G(s) = \dfrac{s + 1}{(s + 2)^2}$

3.3. A white-noise process having a spectral density amplitude of A is applied to the circuit shown. The circuit has been in operation for a long time. Find the mean-square value of the output voltage.

Problem 3.3

3.4. The input to the feedback system shown is a stationary Markov process with an autocorrelation function

$$R_f(\tau) = \sigma^2 e^{-\beta|\tau|}$$

The system is in a stationary condition.
(a) What is the spectral density function of the output?
(b) What is the mean-square value of the output?

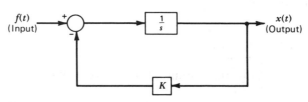

Problem 3.4

3.5. Consider the nonminimum phase filter

$$G(s) = \frac{1 - T_1 s}{1 + T_2 s}$$

driven with a stationary Gauss–Markov process with an autocorrelation function $R(\tau) = \sigma^2 e^{-\beta|\tau|}$. Find:
(a) The spectral density function of the output.
(b) The mean-square value of the output.

3.6. Find the steady-state mean-square value of the output for a first-order low-pass filter [i.e., $G(s) = 1/(1 + Ts)$] if the input has an autocorrelation function of the form

$$R(\tau) = \begin{cases} \sigma^2(1 - \beta|\tau|), & -\dfrac{1}{\beta} \leq \tau \leq \dfrac{1}{\beta} \\ \\ 0, & |\tau| > \dfrac{1}{\beta} \end{cases}$$

(*Hint:* The input spectral function is irrational so the integrals given in Table 3.1 are of no help here. One approach is to write the integral expression for $E(X^2)$ in terms of real ω rather than s and then use conventional integral tables. Also, those familiar with residue theory will find that the integral can be evaluated by the method of residues.)

3.7. Thermal noise in a metallic resistor is sometimes modeled as a white noise voltage source in series with the resistance R of the resistor (11, 12). This is shown in the accompanying figure, along with the parameters describing the spectral amplitude of the noise source. At room temperature the flat spectrum approximation is reasonably accurate from zero frequency out to the infrared range. Clearly, in the idealized model of part (a), the voltage from a to b would be infinity, which is physically impossible. The model is still useful, though, because there is always some shunt capacitance associated with the load connected from a to b. If nothing else, the parasitic capacitance of the resistor leads is sufficient to cause the spectral function to "roll off" at 20 db/decade and thus cause the output to be bounded. This is shown in part (b) of the figure. Now, to demonstrate the validity of the white noise model, consider an example where $R = 1 \times 10^6$ Ω and $C = 1 \times 10^{-12}$ F (a plausible value for parasitic capacitance). Also, assume the temperature is room temperature, about 290 K.

(a) Find the rms voltage across the capacitor C.

(b) Find the half-power frequency in hertz. (Defined as the frequency at which the spectral function is half its value at zero frequency.)

(c) Based on the result of part (b), would you think it reasonable to consider this noise source as being flat (i.e., pure white) in most electronic-circuit applications? Explain briefly.

(a)   (b)

R = Resistance of metallic resistor (ohms)
k = Boltzmann constant $\approx 1.38 \times 10^{-23}$ (joules/deg)
T = Temperature (degrees Kelvin)

Problem 3.7

3.8.   Consider a linear filter whose weighting function is shown in the figure. (This filter is sometimes referred to as a finite-time integrator.) The input to the filter is white noise with a spectral density amplitude A, and the filter has been in operation a long time. What is the mean-square value of the output?

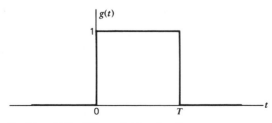

Problem 3.8.   Filter weighting function.

3.9. Find the shaping filter that will shape unity white noise into noise with a spectral function

$$S(j\omega) = \frac{\omega^2 + 1}{\omega^4 + 8\omega^2 + 16}$$

3.10. A series resonant circuit is shown in the figure. Let the resistance R be small such that the circuit is sharply tuned (i.e., high Q or very low damping ratio). Find the noise equivalent bandwidth for this circuit and express it in terms of the damping ratio ζ and the natural undamped resonant frequency ω_r (i.e., $\omega_r = 1/\sqrt{LC}$). Note that the "ideal" response in this case is a unity-gain rectangular pass band centered about ω_r. Also find the usual half-power bandwidth and compare this with the noise equivalent bandwidth. (Half-power bandwidth is defined to be the frequency difference between the two points on the response curve that are "down" by a factor of $1/\sqrt{2}$ from the peak value. It is useful to approximate the resonance curve as being symmetric about the peak for this part of the problem.)

Problem 3.10

3.11. The transfer functions and corresponding bandpass characteristics for first-, second-, and third-order Butterworth filters are shown in the accompanying figure (13). These filters are said to be "maximally flat" at zero frequency with each successive higher-order filter more nearly approaching the ideal curve than the previous one. All three filters have been normalized such that all responses intersect the -3-db point at 1 rad/sec (or $1/2\pi$ Hz).
   (a) Find the noise equivalent bandwidth for each of the filters.
   (b) Insofar as noise suppression is concerned, is there much to be gained by using anything higher order than a third-order Butterworth filter?

3.12. Find the mean-square value of the output (averaged in an ensemble sense) for the following transfer functions. In both cases, the initial conditions are zero and the input $f(t)$ is applied at $t = 0$.
   (a) $G(s) = \dfrac{1}{s^2}$, $R_f(\tau) = A\delta(\tau)$
   (b) $G(s) = \dfrac{1}{s^2 + \omega_0^2}$, $R_f(\tau) = A\delta(\tau)$

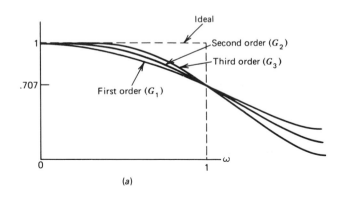

$$G_1(s) = \frac{1}{s+1}$$

$$G_2(s) = \frac{1}{s^2 + \sqrt{2}s + 1}$$

$$G_3(s) = \frac{1}{(s+1)(s^2 + s + 1)}$$

(b)

Problem 3.11. (a) Responses of three Butterworth filters. (b) Transfer functions of Butterworth filters.

3.13. A certain linear system is known to satisfy the following differential equation:

$$\ddot{x} + a\dot{x} = f(t)$$

$$x(0) = \dot{x}(0) = 0$$

where $x(t)$ is the response and $f(t)$ is the input that is applied at $t = 0$. If $f(t)$ is white noise with spectral density amplitude A, what is the mean-square value of the response $x(t)$?

3.14. Consider a simple first-order low-pass filter whose transfer function is

$$G(s) = \frac{1}{1 + 10s}$$

The input to the filter is initiated at $t = 0$, and the filter's initial condition is zero. The input is given by

$$f(t) = Au(t) + n(t)$$

where

$u(t)$ = unit step function
A = random variable with uniform distribution from 0 to 1
$n(t)$ = unity Gaussian white noise

Find:
   (a) The mean, mean square, and variance of the output evaluated at $t = .1$ sec.
   (b) Repeat (a) for the steady-state condition (i.e., for $t = \infty$).

(*Hint:* Since the system is linear, superposition may be used in computing the output. Note that the deterministic component of the input is written explicitly in functional form. Therefore, deterministic methods may be used to compute the portion of the output due to this component. Also, remember that the mean-square value and the variance are not the same if the mean is nonzero.)

3.15. A signal is known to have the following form:

$$s(t) = a_0 + n(t)$$

where a_0 is an unknown constant and $n(t)$ is a stationary noise process with a known autocorrelation function

$$R_n(\tau) = \sigma^2 e^{-\beta|\tau|}$$

It is suggested that a_0 can be estimated by simply averaging $s(t)$ over a finite interval of time T. What would be the rms error in the determination of a_0 by this method? (Note that *root* mean square rather than mean square value is requested in this problem.)

3.16. In the figure shown, $f(t)$ is a time stationary random process whose autocorrelation function is

$$R_f(\tau) = \sigma^2 e^{-\beta|\tau|}$$

The input $f(t)$ is first multiplied by e^{-at} and then integrated, beginning at $t = 0$. The initial value of the integrator is zero. Find the mean-square value of $x(t)$.

Problem 3.16

3.17. Consider an integrator whose initial output at $t = 0$ is a Gaussian random variable with zero mean and variance σ^2. A Gaussian white-noise input with spectral amplitude A is applied at $t = 0$. What is the mean-square value of the output as a function of time?

3.18. Unity Gaussian white noise $f(t)$ is applied to the cascaded combination of integrators shown in the figure. The switch is closed at $t = 0$. The initial condition for the first integrator is zero, and the second integrator has two units as its initial value.

    (a) What is the mean-square value of the output at $t = 2$ sec?

    (b) Sketch the probability density function for the output evaluated at $t = 2$ sec.

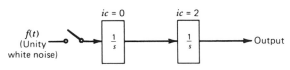

Problem 3.18

3.19. Consider the random process defined by the transfer function shown in the figure. The input $f(t)$ is Gaussian white noise with unity spectral amplitude, and the process is started at $t = 0$ with zero initial conditions. The autocorrelation function of the output $x(t)$ is defined as

$$R_x(t_1, t_2) = E[x(t_1)x(t_2)], \qquad t_1 \text{ and } t_2 > 0$$

Find $R_x(t_1, t_2)$.

Problem 3.19

3.20. Consider again the filter with a rectangular weighting function discussed in Problem 3.8. Consider the filter to be driven with unity Gaussian white noise, which is initiated at $t = 0$ with zero initial conditions.

    (a) Find the mean-square response in the interval from 0 to T.

    (b) Find the mean-square response for $t \geq T$ and compare the result with that obtained in Problem 3.8.

    (c) From the result of (b), would you say the filter's "memory" is finite or infinite?

3.21. The block diagram shown describes the error propagation in one channel of an inertial navigation system with external-velocity-reference damping (14). The inputs shown as $f_1(t)$ and $f_2(t)$ are random driving functions due to the accelerometer and external-velocity-reference instrument errors. These will be assumed to be independent white noise processes with spectral amplitudes A_1 and A_2, respectively. The outputs are labeled x_1, x_2, and x_3, and these physically represent the inertial system's platform tilt, velocity error, and position error. Find the steady-state mean-square value of each of the outputs.

(*Hint:* Since the system is linear, use superposition and note that the driving functions are independent.)

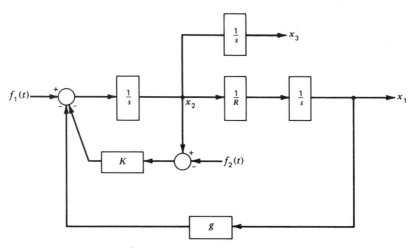

R = earth radius $\approx 2.09 \times 10^7$ ft

g = gravitational constant
≈ 32.2 ft/sec^2

K = feedback constant
(adjustable to yield the
desired damping ratio)

Problem 3.21

3.22. Random processes $x(t)$ and $y(t)$ are generated by passing unity Gaussian white noise through parallel filters as shown below.
(a) Find the autocorrelation functions for each of $x(t)$ and $y(t)$. You may assume that a stationary condition exists.
(b) Find the crosscorrelation function $R_{xy}(\tau)$. (*Hint:* Write x and y as convolution integrals. Then form the appropriate product and take the expectation. Finally, let $t \to \infty$ to get the stationary condition.)

Problem 3.22

3.23. The block diagram in the accompanying figure shows a means of determining the weighting function of a linear system (and thus, indirectly, the system's transfer function). The basic idea is to superimpose a small amount of white noise on the regular input and then crosscorrelate the resulting output with the intentionally added white noise. If the amplitude of the additive noise is relatively small, it is scarcely noticeable, if at all, and this provides a means of continuously monitoring the transfer characteristics of the system without disturbing its normal operation. Show that the output of the crosscorrelator $R_{nx}(\tau)$ is, in fact, proportional to the system weighting function $g(\tau)$. [If you need help on this one, see Truxal (15), p. 437. This idea has "been around" for a long time but has seen only limited application. The reason, of course, lies in the computational effort in determining the crosscorrelation $R_{nx}(\tau)$ or its counterpart in the frequency domain, the cross-spectral density function. Either way, considerable computational effort is involved because the amount of data to be processed needs to be fairly large owing to the presence of a large signal component as well as the random noise. (See Problem 2.30 for more on determining the cross-spectral density function.)]

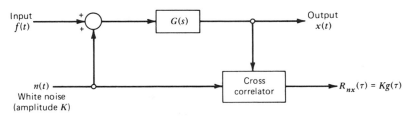

Problem 3.23

3.24. Part (a) of the accompanying figure shows a functional block diagram of an elementary phase-lock loop. Such circuits are used in a wide variety of communication applications (12, 16). Briefly, the phase comparator provides a signal proportional to the difference between the phase of the incoming rf signal and that of the local oscillator. The phase difference is modified by gain factor K_d and is then fed back as a voltage to the local voltage-controlled oscillator (VCO). This voltage causes the frequency of the VCO to shift up or down as necessary to lock onto the phase of the incoming rf signal.

Part (b) of the figure shows the linearized block diagram for a phase-lock loop with the addition of a phase noise input labeled $n(t)$. If the incoming rf signal is frequency modulated (FM), it is the derivative of $\theta_i(t)$ that is proportional to the baseband signal, and thus $\dot{\theta}_i(t)$ is the signal to be recovered (detected) in this case.

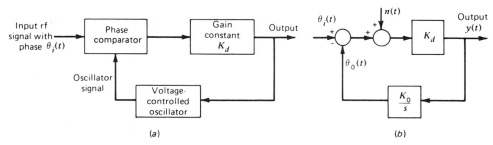

(a)   (b)

Problem 3.24

(a) Show that the transfer function relating $\theta_i(t)$ to the output $y(t)$ is the equivalent of a differentiator in cascade with a first-order, low-pass filter.

(b) Consider the phase noise $n(t)$ to be flat with a spectral amplitude N_0. Also assume that the gain parameters of the phase-lock loop are chosen such that the baseband signal is passed with no distortion, that is, the loop response is flat over the signal frequency range, say from 0 to W Hz. Find the spectral function of the output noise in the 0 to W Hz range, and find its average power (i.e., mean-square value) in this frequency range. (It may help here to think of the noise as being limited to the 0 to W Hz range by a sharp-cutoff postdetector filter.)

3.25.   For two stationary random processes $x(t)$ and $y(t)$, the coherence function is defined to be (see Chapter 2):

$$\gamma_{xy}^2(\omega) = \frac{|S_{xy}(j\omega)|^2}{S_x(j\omega)S_y(j\omega)} \tag{P3.25.1}$$

Consider a special case where $x(t)$ is the input to a linear system, and $y(t)$ is the output as shown in the accompanying figure.

(a) First show that

$$S_{xy}(j\omega) = G(j\omega)S_x(j\omega) \tag{P3.25.2}$$

(b) Then show that the coherence function is unity for all ω. (This shows that x and y do not have to be equal in order to have unity coherence. They only need to be intimately related, as, of course, they are in this situation.)

Problem 3.25

3.26.*   The recursion equation for generating a Markov process from Problem 2.34* is repeated here for convenience:

$$x_{k+1} = e^{-\beta\Delta t}x_k + w_k, \qquad k = 0, 1, 2, \ldots$$

$$w_k = \text{white sequence} \sim N\left[0, \sigma^2(1 - e^{-2\beta\Delta t})\right]$$

$$\sigma^2 = \text{variance of the Markov process}$$

$$\beta = \text{reciprocal time constant of the Markov process}$$

$$\Delta t = \text{time interval between samples}$$

(a) This will be recognized as a simple first-order ARMA model for the x_k process. Using the same parameter values as in Problem 2.34* (i.e., $\sigma^2 = 1$, $\beta = 1$, $\Delta t = .04$ sec), verify analytically that this model generates a stable process. This is easily done by locating the system characteristic pole in the z-plane. What is the system time constant? Express this in terms of both seconds and number of discrete steps.

(b) In order to demonstrate the stability of the process of part (a), generate four sample realizations of the process for $k = 0, 1, 2, \ldots, 500$. The transient phenomenon that takes place and the evolution into a stationary condition will be more obvious if you initialize each realization to be zero at $k = 0$. View the plots of the four sample realizations. (The overall trend toward stationary behavior is more obvious when the four plots are superimposed on the same graph.)

(c) If we let $e^{-\beta \Delta t} = 1$ in the above recursion equation, the ARMA model becomes

$$x_{k+1} - x_k = w_k$$

This model was discussed in Example 3.10, and it was noted to be unstable. To demonstrate this, generate four sample realizations of this process for $k = 1, 2, \ldots, 500$, and let the initial value of x_k be zero just as in part (b). Also, in order to make the initial rate of growth of x_k similar (statistically) to that of part (b), let the variance of w_k be .076884. View the plots of the four realizations and note the instability.

REFERENCES CITED IN CHAPTER 3

1. J. J. D'Azzo and C. H. Houpis, *Linear Control System Analysis and Design*, 3rd ed., New York: McGraw-Hill, 1988.
2. R. C. Dorf, *Modern Control Systems*, 5th ed., Reading, Mass.: Addison-Wesley, 1989.
3. R. A. Gabel and R. A. Roberts, *Signals and Linear Systems*, 2nd ed., New York: Wiley, 1980.
4. H. M. James, N. B. Nichols, and R. S. Phillips, *Theory of Servomechanisms*, Radiation Laboratory Series (Vol. 25), New York: McGraw-Hill, 1947.
5. G. C. Newton, L. A. Gould, and J. F. Kaiser, *Analytical Design of Linear Feedback Controls*, New York: Wiley, 1957.
6. G. R. Cooper and C. D. McGillem, *Probabilistic Methods of Signal and System Analysis*, 2nd ed., New York: Holt, Rinehart, and Winston, 1986.
7. K. S. Shanmugan and A. M. Breipohl, *Random Signals: Detection, Estimation, and Data Analysis*, New York: Wiley, 1988.
8. M. B. Priestley, *Spectral Analysis and Time Series*, New York: Academic Press, 1981.
9. J. M. Mendel, *Optimal Seismic Deconvolution*, New York: Academic Press, 1983.

10. J. S. Bendat and A. G. Piersol, *Random Data: Analysis and Measurement Procedures*, New York: Wiley-Interscience, 1971.
11. A. B. Carlson, *Communication Systems*, 2nd ed., New York: McGraw-Hill, 1975.
12. K. S. Shanmugam, *Digital and Analog Communication Systems*, New York: Wiley, 1979.
13. D. Childers and A. Durling, *Digital Filtering and Signal Processing*, St. Paul: West Publishing, 1975.
14. G. R. Pitman (Ed.), *Inertial Guidance*, New York: Wiley, 1962.
15. J. G. Truxal, *Automatic Feedback Control System Synthesis*, New York: McGraw-Hill, 1955.
16. F. M. Gardner, *Phaselock Techniques*, 2nd ed., New York: Wiley, 1979.

Additional References for General Reading

17. P. Z. Peebles, Jr., *Probability, Random Variables, and Random Signal Principles*, 2nd ed., New York: McGraw-Hill, 1987.
18. H. J. Larson and B. O. Shubert, *Probabilistic Models in Engineering Sciences* (Vols. 1 and 2), New York: Wiley, 1979.

4

Wiener Filtering

In this and subsequent chapters we will consider a particular branch of filter theory that is sometimes referred to as least-squares filtering. Actually, this is an oversimplification because it is the *average* squared error that is minimized and not just the squared error. "Linear minimum mean-square error filtering" would be a more descriptive name for this type of filtering. This is a bit wordy, though, so the name is usually shortened and the theory is included as part of the general theory of least squares. Simply stated, the linear least-squares filter problem is this: Given the spectral characteristics of an additive combination of signal and noise, what linear operation on this input combination will yield the best separation of the signal from the noise? "Best" in this case means minimum mean-square error. This branch of filtering began with N. Wiener's work in the 1940s (1). R. E. Kalman then made an important contribution in the early 1960s by providing an alternative approach to the same problem using state-space methods (2, 3). Kalman's contribution has been especially significant in applied work, because his solution is readily implemented with modern digital methods.

We will consider the Wiener and Kalman theories in their historical order. It should be mentioned that neither is prerequisite material for the other; therefore, they may be studied in either order, or one to the exclusion of the other, for that matter.

4.1
THE WIENER FILTER PROBLEM

The purpose of any filter is to separate one thing from another. In the electric filter case, this usually refers to passing signals in a specified frequency range and rejecting those outside that range; and, historically, filter theory began with the problem of designing a circuit to yield the desired frequency response. This is still an important problem. In many applications in communication and control, one knows intuitively what the ideal frequency response should be. For example, if we want to receive the signal from a particular AM radio station (and do it faithfully), we know that the appropriate filter is one that passes all frequencies within a few kilohertz on either side of the assigned station frequency and rejects all others. Certainly no elaborate theory is needed to determine the desired frequency response in this case. The problem is simply one of circuit design. We will see, though, that this is not always the case. During World War II, Norbert Wiener considered a different sort of filter problem (1). Suppose the signal, as well as the noise, is noiselike in character, and suppose further that there is a significant overlap in the spectra of both the signal and noise. For example, say the signal is a Gauss–Markov process and the corrupting noise is white noise. Their spectral densities are shown in Fig. 4.1. Now, in this case, it should be apparent that no filter is going to yield perfect separation, and the filter that gives the best compromise of passing the signal and, at the same time, suppresses most of the noise is not at all obvious. Neither is it obvious how one should define "best compromise" in order to make the problem mathematically tractable. This is the problem Wiener examined in the 1940s. We note that he was not concerned with filter design in the sense of choosing appropriate resistors, capacitors, and so forth. Instead, his problem was more fundamental; namely, what should the filter's frequency response be in order to give the best possible separation of signal from noise?

The theory that is now loosely referred to as Wiener filter theory is characterized by:

1. The assumption that both signal and noise are random processes with known spectral characteristics or, equivalently, known auto- and cross-correlation functions.

2. The criterion for best performance is minimum mean-square error. This is partially to make the problem mathematically tractable, but it is also a good physical criterion in many applications.

3. A solution based on scalar methods that leads to the optimal filter weighting function (or transfer function in the stationary case).

We now proceed to the filter optimization problem.

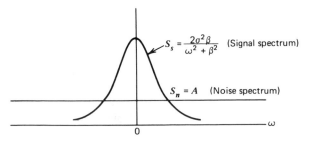

Figure 4.1. Spectral densities of signal and noise.

Figure 4.2. Filter optimization problem.

4.2
OPTIMIZATION WITH RESPECT TO A PARAMETER

One of the early methods of stochastic filter optimization was presented by R. S. Phillips in Volume 25 of the now-famous Radiation Laboratory Series (4). His approach was less general than Wiener's but it is still useful in many applications. Basically, the method is to choose the form of the filter transfer function intuitively, leaving one or more parameters free to vary. One then minimizes the mean-square error with respect to these parameters.

Before looking at an example of Phillip's procedure, we need to derive an expression for the filter's mean-square error. In terms of its Laplace transform, the output of the filter shown in Fig. 4.2. is*

$$X(s) = G(s)[S(s) + N(s)] \qquad (4.2.1)$$

We define the filter error as the difference between the actual output and what we would like it to be ideally—the signal. Therefore let[†]

$$e(t) = s(t) - x(t) \qquad (4.2.2)$$

and

$$E(s) = S(s) - X(s) \qquad (4.2.3)$$

Substituting Eq. (4.2.1) into Eq. (4.2.3) yields

$$E(s) = S(s) - G(s)[S(s) + N(s)] = [1 - G(s)]S(s) - [G(s)]N(s)$$

$$(4.2.4)$$

---

*Where there might be confusion between the signal variable $s(t)$ and the complex variable s, we will show the time dependence explicitly, that is, $s(t)$ is the signal variable. Also, uppercase $S(s)$ without a subscript will denote the Laplace transform of $s(t)$; S with a subscript, for example, $S_s(s)$, denotes spectral density of the subscript variable. This overlap in notation should not be confusing when taken within the context of the subject under consideration.

†Some authors prefer to define the error with the opposite sign. In the subsequent optimization we will always be concerned with minimizing the mean-*square* error. Thus the sign of the error is of no consequence; the resulting optimal filter and mean-square error is the same either way.

It can be seen that the error can be thought of as a superposition of two components, one due to the signal modified by the transfer function $[1 - G(s)]$ and another due to the noise modified by $-G(s)$. If the signal and noise have zero crosscorrelation, the mean-square error is obtained as simply the sum of two terms, that is,

$$E(e^2) = \frac{1}{2\pi j} \int_{-j\infty}^{j\infty} [1 - G(s)][1 - G(-s)] S_s(s) \, ds$$

$$+ \frac{1}{2\pi j} \int_{-j\infty}^{j\infty} G(s) G(-s) S_n(s) \, ds \qquad (4.2.5)$$

We now have an explicit expression for the mean-square error in terms of the spectral functions of $s(t)$ and $n(t)$ (presumably known) and the filter transfer function. If $G(s)$ contains a parameter free to vary, we can now use ordinary differential calculus to minimize $E(e^2)$ with respect to the parameter. We now proceed with an example.

EXAMPLE 4.1

Consider the Gauss–Markov signal and white noise situation shown in Fig. 4.1. It is apparent that some sort of low-pass filter is needed to separate signal from noise. Let us try a simple first-order filter of the form

$$G(s) = \frac{1}{1 + Ts}$$

We have now specified the functional form for $G(s)$, and hence we are ready to use Eq. (4.2.5). The needed quantities are

$$G(s) = \frac{1}{1 + Ts} = \frac{1/T}{s + (1/T)}$$

$$1 - G(s) = 1 - \frac{1}{1 + Ts} = \frac{s}{s + (1/T)}$$

$$S_s(s) = \frac{2\sigma^2 \beta}{-s^2 + \beta^2} = \frac{\sqrt{2\sigma^2 \beta}}{s + \beta} \cdot \frac{\sqrt{2\sigma^2 \beta}}{-s + \beta}$$

$$S_n(s) = A = \sqrt{A} \cdot \sqrt{A}$$

Substituting the above quantities into Eq. (4.2.5) and evaluating $E(e^2)$ using the integral tables of Section 3.4 yields

$$E(e^2) = \frac{\sigma^2 \beta T}{1 + \beta T} + \frac{A}{2T} \qquad (4.2.6)$$

This can now be minimized with respect to T using differential calculus. The result is that $E(e^2)$ is a minimum for

$$T = \frac{\sqrt{A}}{\sigma\sqrt{2\beta} - \beta\sqrt{A}} \qquad (4.2.7)$$

It is interesting to note that this will yield a positive value of T only for certain values of the parameters. A negative solution for T simply means that no *relative* minimum exists within the interval from zero to infinity.

It should be remembered that the minimum obtained is not the absolute minimum possible (unless by coincidence), because the form of the filter transfer function was chosen intuitively. Other functional forms might have done better. ∎

4.3
THE STATIONARY OPTIMIZATION PROBLEM — WEIGHTING FUNCTION APPROACH*

We now consider the filter optimization problem that Wiener first solved in the 1940s (1). Referring to Fig. 4.3, we assume the following:

1. The filter input is an additive combination of signal and noise, both of which are covariance stationary with known auto- and crosscorrelation functions (or corresponding spectral functions).
2. The filter is linear and not time varying. No further assumption is made as to its form.
3. The output is covariance stationary. (A long time has elapsed since any switching operation.)
4. The performance criterion is minimum mean-square error, where the error is defined as $e(t) = s(t + \alpha) - x(t)$.

In addition to the generalization relative of the form of the filter transfer function, we are also generalizing by saying the ideal filter output is to be $s(t + \alpha)$ rather than just $s(t)$. The following terminology has evolved relative to

---

* Two-sided Laplace transform theory is used extensively in this section. See Appendix A for a brief review of two-sided transform theory.

Figure 4.3. Wiener filter problem.

the choice of the α parameter:

1. α positive: This is called the *prediction* problem. (The filter is trying to predict the signal value α units ahead of the present time t.)
2. $\alpha = 0$: This is called the *filter* problem. (The usual problem we have considered before.)
3. α negative: This is called the *smoothing* problem. (The filter is trying to estimate the signal value α units in the past.)

This is an important generalization and there are numerous physical applications corresponding to all three cases. The α parameter is chosen to fit the particular application at hand, and it is fixed in the optimization process.

We begin by defining the filter error as

$$e(t) = s(t + \alpha) - x(t) \tag{4.3.1}$$

The squared error is then

$$e^2(t) = s^2(t + \alpha) - 2s(t + \alpha)x(t) + x^2(t) \tag{4.3.2}$$

We next write $x(t)$ as a convolution integral

$$\int_{-\infty}^{\infty} g(u)[s(t - u) + n(t - u)]\, du \tag{4.3.3}$$

This can then be substituted into Eq. (4.3.2) and both sides averaged to yield*

$$E(e^2) = \int_{-\infty}^{\infty} \int_{-\infty}^{\infty} g(u)g(v) R_{s+n}(u - v)\, du\, dv$$

$$- 2\int_{-\infty}^{\infty} g(u) R_{s+n,s}(\alpha + u)\, du + R_s(0) \tag{4.3.4}$$

---

*We have chosen here to write the mean-square error in terms of the filter weighting function and input autocorrelation functions. In the stationary problem, one can also write $E(e^2)$ in terms of the filter transfer function and input spectral functions, and then proceed with the optimization on that basis (6, 7). We have chosen the time-domain approach because it is easily generalized to the nonstationary problem that is considered in Section 4.4. The frequency-domain approach is not readily generalized.

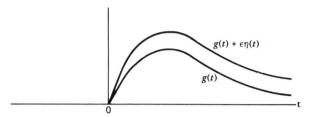

Figure 4.4. Optimal and perturbed weighting functions.

where

R_s = autocorrelation function of $s(t)$
R_{s+n} = autocorrelation function of $s(t) + n(t)$
$R_{s+n,s}$ = crosscorrelation between $s(t) + n(t)$ and $s(t)$

Note that if signal and noise have zero crosscorrelation

$$\left.\begin{array}{l} R_{s+n} = R_s + R_n \\ R_{s+n,s} = R_s \end{array}\right\} \qquad \text{(for zero correlation)} \qquad (4.3.5)$$

We wish to find the function $g(u)$ in Eq. (4.3.4) that minimizes $E(e^2)$. This will be recognized as a problem in calculus of variations (5). Following the usual procedure, we replace $g(u)$ with a perturbed weighting function $g(u) + \varepsilon\eta(u)$

where

$g(u)$ = optimum weighting function [*Note:* From this point on in the solution, $g(u)$ will denote the *optimal* weighting function.]
$\eta(u)$ = an arbitrary perturbing function
ε = small perturbation factor such that the perturbed function approaches the optimum one as ε goes to zero

The optimum and perturbed weighting functions are sketched in Fig. 4.4. Replacing $g(u)$ with $g(u) + \varepsilon\eta(u)$ in Eq. (4.3.4) then leads to

$$E(e^2) = \int_{-\infty}^{\infty}\int_{-\infty}^{\infty} [g(u) + \varepsilon\eta(u)][g(v) + \varepsilon\eta(v)] R_{s+n}(u - v)\, du\, dv$$

$$- 2\int_{-\infty}^{\infty} [g(u) + \varepsilon\eta(u)] R_{s+n,s}(\alpha + u)\, du + R_s(0) \qquad (4.3.6)$$

Note that $E(e^2)$ is now a function of ε, and it is to be a minimum when $\varepsilon = 0$. Now, using differential calculus methods, we differentiate $E(e^2)$ with respect to ε and set the result equal to zero for $\varepsilon = 0$. After interchanging dummy variables of integration freely, the result is

$$\int_{-\infty}^{\infty} \eta(\tau)\left[-R_{s+n,s}(\alpha + \tau) + \int_{-\infty}^{\infty} g(u)R_{s+n}(u - \tau)\,du\right]d\tau = 0 \quad (4.3.7)$$

A subtlety in the solution arises at this point; therefore it is convenient to look at the causal and noncausal cases separately.

Noncausal Solution

If we put no constraint on the filter weighting function, we will very likely obtain a $g(u)$ that is nontrivial for negative as well as positive u. This weighting function is noncausal because it requires the filter to "look ahead" of real time and use data that are not yet available. This is, of course, not possible if the filter is operating on-line. However, in off-line applications, such as postflight analysis of recorded data, the noncausal solution is possible and very much of interest. Thus, it should not be ignored.

If there are no restrictions on $g(u)$, then, similarly, there are no constraints on the perturbation function $\eta(\tau)$. It is arbitrary for all values of its argument. Thus, if the integral with respect to τ in Eq. (4.3.7) is to be zero, the bracketed term must be zero for *all* τ. This leads to

$$\int_{-\infty}^{\infty} g(u)R_{s+n}(u - \tau)\,du = R_{s+n,s}(\alpha + \tau), \qquad -\infty < \tau < \infty \quad (4.3.8)$$

This is an integral equation of the first kind, and in this case it can be solved readily using Fourier transform methods. Since R_{s+n} is symmetric, the term on the left side of Eq. (4.3.8) has the exact form of a convolution integral. Therefore, transforming both sides yields

$$G(s)S_{s+n}(s) = S_{s+n,s}(s)e^{\alpha s} \qquad (4.3.9)$$

or

$$G(s) = \frac{S_{s+n,s}(s)e^{\alpha s}}{S_{s+n}(s)} \qquad (4.3.10)$$

Remember that the transforms indicated in Eq. (4.3.10) are *two-sided* transforms rather than the usual single-sided transforms. Of course, if we wish to find the weighting function $g(u)$, we simply take the inverse transform of the expression given by Eq. (4.3.10).

The filter mean-square error is given by Eq. (4.3.4). If $g(u)$ is the optimal weighting function satisfying Eq. (4.3.8), the mean-square error equation may be simplified as follows. First write the second term of Eq. (4.3.4) as the sum of two equal terms and combine one of these with the double integral term. After rearranging terms, this leads to

$$E(e^2) = R_s(0) - \int_{-\infty}^{\infty} g(u) R_{s+n,s}(\alpha + u)\, du$$

$$+ \int_{-\infty}^{\infty} g(u) \left[-R_{s+n,s}(\alpha + u) + \int_{-\infty}^{\infty} g(v) R_{s+n}(v - u)\, dv \right] du \quad (4.3.11)$$

The bracketed quantity in Eq. (4.3.11) is zero for optimal $g(v)$ for all u. Therefore the mean-square error is

$$E(e^2) = R_s(0) - \int_{-\infty}^{\infty} g(u) R_{s+n,s}(\alpha + u)\, du \quad (4.3.12)$$

EXAMPLE 4.2

Consider the same Markov signal and white noise combination used in Example 4.1. We wish to find the optimal noncausal filter (i.e., $\alpha = 0$). In order to simplify the arithmetic, let $\sigma^2 = \beta = A = 1$. Since the signal and noise have zero crosscorrelation,

$$S_{s+n} = S_s + S_n = \frac{2}{-s^2 + 1} + 1 = \frac{-s^2 + 3}{-s^2 + 1} \quad (4.3.13)$$

$$S_{s+n,s} = S_s = \frac{2}{-s^2 + 1} \quad (4.3.14)$$

$$e^{\alpha s} = 1 \quad (4.3.15)$$

From Eq. (4.3.10) we have

$$G(s) = \frac{\dfrac{2}{-s^2 + 1}}{\dfrac{-s^2 + 3}{-s^2 + 1}} = \frac{2}{-s^2 + 3}$$

Expanding this with a partial fraction expansion yields

$$G(s) = \frac{1/\sqrt{3}}{s + \sqrt{3}} + \frac{1/\sqrt{3}}{-s + \sqrt{3}} \quad (4.3.16)$$

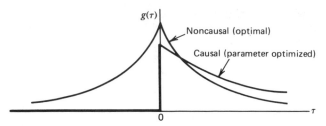

Figure 4.5.   Filter weighting functions for optimal noncausal and parameter-optimized causal filters.

The positive- and negative-time parts of $g(u)$ are given by the first and second terms of Eq. (4.3.16). Thus $g(u)$ is

$$g(u) = \begin{cases} \dfrac{1}{\sqrt{3}} e^{-\sqrt{3}\,u}, & u \geq 0 \\[2mm] \dfrac{1}{\sqrt{3}} e^{\sqrt{3}\,u}, & u < 0 \end{cases} \qquad (4.3.17)$$

This is the optimal noncausal weighting function, and it is sketched in Fig. 4.5 along with the intuitive weighting function of Example 4.1 (evaluated for $\sigma^2 = \beta = A = 1$ and $T = 1 + \sqrt{2}$). Note that since the noncausal filter weights both past and future input data, it can afford to have a smaller time constant than the intuitive filter, which is allowed to weight only past input data.

It is also of interest to compare the mean-square errors for the noncausal optimal filter and the parameter-optimized filter of Example 4.1. These may be computed from Eqs. (4.2.6) and (4.3.12) with the result

$$E(e^2)(\text{parameter-optimized}) \approx 0.914$$

$$E(e^2)(\text{noncausal optimal}) \approx 0.577$$

Notice that the noncausal filter has significantly less error than the causal one. Certainly in off-line applications it would be worthwhile implementing the noncausal filter in preference to the intuitive causal one.   ■

Causal Solution

The calculus of variations procedure led to Eq. (4.3.7), which is repeated here for convenient reference.

$$\int_{-\infty}^{\infty} \eta(\tau)\left[-R_{s+n,s}(\alpha + \tau) + \int_{-\infty}^{\infty} g(u) R_{s+n}(u - \tau)\, du \right] d\tau = 0 \quad (4.3.7)$$

Recall that $\eta(\tau)$ is an arbitrary perturbing function. If we wish to constrain the filter weighting function to be causal, we must place a similar constraint on $\eta(\tau)$ in the variation. Otherwise, we get the unconstrained (noncausal) solution. Thus, for the causal case we require $\eta(\tau)$ to be zero for negative τ, and allow it to be arbitrary for positive τ. The bracketed quantity in Eq. (4.3.7) then needs to be zero only for positive τ. The zero criterion is satisfied for negative τ by virtue of our constraint on $\eta(\tau)$, that is, $\eta(\tau) = 0$ for $\tau < 0$. Therefore, the resulting integral equation is

$$\int_{-\infty}^{\infty} g(u) R_{s+n}(u - \tau)\, du - R_{s+n,\, s}(\alpha + \tau) = 0, \qquad \tau \geq 0 \quad (4.3.18)$$

Equation (4.3.18) is known as the *Wiener–Hopf equation*, and the fact that it is valid only for $\tau \geq 0$ complicates the solution considerably.

One solution of Eq. (4.3.18) that is based on spectral factorization proceeds as follows. First, replace the right side with an unknown negative-time function $a(\tau)$ [i.e., $a(\tau)$ is unknown for negative time, but is known to be zero for positive time.] Eq. (4.3.18) can be written as

$$\int_{-\infty}^{\infty} g(u) R_{s+n}(u - \tau)\, du - R_{s+n,\, s}(\alpha + \tau) = a(\tau), \qquad -\infty < \tau < \infty \quad (4.3.19)$$

Transforming both sides of Eq. (4.3.19) then yields

$$G(s) S_{s+n}(s) - S_{s+n,\, s}(s) e^{\alpha s} \doteq A(s) \tag{4.3.20}$$

Next, use spectral factorization on S_{s+n} and group terms as follows:

$$[G(s) S_{s+n}^{+}(s)] S_{s+n}^{-}(s) - S_{s+n,\, s}(s) e^{\alpha s} = A(s)$$

or

$$G(s) S_{s+n}^{+}(s) = \frac{A(s)}{S_{s+n}^{-}(s)} + \frac{S_{s+n,\, s}(s) e^{\alpha s}}{S_{s+n}^{-}(s)} \tag{4.3.21}$$

In Eq. (4.3.21), the "super $+$" indicates the factored part of the spectral function that has all its poles and zeros in the left half-plane. Similarly, the poles and zeros of S_{s+n}^{-} are mirror images of those of S_{s+n}^{+}. We note here that $g(u)$ is a stable positive-time function; therefore $G(s)$ will have its poles in the left half-plane. Thus, $G(s) S_{s+n}^{+}(s)$ will have all its poles in the left half-plane, and it will be the transform of a positive-time function. Similarly, $A(s)$ is the transform of a negative-time function, so its poles will be in the right half-plane. Also, both the zeros and poles of $S_{s+n}^{-}(s)$ are in the right half-plane. Hence, the three terms of Eq. (4.3.21) translate into words as

$$\begin{bmatrix} \text{Positive-time} \\ \text{function} \end{bmatrix} = \begin{bmatrix} \text{Negative-time} \\ \text{function} \end{bmatrix} + \begin{bmatrix} \text{Both positive- and} \\ \text{negative-time} \\ \text{function} \end{bmatrix}$$

Equating positive-time parts on both sides of Eq. (4.3.21) then leads to

$$G(s)S_{s+n}^{+}(s) = \text{Positive-time part of } \frac{S_{s+n,s}(s)e^{\alpha s}}{S_{s+n}^{-}(s)}$$

or

$$G(s) = \frac{1}{S_{s+n}^{+}(s)} \left[\text{Positive-time part of } \frac{S_{s+n,s}(s)e^{\alpha s}}{S_{s+n}^{-}(s)} \right] \qquad (4.3.22)$$

The bracketed term of Eq. (4.3.22) can be interpreted as follows. First, find the inverse transform of $S_{s+n,s}(s)/S_{s+n}^{-}(s)$. This will normally be nontrivial for both positive and negative time. Next, translate the time function an amount α. (This accounts for $e^{\alpha s}$.) Finally, take the ordinary *single-sided* Laplace transform of the shifted time function, and this will be the bracketed quantity in Eq. (4.3.22). Two examples should be helpful at this point.

EXAMPLE 4.3

Consider the same Markov signal and white noise combination used in Examples 4.1 and 4.2. Again we let $\sigma^2 = \beta = A = 1$ to simplify the arithmetic and, in this example, we are looking for the optimal causal solution. Since the signal and noise are assumed to have zero crosscorrelation, the needed spectral functions are

$$S_{s+n} = S_s + S_n = \frac{2}{-s^2 + 1} + 1 = \frac{-s^2 + 3}{-s^2 + 1} \qquad (4.3.23)$$

$$S_{s+n,s} = S_s = \frac{2}{-s^2 + 1} \qquad (4.3.24)$$

Also, since the prediction time α is assumed to be zero,

$$e^{\alpha s} = 1 \qquad (4.3.25)$$

First, we factor S_{s+n}:

$$S_{s+n} = S_{s+n}^+ S_{s+n}^- = \left[\frac{s + \sqrt{3}}{s + 1} \right]\left[\frac{-s + \sqrt{3}}{-s + 1} \right] \qquad (4.3.26)$$

Next, we form the $S_{s+n,\,s}/S_{s+n}^-$ function

$$\frac{S_{s+n,\,s}}{S_{s+n}^-} = \frac{\dfrac{2}{-s^2 + 1}}{\dfrac{-s + \sqrt{3}}{-s + 1}} = \frac{2}{(-s + \sqrt{3})(s + 1)} \qquad (4.3.27)$$

This, in turn, can be expanded in terms of a partial fraction expansion

$$\frac{S_{s+n,\,s}}{S_{s+n}^-} = \frac{\sqrt{3} - 1}{s + 1} + \frac{\sqrt{3} - 1}{-s + \sqrt{3}} \qquad (4.3.28)$$

Clearly, the first term of Eq. (4.3.28) is the positive-time part. Therefore, $G(s)$ as given by Eq. (4.3.22) is

$$G(s) = \frac{1}{\dfrac{s + \sqrt{3}}{s + 1}} \cdot \frac{\sqrt{3} - 1}{s + 1} = \frac{\sqrt{3} - 1}{s + \sqrt{3}} \qquad (4.3.29)$$

Or, in terms of the filter weighting function,

$$g(t) = \begin{cases} (\sqrt{3} - 1)e^{-\sqrt{3}\,t}, & t \geq 0 \\ 0, & t < 0 \end{cases} \qquad (4.3.30)$$

As before, the mean-square error of the filter can be computed using Eq. (4.3.12). The result is

$$E(e^2) = .732 \qquad (4.3.31)$$

∎

We have now examined three different optimization approaches for the same signal-plus-noise situation. A comparison of the results is shown in Table 4.1, with the most restrictive filter being listed first and the least restrictive one listed last. As should be expected, the mean-square error decreases with each successive relaxation of the constraints on the choice of transfer function. The linear constraint is, of course, present in all three solutions. We will see in later

Table 4.1 Comparison of Results of Examples 4.1, 4.2, and 4.3

Type of Solution	Filter Transfer Function	Weighting Function	Mean-Square Error
Single parameter optimization	$G(s) = \dfrac{1}{1 + (1 + \sqrt{2})\, s}$	$g(t) = \begin{cases} \dfrac{1}{1 + \sqrt{2}}\, e^{-(\sqrt{2}-1)t}, & t \geq 0 \\[2mm] 0, & t < 0 \end{cases}$	.914
Causal Wiener filter	$G(s) = \dfrac{\sqrt{3} - 1}{s + \sqrt{3}}$	$g(t) = \begin{cases} (\sqrt{3} - 1)\, e^{-\sqrt{3}\, t}, & t \geq 0 \\[2mm] 0, & t < 0 \end{cases}$	.732
Noncausal Wiener filter	$G(s) = \dfrac{2}{(s + \sqrt{3})(-s + \sqrt{3})}$	$g(t) = \begin{cases} \dfrac{1}{\sqrt{3}}\, e^{-\sqrt{3}\, t}, & t \geq 0 \\[2mm] \dfrac{1}{\sqrt{3}}\, e^{\sqrt{3}\, t}, & t < 0 \end{cases}$	.577

chapters that this is not as serious as one might think at first glance. The explanation of this, though, will be deferred until Chapter 5.

EXAMPLE 4.4

There is a classical problem in random process theory known as the *pure prediction* problem. Here we assume the additive noise corrupting the signal is zero, and we pose the problem of looking ahead and finding the best estimate of the signal α units ahead of the present time t. To demonstrate the applicability of Wiener filter theory to this problem, let the signal be Markov with a known autocorrelation function

$$R_s(\tau) = \sigma^2 e^{-\beta|\tau|} \quad \text{or} \quad S_s(s) = \frac{2\sigma^2\beta}{-s^2 + \beta^2} \quad (4.3.32)$$

Also,

$$R_n(\tau) = 0$$

We first factor $S_{s+n}(s)$.

$$S_{s+n} = S_s = S_s^+ \cdot S_s^- = \frac{\sqrt{2\sigma^2\beta}}{s + \beta} \cdot \frac{\sqrt{2\sigma^2\beta}}{-s + \beta} \quad (4.3.33)$$

Next, we form the $S_{s+n,s}/S_{s+n}^-$ function

$$\frac{S_{s+n,s}}{S_{s+n}^-} = \frac{S_s}{S_s^-} = S_s^+ = \frac{\sqrt{2\sigma^2\beta}}{s + \beta} \quad (4.3.34)$$

In this problem α is not zero, and hence we must first multiply Eq. (4.3.34) by $e^{\alpha s}$ and then find the positive-time part of the result. This is readily accomplished by appropriate shifting in the time domain as shown in Fig. 4.6. Finally,

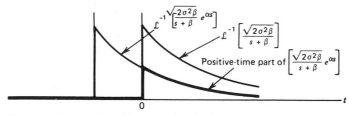

Figure 4.6. The shifted time functions for Example 4.4.

substituting the proper positive-time part into Eq. (4.3.22) yields

$$G(s) = \frac{1}{\dfrac{\sqrt{2\sigma^2\beta}}{s+\beta}} \cdot e^{-\alpha\beta} \frac{\sqrt{2\sigma^2\beta}}{s+\beta} = e^{-\alpha\beta} \qquad (4.3.35)$$

Or the corresponding weighting function is

$$g(t) = e^{-\alpha\beta}\delta(t) \qquad (4.3.36)$$

The Wiener solution says that the best one can hope to do (in the least-squares sense) is to multiply the present value of the input by an attenuation factor $e^{-\alpha\beta}$, and this will yield the best estimate of the process α units ahead of the present time. Observe that the predictive estimate is dependent only on the present value of the input signal and not on its past history. If the Markov signal under consideration is also Gaussian, then the predictive estimate given by Wiener theory is also identical to the conditional mean of the process at the predicted time, conditioned on all the prior input data. The use of the term Markov for a process with an exponential autocorrelation function should now be apparent. This will be expanded on later, in Chapter 5. ■

4.4
THE NONSTATIONARY PROBLEM

In the preceding discussion it was assumed that the filter was "turned on" at $t = -\infty$, which made the entire past history of $s(t) + n(t)$ available for weighting. This led to the steady-state or stationary solution. In the transient or nonstationary problem, we consider the signal and noise to be covariance stationary processes with known spectral characteristics as before, but we consider the input to be applied at $t = 0$ rather than $-\infty$. This is indicated in Fig. 4.7 as a switching operation at the input. We assume zero initial conditions; hence the filter output can be written as

$$x(t) = \int_0^t g(\tau)[s(t-\tau) + n(t-\tau)]\, d\tau, \qquad t \geq 0 \qquad (4.4.1)$$

Figure 4.7. Block diagram for the nonstationary problem.

As before, the ideal output would be $s(t + \alpha)$. Thus, the filter error is

$$e(t) = s(t + \alpha) - x(t) \qquad (4.4.2)$$

Substituting Eq. (4.4.1) into Eq. (4.4.2), squaring, and then taking the expectation of both sides yields

$$E(e^2) = \int_0^t \int_0^t g(u)g(v)R_{s+n}(u - v)\, du\, dv$$

$$- 2\int_0^t g(u)R_{s+n,s}(\alpha + u)\, du + R_s(0) \qquad (4.4.3)$$

The problem is to choose $g(u)$ such as to minimize $E(e^2)$. The variational procedure to be followed is essentially the same as for the stationary case, so it will not be repeated. The only difference is in the limits on the integration, which, in turn, traces back to the expression for $x(t)$ given by Eq. (4.4.1). This is an important difference, though. We need not worry about placing a causality constraint on $g(\tau)$ or its perturbation, because the range of integration is limited to be from 0 to t. We can arbitrarily truncate $g(\tau)$ to zero outside this range.

As before, the variational procedure leads to an integral equation in $g(\tau)$. For the nonstationary problem, it is

$$\int_0^t g(u)R_{s+n}(\tau - u)\, du = R_{s+n,s}(\alpha + \tau), \qquad 0 \le \tau \le t \qquad (4.4.4)$$

This equation is similar to the Wiener–Hopf equation except for the limits on the integral and the range of τ for which the equation is valid. These seemingly small differences complicate the solution even further, though. The spectral factorization method cannot be applied to Eq. (4.4.4) because the range of τ is finite rather than semi-finite as in the Wiener–Hopf equation. Equation (4.4.4) is amenable to solution, though, by another method that involves transforming the integral equation to a differential equation and then solving the differential equation for $g(\tau)$ (8). This technique will work only for the case where the spectral function for $s(t) + n(t)$ is rational, that is, a ratio of polynomials in s^2. However, we note that this same restriction is necessary for spectral factorization, and therefore only limited types of spectral situations can be handled for either the stationary or nonstationary case.

The solution of Eq. (4.4.4) proceeds as follows. We first write R_{s+n} as a Fourier integral

$$\int_0^t g(u)\left[\frac{1}{2\pi j} \int_{-j\infty}^{j\infty} S_{s+n}(s)e^{s(\tau-u)}\, ds \right] du = R_{s+n,s}(\alpha + \tau) \qquad (4.4.5)$$

Next, we write $S_{s+n}(s)$ as a ratio of polynomials in s^2

$$S_{s+n}(s) = \frac{N(s^2)}{D(s^2)} \tag{4.4.6}$$

Now we note that operating on $e^{s(\tau-u)}$ with the differential operator $D(d^2/d\tau^2)$ will generate $D(s^2)e^{s(\tau-u)}$.* Thus the denominator of S_{s+n} in Eq. (4.4.5) can be canceled by operating on both sides of the equation with $D(d^2/d\tau^2)$. Similarly, algebraic multiplication within the integral by $N(s^2)$ is equivalent to operating in front of the first integral with $N(d^2/d\tau^2)$. Inserting these equivalent operations in Eq. (4.4.5) yields

$$N\left(\frac{d^2}{d\tau^2}\right)\int_0^t g(u)\left[\frac{1}{2\pi j}\int_{-j\infty}^{j\infty} e^{s(\tau-u)}\,ds\right]du = D\left(\frac{d^2}{d\tau^2}\right)R_{s+n,s}(\alpha+\tau) \tag{4.4.7}$$

We now note that the Fourier integral in brackets in Eq. (4.4.7) is just the Dirac delta function $\delta(\tau-u)$. Inserting this and using the sifting property of the impulse function leads to

$$N\left(\frac{d^2}{d\tau^2}\right)g(\tau) = D\left(\frac{d^2}{d\tau^2}\right)R_{s+n,s}(\alpha+\tau), \qquad 0 < \tau < t \tag{4.4.8}$$

We now have a differential equation in $g(\tau)$ rather than an integral equation. Furthermore, it is a linear differential equation with constant coefficients, and the solution of this type of equation is well known. Before proceeding, note that the interval on τ in Eq. (4.4.8) is the open interval $(0, t)$ rather than the closed interval $[0, t]$ associated with the integral equation, Eq. (4.4.4). This is intentional and arises because of the problem of continuity and differentiation at the end points. In other words, we may safely assume only that the differential equation is valid in the interior region of the interval.

The solution of Eq. (4.4.8) will, of course, contain arbitrary constants of integration. Furthermore, because of the end-point problem, impulse functions with undetermined amplitudes must be added at $\tau = 0+$ and $\tau = t-$. (The + and − indicate that the impulses are placed at the inside edges of the interval.) The rule for adding the impulses is as follows: If the order of $D(s^2)$ is:

1. The same as $N(s^2)$, add no impulses.
2. If it is two greater than $N(s^2)$, add simple impulses.
3. If it is four greater than $N(s^2)$, add simple impulses plus doublet impulses.
4. Etc.

---

*Remember that D refers to the polynomial in the denominator of $S_{s+n}(s)$ and not the "D-operator" used in differential equation theory. That is, $D(d^2/d\tau^2)$ is $D(s^2)$ with s^2 replaced by $d^2/d\tau^2$. Similarly, $N(d^2/d\tau^2)$ is the numerator polynomial $N(s^2)$ with s^2 replaced with $d^2/d\tau^2$.

The unknown coefficients in the general solution are evaluated by substituting the assumed solution into the original integral equation and demanding equality on both sides of the equation. This is much the same as using the initial conditions to evaluate the constants of integration in the usual initial-condition problem.

Remember that the procedure just described is highly specialized and applies only to the case where $S_{s+n}(s)$ can be written as a ratio of polynomials in s^2. The justification of the procedure in any particular case lies in the final substitution of the solution into the integral equation. We can ask no more of the solution than to satisfy the original integral equation.

EXAMPLE 4.5

We again look at the situation where the signal is Markov and the noise is white. The additive combination forms the input that is applied to the filter at $t = 0$. Let

$$R_s(\tau) = e^{-|\tau|} \quad \text{or} \quad S_s(s) = \frac{2}{-s^2 + 1} \tag{4.4.9}$$

$$R_n(\tau) = \delta(\tau) \quad \text{or} \quad S_n(s) = 1 \tag{4.4.10}$$

If we assume the signal and noise have zero crosscorrelation,

$$S_{s+n}(s) = \frac{2}{-s^2 + 1} + 1 = \frac{-s^2 + 3}{-s^2 + 1} = \frac{N(s^2)}{D(s^2)} \tag{4.4.11}$$

Also, if $\alpha = 0$

$$R_{s+n,s}(\alpha + \tau) = R_s(\tau) = e^{-\tau}, \qquad \tau \geq 0 \tag{4.4.12}$$

Note that the absolute magnitude signs around τ may be dropped because τ is always positive. Using the polynomials $N(s^2)$ and $D(s^2)$ as given by Eq. (4.4.11), the differential equation (4.4.8) becomes

$$\left(-\frac{d^2}{d\tau^2} + 3 \right) g(\tau) = \left(-\frac{d^2}{d\tau^2} + 1 \right) e^{-\tau} \tag{4.4.13}$$

and this reduces to

$$-\frac{d^2 g(\tau)}{d\tau^2} + 3g(\tau) = 0 \tag{4.4.14}$$

The general solution of this equation is recognized to be

$$g(\tau) = ae^{-\sqrt{3}\tau} + be^{\sqrt{3}\tau} \qquad (4.4.15)$$

We do not need to add impulses in this case because $N(s^2)$ and $D(s^2)$ are both second order. Thus, we know the filter weighting function is of the form given by Eq. (4.4.15) without additional impulses. The a and b coefficients may be evaluated by substituting the known form of solution [i.e., Eq. (4.4.15)] into the original integral equation, Eq. (4.4.4), and then choosing a and b such that the resulting functions on left and right sides of the equation are identical functions of τ. This is straightforward, but considerable algebra is involved, which will be omitted. The end result is

$$a(t) = \frac{2(\sqrt{3} + 1)e^{\sqrt{3}t}}{(\sqrt{3} + 1)^2 e^{\sqrt{3}t} - (-\sqrt{3} + 1)^2 e^{-\sqrt{3}t}} \qquad (4.4.16)$$

$$b(t) = \frac{-2(-\sqrt{3} + 1)e^{-\sqrt{3}t}}{(\sqrt{3} + 1)^2 e^{\sqrt{3}t} - (-\sqrt{3} + 1)^2 e^{-\sqrt{3}t}} \qquad (4.4.17)$$

Note that the "constants" are functions of the running time variable t. The final solution for weighting function is then

$$g(\tau; t) = a(t)e^{-\sqrt{3}\tau} + b(t)e^{\sqrt{3}\tau} \qquad (4.4.18)$$

where $a(t)$ and $b(t)$ are given by Eqs. (3.4.16) and (3.4.17). The semicolon in $g(\tau; t)$ is used to emphasize the fact that τ is the usual age variable in the weighting function and t is just a parameter. The resulting filter is, of course, a time-variable filter.

It is readily verified that as t approaches ∞, the solution for $g(\tau)$ becomes

$$g(\tau) = (\sqrt{3} - 1)e^{-\sqrt{3}\tau} \qquad (4.4.19)$$

As should be expected, this is the same steady-state solution that was obtained previously using spectral factorization methods. It is of interest to note that the differential-equation approach provides an alternative method of solving the stationary problem. ∎

In Example 4.5 it is worth noting that the running time variable t came into the weighting-function solution naturally (i.e., without any conscious effort) because we chose to write the superposition integral in the form

$$x(t) = \int_0^t g(\tau; t) f(t - \tau) \, d\tau \tag{4.4.20}$$

The other form, which is equally valid, and sometimes preferred in books on linear systems theory (9), is

$$x(t) = \int_0^t h(t, \tau) f(\tau) \, d\tau \tag{4.4.21}$$

In Eq. (4.4.21), $h(t, \tau)$ has the physical meaning of the system response to a unit impulse applied at time τ. The relationship between the impulsive response and weighting function is obtained by making the appropriate change of variable in either Eq. (4.4.20) or Eq. (4.4.21) and then comparing the two integrals. The result is

$$h(t, \tau) = g(t - \tau; t) \tag{4.4.22}$$

4.5 ORTHOGONALITY

It was shown in Section 4.4 that the filter weighting function that minimizes the mean-square error must satisfy the integral equation

$$\int_0^t g(u) R_{s+n}(\tau - u) \, du = R_{s+n, s}(\alpha + \tau), \qquad 0 \le \tau \le t \tag{4.5.1}$$

Also, the filter error is given by

$$e(t) = s(t + \alpha) - x(t) = s(t + \alpha) - \int_0^t g(u)[s(t - u) + n(t - u)] \, du \tag{4.5.2}$$

We wish to examine the expectation of the product of the filter error at time t and input at some time t_1 where $0 \le t_1 \le t$. Let the input be denoted as $z(t)$. Then

$$z(t_1) = s(t_1) + n(t_1) \tag{4.5.3}$$

and

$$E[z(t_1)e(t)] = E\left[\{s(t_1) + n(t_1)\}\right.$$

$$\left.\times \left\{s(t + \alpha) - \int_0^t g(u)[s(t - u) + n(t - u)]\, du\right\}\right] \quad (4.5.4)$$

Moving $s(t_1) + n(t_1)$ inside the integration and carrying out the expectation operation yields

$$E[z(t_1)e(t)] = R_{s+n,s}(t - t_1 + \alpha) - \int_0^t g(u)R_{s+n}(t - t_1 - u)\, du \quad (4.5.5)$$

However, $g(u)$ must satisfy the integral equation, Eq. (4.5.1). Thus, the above must be zero for $0 \le (t - t_1) \le t$. Since the time t_1 was assumed to lie between 0 and t, this is equivalent to saying

$$E[z(t_1)e(t)] = 0, \qquad 0 \le t_1 \le t \quad (4.5.6)$$

If the expectation of the product of two random variables is zero, the variables are said to be *orthogonal*. Equation (4.5.6) states that the filter error at the current time t is not only orthogonal to the input at the same time t, but it is also orthogonal to the input evaluated at any previous time during the past history of the filter operation. This is a consequence of minimizing the mean-square error. Furthermore, it should be apparent from the derivation that the argument can be reversed. That is, if we begin by assuming that $e(t)$ is orthogonal to $z(t_1)$, we can then conclude that the integral equation, Eq. (4.5.1), is satisfied. It is important to recognize this equivalence because some authors prefer to begin their optimality arguments with the orthogonality relationship rather than the minimization of the mean-square error (2, 3, 8).

4.6
COMPLEMENTARY FILTER

Applications of Wiener filter theory are not as commonplace as one might expect. Perhaps one reason for this is that Wiener theory demands that the signal be noiselike in character. In the usual communication problem, this is not the case. The signal usually has at least some deterministic structure, and it is often not reasonable to assume it to be completely random. Thus, the typical filtering problem encountered in communication engineering simply does not fit the Wiener mold. There is an instrumentation application, though, where Wiener methods have been used extensively. In this application, redundant

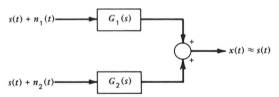

Figure 4.8. General two-input Wiener problem.

measurements of the same signal are available, and the problem is to combine all the information in such a way as to minimize the instrumentation errors. In order to keep the discussion as simple as possible, the two-input case will be considered first.

Consider the general problem of combining two independent noisy measurements of the same signal as depicted in Fig. 4.8. In the context of instrumentation, the measurements might come from two completely different types of instruments, each with its own particular error characteristic. We wish to blend the two measurements together in such a way as to eliminate as much of the error as possible. If the signal $s(t)$ is noiselike, Wiener methods may be used to determine the transfer functions $G_1(s)$ and $G_2(s)$ that minimize the mean-square error. This is the two-input Wiener problem, and more will be said of it in the next section. However, more often than not the signal may not be properly modeled as a random process with known spectral characteristics. For example, if $s(t)$ represents the position of an airplane in flight along a prescribed air route, certainly the signal is not random. Furthermore, in this case we would not want to delay or distort the signal in any way in the process of filtering the measurement errors.

A method of filtering the noise without distorting the signal is shown in Fig. 4.9. From the block diagram, it should be apparent that the Laplace transform of the output may be written as

$$X(s) = \underbrace{S(s)}_{\text{Signal term}} + \underbrace{N_1(s)[1 - G(s)] + N_2(s)G(s)}_{\text{Noise term}} \qquad (4.6.1)$$

Clearly, the signal term $S(s)$ is not affected by our choice of $G(s)$ in any way. On the other hand, the two noise inputs are modified by the complementary transfer functions $[1 - G(s)]$ and $G(s)$. If the two noises have complementary

Figure 4.9. Complementary filter.

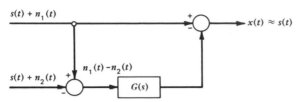

Figure 4.10. Differencing and feedforward configuration for a complementary filter.

spectral characteristics, $G(s)$ may be chosen to mitigate the noise in both channels. For example, if n_1 is predominantly low-frequency noise and n_2 high-frequency, then choosing $G(s)$ to be a low-pass filter will automatically attenuate n_1 as well as n_2.

We now note that the noise term in Eq. (4.6.1) has the same form as seen before [Eq. (4.2.4)] except for the sign on the $N_2(s)$ term. Thus, we would expect to be able to use Wiener methods in minimizing this term. This is perhaps even more evident from Fig. 4.10. It can be easily verified that the input–output relationships are identical for the systems of Figs. 4.9 and 4.10, and thus they are equivalent. From Fig. 4.10 we see that the purpose of the filter $G(s)$ is to give the best possible estimate of $n_1(t)$, and this, in turn, is subtracted from $s(t) + n_1(t)$ in order to give an improved estimate of $s(t)$. The input to $G(s)$ is $n_1(t) - n_2(t)$, and hence the filter must separate one *noiselike* signal from another. Clearly, if we let $n_1(t)$ play the role of signal and $-n_2(t)$ the role of the noise, this problem fits the Wiener theory perfectly. An example will illustrate an engineering application of the complementary-filter method of combining redundant measurement data.

EXAMPLE 4.6*

In the standard aircraft instrument landing system (ILS), the aircraft descends along a radio beam that provides the pilot with steering signals proportional to the plane's deviation from the desired glide path. The pilot uses these to correct the plane's course in order to follow the beam as closely as possible. In automating the system, say in the lateral channel, it is desirable to have a rate signal to aid in stabilizing the closed-loop control system. One such signal can be obtained by differentiating (approximately) the lateral-deviation radio signal. However, this gives rise to high-frequency noise in the rate signal. Another completely independent lateral rate signal may be obtained from the aircraft heading. If the aircraft speed is constant, an incremental change in heading, when multiplied by speed, will yield a corresponding lateral rate change. This signal, however, is also contaminated with error, the major source in this case being wind changes that occur during the plane's descent. These errors are

---

*This example is a simplified version of a system described in a patent granted to W. H. Wirkler (10). Even though Wirkler did not use the term *complementary filter* in his disclosure, the principle is there. The system patented by Wirkler (with some modifications) was described later by Anderson and Fritze (11), and they introduced the term *complementary filter* in their paper.

primarily low-frequency in nature. Thus, we have the ideal setting here for a complementary filter. We have two independent measurements of the same quantity, one contaminated with high-frequency noise and the other with low-frequency noise.

At this point, we assume the following spectral characteristics for the two noises:

1. *Heading-derived rate signal.* Markov noise (primarily low-frequency):

$$S_{n_1}(s) = \frac{2\sigma_1^2 \omega_1}{-s^2 + \omega_1^2} \tag{4.6.2}$$

2. *Radio-derived rate signal.* Differentiated Markov noise (primarily high-frequency):

$$S_{n_2}(s) = \frac{2\sigma_2^2 \omega_2(-s^2)}{-s^2 + \omega_2^2} \tag{4.6.3}$$

We can now take either of two approaches. The filter form may be assumed, and then it can be optimized with respect to a parameter (e.g., the time constant); or the more general Wiener approach may be used, treating n_1 as the signal and n_2 as the noise. We take the latter approach and, in order to keep the algebra reasonable, we assume plausible values for the various parameters. Let

$$
\begin{array}{ll}
\sigma_1 = 2 \text{ m/sec} & \left\{ \begin{array}{l} \text{Heading-derived} \\ \text{velocity noise} \end{array} \right. \\
\omega_1 = 1/20 \text{ rad/sec} & \\[2mm]
\sigma_2 = 20 \text{ m} & \left\{ \begin{array}{l} \text{Radio-derived} \\ \text{position noise} \end{array} \right. \\
\omega_2 = 1/2 \text{ rad/sec} &
\end{array}
$$

Then

$$S_{n_1+n_2}(s) = \frac{2\sigma_1^2 \omega_1}{-s^2 + \omega_1^2} + \frac{2\sigma_2^2 \omega_2(-s^2)}{-s^2 + \omega_2^2} \tag{4.6.4}$$

$$= 20 \frac{(s^2 + .187s + .0158)}{(s + .05)(s + .5)}$$

$$\cdot 20 \frac{(-s)^2 + .187(-s) + .0158}{(-s + .05)(-s + .5)} \tag{4.6.5}$$

Spectral factorization of $S_{n_1+n_2}$ is obvious from the way Eq. (4.6.5) has been written. We can now form $S_{n_1}/S^-_{n_1+n_2}$ as

$$\frac{S_{n_1}(s)}{S^-_{n_1+n_2}(s)} = \frac{\dfrac{.4}{(s+.05)(-s+.05)}}{20\,\dfrac{(-s)^2 + .187(-s) + .0158}{(-s+.05)(-s+.5)}}$$

$$= .02\,\frac{-s+.5}{(s+.05)\left[(-s)^2 + .187(-s) + .0158\right]} \tag{4.6.6}$$

Clearly, the positive-time part of this will contain only a $K/(s+.05)$ term. Evaluating this yields

$$\text{Positive-time part of } \frac{S_{n_1}}{S^-_{n_1+n_2}} = \frac{.3976}{s+.05} \tag{4.6.7}$$

The transfer function $G(s)$ is then given by Eq. (4.3.22):

$$G(s) = \frac{1}{20\,\dfrac{s^2 + .187s + .0158}{(s+.05)(s+.5)}}\,\frac{.3976}{s+.05}$$

$$= .0199\,\frac{s+.5}{s^2 + .187s + .0158} \tag{4.6.8}$$

This is obviously a low-pass filter with a break point of about .1 rad/sec. It is interesting to note that the zero-frequency response of $G(s)$ is about .63 rather than unity, a might be expected intuitively. This means that even at very low frequencies the heading-derived signal will be given significant weight—.37 relative to the weight of .63 given the radio signal. Obviously, this will depend on the numerical values assigned to the various parameters. However, in any event, optimal filter theory says the heading-derived signal has at least some value at low frequencies. On the other hand, the weight shifts entirely to the heading signal at high frequencies. ∎

The principal of complementary filtering is easily extended to the case of more than two signals. All one has to do is let one of the transfer characteristics be the complement of the sum of the others. For example, for the three-input case, let

$$G_1(s) = \text{transfer function for noisy measurement 1}$$

$$G_2(s) = \text{transfer function for noisy measurement 2}$$

$$1 - G_1(s) - G_2(s) = \text{transfer function for noisy measurement 3}$$

Then the signal component passes through the system undistorted, and one chooses $G_1(s)$ and $G_2(s)$ such as to give the best suppression of the noise. The problem of determining the optimal $G_1(s)$ and $G_2(s)$ is a multiple-input Wiener problem and this will be discussed later.

It is important to note that the choice of complementary filter transfer function does not depend on any prior assumptions about the *signal* structure. It can be either noiselike or deterministic, and the complementary feature assures that the signal will not be distorted in any way by the filtering action. Thus, philosophically, it is a "safe" design and is particularly applicable in situations where the designer wants a filter that will cope reasonably well with statistically unusual situations without giving disastrously large errors. A strict Wiener design with no complementary constraint is, of course, chosen to minimize the *average* squared error. So in the unusual situation, the error may be quite large. This can be disastrous in some applications.

4.7
MULTIPLE-INPUT WIENER
FILTER PROBLEM

Figure 4.11 shows the general two-input Wiener problem that will be used to illustrate the multiple-input problem. Since the extension to more than two inputs is fairly obvious, we confine our attention to the two-input case. We assume that $s(t)$ as well as $n_1(t)$ and $n_2(t)$ are stationary random processes with known spectral characteristics, and in this case, we do *not* wish to constrain one transfer function to be the complement of the other. The problem is to find the causal transfer functions $G_1(s)$ and $G_2(s)$ that will minimize the mean-square error.

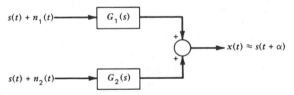

Figure 4.11.   General two-input Wiener problem.

Since the solution for the two-input case proceeds much the same as for the single-input problem, some of the details will be omitted. Also, in order to keep the solution as simple as possible, the signal and noises will be assumed to have zero crosscorrelation. We first write the expression for the error as

$$e(t) = s(t + \alpha) - x(t)$$

$$= s(t + \alpha) - \int_{-\infty}^{\infty} g_1(\tau)[s(t - \tau) + n_1(t - \tau)] \, d\tau$$

$$- \int_{-\infty}^{\infty} g_2(\tau)[s(t - \tau) + n_2(t - \tau)] \, d\tau \qquad (4.7.1)$$

The expectation of the square of $e(t)$ is then formed, and this will involve the two weighting functions $g_1(\tau)$ and $g_2(\tau)$. Next, following the usual calculus-of-variations procedure, g_1 and g_2 are replaced with perturbed weighting functions $g_1(\tau) + \varepsilon_1 \eta_1(\tau)$ and $g_2(\tau) + \varepsilon_2 \eta_2(\tau)$. This allows us to consider independent variations of both weighting functions by varying ε_1 and ε_2. After the indicated substitution, the mean-square error becomes a function of ε_1 and ε_2. The partial derivatives of $E(e^2)$ with respect to ε_1 and ε_2 may then be formed and set equal to zero for $\varepsilon_1 = \varepsilon_2 = 0$. Finally, noting that $\eta_1(\tau)$ and $\eta_2(\tau)$ are arbitrary for $\tau \geq 0$ leads to the following pair of integral equations.

$$\left. \begin{array}{l} \int_{-\infty}^{\infty} g_1(u) R_{s+n_1}(\tau - u) \, du \\ \qquad + \int_{-\infty}^{\infty} g_2(u) R_s(\tau - u) \, du - R_s(\tau + \alpha) = 0 \\ \int_{-\infty}^{\infty} g_1(u) R_s(\tau - u) \, du \\ \qquad + \int_{-\infty}^{\infty} g_2(u) R_{s+n_2}(\tau - u) \, du - R_s(\tau + \alpha) = 0 \end{array} \right\} \quad \tau \geq 0$$

$$(4.7.2)$$

Note that these are *coupled* integral equations and must be solved simultaneously. Also, because of the causality constraint on g_1 and g_2, the equations are valid only for $\tau \geq 0$.

The solution of Eq. (4.7.2) is not as straightforward as one might think at first glance. If we first transform the equations and then try using spectral factorization as before, we find that the positive- and negative-time parts are not readily separated and identified. Thus, we must look for another approach. A method of solution that is somewhat cumbersome, but workable, proceeds as follows. First, use differential operators similar to those used in Section 4.4 to convert the two integral equations into differential equations. These equations may then be solved using standard methods of ordinary differential equations.

Impulses may then be added at the end points, if needed, and the general solution may then be substituted into the integral equations to determine the undetermined coefficients. This is a long and laborious method of solution, though. An alternative approach, which is less general but may be used in simple situations, is given by Brown and Nilsson (12). This method is summarized in Problem 4.17. Also, we will see later, in Chapter 7, that multiple-input problems may be solved using continuous Kalman filter methods.

As mentioned in Section 4.6, we often wish to use the complementary constraint in instrumentation applications where redundant measurements are available. The complementary constraint is especially important in those situations where we want to avoid making questionable assumptions about the nature of the signal. If there are three or more redundant measurements, the complementary filter problem always reduces to a general multiple-input Wiener problem of one less order than the original problem (see Problem 4.16). For example, if three measurements are available, the problem reduces to the general two-input Wiener problem just considered. In effect, one of the noises, say n_1, plays the role of signal and the problem is to separate it from the other two noises n_2 and n_3. Once this is done, the filter's best estimate of n_1 may be subtracted from the first noisy measurement $s + n_1$ to yield an improved estimate of the signal s.

4.8
WEIGHTING FUNCTION APPROACH
TO THE SAMPLED-DATA FILTER

The Wiener approach to least-squares filtering is basically a weighting function approach. When viewed this way, the basic problem always reduces to: How should the past history of the input be weighted in order to yield the present best estimate of the variable of interest? It is instructive to see how this approach extends to the discrete-measurement situation.

Consider the filter input to be a sequence of discrete noisy measurements $z_1, z_2, \ldots, z_n$ as shown in Fig. 4.12. These are additive combinations of signal and noise; hence $z_1 = s_1 + n_1$, $z_2 = s_2 + n_2$, and so on. As before, we denote the filter output as x, and therefore the corresponding samples of the output are $x_1, x_2, \ldots, x_n$. We now write the output at time t_n as a general linear

Figure 4.12. Discrete measurement situation.

combination of the past measurements

$$x_n = k_1 z_1 + k_2 z_2 + \cdots + k_n z_n \tag{4.8.1}$$

The filter error may then be written as

$$e_n = s_n - x_n$$

$$= s_n - (k_1 z_1 + k_2 z_2 + \cdots + k_n z_n) \tag{4.8.2}$$

The mean-square error is then

$$E(e_n^2) = E[s_n - (k_1 z_1 + k_2 z_2 + \cdots + k_n z_n)]^2$$

$$= E(s_n^2) + \left[k_1^2 E(z_1^2) + k_2^2 E(z_2^2) + \cdots + k_n^2 E(z_n^2) \right.$$

$$\left. + 2k_1 k_2 E(z_1 z_2) + 2k_1 k_3 E(z_1 z_3) + \cdots \right]$$

$$- \left[2k_1 E(z_1 s_n) + 2k_2 E(z_2 s_n) + \cdots + 2k_n E(z_n s_n) \right] \tag{4.8.3}$$

We now wish to find $k_1, k_2, \ldots, k_n$ such as to minimize $E(e_n^2)$. The usual methods of differential calculus lead to the following set of linear equations:

$$\begin{bmatrix} E(z_1^2) & E(z_1 z_2) & \cdots & \cdots \\ E(z_2 z_1) & & & \\ & & \ddots & \\ \vdots & & & \ddots \\ E(z_n z_1) & & & E(z_n^2) \end{bmatrix} \begin{bmatrix} k_1 \\ k_2 \\ \vdots \\ \\ k_n \end{bmatrix} = \begin{bmatrix} E(z_1 s_n) \\ E(z_2 s_n) \\ \vdots \\ \\ E(z_n s_n) \end{bmatrix} \tag{4.8.4}$$

Just as in the continuous problem, we assume that the auto- and crosscorrelation function of the signal and noise are known, so that all the expectations indicated in Eq. (4.8.4) are available and the equations may be solved for the weighting factors $k_1, k_2, \ldots, k_n$. Thus, in principle at least, we have a formal solution. The indicated solution may even be practical in off-line situations where the amount of measurement data is limited. However, in on-line situations, the solution keeps growing in size as time evolves. With each new measurement, we have to add an additional equation, compute a new row of covariances, and re-solve the resultant equations for a new set of k's. The solution gets out of hand in a hurry! Even with clever programming in the solution of Eq. (4.8.4), one cannot avoid the "growing memory" problem caused by the need to store all the past measurement data. One has no right, a priori at least, to discard any of the past input data, no matter how old it may be. There are many situations where "old" data are just as good as "new" data. The net result is that the weighting function approach is, at best, unwieldy in on-line applications. We will see in Chapter 5 that the discrete Kalman filter provides a solution that is readily implemented on a computer, and usually it is preferred over the direct weighting-function solution given by Eq. (4.8.4). (See Problem 5.7, Chapter 5, for an example that compares the weighting-function and Kalman-filter solutions.)

4.9
PERSPECTIVE

A Wiener filter minimizes the mean-square estimation error subject to certain constraints and assumptions. It is important to remember that this optimization is only intended to apply to the problem of separating one *noiselike* signal from another, which is a very restricted class of filtering problems. Also the assumption of *linear* filtering was built into the derivation from the start. We will see later, in Chapter 5, that this is not a serious restriction if all the random processes involved are Gaussian. In this one case the linear filter is optimum by almost any reasonable criterion of performance (13). However, the non-Gaussian case is another matter. A nonlinear filter may be better, and the Wiener filter is optimal only within the restricted class of linear filters.

Sometimes, minimization of the mean-square error can lead to seemingly strange physical results; therefore, the results of Wiener filtering should be viewed with a degree of caution. An example will illustrate this. It was mentioned in Section 2.11 that both the random telegraph wave and the Gauss–Markov process have exponential autocorrelation functions. Since the Wiener filter design depends only on the auto- and crosscorrelation functions, the solution for the pure prediction problem will be the same for both random-telegraph and Gauss–Markov signals with the same autocorrelation functions. It was found in Example 4.4 that the solution in this case is a simple attenuator and, for large prediction time, the predictor output is approximately zero. This makes good sense for the Gauss–Markov signal, because it is noiselike with a central tendency toward zero; and, in the absence of relevant (i.e., "recent") measurement information, one should just pick the process mean as the best estimate. This way the estimate is only rarely in error by a gross amount, and it is often close to the signal value. On the other hand, to estimate the random telegraph signal to be zero is pure nonsense. We know a priori that the signal is either $+1$ or -1, and it is *never* zero. That is, the Wiener predictor never predicts the correct answer, nor is it even close to the correct answer! We might better arbitrarily choose $+1$ as our estimate. Then, at least we would be correct half the time! Think of the many game (and more serious) situations where you would be better off to be exactly correct half the time (and grossly in error the other half) than to be significantly (but perhaps not grossly) in error *all* the time.

The reason for the strange result in the random-telegraph-wave predictor is that the optimization procedure did not account for the higher-order "statistics" of the process. The mean-square-error criterion is simplistic and only calls for knowledge of the correlation functions of the processes involved. This is, of course, convenient and it also happens to fit the Gaussian process case quite well (for reasons that may not be apparent yet). However, as has just been demonstrated, the mean-square-error criterion of performance can lead to strange results when dealing with non-Gaussian processes.

PROBLEMS

4.1 The first-order filter assumed in Example 4.1 will be recognized as a first-order Butterworth filter. Variation of the time constant T in the optimization process is the equivalent of varying the -3-db cutoff frequency of the filter. It was noted later in Example 4.2 that for $\sigma^2 = \beta = A = 1$, the Phillips procedure led to a cutoff frequency of $1/(1 + \sqrt{2})$ rad/sec and a mean-square error of .914.

In this problem, consider a second-order Butterworth filter of the form

$$G(s) = \frac{1}{\left(\dfrac{s}{\omega_c}\right)^2 + \sqrt{2}\left(\dfrac{s}{\omega_c}\right) + 1} \tag{P4.1}$$

Just as in the first-order case, only one parameter is free to vary—the -3-db cutoff frequency ω_c. Using the same noise parameters as before (i.e., $\sigma^2 = \beta = A = 1$), find the optimum cutoff frequency for the second-order Butterworth filter. Also find the associated mean-square error and compare the results with those obtained for the first-order filter. (*Note:* Even a second-order filter with only one adjustable parameter involves a considerable amount of algebraic effort. It might be easier to find the minimum by trial-and-error, numerical means rather than by analytical methods. Also note that the resulting second-order filter may not be better than the first-order filter because of the constraints placed on the form of the filter.)

4.2 A closed-loop position control system has the form shown in the block diagram. The spectral density function of the *derivative* of the signal $s(t)$ is given by

$$S_{\dot s}(j\omega) = \frac{1000}{\omega^2 + .01}$$

Problem 4.2

and the noise is unity-amplitude Gaussian white noise. Find the value of the gain constant K that will minimize the mean-square error, and find the damping ratio corresponding to this gain. (*Hint:* The error term that involves the signal spectrum is of the form $S(s)[1 - G(s)]$ [see Eq. (4.2.3)]. In this problem $[1 - G(s)]$ contains an s factor in the numerator that may be linked with $S(s)$. Since this has the interpretation of *derivative* of $s(t)$ in the time domain, the mean-square error term due to signal may be written in terms of the spectrum of the derivative of the signal, which is the spectral function given in the problem. This problem is worked out as an example in Truxal (14).)

4.3 The figure for Problem 3.21 is repeated here for convenience. Let $f_1(t)$ and $f_2(t)$ be independent white noise inputs with spectral amplitudes A_1 and A_2, just as in Problem 3.21. Consider x_3 (the inertial system position error) as the output and find the value of K that minimizes the mean-square value of x_3.

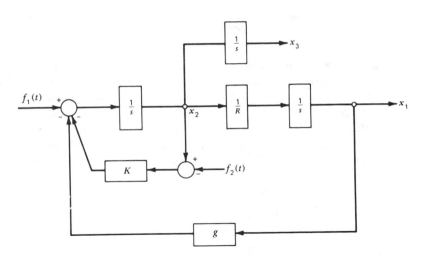

R = earth radius $\approx 2.09 \times 10^7$ ft

g = gravitational constant
≈ 32.2 ft/sec^2

K = feedback constant
(adjustable to yield the
desired damping ratio)

Problem 4.3

4.4 One facet of biomedical electronic instrumentation work is that of implanting miniature, telemetering transducers in live animals. This is done in order that various body functions may be observed under normal conditions of activity and environment. When considering such an implant, one is immediately confronted with the problem of supplying energy to the transducer. Batteries are an obvious possibility, but they are bulky and have finite life. Another possibility is to take advantage of the random motion of the animal, and a

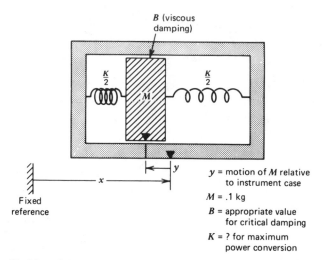

Problem 4.4

device for accomplishing this is shown in the simplified diagram of the accompanying figure. The energy conversion takes place in the damping mechanism. This is shown as simple viscous damping, but, in fact, would have to be some sort of electromechanical conversion device. Assuming that all the power indicated as being dissipated in damping can be converted to electrical form, what is the optimum value for the spring constant K and how much power is converted for this optimum condition? The spring-mass arrangement is to be critically damped, the mass is .1 kg, and the autocorrelation function for the velocity $\dot{x}$ is estimated to be

$$R_{\dot{x}}(\tau) = 1e^{-2\pi|\tau|} \ (\text{ft/sec})^2$$

[For more details see Long (15).]

4.5   Find the optimal causal transfer function for the case where the autocorrelation functions of the signal and noise are

$$R_s(\tau) = 2e^{-2|\tau|}$$

$$R_n(\tau) = e^{-|\tau|}$$

Also find the mean-square error. The signal $s(t)$ and the noise $n(t)$ may be assumed to be independent of each other and both are time stationary. Also let the prediction time be zero. In brief, this is the classical Wiener filter problem with zero prediction time.

4.6   Find the optimal noncausal $G(s)$ for the signal and noise situation given in Problem 4.5. Also, find the corresponding weighting function and the resultant mean-square error.

4.7 Consider a stationary Gaussian signal whose spectral density function is

$$S_s(j\omega) = \frac{\omega^2 + 1}{\omega^4 + 8\omega^2 + 16}$$

Find the optimal predictor for the stationary case and $\alpha = 1$. (*Note:* The causal solution is desired, and the predictor may be specified in terms of either a transfer function or a weighting function.)

4.8 Consider an additive combination of signal and noise where the spectral densities are given by

$$S_s(s) = \frac{2}{-s^2 + 1}$$

$$S_n(s) = \frac{4}{-s^2 + 4}$$

Both the signal and noise are stationary Gaussian processes, and they are statistically independent.

(a) Without regard to causality, what is the optimal linear filter for the stationary case? The answer may be specified in terms of either a transfer function or a weighting function.

(b) Does the noncausal result of (a) have any physical significance? That is, is there a corresponding estimation problem where the computed theoretical mean-square error would have the significance of actual mean-square estimation error? Explain briefly.

4.9 Consider the following autocorrelation functions of the signal and noise in a stationary Wiener prediction problem

$$R_s(\tau) = 4e^{-4|\tau|}$$

$$R_n(\tau) = e^{-|\tau|}$$

The signal and noise are independent and the prediction time is .25 sec. Find the optimal causal weighting function.

4.10 Show that the Wiener–Hopf equation [Eq. (4.3.18)] is a sufficient as well as a necessary condition for minimizing the mean-square error. [*Hint:* Differentiate Eq. (4.3.6) twice with respect to ε and then examine $d^2E(e^2)/d\varepsilon^2$. Recall from elementary calculus that the extremum is a relative minimum if this quantity is positive at the extremum point, which, in this case is $\varepsilon = 0$.]

4.11 The orthogonality principle states that the filter error at the current time t is orthogonal to the filter input evaluated at any previous time since initiation of the input. In Section 4.5 this principle was derived from the nonstationary version of the Wiener–Hopf equation [Eq. (4.4.4)]. Reverse the arguments and show that the integral equation [i.e., Eq. (4.4.4)] may be derived from the orthogonality principle. (If you need help on this one, see Davenport and Root (8) p. 240.

4.12 Consider a noisy measurement of the form

$$z(t) = a_0 + n(t)$$

where a_0 is an unknown random constant with zero-mean normal distribution and a variance of σ^2. The additive noise may be assumed to be white with spectral amplitude A. Find the optimal time-variable filter for estimating a_0. The filter is turned on at $t = 0$. (*Hint:* A random constant may be thought of as a limiting case of a Markov process with a very large time constant.)

4.13 Verify that the $a(t)$ and $b(t)$ coefficients given in Example 4.5 are correct.

4.14 The accompanying figure was taken from Kayton and Fried (16) p. 318. It describes a means of blending together barometrically derived and inertially derived altitude signals in such a way as to take advantage of the best properties of both signals. The barometric signal, when taken by itself, has large inherent time lag that is undesirable. It is, however, stable and reasonably accurate in the steady-state. On the other hand, integrated vertical acceleration has an unbounded steady-state error because of accelerometer bias error. In effect, the high-frequency response of the accelerometer is good, but its low-frequency response is poor; just the reverse is true for the barometric instrument. Thus this is an ideal setting for a complementary filter application.

Let $G_1(s)$ and $G_2(s)$ in the figure be constant gains G_1 and G_2 and show that:

(a) The system fits the form of a complementary filter as discussed in Section 4.6.

(b) The system error has second-order characteristics with the natural frequency and damping ratio being controlled by the designer's choice of G_1 and G_2.

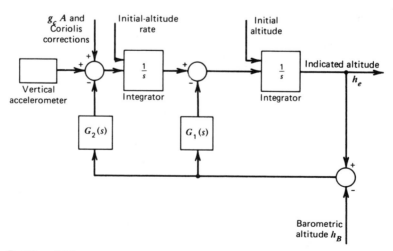

Problem 4.14

[*Note:* In order to show the complementary-filter property, you must conceptually think of directly integrating the accelerometer signal twice ahead of the summer to obtain an inertially derived altitude signal contaminated with error. However, direct integration of the accelerometer output is not required in the final implementation because of cancellation of s's in numerator and denominator of the transfer function. This will be apparent after carrying through the details of the problem. Also note that the g_c, A, and Coriolis corrections indicated in the figure are simply the gravity, accelerometer-bias, and Coriolis corrections required in order that the accelerometer output be $\ddot{h}$ (as best possible). Also, the initial conditions indicated in the figure may be ignored because the system is stable and we are interested in only the steady-state condition here.

4.15 It is desired to add rate feedback in a control system. The rate signal is to come from a permanent-magnet, dc tachometer on the output shaft. However, it is noticed that the tachometer output is noisy due to the combined commutator-brush action. Filtering the tachometer output is a possibility, but this introduces unwanted delay in the rate signal. Someone suggests that if we add an angular accelerometer as well as the dc tachometer on the output shaft, we can filter the rate signal without paying the price of time delay. The proposed scheme is shown in the accompanying figure.
    (a) Show that this is a form of complementary filtering.
    (b) Assume that the errors associated with the tachometer and accelerometer signals are independent white noise processes with spectral amplitudes K_T and K_A, respectively. Find the value of T that minimizes the mean-square error.

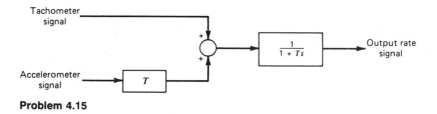

Problem 4.15

4.16 Consider the situation where N redundant noisy measurements of the same signal are available. The noisy signals are to be summed after passing each through its respective filter transfer function $G_1(s), G_2(s), \ldots, G_N(s)$ and the sum is to be an estimate of the signal $s(t)$. Show that if the last filter $G_N(s)$ is constrained to be $G_N(s) = 1 - [G_1(s) + G_2(s) + \cdots + G_{N-1}(s)]$, then the remaining optimization problem is a general $(N-1)$ input Wiener filter problem.

4.17 With reference to the general two-input Wiener filter problem shown in Fig. 4.11, let the signal and noises have zero crosscorrelation and let their respective autocorrelation functions be

$$R_s(\tau) = e^{-|\tau|}$$

$$R_{n_1}(\tau) = e^{-2|\tau|}$$

$$R_{n_2}(\tau) = \delta(\tau)$$

Also, let the prediction time α be zero. Find the optimal causal $G_1(s)$ and $G_2(s)$. *Hint:* Follow these steps:

1. Replace the right sides of Eqs. (4.7.2) with negative-time functions $a_1(\tau)$ and $a_2(\tau)$.
2. Take the Fourier transform of both sides of Eqs. (4.7.2) just as if $a_1(\tau)$ and $a_2(\tau)$ were known; then solve these equations algebraically for $G_1(s)$ and $G_2(s)$.
3. Note that $A_1(s)$ and $A_2(s)$ correspond to negative-time functions, and thus their poles are in the right half-plane. Hence, from the solution obtained in step 2, you can find the characteristic poles of $G_1(s)$ and $G_2(s)$, which must be in the left half-plane.
4. Next, write out the partial fraction expansion form of $G_1(s)$ and $G_2(s)$ leaving the coefficients undetermined. Direct feed-through terms (i.e., constants) must be inserted where appropriate. Note that a constant need not be included in $G_2(s)$, because the white noise input in this channel would lead to an output with infinite variance.
5. Substitute the assumed solution from step 4 into the transformed equations obtained in step 2, and then determine the unknown coefficients of $G_1(s)$ and $G_2(s)$.

Answer:

$$G_1(s) = \frac{1 + \sqrt{\dfrac{10}{3}}}{7} + \frac{\dfrac{1}{7}\left(-\dfrac{4}{3} + \sqrt{\dfrac{10}{3}}\right)}{s + \sqrt{\dfrac{10}{3}}}$$

$$G_2(s) = \frac{\dfrac{1}{7}\left(-8 + 6\sqrt{\dfrac{10}{3}}\right)}{s + \sqrt{\dfrac{10}{3}}}$$

REFERENCES CITED IN CHAPTER 4

1. N. Wiener, *Extrapolation, Interpolation, and Smoothing of Stationary Time Series*, New York: Wiley, 1949.
2. R. E. Kalman, "A New Approach to Linear Filtering and Prediction Problems," *Trans. of the ASME—Journal of Basic Engineering*, 35–45 (March 1960).
3. R. E. Kalman and R. Bucy, "New Results in Linear Filtering and Prediction," *Trans, of the ASME—Journal of Basic Engineering*, *83*, 95 (1961).
4. H. M. James, N. B. Nichols, and R. S. Phillips, *Theory of Servomechanisms* (Vol. 25), Radiation Laboratory Series, New York: McGraw-Hill, 1947.
5. R. Weinstock, *Calculus of Variations*, New York: McGraw-Hill, 1952.
6. H. W. Bode and C. E. Shannon, "A Simplified Derivation of Linear Least Squares Smoothing and Prediction Theory," *Proc. I.R.E.*, *38*, 417–424 (April 1950).
7. G. R. Cooper and C. D. McGillem, *Probabilistic Methods of Signal and System Analysis*, 2nd ed., New York: Holt, Rinehart, and Winston, 1986.
8. W. B. Davenport, Jr., and W. L. Root, *Introduction to Random Signals and Noise*, New York: McGraw-Hill, 1958.
9. C. T. Chen, *Linear System Theory and Design*, New York: Holt, Rinehart, and Winston, 1984.
10. W. H. Wirkler, "Aircraft Course Stabilization Means," U.S. Patent 2,548,278, issued April 10, 1951.
11. W. G. Anderson and E. H. Fritze, "Instrument Approach System Steering Computer," *Proc. of I.R.E.*, *41* (February 1953).
12. R. G. Brown and J. W. Nilsson, *Introduction to Linear Systems Analysis*, New York: Wiley, 1962.
13. J. S. Meditch, *Stochastic Optimal Linear Estimation and Control*, New York: McGraw-Hill, 1969.
14. J. G. Truxal, *Automatic Feedback Control System Synthesis*, New York: McGraw-Hill, 1955, pp. 472–477.
15. F. M. Long, *Biological Power Sources*, Ph.D. Dissertation, Ames, IA: Iowa State University, 1961.
16. M. Kayton and W. R. Fried (Eds.), *Avionics Navigation Systems*, New York: Wiley, 1969, p. 318.

5

The Discrete Kalman Filter

The end result of the Wiener solution of the optimal filter problem is a filter weighting function.* In effect this tells how the past values of the input should be weighted in order to determine the present value of the output, that is, the optimal estimate. Unfortunately, the Wiener solution does not lend itself very well to the corresponding discrete-data problem nor is it easily extended to more complicated time-variable, multiple-input/output problems. In 1960, R. E. Kalman provided an alternative way of formulating the least squares filtering problem using state-space methods (1). Engineers, especially in the field of navigation, were quick to see the Kalman technique as a practical solution to a number of problems that were previously considered intractable using Wiener methods. The two main features of the Kalman formulation and solution of the problem are (1) vector modeling of the random processes under consideration and (2) recursive processing of the noisy measurement (input) data. We now proceed to the details of the recursive solution of the discrete-data linear filter problem.

5.1
A SIMPLE RECURSIVE EXAMPLE

When working with practical problems involving discrete data, it is important that our methods be computationally feasible as well as mathematically correct. A simple example will illustrate this. Consider the problem of estimating the mean of some random constant based on a sequence of noisy measurements. Let us assume that our estimate is to be the sample mean and that we wish to refine our estimate with each new measurement as it becomes available. That is, think of processing the data on-line. Let the measurement sequence be denoted

---

* Even though the details of Wiener filter theory are not needed for this and subsequent chapters, the reader should at least browse through Chapter 4 in order to understand the problem statement and assumptions used in the Wiener solution.

as $z_1, z_2, \ldots, z_n$, where the subscript denotes the time at which the measurement is taken. One method of processing the data would be to store each measurement as it becomes available and then compute the sample mean in accordance with the following algorithm (in words):

1. First measurement z_1: Store z_1 and estimate the mean as

$$\hat{m}_1 = z_1$$

2. Second measurement z_2: Store z_2 along with z_1 and estimate the mean as

$$\hat{m}_2 = \frac{z_1 + z_2}{2}$$

3. Third measurement z_3: Store z_3 along with z_1 and z_2 and estimate the mean as

$$\hat{m}_3 = \frac{z_1 + z_2 + z_3}{3}$$

4. And so forth.

Clearly, this would yield the correct sequence of sample means as the experiment progresses. It should also be clear that the amount of memory needed to store the measurements keeps increasing with time, and also the number of arithmetic operations needed to form the estimate increases correspondingly. This would lead to obvious problems when the total amount of data is large. Thus, consider a simple variation in the computational procedure in which each new estimate is formed as a blend of the old estimate and the current measurement. To be specific, consider the following algorithm:

1. First measurement z_1: Compute the estimate as

$$\hat{m}_1 = z_1$$

   Store $\hat{m}_1$ and discard z_1.

2. Second measurement z_2: Compute the estimate as a weighted sum of the previous estimate $\hat{m}_1$ and the current measurement z_2:

$$\hat{m}_2 = \tfrac{1}{2}\hat{m}_1 + \tfrac{1}{2}z_2$$

   Store $\hat{m}_2$ and discard z_2 and $\hat{m}_1$.

3. Third measurement z_3: Compute estimate as a weighted sum of $\hat{m}_2$ and z_3:

$$\hat{m}_3 = \tfrac{2}{3}\hat{m}_2 + \tfrac{1}{3}z_3$$

   Store $\hat{m}_3$ and discard z_3 and $\hat{m}_2$.

4. And so forth. It should be obvious that at the nth stage the weighted sum is

$$\hat{m}_n = \left(\frac{n-1}{n}\right)\hat{m}_{n-1} + \left(\frac{1}{n}\right)z_n$$

Clearly, the above procedure yields the same identical sequence of estimates as before, but without the need to store all the previous measurements. We simply use the result of the previous step to help obtain the estimate at the current step of the process. In this way, the previous computational effort is used to good advantage and not wasted. The second algorithm can proceed on ad infinitum without a growing memory problem. Eventually, of course, as n becomes extremely large, a round-off problem might be encountered. However, this is to be expected with either of the two algorithms.

The second algorithm is a simple example of a *recursive* mode of operation. The key element in any recursive procedure is the use of the results of the previous step to aid in obtaining the desired result for the current step. This is one of the main features of Kalman filtering, and one that clearly distinguishes it from the weighting-function (Wiener) approach, which requires arithmetic operations on all the past data.

In order to apply the recursive philosophy to estimation of a random process, it is first necessary that both the process and the measurement noise be modeled in vector form. Thus we digress for a moment and look at the vector description of a random process. We will then return to the recursive filtering problem in Section 5.4.

5.2
VECTOR DESCRIPTION OF A
CONTINUOUS-TIME RANDOM PROCESS

In the material to follow, it is desirable to have the random process description written in vector form. We begin with a scalar process $y(t)$ for which we have a description as discussed in Chapter 2. We now wish to rewrite the description in the following format:

$$\dot{\mathbf{x}} = \mathbf{Fx} + \mathbf{Gu} \qquad (5.2.1)$$

$$\mathbf{y} = \mathbf{Bx} \qquad (5.2.2)$$

The vector process under consideration is denoted as $\mathbf{x}$ and is an $(n \times 1)$ column vector. $\mathbf{F}$, $\mathbf{G}$, and $\mathbf{B}$ are rectangular matrices whose elements may be time varying, and $\mathbf{u}$ is a $(p \times 1)$ column vector whose elements are unity white

Figure 5.1. Shaping filter ($m = n - 1$).

noise processes. This is called the *state model* of the process, and, in words, it simply says that we want to think of our process as the result of passing white noise through some system with linear dynamics.* The original process $y(t)$ is required to be a linear combination of the system state variables via Eq. (5.2.2) and, thus, the old process can be recovered from the new model with the appropriate additive combination of state variables.

In general, it is not possible to model all random processes in form of Eqs. (5.2.1) and (5.2.2). However, many of the processes encountered in physical applications may be so modeled; and, in particular, all processes with rational spectral density functions have state models with finite dimensionality. A procedure for forming the state model for such processes follows.

State Model for a Process with a Rational Spectral Function

Assume we know the spectral density function $S_y(s)$ for the process $y(t)$. We first factor the spectral function using the same spectral factorization methods used in Chapters 3 and 4. This leads to

$$S_y(s) = S_y^+(s) S_y^-(s) \qquad (5.2.3)$$

where we are assured that both the poles and zeros of S_y^+ lie in the left half-plane. Now, it is clear from Section 3.6 that $S_y^+(s)$ is just the shaping filter required to shape unity white noise into the process $y(t)$. The assumed form of the shaping-filter transfer function [i.e., $S_y^+(s)$] is shown in Fig. 5.1, and it is assumed that the order of the denominator is at least one greater than the numerator. This is necessary in order for the $y(t)$ process to have finite variance.

It is now convenient to decompose the block diagram of Fig. 5.1 into the equivalent form shown in Fig. 5.2. With the state variables chosen as the usual phase variables, we are assured that each will have bounded variance. The final

---

*Only an elementary knowledge of state-space methods is needed for our discussion of Kalman filtering. The treatment of state-space methods given in most introductory books on linear control theory (e.g., Refs. 2 and 3) is quite adequate for our purposes here.

Figure 5.2. Block diagram showing state variables ($m = n - 1$).

state model for the process $y(t)$ is then

$$
\begin{bmatrix} \dot{x}_1 \\ \dot{x}_2 \\ \cdot \\ \cdot \\ \cdot \\ \dot{x}_n \end{bmatrix} = \begin{bmatrix} 0 & 1 & 0 & \cdots \\ 0 & 0 & 1 & \cdots \\ & & & \\ & & & \\ -a_0 & -a_1 & \cdots & -a_{n-1} \end{bmatrix} \begin{bmatrix} x_1 \\ x_2 \\ \cdot \\ \cdot \\ \cdot \\ x_n \end{bmatrix} + \begin{bmatrix} 0 \\ 0 \\ \cdot \\ \cdot \\ \cdot \\ 1 \end{bmatrix} u(t) \quad (5.2.4)
$$

$$
y = [b_0 b_1 \cdots b_{n-1}] \begin{bmatrix} x_1 \\ x_2 \\ \cdot \\ \cdot \\ \cdot \\ x_n \end{bmatrix} \quad (5.2.5)
$$

The special form of the state equations given by Eqs. (5.2.4) and (5.2.5) is known as the controllable canonical form in control theory. Note that the process $y(t)$ is not one of the system state variables directly. Rather, it is reconstructed as a linear combination of state variables via Eq. (5.2.5). Two examples illustrate this modeling procedure.

EXAMPLE 5.1

Consider a random process with a spectral function given by

$$
S_y(j\omega) = \frac{16}{\omega^4 - 4\omega^2 + 16} \quad (5.2.6)
$$

A sketch of this spectral function is shown in Fig. 5.3 and it shows a mild resonance phenomenon occurring at $\sqrt{2}$ rad/sec. We first need to convert

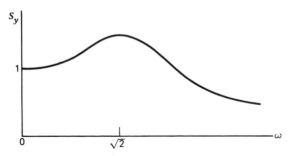

Figure 5.3. Spectral function for Example 5.1.

$S_y(j\omega)$ to the s domain by replacing $j\omega$ with s. The result is

$$S_y(s) = \frac{16}{s^4 + 4s^2 + 16} \qquad (5.2.7)$$

Since the poles of $S_y(s)$ are at $s = -1 \pm j\sqrt{3}$ and $s = 1 \pm j\sqrt{3}$, $S_y(s)$ can be factored in the form

$$S_y(s) = S_y^+(s) \cdot S_y^-(s) = \frac{4}{s^2 + 2s + 4} \cdot \frac{4}{(-s)^2 + 2(-s) + 4} \qquad (5.2.8)$$

A block diagram showing the state and output relationships can now be formed as shown in Fig. 5.4.

The complete state model is then

$$\begin{bmatrix} \dot{x}_1 \\ \dot{x}_2 \end{bmatrix} = \begin{bmatrix} 0 & 1 \\ -4 & -2 \end{bmatrix}\begin{bmatrix} x_1 \\ x_2 \end{bmatrix} + \begin{bmatrix} 0 \\ 1 \end{bmatrix}u(t) \qquad (5.2.9)$$

$$y = \begin{bmatrix} 4 & 0 \end{bmatrix}\begin{bmatrix} x_1 \\ x_2 \end{bmatrix} \qquad (5.2.10)$$

∎

Figure 5.4. Block diagram for state model for Example 5.1.

EXAMPLE 5.2

The Gauss–Markov process model is frequently used in engineering applications. This is partly because of its simplicity and partly because the analyst often lacks the detailed knowledge of the process under consideration that would be needed to devise a more complicated model. The Gauss–Markov process has an autocorrelation function given by

$$R_y(\tau) = \sigma^2 e^{-\beta|\tau|}$$

The corresponding spectral function is

$$S_y(s) = \frac{2\sigma^2\beta}{-s^2 + \beta^2} = \frac{\sqrt{2\sigma^2\beta}}{s + \beta} \; \frac{\sqrt{2\sigma^2\beta}}{-s + \beta}$$

Clearly, the shaping filter is

$$S_y^+(s) = \frac{\sqrt{2\sigma^2\beta}}{s + \beta}$$

and the corresponding differential equation for y is

$$\dot{y} = -\beta y + \sqrt{2\sigma^2\beta}\, u(t) \tag{5.2.11}$$

where $u(t)$ is unity white noise. Equation (5.2.11) is first order, so this is the state model for the process. ∎

Nonstationary and Deterministic Processes

The state model of a random process is reasonably general in form and will accommodate a wide variety of situations. All that is required is that the process under consideration be related to white noise via a linear differential equation. A few examples of processes that do not have rational spectral functions will illustrate this further.

　　1. *Wiener process (Brownian motion):* The Wiener process is defined as the integral of Gaussian white noise with zero initial condition. Thus, the

appropriate state model is first order and is

$$\dot{y} = ku(t) \qquad y(0) = 0 \tag{5.2.12}$$

where $u(t)$ is unity Gaussian white noise and k is a scale factor.

2. *Random bias:* A random constant satisfies the differential equation

$$\dot{y} = 0 \qquad y(0) = a_0 \tag{5.2.13}$$

The initial condition a_0 is a random variable whose distribution is presumed to be known. (See Problems 5.5 and 5.6 for more on modeling random bias and Wiener processes.)

3. *Random ramp:* A random ramp process with random initial value and slope may be written as

$$y(t) = a_0 + a_1 t \tag{5.2.14}$$

where a_0 and a_1 are random variables. The differential equation corresponding to Eq. (5.2.14) is

$$\ddot{y} = 0 \qquad y(0) = a_0 \qquad \dot{y}(0) = a_1 \tag{5.2.15}$$

This is a second-order differential equation, so the state vector for the process y must be a two-tuple. Using phase variables in the vector model leads to

$$\begin{bmatrix} \dot{x}_1 \\ \dot{x}_2 \end{bmatrix} = \begin{bmatrix} 0 & 1 \\ 0 & 0 \end{bmatrix} \begin{bmatrix} x_1 \\ x_2 \end{bmatrix} + \begin{bmatrix} 0 \\ 0 \end{bmatrix} u, \quad \begin{bmatrix} x_1(0) \\ x_2(0) \end{bmatrix} = \begin{bmatrix} a_0 \\ a_1 \end{bmatrix} \tag{5.2.16}$$

$$y = \begin{bmatrix} 1 & 0 \end{bmatrix} \begin{bmatrix} x_1 \\ x_2 \end{bmatrix} \tag{5.2.17}$$

4. *Harmonic motion with random amplitude and phase:* A process having harmonic motion with known frequency ω_0 rad/sec satisfies the differential equation

$$\ddot{y} + \omega_0^2 y = 0 \tag{5.2.18}$$

The solution of this equation is

$$y(t) = y(0) \cos \omega_0 t + \frac{\dot{y}(0)}{\omega_0} \sin \omega_0 t \tag{5.2.19}$$

To specify "random amplitude and phase" is equivalent to saying "random sine and cosine components." Thus, choosing phase variables as the state variables in the vector model, we have

$$\begin{bmatrix} \dot{x}_1 \\ \dot{x}_2 \end{bmatrix} = \begin{bmatrix} 0 & 1 \\ -\omega_0^2 & 0 \end{bmatrix} \begin{bmatrix} x_1 \\ x_2 \end{bmatrix} + \begin{bmatrix} 0 \\ 0 \end{bmatrix} u, \quad \begin{bmatrix} x_1(0) \\ x_2(0) \end{bmatrix} = \begin{bmatrix} A \\ B \end{bmatrix} \quad (5.2.20)$$

$$y = \begin{bmatrix} 1 & 0 \end{bmatrix} \begin{bmatrix} x_1 \\ x_2 \end{bmatrix} \quad (5.2.21)$$

The random variables A and B are assigned probability density functions in accordance with the given distributions of the sine and cosine components of y. A common assumption is to let A and B be independent zero-mean normal random variables with variances σ^2 and $\omega_0^2\sigma^2$. The resultant harmonic motion then has Rayleigh amplitude and uniform phase distributions.

The preceding examples are just a few of the many processes that can be modeled in vector form. Stationarity is not necessary. The only requirement is that the process must somehow or other be related to white noise through a linear differential equation. It is especially important to note that one does not choose the number of state variables in the model at will; rather, the number of elements in the state vector is fixed by the order of the differential equations relating the various processes in the model to the white-noise driving functions. This is a basic fact of life and must be adhered to if the state model is to faithfully represent the situation at hand. Engineering literature abounds with examples of sloppy modeling and the reader is cautioned accordingly.

Also, remember that the vector model of a random process is not unique. You can always perform any nonsingular linear transformation on a set of state variables and obtain another valid set. Of course, since the **F**, **G**, and **B** matrices [see Eqs. (5.2.1) and (5.2.2)] for the transformed set will be different from those of the original set, the new model will look different. It will, however, be a perfectly proper model if the transformation is done correctly. Problems 5.8, 5.9, and 5.10 elaborate on this further.

5.3
DISCRETE-TIME MODEL

Discrete-time processes may arise in either of two ways. First, there is the situation where a sequence of events takes place naturally in discrete steps. This might be the result of a sequence of chance experiments such as the discrete random-walk problems of statistics. Recall in this problem that we imagine the walker taking a number of steps at random, either forward or backward. The length of each step may be either a fixed or a random variable. In either case, the random variable of interest is the distance from the origin after taking n

steps. In this problem, there is no such thing as fractional steps. The time variable moves in discrete jumps!

Discrete-time processes may also arise from sampling a continuous process at discrete times. The sampling may be intentional and may be under the control of the designer, as is the case when analog data are converted to digital form. Or sometimes the sampling may be unintentional and forced on us by a measurement constraint that allows observation of the process only at discrete points in time. For example, the TRANSIT satellite navigation system provides the user with only a single position fix with each satellite pass (4). Thus the user is allowed to observe his or her position only at discrete points in time. Furthermore, in this case the observation times are not equally spaced because the various satellite orbits do not have perfect symmetry.

Irrespective of how the discretization arises in the physical problem, we wish to fit all situations into the following format:

$$\mathbf{x}_{k+1} = \boldsymbol{\phi}_k \mathbf{x}_k + \mathbf{w}_k \tag{5.3.1}$$

$$\mathbf{y}_k = \mathbf{B}_k \mathbf{x}_k \tag{5.3.2}$$

where

$\mathbf{x}_k$ = vector state of the process at time t_k, i.e., $\mathbf{x}_k = \mathbf{x}(t_k)$
$\boldsymbol{\phi}_k$ = matrix that relates $\mathbf{x}_k$ to $\mathbf{x}_{k+1}$ in the absence of a forcing function (In the sampled version of a continuous process this is the state transition matrix.)
$\mathbf{w}_k$ = vector whose elements are white sequences
$\mathbf{B}_k$ = linear connection matrix between output $\mathbf{y}_k$ and state $\mathbf{x}_k$

Recall from Chapter 3 that a white sequence is a sequence of zero-mean random variables that are uncorrelated timewise. Note, however, that the elements of $\mathbf{w}_k$ may have a mutual nontrivial correlation at any point in time t_k. The covariance matrix associated with $\mathbf{w}_k$ is assumed to be known, and it will be denoted as $\mathbf{Q}_k$. Thus we have

$$E\left[\mathbf{w}_k \mathbf{w}_i^T\right] = \begin{cases} \mathbf{Q}_k, & i = k \\ 0, & i \neq k \end{cases} \tag{5.3.3}$$

where "super T" denotes transpose. (We will reserve "prime" for other uses.)

Sampled Continuous-Time Systems

As mentioned previously, the discrete model of Eqs. (5.3.1) and (5.3.2) need not come from a sampled continuous situation, but this is so common in applied work that it warrants amplification. Let us say we begin with a continuous process described by

$$\dot{\mathbf{x}} = \mathbf{F}\mathbf{x} + \mathbf{G}\mathbf{u} \qquad (5.3.4)$$

where $\mathbf{u}$ is a vector forcing function whose elements are white noise. Let us consider samples of this process at discrete times $t_0, t_1, \ldots, t_k, \ldots$. State-space methods may be used to obtain the difference equation relating the samples of $\mathbf{x}$. Specifically, the solution of Eq. (5.3.4) at time t_{k+1} may be written as

$$\mathbf{x}(t_{k+1}) = \boldsymbol{\phi}(t_{k+1}, t_k)\mathbf{x}(t_k) + \int_{t_k}^{t_{k+1}} \boldsymbol{\phi}(t_{k+1}, \tau)\mathbf{G}(\tau)\mathbf{u}(\tau)\, d\tau \quad (5.3.5)$$

Or, in our abbreviated notation,

$$\mathbf{x}_{k+1} = \boldsymbol{\phi}_k \mathbf{x}_k + \mathbf{w}_k$$

Clearly, $\boldsymbol{\phi}_k$ is the state transition matrix for the step from t_k to t_{k+1}, and $\mathbf{w}_k$ is the driven response at t_{k+1} due to the presence of the white noise input during the (t_k, t_{k+1}) interval. Note that the white-noise input requirement in the continuous model automatically assures that $\mathbf{w}_k$ will be a white sequence in the discrete model.

Analytical methods for finding the state transition matrix are well known, and these may be used in systems with low dimensionality. However, evaluation of the $\mathbf{Q}_k$ matrix that describes $\mathbf{w}_k$ may not be so obvious. Formally, we can write $\mathbf{Q}_k$ in integral form as

$$\mathbf{Q}_k = E\left[\mathbf{w}_k \mathbf{w}_k^T\right]$$

$$= E\left\{\left[\int_{t_k}^{t_{k+1}} \boldsymbol{\phi}(t_{k+1}, \xi)\mathbf{G}(\xi)\mathbf{u}(\xi)\, d\xi\right]\left[\int_{t_k}^{t_{k+1}} \boldsymbol{\phi}(t_{k+1}, \eta)\mathbf{G}(\eta)\mathbf{u}(\eta)\, d\eta\right]^T\right\}$$

$$= \int_{t_k}^{t_{k+1}} \int_{t_k}^{t_{k+1}} \boldsymbol{\phi}(t_{k+1}, \xi)\mathbf{G}(\xi)E\left[\mathbf{u}(\xi)\mathbf{u}^T(\eta)\right]\mathbf{G}^T(\eta)\boldsymbol{\phi}^T(t_{k+1}, \eta)\, d\xi\, d\eta$$

$$(5.3.6)$$

The matrix $E[\mathbf{u}(\xi)\mathbf{u}^T(\eta)]$ is a matrix of Dirac delta functions that, presumably, is known from the continuous model. Thus, in principle, $\mathbf{Q}_k$ may be evaluated from Eq. (5.3.6). This is not a trivial task, though, even for low-order systems. If the continuous system giving rise to the discrete situation has constant parameters and if the various white noise inputs have zero crosscorrelation, some simplification is possible and the transfer function methods of Chapter 3 may be applied. This is best illustrated with an example rather than in general terms.

EXAMPLE 5.3

The integrated Gauss–Markov process shown in Fig. 5.5 is frequently encountered in engineering applications. The continuous model in this case is

$$\begin{bmatrix} \dot{x}_1 \\ \dot{x}_2 \end{bmatrix} = \begin{bmatrix} 0 & 1 \\ 0 & -\beta \end{bmatrix} \begin{bmatrix} x_1 \\ x_2 \end{bmatrix} + \begin{bmatrix} 0 \\ \sqrt{2\sigma^2\beta} \end{bmatrix} u(t) \qquad (5.3.7)$$

$$y = \begin{bmatrix} 1 & 0 \end{bmatrix} \begin{bmatrix} x_1 \\ x_2 \end{bmatrix} \qquad (5.3.8)$$

Let us say the sampling interval is Δt and we wish to find the corresponding discrete model. This amounts to the determination of ϕ_k, $\mathbf{Q}_k$, and $\mathbf{B}_k$. The transition matrix is easily determined as

$$\phi_k = \left[\mathscr{L}^{-1}\left[(s\mathbf{I} - \mathbf{F})^{-1} \right] \right]_{t=\Delta t}$$

$$= \mathscr{L}^{-1}\begin{bmatrix} s & -1 \\ 0 & s+\beta \end{bmatrix}^{-1} = \mathscr{L}^{-1}\begin{bmatrix} \dfrac{1}{s} & \dfrac{1}{s(s+\beta)} \\ 0 & \dfrac{1}{s+\beta} \end{bmatrix}$$

$$= \begin{bmatrix} 1 & \dfrac{1}{\beta}(1 - e^{-\beta\Delta t}) \\ 0 & e^{-\beta\Delta t} \end{bmatrix} \qquad (5.3.9)$$

Next, rather than using Eq. (5.3.6) directly to determine $\mathbf{Q}_k$, we use the transfer function approach. From the block diagram of Fig. 5.5 we observe the following transfer functions:

$$G(u \text{ to } x_1) = G_1 = \frac{\sqrt{2\sigma^2\beta}}{s(s+\beta)} \qquad (5.3.10)$$

$$G(u \text{ to } x_2) = G_2 = \frac{\sqrt{2\sigma^2\beta}}{s+\beta} \qquad (5.3.11)$$

Figure 5.5. Integrated Gauss–Markov process.

The corresponding weighting functions are

$$g_1(t) = \sqrt{\frac{2\sigma^2}{\beta}} \, (1 - e^{-\beta t}) \qquad (5.3.12)$$

$$g_2(t) = \sqrt{2\sigma^2 \beta} \, e^{-\beta t} \qquad (5.3.13)$$

We can now use the methods of Chapter 3 to find the needed mean-square responses

$$E[x_1 x_1] = \int_0^{\Delta t} \int_0^{\Delta t} g_1(\xi) g_1(\eta) E[u(\xi) u(\eta)] \, d\xi \, d\eta$$

$$= \int_0^{\Delta t} \int_0^{\Delta t} \frac{2\sigma^2}{\beta} (1 - e^{-\beta \xi})(1 - e^{-\beta \eta}) \delta(\xi - \eta) \, d\xi \, d\eta$$

$$= \frac{2\sigma^2}{\beta} \left[\Delta t - \frac{2}{\beta}(1 - e^{-\beta \Delta t}) + \frac{1}{2\beta}(1 - e^{-2\beta \Delta t}) \right] \qquad (5.3.14)$$

$$E[x_1 x_2] = \int_0^{\Delta t} \int_0^{\Delta t} g_1(\xi) g_2(\eta) E[u(\xi) u(\eta)] \, d\xi \, d\eta$$

$$= \int_0^{\Delta t} \int_0^{\Delta t} 2\sigma^2 e^{-\beta \xi}(1 - e^{-\beta \eta}) \delta(\xi - \eta) \, d\xi \, d\eta$$

$$= 2\sigma^2 \left[\frac{1}{\beta}(1 - e^{-\beta \Delta t}) - \frac{1}{2\beta}(1 - e^{-2\beta \Delta t}) \right] \qquad (5.3.15)$$

$$E[x_2 x_2] = \int_0^{\Delta t} \int_0^{\Delta t} g_2(\xi) g_2(\eta) E[u(\xi) u(\eta)] \, d\xi \, d\eta$$

$$= \int_0^{\Delta t} \int_0^{\Delta t} 2\sigma^2 \beta e^{-\beta \xi} e^{-\beta \eta} \delta(\xi - \eta) \, d\xi \, d\eta$$

$$= \sigma^2(1 - e^{-2\beta \Delta t}) \qquad (5.3.16)$$

Thus the $\mathbf{Q}_k$ matrix is

$$\mathbf{Q}_k = \begin{bmatrix} E[x_1 x_1] & E[x_1 x_2] \\ E[x_1 x_2] & E[x_2 x_2] \end{bmatrix} = \begin{bmatrix} \text{Eq. (5.3.14)} & \text{Eq. (5.3.15)} \\ \text{Eq. (5.3.15)} & \text{Eq. (5.3.16)} \end{bmatrix} \qquad (5.3.17)$$

The $\mathbf{B}_k$ matrix is the same for both the continuous and discrete models and is

$$\mathbf{B}_k = \begin{bmatrix} 1 & 0 \end{bmatrix} \tag{5.3.18}$$

The discrete model is now complete with the specification of $\boldsymbol{\phi}_k$, $\mathbf{Q}_k$, and $\mathbf{B}_k$ as given by Eqs. (5.3.9), (5.3.17), and (5.3.18). Note that the k subscript could have been dropped in this example because the sampling interval is constant. ■

Numerical Evaluation of $\boldsymbol{\phi}_k$ and $\mathbf{Q}_k$

Analytical methods for finding $\boldsymbol{\phi}_k$ and $\mathbf{Q}_k$ work quite well for constant parameter systems with just a few elements in the state vector. However, the dimensionality does not have to be very large before it becomes virtually impossible to work out explicit expressions for $\boldsymbol{\phi}_k$ and $\mathbf{Q}_k$. This is especially true if the system **F** matrix contains time-varying terms. Thus, we often need to resort to numerical methods.

The state transition matrix (STM) tells how the dynamical system naturally evolves from one state to another in the absence of a driving function. Thus the elements of the STM are obtained as the solutions of certain initial-condition problems. For example, let the system input be zero and let the initial state at t_k be

$$\mathbf{x}(t_k) = \begin{bmatrix} 1 \\ 0 \\ 0 \\ \cdot \\ \cdot \\ \cdot \\ 0 \end{bmatrix} \tag{5.3.19}$$

The solution for $\mathbf{x}(t)$ for $t \geq t_k$ is then

$$\mathbf{x}(t) = \begin{bmatrix} \phi_{11}(t,t_k) & \phi_{12}(t,t_k) & \cdots \\ \phi_{21}(t,t_k) & & \cdots \\ & \cdots & \end{bmatrix} \begin{bmatrix} 1 \\ 0 \\ 0 \\ \cdot \\ \cdot \end{bmatrix} = \begin{bmatrix} \phi_{11}(t,t_k) \\ \phi_{21}(t,t_k) \\ \cdot \\ \cdot \end{bmatrix} \tag{5.3.20}$$

Thus the first column of the STM is obtained by solving for the system response with the initial value of **x** set at $\begin{bmatrix} 1 & 0 & 0 & \cdots \end{bmatrix}^T$. Similarly, the second column of $\boldsymbol{\phi}$ is obtained by letting the initial condition be $\begin{bmatrix} 0 & 1 & 0 & \cdots \end{bmatrix}^T$, and so forth. Thus, in effect, we must solve a family of initial condition problems in order to obtain all the elements of the state transition matrix. This is laborious, but fortunately numerical methods for solving the initial-condition problem of ordinary differential equations are readily available. Whole books have been written on the subject (5); therefore it will suffice to say here that the STM can usually be evaluated in a routine manner.

One special case of evaluating ϕ_k is worthy of mention at this point. If the dynamical system has fixed parameters (i.e., **F** is a constant), the STM may be written as an exponential series.

$$\phi_k = e^{\mathbf{F}\Delta t} = \mathbf{I} + (\mathbf{F}\Delta t) + \frac{(\mathbf{F}\Delta t)^2}{2!} + \cdots \qquad (5.3.21)$$

where Δt is the step size. This is probably the easiest way of evaluating the STM (from the analyst's viewpoint), because Eq. (5.3.21) is easily programmed and involves only matrix-add and matrix-multiply subroutines. Equation (5.3.21) is valid for any step size Δt. However, a large number of terms in the series may be needed if Δt is large, and this may lead to computational problems. Therefore, simply look at each individual case on its own merits, and then decide as to the best method of evaluating the state transition matrix.

The numerical procedure for evaluating $\mathbf{Q}_k$ is straightforward, but laborious. In general this calls for some sort of numerical procedure for evaluating the double integral given by Eq. (5.3.6). This can be a difficult chore if the system has time-varying parameters and its dimensionality is high. An alternative approach that may be applied in some cases (and one that is easily programmed) is to subdivide the Δt interval into smaller δt steps such that only first-order terms in δt need be retained in the $\mathbf{Q}_k$ expression. The contributions of each of the subintervals may then be projected ahead via a recursive equation to yield the total $\mathbf{Q}_k$ for the whole Δt interval. An example will illustrate this procedure. (Also see Problem 7.13* for another method of finding $\mathbf{Q}_k$.)

EXAMPLE 5.4

Let us reconsider the integrated Gauss–Markov process used in Example 5.3. Suppose $\beta = 10$ rad/sec and $\Delta t = 0.1$ sec. The product $\beta\Delta t$ (which is critical in this case) is certainly not small, and higher-order terms in Δt would have to be included in evaluating $\mathbf{Q}_k$ terms such as $e^{-\beta\Delta t}$ and $e^{-2\beta\Delta t}$. Hence, in this case, we divide Δt into 1000 subintervals of equal length. The $\beta\delta t$ of each of these is .001, and it is now reasonable to neglect higher-order terms in $\beta\delta t$.

We first need to work out the "Q" for the first subinterval. Referring to Eq. (5.3.12) and Eq. (5.3.13), we note that the weighting functions g_1 and g_2 may be approximated as being zero and $\sqrt{2\sigma^2\beta}$ over the δt interval. This leads to

$$E[x_1x_1] \approx 0$$

$$E[x_1x_2] \approx 0$$

$$E[x_2x_2] \approx (2\sigma^2\beta)\,\delta t \qquad (5.3.22)$$

Figure 5.6. Subinterval notation for Example 5.4.

The **Q** matrix for each subinterval is then*

$$\mathbf{Q}_{\text{sub}} = \begin{bmatrix} 0 & 0 \\ 0 & (2\sigma^2\beta)\,\delta t \end{bmatrix} \tag{5.3.23}$$

Ultimately, we want $\mathbf{Q}_k$ for the whole interval Δt. We obtain this as the result of taking 1000 steps as follows (refer to Fig. 5.6 for notation):

Step 1 (Determination of $E[\mathbf{x}_b\mathbf{x}_b^T]$).
We have just determined that the **Q** for a single subinterval is given by Eq. (5.3.23). Thus

$$E\left[\mathbf{x}_b\mathbf{x}_b^T\right] = \mathbf{Q}_{\text{sub}} \tag{5.3.24}$$

Step 2 (Determination of $E[\mathbf{x}_c\mathbf{x}_c^T]$).
We first note that

$$\mathbf{x}_c = \boldsymbol{\phi}_{\text{sub}}\mathbf{x}_b + (\text{response during second interval}) \tag{5.3.25}$$

where $\boldsymbol{\phi}_{\text{sub}}$ is the state transition matrix for the subinterval. Here we can approximate $\boldsymbol{\phi}_{\text{sub}}$ with just the first two terms of the series expansion given by Eq. (5.3.21) because δt is small. Therefore,

$$E\left[\mathbf{x}_c\mathbf{x}_c^T\right] = \boldsymbol{\phi}_{\text{sub}}E\left[\mathbf{x}_b\mathbf{x}_b^T\right]\boldsymbol{\phi}_{\text{sub}}^T + \mathbf{Q}_{\text{sub}} \tag{5.3.26}$$

---

*The expression for $\mathbf{Q}_{\text{sub}}$ can also be obtained directly from Eq. (5.3.6) by noting that the range of integration is small and retaining only first-order terms in δt. Toward this end we approximate $\boldsymbol{\phi}$ as the identity matrix and note that **G** is given in Eq. (5.3.7). Therefore,

$$\mathbf{Q}_{\text{sub}} \approx \begin{bmatrix} 0 \\ \sqrt{2\sigma^2\beta} \end{bmatrix} \int_t^{t+\delta t}\int_t^{t+\delta t} E[u(\xi)u(\eta)]\,d\xi\,d\eta \begin{bmatrix} 0 & \sqrt{2\sigma^2\beta} \end{bmatrix}$$

Now note that u is unity white noise, so the first integral is 1 and the second integration yields δt. Finally, $\mathbf{Q}_{\text{sub}}$ is obtained by performing the indicated matrix multiply operation, and the result is the same as that given by Eq. (5.3.23).

Step 3 (Determination of $E[\mathbf{x}_d\mathbf{x}_d^T]$).
Similarly, $E[\mathbf{x}_d\mathbf{x}_d^T]$ can be obtained using the result of the previous step. Thus

$$E[\mathbf{x}_d\mathbf{x}_d^T] = \boldsymbol{\phi}_{\text{sub}}E[\mathbf{x}_c\mathbf{x}_c^T]\boldsymbol{\phi}_{\text{sub}}^T + \mathbf{Q}_{\text{sub}} \qquad (5.3.27)$$

The recursive process is now clear. We simply take the previous covariance matrix, pre- and postmultiply it by $\boldsymbol{\phi}_{\text{sub}}$ and $\boldsymbol{\phi}_{\text{sub}}^T$, and then add $\mathbf{Q}_{\text{sub}}$. After repeating this 1000 times, we have the $\mathbf{Q}_k$ for the whole Δt interval.

If we should be in doubt about the first-order approximation in δt, we can always halve the subinterval and then carry the process through 2000 steps instead of 1000. This should lead to the same result as before. If not, we can keep halving the subinterval (and perhaps use double or triple precision) until results are consistent. This may seem to be a bit crude and wasteful of computer time. Without apology, it is. It is quite effective, though, and the programming is trivial for the analyst. In this day and age, the trade-off of computer time for analyst's time is usually a good one! ■

State Model from the ARMA Model

It was mentioned previously that sometimes processes are intrinsically discrete and have nothing to do with a continuous evolution of time. Such processes are often described in terms of an ARMA model, which was touched on briefly in Section 3.9 (6, 7, 8). We will now continue that discussion and show how the ARMA model can be converted to state-space form. We begin by repeating the ARMA model equation given in Section 3.9:

$$y(k + n) + \alpha_{n-1}y(k + n - 1) + \alpha_{n-2}y(k + n - 2) + \cdots + \alpha_0 y(k)$$

$$= \beta_m w(k + m) + \beta_{m-1}w(k + m - 1) + \cdots + \beta_0 w(k),$$

$$m = n - 1 \quad \text{and} \quad k = 0, 1, 2, \ldots \qquad (5.3.28)$$

Note that the model is single-sided in k and the order of the MA part is at least one less than the AR part. The reasons for these restrictions will be apparent shortly.

Conversion of the ARMA model to the controllable canonical form is effected by following a procedure analogous to that used in the continuous differential equation case. First, we define an intermediate variable $r(k)$, which is the solution of

$$r(k + n) + \alpha_{n-1}r(k + n - 1)$$

$$+ \alpha_{n-2}r(k + n - 2) + \cdots + \alpha_0 r(k) = w(k) \qquad (5.3.29)$$

Next, we define state variables that are analogous to the phase variables in the continuous case:

$$x_1(k) = r(k)$$

$$x_2(k) = r(k + 1)$$

$$\vdots$$

$$x_n(k) = r(k + n - 1) \tag{5.3.30}$$

Using state variables as defined by Eq. (5.3.30) and the difference equation, Eq. (5.3.29), leads to the matrix equation

$$
\begin{bmatrix} x_1(k+1) \\ x_2(k+1) \\ \vdots \\ x_n(k+1) \end{bmatrix} =
\begin{bmatrix} 0 & 1 & 0 & 0 & \cdots \\ 0 & 0 & 1 & 0 & \cdots \\ \vdots & & & & \\ -\alpha_0 & -\alpha_1 & -\alpha_2 & \cdots & -\alpha_{n-1} \end{bmatrix}
\begin{bmatrix} x_1(k) \\ x_2(k) \\ \vdots \\ x_n(k) \end{bmatrix} +
\begin{bmatrix} 0 \\ 0 \\ \vdots \\ 1 \end{bmatrix} w(k)
$$

$$\tag{5.3.31}$$

We now need to reconstruct the original $y(k)$ variable as a linear combination of the elements of the state vector. This can be done by considering a superposition of equations of the type given by Eq. (5.3.29) with the index k advanced appropriately. However, it is a bit easier to do this with a block diagram by using z transforms as shown in Fig. 5.7. The analogy between the block diagrams of Figs. 5.2 and 5.7 should be apparent. It should be clear now that the output equation that takes us from the state space back to $y(k)$ is

$$
y(k) = \begin{bmatrix} \beta_0 & \beta_1 & \beta_2 & \cdots & \beta_n \end{bmatrix}
\begin{bmatrix} x_1(k) \\ x_2(k) \\ \vdots \\ x_n(k) \end{bmatrix} \tag{5.3.32}
$$

We will illustrate the procedure with an example.

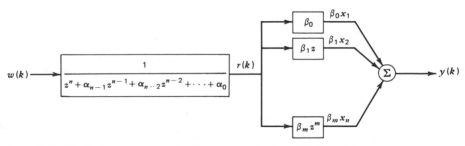

Figure 5.7. Block diagram for constructing state model from ARMA model.

EXAMPLE 5.5

We will use the same ARMA model here that was introduced in Example 3.11 in Chapter 3. The difference equation is

$$y(k+2) - y(k+1) + .5y(k) = .5w(k+1) + .25w(k), \qquad k = 0, 1, 2, \ldots$$
$$(5.3.33)$$

The block diagram that leads to the controllable canonical form is shown in Fig. 5.8. The state model is then

$$\begin{bmatrix} x_1(k+1) \\ x_2(k+1) \end{bmatrix} = \begin{bmatrix} 0 & 1 \\ -.5 & 1 \end{bmatrix} \begin{bmatrix} x_1(k) \\ x_2(k) \end{bmatrix} + \begin{bmatrix} 0 \\ 1 \end{bmatrix} w(k) \qquad (5.3.34)$$

$$y(k) = \begin{bmatrix} .25 & .5 \end{bmatrix} \begin{bmatrix} x_1(k) \\ x_2(k) \end{bmatrix} \qquad (5.3.35)$$

Eqs. (5.3.34) and (5.3.35) now compose the model.

We can use this same example to demonstrate why we need to put certain restrictions on our ARMA model. First, we restrict k to integer increments beginning at $k = 0$, that is, $k = 0, 1, 2, \ldots$. This is done simply because in the filter theory that follows in Section 5.4 we wish to consider measurement sequences that begin at some finite real time, and it is a matter of convenience to let point of initiation be $k = 0$ (equivalent to $t = 0$ in the continuous case). We have no problem with letting k (or t) proceed on indefinitely in a positive sense, but time prior to starting the process will be of no consequence in the material that follows. The second restriction put on our ARMA model has to do with the order of the right side of the ARMA equation. Suppose we were to let the right side be of the same order as the left side. This would lead to a transfer function in the z-domain that has the same order polynomial in the numerator as in the denominator. For example, we might then have something like

$$\text{Transfer function} = \frac{z^2 + 3z + 1}{z^2 - z + .5}$$

Figure 5.8. Block diagram for state model.

It can be seen by long division that there will be a direct feedthrough of the input $w(k)$ into the output. We do not wish to allow this in the Kalman filter model that follows in Section 5.4 (at least in its most elementary form), so we will require the order of the MA part to be at least one less than the AR part. There is also a problem of causality when the order of the right side is equal to or greater than the left side, and this will be elaborated on further in Section 5.4.

To summarize, we put the restrictions on our ARMA model because we only wish to consider processes that begin at a finite time, and ones that are causal with no direct feedthrough of the input into the output.

To demonstrate further that the model specified by Eqs. (5.3.34) and (5.3.35) is, in fact, causal and has no direct feedthrough, we will go through three steps of a simple simulation. Suppose that the initial conditions on the two state variables are

$$x_1(0) = x_2(0) = 0$$

These can be specified arbitrarily, because the system is second order. Now suppose that the first three values of the input sequence are $w(0) = 1.0$, $w(1) = -.5$, and $w(2) = .2$ (by chance, of course). We begin at $k = 0$.

At k = 0:

$$\begin{bmatrix} x_1(0) \\ x_2(0) \end{bmatrix} = \begin{bmatrix} 0 \\ 0 \end{bmatrix}, \quad \text{(This is the given initial condition.)}$$

$$y(0) = \begin{bmatrix} .25 & .5 \end{bmatrix} \begin{bmatrix} 0 \\ 0 \end{bmatrix} = 0$$

Note that the state and output at $k = 0$ do not depend on $w(0)$.

At k = 1:

$$\begin{bmatrix} x_1(1) \\ x_2(1) \end{bmatrix} = \begin{bmatrix} 0 & 1 \\ -.5 & 1 \end{bmatrix} \begin{bmatrix} 0 \\ 0 \end{bmatrix} + \begin{bmatrix} 0 \\ 1 \end{bmatrix} (1.0) = \begin{bmatrix} 0 \\ 1 \end{bmatrix}$$

$$y(1) = \begin{bmatrix} .25 & .5 \end{bmatrix} \begin{bmatrix} 0 \\ 1 \end{bmatrix} = .5$$

Note that the state and output at $k = 1$ depend on $w(0)$ but not on $w(1)$.

At k = 2:

$$\begin{bmatrix} x_1(2) \\ x_2(2) \end{bmatrix} = \begin{bmatrix} 0 & 1 \\ -.5 & 1 \end{bmatrix} \begin{bmatrix} 0 \\ 1 \end{bmatrix} + \begin{bmatrix} 0 \\ 1 \end{bmatrix} (-.5) = \begin{bmatrix} 1 \\ .5 \end{bmatrix}$$

$$y(2) = \begin{bmatrix} .25 & .5 \end{bmatrix} \begin{bmatrix} 1 \\ .5 \end{bmatrix} = .5$$

As before, note that the state and output at $k = 2$ do not depend on the "current" value of the input, $w(2)$. This is largely a matter of notation, but it is important in the material that follows in Section 5.4. Conceptually, we can think of the input white sequence unfolding with time one step back of the state and output sequences. That is, when $y(1)$ unfolds, only $w(0)$ is known; when $y(2)$ unfolds, only $w(0)$ and $w(1)$ are known; when $y(3)$ unfolds only $w(0)$, $w(1)$, and $w(2)$ are known; and so forth. Thinking in terms of real time, at time k nothing is known about $w(k)$ other than the fact that it has zero mean and a known variance. It does not unfold and reveal itself until step $k + 1$.

Before leaving this example, it is of interest to check to see if the original ARMA model, Eq. (5.3.33), is satisfied. We have worked through enough steps to check this at $k = 0$. Substituting the computed values for $y(0)$, $y(1)$, $y(2)$ and the given values of $w(0)$ and $w(1)$ yields

$$(.5) - (.5) + (.5)(0) = (.5)(-.5) + (.25)(1.0)$$

which checks. It might also be noted that the "initial conditions" on the scalar difference equation in y, that is, $y(0)$ and $y(1)$, are not the same as those of the state variables $x_1(0)$ and $x_2(0)$, nor should they be. They are algebraically related, but there is not a one-to-one correspondence, because the state variables are intermediate phase variables and not just $y(k)$ and $y(k + 1)$. ■

In closing it should be mentioned that the state-space model for a given system is not unique. This should be obvious from the fact that we can make any nonsingular transformation on any set of state variables and get another equally legitimate set. The choice of state variables is largely a matter of convenience and may vary from one application to another. The controllable canonical form that was used here is especially convenient in making the transition from a differential (or difference) equation model to a vector model, because the coefficients in the scalar equation transfer directly without any algebraic manipulation. It might also be mentioned that one can also go the other direction in modeling, that is, go from the state model to the scalar model. The conversion is relatively easy and the details are discussed in Problem 5.15.

5.4
THE DISCRETE KALMAN FILTER

R. E. Kalman's paper describing a recursive solution of the discrete-data linear filtering problem was published in 1960 (1). About this same time, advances in digital computer technology made it possible to consider implementing his recursive solution in a number of real-time applications. This was a fortuitous circumstance, and Kalman filtering caught hold almost immediately. We will consider some examples shortly, but we must first derive the Kalman filter recursive equations, which are, in effect, the "filter."

We begin by assuming the random process to be estimated can be modeled in the form

$$\mathbf{x}_{k+1} = \boldsymbol{\phi}_k \mathbf{x}_k + \mathbf{w}_k \tag{5.4.1}$$

The observation (measurement) of the process is assumed to occur at discrete points in time in accordance with the linear relationship

$$\mathbf{z}_k = \mathbf{H}_k \mathbf{x}_k + \mathbf{v}_k \tag{5.4.2}$$

Some elaboration on notation and the various terms of Eqs. (5.4.1) and (5.4.2) is in order:

$\mathbf{x}_k = (n \times 1)$ process state vector at time t_k

$\boldsymbol{\phi}_k = (n \times n)$ matrix relating $\mathbf{x}_k$ to $\mathbf{x}_{k+1}$ in the absence of a forcing function (If $\mathbf{x}_k$ is a sample of continuous process, $\boldsymbol{\phi}_k$ is the usual state transition matrix.)

$\mathbf{w}_k = (n \times 1)$ vector—assumed to be a white sequence with known covariance structure

$\mathbf{z}_k = (m \times 1)$ vector measurement at time t_k

$\mathbf{H}_k = (m \times n)$ matrix giving the ideal (noiseless) connection between the measurement and the state vector at time t_k

$\mathbf{v}_k = (m \times 1)$ measurement error—assumed to be a white sequence with known covariance structure and having zero crosscorrelation with the $\mathbf{w}_k$ sequence

The covariance matrices for the $\mathbf{w}_k$ and $\mathbf{v}_k$ vectors are given by

$$E\left[\mathbf{w}_k \mathbf{w}_i^T\right] = \begin{cases} \mathbf{Q}_k, & i = k \\ 0, & i \neq k \end{cases} \tag{5.4.3}$$

$$E\left[\mathbf{v}_k \mathbf{v}_i^T\right] = \begin{cases} \mathbf{R}_k, & i = k \\ 0, & i \neq k \end{cases} \tag{5.4.4}$$

$$E\left[\mathbf{w}_k \mathbf{v}_i^T\right] = 0, \quad \text{for all } k \text{ and } i \tag{5.4.5}$$

We assume at this point that we have an initial estimate of the process at some point in time t_k, and that this estimate is based on all our knowledge about the process prior to t_k. This prior (or *a priori*) estimate will be denoted as $\hat{\mathbf{x}}_k^-$ where the "hat" denotes estimate, and the "super minus" is a reminder that this is our best estimate prior to assimilating the measurement at t_k. (Note that super minus as used here is not related in any way to the super minus notation used in spectral factorization.) We also assume that we know the error covariance matrix associated with $\hat{\mathbf{x}}_k^-$. That is, we define the estimation error to be

$$\mathbf{e}_k^- = \mathbf{x}_k - \hat{\mathbf{x}}_k^- \tag{5.4.6}$$

and the associated error covariance matrix is*

$$\mathbf{P}_k^- = E\left[\mathbf{e}_k^- \mathbf{e}_k^{-T}\right] = E\left[(\mathbf{x}_k - \hat{\mathbf{x}}_k^-)(\mathbf{x}_k - \hat{\mathbf{x}}_k^-)^T\right] \qquad (5.4.7)$$

In many cases, we begin the estimation problem with no prior measurements. Thus, in this case, if the process mean is zero, the initial estimate is zero, and the associated error covariance matrix is just the covariance matrix of $\mathbf{x}$ itself.

With the assumption of a prior estimate $\hat{\mathbf{x}}_k^-$, we now seek to use the measurement $\mathbf{z}_k$ to improve the prior estimate. We choose a linear blending of the noisy measurement and the prior estimate in accordance with the equation

$$\hat{\mathbf{x}}_k = \hat{\mathbf{x}}_k^- + \mathbf{K}_k(\mathbf{z}_k - \mathbf{H}_k \hat{\mathbf{x}}_k^-) \qquad (5.4.8)$$

where
  $\hat{\mathbf{x}}_k$ = updated estimate
  $\mathbf{K}_k$ = blending factor (yet to be determined)

The justification of the special form of Eq. (5.4.8) will be deferred until Section 5.6. The problem now is to find the particular blending factor $\mathbf{K}_k$ that yields an updated estimate that is optimal in some sense. Just as in the Wiener solution, we use minimum mean-square error as the performance criterion. Toward this end we first form the expression for the error covariance matrix associated with the updated (*a posteriori*) estimate.

$$\mathbf{P}_k = E\left[\mathbf{e}_k \mathbf{e}_k^T\right] = E\left[(\mathbf{x}_k - \hat{\mathbf{x}}_k)(\mathbf{x}_k - \hat{\mathbf{x}}_k)^T\right] \qquad (5.4.9)$$

Next, we substitute Eq. (5.4.2) into Eq. (5.4.8) and then substitute the resulting expression for $\hat{\mathbf{x}}_k$ into Eq. (5.4.9). The result is

$$\mathbf{P}_k = E\Big\{\left[(\mathbf{x}_k - \hat{\mathbf{x}}_k^-) - \mathbf{K}_k(\mathbf{H}_k \mathbf{x}_k + \mathbf{v}_k - \mathbf{H}_k \hat{\mathbf{x}}_k^-)\right]$$

$$\left[(\mathbf{x}_k - \hat{\mathbf{x}}_k^-) - \mathbf{K}_k(\mathbf{H}_k \mathbf{x}_k + \mathbf{v}_k - \mathbf{H}_k \hat{\mathbf{x}}_k^-)\right]^T\Big\} \qquad (5.4.10)$$

Now, performing the indicated expectation and noting the $(\mathbf{x}_k - \hat{\mathbf{x}}_k^-)$ is the a priori estimation error that is uncorrelated with the measurement error $\mathbf{v}_k$, we have

$$\mathbf{P}_k = (\mathbf{I} - \mathbf{K}_k \mathbf{H}_k)\mathbf{P}_k^-(\mathbf{I} - \mathbf{K}_k \mathbf{H}_k)^T + \mathbf{K}_k \mathbf{R}_k \mathbf{K}_k^T \qquad (5.4.11)$$

---

* We tacitly assume here that the estimation error has zero mean, and thus it is proper to refer to $E[\mathbf{e}_k^- \mathbf{e}_k^{-T}]$ as a covariance matrix. It is also, of course, a moment matrix, but it is usually not referred to as such.

Notice here that Eq. (5.4.11) is a perfectly general expression for the updated error covariance matrix, and it applies for any gain $\mathbf{K}_k$, suboptimal or otherwise.

Returning to the optimization problem, we wish to find the particular $\mathbf{K}_k$ that minimizes the individual terms along the major diagonal of $\mathbf{P}_k$, because these terms represent the estimation error variances for the elements of the state vector being estimated. The optimization can be done in a number of ways. We will do this using a straightforward differential calculus approach, and to do so we need two matrix differentiation formulas. They are

$$\frac{d[\text{trace}(\mathbf{AB})]}{d\mathbf{A}} = \mathbf{B}^T, \quad (\mathbf{AB} \text{ must be square}) \tag{5.4.12}$$

$$\frac{d[\text{trace}(\mathbf{ACA}^T)]}{d\mathbf{A}} = 2\mathbf{AC}, \quad (\mathbf{C} \text{ must be symmetric}) \tag{5.4.13}$$

where the derivative of a scalar with respect to a matrix is defined as

$$\frac{ds}{d\mathbf{A}} = \begin{bmatrix} \dfrac{ds}{da_{11}} & \dfrac{ds}{da_{12}} & \cdots \\ \dfrac{ds}{da_{21}} & \dfrac{ds}{da_{22}} & \cdots \\ \vdots & & \end{bmatrix} \tag{5.4.14}$$

Proof of these two differentiation formulas will be left as an exercise (see Problem 5.11). We will now temporarily drop the subscripts in the $\mathbf{P}_k$ expression, Eq. (5.4.11), and expand it out and rewrite it in the form:

$$\mathbf{P} = \mathbf{P}^- - \mathbf{KHP}^- - \mathbf{P}^-\mathbf{H}^T\mathbf{K}^T + \mathbf{K}(\mathbf{HP}^-\mathbf{H}^T + \mathbf{R})\mathbf{K}^T \tag{5.4.15}$$

Notice that the second and third terms are linear in $\mathbf{K}$ and that the fourth term is quadratic in $\mathbf{K}$. The two matrix differentiation formulas may now be applied to Eq. (5.4.15). We wish to minimize the trace of $\mathbf{P}$ because it is the sum of the mean-square errors in the estimates of all the elements of the state vector. We can use the argument here that the individual mean-square errors are also minimized when the total is minimized, provided that we have enough degrees of freedom in the variation of $\mathbf{K}$, which we do in this case. We proceed now to differentiate the trace of $\mathbf{P}$ with respect to $\mathbf{K}$, and we note that the trace of $\mathbf{P}^-\mathbf{H}^T\mathbf{K}^T$ is equal to the trace of its transpose $\mathbf{KHP}^-$. The result is

$$\frac{d(\text{trace } \mathbf{P})}{d\mathbf{K}} = -2(\mathbf{HP}^-)^T + 2\mathbf{K}(\mathbf{HP}^-\mathbf{H}^T + \mathbf{R}) \tag{5.4.16}$$

We now set the derivative equal to zero and solve for the optimal gain. The result is (with the subscripts reinserted)

$$\mathbf{K}_k = \mathbf{P}_k^- \mathbf{H}_k^T \left(\mathbf{H}_k \mathbf{P}_k^- \mathbf{H}_k^T + \mathbf{R}_k \right)^{-1} \tag{5.4.17}$$

This particular $\mathbf{K}_k$, namely the one that minimizes the mean-square estimation error, is called the *Kalman gain*.

The covariance matrix associated with the optimal estimate may now be computed. Referring to Eq. (5.4.11), we have

$$\mathbf{P}_k = (\mathbf{I} - \mathbf{K}_k \mathbf{H}_k) \mathbf{P}_k^- (\mathbf{I} - \mathbf{K}_k \mathbf{H}_k)^T + \mathbf{K}_k \mathbf{R}_k \mathbf{K}_k^T \tag{5.4.18}$$

$$= \mathbf{P}_k^- - \mathbf{K}_k \mathbf{H}_k \mathbf{P}_k^- - \mathbf{P}_k^- \mathbf{H}_k^T \mathbf{K}_k^T + \mathbf{K}_k (\mathbf{H}_k \mathbf{P}_k^- \mathbf{H}_k^T + \mathbf{R}_k) \mathbf{K}_k^T \tag{5.4.19}$$

Routine substitution of the optimal gain expression, Eq. (5.4.17), into Eq. (5.4.19) leads to

$$\mathbf{P}_k = \mathbf{P}_k^- - \mathbf{P}_k^- \mathbf{H}_k^T \left(\mathbf{H}_k \mathbf{P}_k^- \mathbf{H}_k^T + \mathbf{R}_k \right)^{-1} \mathbf{H}_k \mathbf{P}_k^- \tag{5.4.20}$$

or

$$\mathbf{P}_k = \mathbf{P}_k^- - \mathbf{K}_k (\mathbf{H}_k \mathbf{P}_k^- \mathbf{H}_k^T + \mathbf{R}_k) \mathbf{K}_k^T \tag{5.4.21}$$

or

$$\mathbf{P}_k = (\mathbf{I} - \mathbf{K}_k \mathbf{H}_k) \mathbf{P}_k^- \tag{5.4.22}$$

Of the three expressions for $\mathbf{P}_k$, the latter one given by Eq. (5.4.22) is the simplest, so it is used more frequently than the others. Again, note that Eq. (5.4.18) is valid for any gain, suboptimal or otherwise, whereas Eqs. (5.4.20), (5.4.21), and (5.4.22) are valid only for the Kalman (optimal) gain.

We now have a means of assimilating the measurement at t_k by the use of Eq. (5.4.8) with $\mathbf{K}_k$ set equal to the Kalman gain as given by Eq. (5.4.17). Note that we need $\hat{\mathbf{x}}_k^-$ and $\mathbf{P}_k^-$ to accomplish this, and we can anticipate a similar need at the next step in order to make optimal use of the measurement $\mathbf{z}_{k+1}$. The updated estimated $\hat{\mathbf{x}}_k$ is easily projected ahead via the transition matrix. We are justified in ignoring the contribution of $\mathbf{w}_k$ in Eq. (5.4.1) because it has zero mean and is not correlated with any of the previous $\mathbf{w}$'s.* Thus we have

$$\hat{\mathbf{x}}_{k+1}^- = \boldsymbol{\phi}_k \hat{\mathbf{x}}_k \tag{5.4.23}$$

---

* Recall that in our notation $\mathbf{w}_k$ is the process noise that accumulates during the step ahead from t_k to t_{k+1}. This is purely a matter of notation (but an important one), and in some books it is denoted as $\mathbf{w}_{k+1}$ rather than $\mathbf{w}_k$ (6, 9). Consistency in notation is the important thing here. Conceptually, we are thinking of doing real-time filtering in contrast to smoothing, which we usually think of doing offline (see Chapter 8). Therefore, if we begin with a discrete ARMA model, the white forcing sequence $w(k)$ must conform to this same step-ahead notation, and we need to restrict the order of the MA part of the model to be at least one less than the AR part (see Section 5.3). This then assures us that $w(k)$ contributes only to the state vector *after* time t_k and not before.

The error covariance matrix associated with $\hat{\mathbf{x}}_{k+1}^-$ is obtained by first forming the expression for the a priori error

$$
\begin{aligned}
\mathbf{e}_{k+1}^- &= \mathbf{x}_{k+1} - \hat{\mathbf{x}}_{k+1}^- \\
&= (\boldsymbol{\phi}_k \mathbf{x}_k + \mathbf{w}_k) - \boldsymbol{\phi}_k \hat{\mathbf{x}}_k \\
&= \boldsymbol{\phi}_k \mathbf{e}_k + \mathbf{w}_k
\end{aligned}
\tag{5.4.24}
$$

We now note that $\mathbf{w}_k$ and $\mathbf{e}_k$ have zero crosscorrelation, because $\mathbf{w}_k$ is the process noise for the step ahead of t_k. Thus we can write the expression for $\mathbf{P}_{k+1}^-$ as

$$
\begin{aligned}
\mathbf{P}_{k+1}^- &= E\left[\mathbf{e}_{k+1}^- \mathbf{e}_{k+1}^{-T}\right] = E\left[(\boldsymbol{\phi}_k \mathbf{e}_k + \mathbf{w}_k)(\boldsymbol{\phi}_k \mathbf{e}_k + \mathbf{w}_k)^T\right] \\
&= \boldsymbol{\phi}_k \mathbf{P}_k \boldsymbol{\phi}_k^T + \mathbf{Q}_k
\end{aligned}
\tag{5.4.25}
$$

We now have the needed quantities at time t_{k+1}, and the measurement $\mathbf{z}_{k+1}$ can be assimilated just as in the previous step.

Equations (5.4.8), (5.4.17), (5.4.22), (5.4.23), and (5.4.25) comprise the Kalman filter recursive equations. It should be clear that once the loop is entered, it can be continued ad infinitum. The pertinent equations and the sequence of computational steps are shown pictorially in Fig. 5.9. This summarizes what is now known as the *Kalman filter*.

Before proceeding to some examples, it is interesting to reflect on the Kalman filter in perspective. If you were to stumble onto the recursive process of Fig. 5.9 without benefit of previous history, you might logically ask, "Why in the world did somebody call that a filter? It looks more like a computer algorithm." You would, of course, be quite right in your observation. The Kalman filter is just a computer algorithm for processing discrete measurements

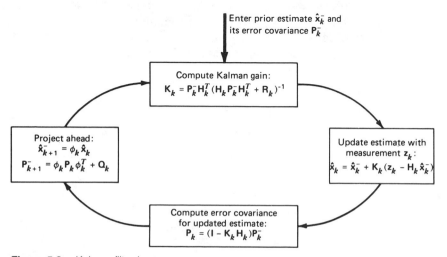

Figure 5.9. Kalman filter loop.

(the input) into optimal estimates (the output). Its roots, though, go back to the days when filters were made of electrical elements wired together in such a way as to yield the desired frequency response. The design was often heuristic. Wiener then came on the scene in the 1940s and added a more sophisticated type of filter problem. The end result of his solution was a filter weighting function or a corresponding transfer function in the complex domain. Implementation in terms of electrical elements was left as a further exercise for the designer. The sampled-data version of the Wiener problem remained unsolved (in a practical sense, at least) until Kalman's paper of 1960. Even though his presentation appeared to be quite abstract at first glance, engineers soon realized that this work provided a practical solution to a number of unsolved filtering problems, especially in the field of navigation. More than 30 years have elapsed since Kalman's original paper, and there are still numerous current papers dealing with new applications and variations on the basic Kalman filter. It has withstood the test of time!

5.5
EXAMPLES

The basic recursive equations for the Kalman filter were presented in Section 5.4. Two simple scalar examples illustrate the use of these equations. Examples of higher-order systems will be deferred until Chapter 6.

EXAMPLE 5.6

Wiener (Brownian Motion) Process   Consider the Wiener process described in Fig. 5.10. Recall that this process has Gaussian statistics and is assumed to be zero at $t = 0$. Assume that we have a sequence of independent noisy measurements taken at unit intervals as shown in Fig. 5.10 and let the standard deviation of the measurement error be one-half. We wish to determine the optimal estimate of the process at the sample times $0, 1, 2, 3, \ldots$, etc.

The model parameters for the Kalman filter may be computed as follows: The process differential equation for this case is

$$\dot{x} = u(t) \tag{5.5.1}$$

Therefore, the transition matrix is

$$\phi_k = 1 \tag{5.5.2}$$

Figure 5.10.   Block diagram of Wiener process and typical sample function of the process.

The Q_k matrix is computed as

$$Q_k = E[w_k, w_k] = E\left[\int_0^1 1 \cdot u(\xi)\, d\xi \int_0^1 1 \cdot u(\eta)\, d\eta\right]$$

$$= \int_0^1 \int_0^1 E[u(\xi)u(\eta)]\, d\xi\, d\eta = \int_0^1 \int_0^1 \delta(\xi - \eta)\, d\xi\, d\eta = 1 \quad (5.5.3)$$

Since the measurement has a direct one-to-one correspondence to the process $x(t)$, the H_k matrix is

$$H_k = 1 \quad\quad (5.5.4)$$

The variance of the measurement error is

$$R_k = (\text{Standard deviation})^2 = \left(\tfrac{1}{2}\right)^2 = \tfrac{1}{4} \quad\quad (5.5.5)$$

Our initial estimate at $t = 0$ is $\hat{x}_0^-$ and is zero because the Wiener process is defined to begin at zero. Furthermore, because of this prior knowledge about the process, the error associated with our initial estimate is zero; that is, the a priori estimate, $\hat{x}_0^- = 0$, is perfect by definition of the process. Thus

$$P_0^- = 0 \quad\quad (5.5.6)$$

We now have all the parameters needed for the Kalman filter and are ready to begin the recursive process. Referring to Fig. 5.9, we enter the loop at $t = 0$ and process the first measurement.

Step 1: $t = 0$ (Subscripts will be dropped for constant parameters.)
Compute gain:

$$K_0 = P_0^- H^T \left(H P_0^- H^T + R\right)^{-1}$$

$$= \frac{0}{.25} = 0 \quad\quad (5.5.7)$$

Update estimate:

$$\hat{x}_0 = \hat{x}_0^- + K_0(z_0 - H\hat{x}_0^-)$$

$$= \hat{x}_0^- = 0 \quad\quad (5.5.8)$$

Update P:

$$P_0 = (I - K_0 H) P_0^-$$

$$= 1 \cdot P_0^- = 0 \quad\quad (5.5.9)$$

Project ahead:

$$\hat{x}_1^- = \phi \hat{x}_0 = 1 \cdot 0 = 0$$

$$P_1^- = \phi P_0 \phi^T + Q \quad\quad (5.5.10)$$

$$= 1 \cdot 0 \cdot 1 + 1 = 1 \quad\quad (5.5.11)$$

Note that the measurement at $t = 0$ is given zero weight relative to the prior estimate. This makes good sense because the initial estimate is known to be perfect, whereas the measurement is known to be noisy.

Step 2: $t = 1$

Compute gain:

$$K_1 = P_1^- H^T \left(H P_1^- H^T + R \right)^{-1}$$

$$= 1 \cdot 1 \left(1 \cdot 1 \cdot 1 + \tfrac{1}{4} \right)^{-1} = \tfrac{4}{5} \qquad (5.5.12)$$

Update estimate:

$$\hat{x}_1 = \hat{x}_1^- + K_1 (z_1 - H\hat{x}_1^-)$$

$$= 0 + \tfrac{4}{5}(z_1) \qquad (5.5.13)$$

Update P:

$$P_1 = (I - K_1 H) P_1^-$$

$$= \left(1 - \tfrac{4}{5} \cdot 1 \right) 1 = \tfrac{1}{5} \qquad (5.5.14)$$

Project ahead:

$$\hat{x}_2^- = \phi \hat{x}_1 = 1 \cdot \hat{x}_1$$

$$P_2^- = \phi P_1 \phi^T + Q \qquad (5.5.15)$$

$$= 1 \cdot \tfrac{1}{5} \cdot 1 + 1 = \tfrac{6}{5} \qquad (5.5.16)$$

This could now be carried on ad infinitum. Note that with step 2 the filter begins to give the measurement some weight in determining the optimal estimate. As times goes on, the filter depends more and more on the measurements and less on the initial assumptions. ∎

EXAMPLE 5.7

Scalar Gauss–Markov Process   Consider a stationary Gauss–Markov process whose autocorrelation function is

$$R_x(\tau) = 1 \cdot e^{-|\tau|} \qquad (5.5.17)$$

Clearly, the correlation time and variance for this process are both unity.

Figure 5.11. Shaping filter for Gauss–Markov process.

Therefore, the spectral function for the process is

$$S_x(s) = \frac{2}{-s^2 + 1} = \frac{\sqrt{2}}{s + 1} \cdot \frac{\sqrt{2}}{-s + 1} \qquad (5.5.18)$$

and the shaping filter that shapes white noise into the process is shown in Fig. 5.11. Thus, the state equation for this process is

$$\dot{x} = -x + \sqrt{2}\,u(t) \qquad (5.5.19)$$

Suppose we have a sequence of noisy measurements of this process taken .02 sec apart beginning at $t = 0$. The measurement error will be assumed to have a variance of unity. We wish to process these via a Kalman filter and obtain an optimal estimate of $x(t)$. First we need to determine the filter parameters ϕ_k, H_k, Q_k, and R_k.

The state transition matrix (scalar in this case) is

$$\phi_k = e^{-.02} \approx .9802 \qquad (5.5.20)$$

The measurement relationship to x is

$$H_k = 1 \qquad (5.5.21)$$

The input noise sequence is

$$Q_k = E[w_k^2] = E\left[\int_0^{.02} \sqrt{2}\,e^{-\xi}u(\xi)\,d\xi \int_0^{.02} \sqrt{2}\,e^{-\eta}u(\eta)\,d\eta\right]$$

$$= \int_0^{.02}\left(\sqrt{2}\,e^{-\eta}\right)^2 d\eta = 1 - e^{-2(.02)} \approx .03921 \qquad (5.5.22)$$

The measurement error is

$$R_k = 1 \qquad (5.5.23)$$

We also need the initial conditions $\hat{x}_0^-$ and P_0^- to enter the recursive loop. In this case, we have assumed that the process is Gauss–Markov and stationary with a known autocorrelation function. We have no measurements prior to $t = 0$, but the assumed knowledge of the process autocorrelation function tells us the process has zero mean and a variance of unity. This is important information and enables us to start the recursive process at $t = 0$ with initial conditions

$$\hat{x}_0^- = 0 \qquad\qquad (5.5.24)$$

$$P_0^- = 1 \qquad\qquad (5.5.25)$$

Now that the four filter parameters and initial conditions are determined, it is a routine matter to cycle through the Kalman filter loop shown in Fig. 5.9 as many times as desired.

In order to add a note of realism to this example, the Markov process and noisy measurement situation just described were simulated using random numbers selected by chance. The results for the first 51 steps of the simulation are shown in Fig. 5.12. The discrete measurements z_k are shown as dots, and it should be obvious that we have postulated a very noisy measurement situation for this example. In spite of this, though, the filter does a reasonably good job of tracking $x(t)$. After about the first 20 steps, the filter settles down to a steady-state condition where the Kalman gain is about .165 and the standard deviation of the filter error is about .4. That is, $\sqrt{P_k}$ in the steady state is about .4 units. This is consistent, qualitatively at least, with what we see in the results shown in Fig. 5.12. It is comforting to know that our single sample realization of

Figure 5.12.   Simulation for Gauss–Markov example.

the process is typical rather than atypical. (See Problem 5.16* for a continuation of this example.) ■

5.6
THE CONDITIONAL DENSITY VIEWPOINT

In our discussion thus far, we have used minimum mean-square error as the performance criterion, and we have assumed a linear form for the filter. This was partly a matter of convenience, but not entirely so, because we will see presently that the resulting linear filter has more far-reaching consequences than are apparent at first glance. This is especially so in the Gaussian case. We now elaborate on this.

We first show that if we choose as our estimate the mean of $\mathbf{x}_k$ conditioned on the available measurement stream, then this estimate will minimize the mean-square error. This is a somewhat restrictive form of what is sometimes called the fundamental theorem of estimation theory (7, 10). The same notation and model assumptions that were used in Section 5.4 will be used here, and our derivation follows closely that given by Mendel (7). Also, to save writing we will temporarily drop the k subscripts, and we will denote the complete measurement stream $\mathbf{z}_0, \mathbf{z}_1, \ldots, \mathbf{z}_k$ simply as $\mathbf{z}^*$. We first write the mean-square estimation error of $\mathbf{x}$, conditioned on $\mathbf{z}^*$, as

$$E\left[(\mathbf{x} - \hat{\mathbf{x}})^T(\mathbf{x} - \hat{\mathbf{x}})|\mathbf{z}^*\right] = E\left[(\mathbf{x}^T\mathbf{x} - \mathbf{x}^T\hat{\mathbf{x}} - \hat{\mathbf{x}}^T\mathbf{x} + \hat{\mathbf{x}}^T\hat{\mathbf{x}})|\mathbf{z}^*\right]$$

$$= E(\mathbf{x}^T\mathbf{x}|\mathbf{z}^*) - E(\mathbf{x}^T|\mathbf{z}^*)\hat{\mathbf{x}} - \hat{\mathbf{x}}^T E(\mathbf{x}|\mathbf{z}^*) + \hat{\mathbf{x}}^T\hat{\mathbf{x}} \quad (5.6.1)$$

Factoring $\hat{\mathbf{x}}$ away from the expectation operator in Eq. (5.6.1) is justified, because $\hat{\mathbf{x}}$ is a function of $\mathbf{z}^*$, which is the conditioning on the random variable $\mathbf{x}$. We now complete the square of the last three terms in Eq. (5.6.1) and obtain

$$E\left[(\mathbf{x} - \hat{\mathbf{x}})^T(\mathbf{x} - \hat{\mathbf{x}})|\mathbf{z}^*\right]$$

$$= E(\mathbf{x}^T\mathbf{x}|\mathbf{z}^*) + \left[\hat{\mathbf{x}} - E(\mathbf{x}|\mathbf{z}^*)\right]^T\left[\hat{\mathbf{x}} - E(\mathbf{x}|\mathbf{z}^*)\right] - E(\mathbf{x}^T|\mathbf{z}^*)E(\mathbf{x}|\mathbf{z}^*)$$

$$(5.6.2)$$

Clearly, the first and last terms on the right side of Eq. (5.6.2) do not depend on our choice of the estimate $\hat{\mathbf{x}}$. Therefore, in our search among the admissible estimators (both linear and nonlinear), it should be clear that the best we can do is to force the middle term to be zero. We do this by letting

$$\hat{\mathbf{x}}_k = E(\mathbf{x}_k|\mathbf{z}_k^*) \quad (5.6.3)$$

where we have now reinserted the k subscripts. We have tacitly assumed here that we are dealing with the filter problem, but a similar line of reasoning would also apply to the predictive and smoothed estimates of the **x** process. (See Section 4.4 for the definition and a brief discussion of prediction and smoothing.)

Equation (5.6.3) now provides us with a general formula for finding the estimator that minimizes the mean-square error, and it is especially useful in the Gaussian case because it enables us to write out an explicit expression for the optimal estimate in recursive form. Toward this end, we will now assume Gaussian statistics throughout. We will further assume that we have, by some means, an optimal prior estimate $\hat{\mathbf{x}}_k^-$ and its associated error covariance $\mathbf{P}_k^-$. Now, at this point we will stretch our notation somewhat and let $\mathbf{x}_k$ denote the **x** random variable at t_k *conditioned* on the measurement stream $\mathbf{z}_{k-1}^*$. We know that the form of the probability density of $\mathbf{x}_k$ is then

$$f_{\mathbf{x}_k} \sim N(\hat{\mathbf{x}}_k^-, \mathbf{P}_k^-) \tag{5.6.4}$$

Now, from our measurement model we know that $\mathbf{x}_k$ is related to $\mathbf{z}_k$ by

$$\mathbf{z}_k = \mathbf{H}_k \mathbf{x}_k + \mathbf{v}_k \tag{5.6.5}$$

Therefore, we can immediately write the density function for $\mathbf{z}_k$ as

$$f_{\mathbf{z}_k} \sim N\big(\mathbf{H}_k \hat{\mathbf{x}}_k^-, \mathbf{H}_k \mathbf{P}_k^- \mathbf{H}_k^T + \mathbf{R}_k\big) \tag{5.6.6}$$

(Again, remember that conditioning on $\mathbf{z}_{k-1}^*$ is implied.) Also, from Eq. (5.6.5) we can write out the form for the conditional density of $\mathbf{z}_k$, given $\mathbf{x}_k$. It is

$$f_{\mathbf{z}_k|\mathbf{x}_k} \sim N(\mathbf{H}_k \mathbf{x}_k, \mathbf{R}_k) \tag{5.6.7}$$

Finally, we can now use Bayes formula and write

$$f_{\mathbf{x}_k|\mathbf{z}_k} = \frac{f_{\mathbf{z}_k|\mathbf{x}_k} f_{\mathbf{x}_k}}{f_{\mathbf{z}_k}} \tag{5.6.8}$$

where the terms on the right side of the equation are given by Eqs. (5.6.4), (5.6.6), and (5.6.7). But recall that $\mathbf{x}_k$ itself was conditioned on $\mathbf{z}_0, \mathbf{z}_1, \ldots, \mathbf{z}_{k-1}$. Thus the density function on the left side of Eq. (5.6.8) is actually the density of the usual random variable $\mathbf{x}_k$, conditioned on the whole measurement stream up through $\mathbf{z}_k$. Thus we will change the notation slightly and rewrite Eq. (5.6.8) as

$$f_{\mathbf{x}_k|\mathbf{z}_k} = \frac{[N(\mathbf{H}_k \mathbf{x}_k, \mathbf{R}_k)][N(\hat{\mathbf{x}}_k^-, \mathbf{P}_k^-)]}{[N(\mathbf{H}_k \hat{\mathbf{x}}_k^-, \mathbf{H}_k \mathbf{P}_k^- \mathbf{H}_k^T + \mathbf{R}_k)]} \tag{5.6.9}$$

where it is implied that we substitute the indicated normal functional expressions into the right side of the equation (see Section 1.15 for the vector normal form). It is a routine matter now to make the appropriate substitutions in Eq. (5.6.9) and determine the mean and covariance by inspection of the exponential term. The algebra is routine, but a bit laborious, so we will not pursue it further here (see Problem 5.12). The resulting mean and covariance for $\mathbf{x}_k | \mathbf{z}_k^*$ are

$$\text{Mean} = \hat{\mathbf{x}}_k^- + \mathbf{P}_k^- \mathbf{H}_k^T \left(\mathbf{H}_k \mathbf{P}_k^- \mathbf{H}_k^T + \mathbf{R}_k \right)^{-1} \left(\mathbf{z}_k - \mathbf{H}_k \hat{\mathbf{x}}_k^- \right) \quad (5.6.10)$$

$$\text{Covariance} = \left[\left(\mathbf{P}_k^- \right)^{-1} + \mathbf{H}_k^T \mathbf{R}_k^{-1} \mathbf{H}_k \right]^{-1} \quad (5.6.11)$$

Note that the expression for the mean is identical to the optimal estimate previously derived by other methods and given by Eqs. (5.4.8) and (5.4.17). The expression for the covariance given by Eq. (5.6.11) may not look familiar, but in Chapter 6 it will be shown to be identically equal to the usual $\mathbf{P} = (\mathbf{I} - \mathbf{KH})\mathbf{P}^-$ expression, provided that $\mathbf{K}$ is the Kalman gain.

In summary, we see that in the Gaussian case the conditional density viewpoint leads to the same identical result that was obtained in Section 5.4, where we assumed a special linear form for our estimator. There are some far-reaching conclusions that can be drawn from the conditional density viewpoint:

1. Note that in the conditional density function approach, we did not need to *assume* a linear relationship between the estimate and the measurements. Instead, this came out naturally as a consequence of the Gaussian assumption and our choice of the conditional mean as our estimate. Thus, in the Gaussian case, we know that we need not search among nonlinear filters for a better one; it cannot exist. Thus, our earlier linear assumption in the derivation of both the Wiener and Kalman filters turns out to be a fortuitous one. That is, in the Gaussian case, the Wiener–Kalman filter is not just best within a class of linear filters; it is best within a class of all filters, linear or nonlinear.

2. For the Gaussian case, the conditional mean is also the "most likely" value in that the maximum of the density function occurs at the mean. Also, it can be shown that the conditional mean minimizes the expectation of almost any reasonable nondecreasing function of the magnitude of the error (as well as the squared error). [See Meditch (10) for a more complete discussion of this.] Thus, *in the Gaussian case, the Kalman filter is best by almost any reasonable criterion.*

3. In physical problems, we often begin with incomplete knowledge of the process under consideration. Perhaps only the covariance structure of the process is known. In this case we can always imagine a corresponding Gaussian process with the same covariance structure. This process is then completely defined and conclusions for the equivalent Gaussian

process can be drawn. It is, of course, a bit risky to extend these conclusions to the original process, not knowing it to be Gaussian. However, even risky conclusions are better than none if viewed with proper caution.

PROBLEMS

5.1 A stationary Gaussian random process is known to have a spectral density function of the form

$$S_y(j\omega) = \frac{\omega^2 + 1}{\omega^4 + 8\omega^2 + 16}$$

Assume that discrete noisy measurements of y are available at $t = 0, 1, 2, 3, \ldots$ and that the measurement errors are uncorrelated and have a variance of two units. Model this process in appropriate form for a Kalman filter. That is, find ϕ_k, $\mathbf{H}_k$, $\mathbf{Q}_k$, $\mathbf{R}_k$ and the initial conditions $\hat{\mathbf{x}}_0^-$ and $\mathbf{P}_0^-$. Assume that there is no measurement information prior to $t = 0$. (The integrations involved in evaluating the $\mathbf{Q}_k$ matrix are a bit laborious, so the terms of $\mathbf{Q}_k$ may be left in integral form in this problem. See Problem 5.18* for a continuation of the $\mathbf{Q}_k$ determination problem.)

5.2. A classical problem in Wiener-filter theory is one of separating signal from noise when both the signal and noise have exponential autocorrelation functions. Let the noisy measurement be

$$z(t) = s(t) + n(t)$$

and let signal and noise be stationary independent processes with autocorrelation functions

$$R_s(\tau) = \sigma_s^2 e^{-\beta_s|\tau|}$$

$$R_n(\tau) = \sigma_n^2 e^{-\beta_n|\tau|}$$

Assume that we have discrete samples of $z(t)$ spaced Δt apart and wish to form the optimal estimate of $s(t)$ at the sample points. Let $s(t)$ and $n(t)$ be the first and second elements of the process state vector, and then find the parameters of the Kalman filter for this situation. That is, find ϕ_k, $\mathbf{H}_k$, $\mathbf{Q}_k$, $\mathbf{R}_k$ and the initial conditions $\hat{\mathbf{x}}_0^-$ and $\mathbf{P}_0^-$. Assume that the measurement sequence begins at $t = 0$.

5.3. Consider the two cascaded integrators as shown in the accompanying figure. Integrator scale factors are shown in the "boxes." The initial conditions for both integrators are zero, and state variables x_1 and x_2 are defined as shown on the figure. State variable x_1 is the quantity being measured and it is sampled at 2-sec intervals. The measurement error variance is four units. Find

Problem 5.3

the Kalman filter parameters ϕ_k, $\mathbf{H}_k$, $\mathbf{Q}_k$, and $\mathbf{R}_k$ and the appropriate initial conditions $\hat{\mathbf{x}}_0^-$, $\mathbf{P}_0^-$ for this situation.

5.4. The system shown is driven by two independent Gaussian white sources $u_1(t)$ and $u_2(t)$. Their spectral functions are given by

$$S_{u1} = 4 \left(\text{ft/sec}^2\right)^2/(\text{rad})/(\text{sec})$$

$$S_{u2} = 16 \left(\text{ft/sec}\right)^2/(\text{rad})/(\text{sec})$$

Let state variables be chosen as shown on the diagram, and assume that noisy measurements of x_1 are obtained at unit intervals of time. A discrete Kalman filter model is desired. Find $\mathbf{Q}_k$ for this model.

Problem 5.4

5.5. In Example 5.6, Section 5.5, the random walk process was assumed to have an initial value of zero, that is, it was a Wiener process. Consider a similar process except that the initial value is a random variable described as $N(0, 1)$.

    (a) How will this change in initial condition affect the discrete Kalman filter model?

    (b) Cycle through two recursive steps and find the Kalman gain at $t = 1$. Compare this with the corresponding gain in Example 5.6.

    (c) Would you expect the two gains, that is, for Example 5.6 and this problem, to approach the same value as the number of steps becomes large? (See Problem 5.17* for a continuation of this question.)

5.6. Suppose we have a scalar process $y(t)$ that is an additive combination of a random bias and a Wiener process (see Section 5.2). The $y(t)$ process can be modeled either as the sum of two separate state variables or as a single variable with a random initial condition as was done in Problem 5.5. Assume that the measurements of $y(t)$ occur at unit intervals beginning at $t = 0$ and that the measurement error has a variance R. Also assume that the random bias component is described as $N(0, \sigma^2)$, and that the Wiener process has a variance that increases in accordance with At. Both σ^2 and A are known parameters.

    (a) Demonstrate that both the single-state and two-state models yield the same optimal estimate of y by cycling through two recursive steps for each model.

    (b) Assume that only the sum of the random constant and Wiener process is of interest. Which of the two models would you prefer? Why?

5.7. Consider the same Gauss–Markov process and measurement situation discussed in Example 5.7.

    (a) Carry out two recursive steps of the Kalman filter and write the estimate at $t = .02$ sec explicitly in terms of z_0 and z_1.

    (b) Using the weighting function approach of Section 4.8, write the optimal estimate after two measurements (i.e., at $t = .02$ sec) in terms of z_0 and z_1. [Note that the answer should be the same as that obtained in part (a)].

5.8. A common process encountered in physical applications is harmonic motion with random amplitude and phase. If the motion is truly harmonic, it satisfies the undriven differential equation

$$\ddot{x} + \omega_r^2 x = 0$$

and it can be seen to be a deterministic random process. Suppose we have a sequence of discrete noisy measurements of x beginning at $t = 0$, and we wish to estimate x. Assume that we have no prior knowledge of either amplitude or phase, and that the measurements are equally spaced by Δt and each has a measurement variance of R.

    (a) Choose phase variables (i.e., x and $\dot{x}$) as the state variables and find the Kalman filter parameters for this situation. (*Partial answer*: In the absence of better information, it is appropriate to let $\hat{\mathbf{x}}_0^- = \mathbf{0}$ and $\mathbf{P}_0^- = \infty$. As a practical matter, $\mathbf{P}_0^-$ can be set at a very large value relative to the measurement variance and thus the same result as "∞" is accomplished, approximately. There is also an alternative approach discussed in Chapter 6.)

    (b) It will be noted that the solution for x is of the form

$$x(t) = x(0) \cos \omega_r t + \frac{\dot{x}(0)}{\omega_r} \sin \omega_r t$$

This suggests that we might choose $x(0)$ and $\dot{x}(0)$ as our state variables and then model the Kalman filter accordingly. Work out the filter parameters for this model.

    (c) Find the linear transformation relating the state variables of parts (a) and (b) (*Note:* The elements of the transformation matrix will be time-varying in this case.)

5.9. Suppose that we make the following linear transformation on the process state vector of Problem 5.2:

$$\begin{bmatrix} x_1' \\ x_2' \end{bmatrix} = \begin{bmatrix} 1 & 1 \\ 0 & 1 \end{bmatrix}\begin{bmatrix} x_1 \\ x_2 \end{bmatrix}$$

This transformation is nonsingular, and hence we should be able to consider $\mathbf{x}'$ as the state vector to be estimated and write out the Kalman filter equations accordingly. Specify the Kalman filter parameters for the transformed problem. (Note that the specified transformation yields a simplification of the measurement matrix, but this is at the expense of complicating the model elsewhere.)

5.10. It is almost self-evident that if the estimation errors are minimized in one set of state variables, this also will minimize the error in any linear combination of those state variables. This can be shown formally by considering a new state vector $\mathbf{x}'$ to be related to the original state vector $\mathbf{x}$ via a general nonsingular transformation $\mathbf{x}' = \mathbf{A}\mathbf{x}$. Proceeding in this manner, show that the Kalman estimate obtained in the transformed domain is the same as would be obtained by performing the update [i.e., Eq. (5.4.8)] in the original $\mathbf{x}$ domain and then transforming this estimate via the $\mathbf{A}$ matrix.

5.11. The matrix differentiation formulas used in Section 5.4 are repeated here for convenience.

$$\frac{d[\text{trace}(\mathbf{AB})]}{d\mathbf{A}} = \mathbf{B}^T, \qquad (\mathbf{AB} \text{ must be square})$$

$$\frac{d[\text{trace}(\mathbf{ACA}^T)]}{d\mathbf{A}} = 2\mathbf{AC}, \qquad (\mathbf{C} \text{ must be symmetric})$$

Verify that these formulas are correct. [A straightforward way of doing this is to write out a few terms of the indicated matrices, perform the products, and then differentiate term-by-term as indicated in Eq. (5.4.14).]

5.12. Show that the mean and covariance associated with conditional density function $f_{\mathbf{x}_k|z_k}$ are as given by Eqs. (5.6.10) and (5.6.11). (*Hint:* When the expression for $f_{\mathbf{x}_k|z_k}$ is written out explicitly in terms of $\mathbf{x}_k$, it will be found that the quantity in the exponent is quadratic in $\mathbf{x}_k$. Next, you will need to do the matrix equivalent of completing-the-square in order to recognize the mean and covariance of the density function. Also, you will find the alternative gain and error-covariance expressions derived in Chapter 6 to be useful here. They are, with subscripts omitted,

$$\mathbf{K} = \mathbf{P}\mathbf{H}^T\mathbf{R}^{-1}$$

$$\mathbf{P} = \left[(\mathbf{P}^-)^{-1} + \mathbf{H}^T\mathbf{R}^{-1}\mathbf{H}\right]^{-1}$$

Assume these to be valid and use them wherever they may be helpful in this problem.)

5.13. It is desired to simulate a Wiener process at discrete sample points separated by .1 sec. Call the sequence of samples of the Wiener process $x_0, x_1, x_2, \ldots$ with x_0 being the sample at $t = 0$. The Wiener process under consideration may be thought of as integrated unity white noise. To accomplish the simulation, one has available a sequence of samples of a normal distribution $N(0, 1)$. Call this sequence $w_0, w_1, \ldots$ and write the difference equation for generating the $x_0, x_1, x_2, \ldots$ sequence from the $w_0, w_1, \ldots$ sequence.

5.14. Occasionally we need to transform the process state model back to a corresponding scalar differential equation (continuous case) or an ARMA model (discrete case). In either case we are looking for the direct relationship between the scalar input and the scalar output. In the continuous case the state equation has the following form:

$$\dot{\mathbf{x}} = \mathbf{F}\mathbf{x} + \mathbf{G}u, \quad (u \text{ is unity white noise})$$

$$y = \mathbf{B}\mathbf{x}, \quad (y \text{ is scalar})$$

If we transform to the s-domain and solve for $Y(s)$, we get

$$Y(s) = \left[\mathbf{B}(s\mathbf{I} - \mathbf{F})^{-1}\mathbf{G}\right]U(s)$$

Clearly, the denominator of $[\mathbf{B}(s\mathbf{I} - \mathbf{F})^{-1}\mathbf{G}]$ provides the characteristic polynomial of the desired differential equation, and the numerator gives the coefficients of the forcing function part. Derive the corresponding input–output transfer function relationship in the z-domain for the discrete model and relate the denominator and the numerator to the respective AR and MA parts of the ARMA model. In other words, develop an ARMA model from the discrete state model. Demonstrate that this procedure does yield the correct ARMA equation by applying it to Eqs. (5.3.33) and (5.3.34) of Example 5.5.

5.15. Most Kalman filter simulation exercises are best done with the help of computer software that was developed especially for this purpose. However, it is still instructive to cycle through a few recursive steps of a simple example using just a hand-held calculator. A set of samples from a normal distribution are included here for your convenience in this problem.

    (a) Using the same parameters as in Example 5.7, generate a five-step sample realization of a typical Markov process x_k and associated measurement sequence z_k. Use the five-step measurement sequence as the input to the Kalman filter for this situation and compute the sequence of optimal estimates. Note that the Kalman gain varies with each step, so this is truly a time-variable filter.

    (b) This problem is easily programmed with just a few lines of code using one of the scientific languages such as FORTRAN, BASIC, and so

102 Samples from a Normal Distribution $N(0, 1)$

− 1.706	.518	.737	.048	−.716	− 1.728
.410	−.153	−.128	.649	−.927	−.493
.703	.197	.075	−1.477	.916	
−.923	.131	.848	−.182	−1.386	
.890	−.837	−.402	.668	−.919	
−.202	2.511	.365	.210	−.559	
.995	−.114	1.143	.628	−1.137	
.810	−.005	1.115	.285	.987	
−1.598	−.355	.703	1.055	.332	
−.235	−.967	1.080	−.105	1.190	
.018	−.233	.968	.650	−.920	
−.215	1.675	1.392	−.068	.348	
−.121	−.842	−.610	.683	.510	
−.128	.977	1.675	−.027	.719	
.974	−.365	−.817	−1.360	−.623	
1.911	−.668	−1.685	−.713	−.017	
.367	−1.054	−.451	−1.599	−1.160	
−.194	−.997	−.572	.752	1.927	
1.175	−.403	.144	.075	1.041	
−.593	1.190	−.028	−.156	−.105	

forth. Try programming this problem using the language of your choice, and run the problem for 51 steps just as in Example 5.7. (Students with access to a personal computer may wish to skip this part and go directly to Problem 5.16*.)

5.16*. The Gauss–Markov process and the accompanying estimate given in Example 5.7 represent sample realizations of just one Monte Carlo simulation. With just one simulation there is always a chance that a statistically unusual situation will be encountered. With appropriate computer software it is relatively easy to run an ensemble of such simulations; and, in doing so, one gains some insight into what is typical for a given set of parameters.

(a) Using the same set of parameters as given in Example 5.7, compute the sequence of Kalman filter gains and error covariances for $k = 0, 1, 2, \ldots, 50$. Note that a stationary condition is reached well before the end of the run.

(b) Run three Monte Carlo simulations for the situation of Example 5.7. Make files for both the true x_k and the Kalman filter estimate $\hat{x}_k$ for each run, and then view the plots of x_k and $\hat{x}_k$ superimposed on one graph. Do this for each Monte Carlo simulation. On the basis of the observed plots, would you say that the "typical" assertion made in Example 5.7 is correct?

(c) The measurement situation in Example 5.7 is obviously quite noisy, and, as a result, the filter performance is not spectacular. Now consider a similar model except that the measurement error variance in the new model is to be .01 rather than 1.0. Run three Monte Carlo simulations for this new situation and observe the plots of the resulting x_k and $\hat{x}_k$ just as in part (b). You should see a marked improvement in filter performance because of the reduced measurement noise. Note that the R_k parameter must be changed to .01 in the Kalman filter as well as in the generation of the measurement z_k. Otherwise, the filter will not be optimal.

5.17*.   In Problem 5.5, part (c), you were asked to speculate about the steady-state Kalman gains for two different initial-condition situations. Questions such as this are easily answered with a computer solution. Cycle each of the two Kalman filters (each with a different P_0^-) through a sufficient number of steps to reach steady state and, in this way, determine the answer to part (c) empirically.

Suppose the standard deviation of the measurement error is 2.0 rather than .5 as given in Problem 5.5 (and Example 5.6). How would this affect the rapidity of convergence to the steady-state condition? Verify your answer empirically by cycling the Kalman filter through an appropriate number of steps with $R = 4$ rather than .25.

5.18*.   A suitable state model for the y process of Problem 5.1 is as follows:

$$\begin{bmatrix} \dot{x}_1 \\ \dot{x}_2 \end{bmatrix} = \begin{bmatrix} 0 & 1 \\ -4 & -4 \end{bmatrix} \begin{bmatrix} x_1 \\ x_2 \end{bmatrix} + \begin{bmatrix} 0 \\ 1 \end{bmatrix} u(t)$$

$$y = \begin{bmatrix} 1 & 1 \end{bmatrix} \begin{bmatrix} x_1 \\ x_2 \end{bmatrix}$$

This model has phase variables as its state variables, and it can be developed using the methods given in Section 5.2. The **F** matrix is relatively simple in this case, so the state transition matrix ϕ_k can be readily found in general terms analytically. However, this is not so with the process noise matrix $\mathbf{Q}_k$. Its evaluation involves integration by parts, and there are many opportunities for mistakes. Thus, evaluating $\mathbf{Q}_k$ for a general step size Δt is an onerous task, at best.

(a) Using appropriate computer software, evaluate the state transition matrix numerically for $\Delta t = 1$ sec. Check your result term by term with the values obtained from the analytical expression for ϕ_k.

(b) The $\mathbf{Q}_k$ matrix is easily evaluated for small Δt using the methods given in Example 5.4. Also, the elements of $\mathbf{Q}_k$ can be found for very large Δt by noting that the limits on the integrals to be evaluated are the same as for the Laplace transform. Thus, Laplace transform tables with s set equal to zero can be used to evaluate the terms of $\mathbf{Q}_k$ for large Δt. Omitting the details, the results are

$$
\underbrace{\mathbf{Q}_k \approx \begin{bmatrix} 0 & 0 \\ 0 & \Delta t \end{bmatrix}}_{\text{For small } \Delta t} \qquad \underbrace{\mathbf{Q}_k \approx \begin{bmatrix} .03125 & 0 \\ 0 & .125 \end{bmatrix}}_{\text{For large } \Delta t}
$$

Using appropriate software for evaluating $\mathbf{Q}_k$ numerically, investigate the range of Δt for which each of the above approximations is reasonably accurate. Use 1 percent as a rough figure of merit for "reasonable" accuracy. Note that the time constant in this system is .5 sec. (It is second order and critically damped.) Express your bounds on Δt in relative terms, that is, relative to the system time constant. This problem is intended to give you at least a rough idea about the range of validity of the small and large Δt approximations.

5.19*. Consider two different measurement situations for the same random walk dynamical process:

Process model:

$$x_{k+1} = x_k + w_k$$

Measurement model 1:

$$z_k = .5x_k + v_k$$

Measurement model 2:

$$z_k = (\cos \theta_k)x_k + v_k, \qquad \theta_k = 1 + \frac{k}{120} \text{ rad}$$

Using $Q = 4$, $R = 1$, and $P_0^- = 100$, run error covariance analyses for each measurement model for $k = 0, 1, 2, \ldots, 200$. Plot the estimation error variance for the scalar state x against the time index k for each case. Explain the difference seen between the two plots, particularly as the iterative process approaches and passes $k \approx 70$.

REFERENCES CITED IN CHAPTER 5

1. R. E. Kalman, "A New Approach to Linear Filtering and Prediction Problems," *Trans. of the ASME—J. of Basic Engr.*, 35–45 (March 1960).
2. J. J. D'Azzo and C. H. Houpis, *Linear Control System Analysis and Design*, 3rd ed., New York: McGraw-Hill, 1988.
3. R. C. Dorf, *Modern Control Systems*, 5th ed., Reading, Mass.: Addison-Wesley, 1989.

4.   T. A. Stansell, Jr., "The Many Faces of Transit," *Navigation*, Jour. of the Inst. of Navigation, 25:1, 55–70 (Spring 1978).

5.   P. Henrici, *Discrete Variable Methods in Ordinary Differential Equations*, New York: Wiley, 1962.

6.   K. S. Shanmugan and A. M. Breipohl, *Random Signals: Detection, Estimation, and Data Analysis*, New York: Wiley, 1988.

7.   J. M. Mendel, *Optimal Seismic Deconvolution*, New York: Academic Press, 1983.

8.   M. B. Priestly, *Spectral Analysis and Time Series*, New York: Academic Press, 1981.

9.   A. H. Jazwinski, *Stochastic Processes and Filtering Theory*, New York: Academic Press, 1970.

10.   J. S. Meditch, *Stochastic Optimal Linear Estimation and Control*, New York: McGraw-Hill, 1969.

Additional References on Kalman Filtering

11.   A. Gelb (Ed.), *Applied Optimal Estimation*, Cambridge, Mass.: MIT Press, 1974.

12.   P. S. Maybeck, *Stochastic Models, Estimation and Control* (Vol. 1), New York: Academic Press, 1979.

13.   A. P. Sage and J. L. Melsa, *Estimation Theory with Applications to Communications and Control*, New York: McGraw-Hill, 1971.

14.   S. M. Bozic, *Digital and Kalman Filtering*, London: E. Arnold, Publisher, 1979.

15.   B. D. O. Anderson and J. B. Moore, *Optimal Filtering*, Englewood Cliffs, N. J.: Prentice-Hall, 1979.

16.   C. T. Leondes, (Ed.), *Theory and Application of Kalman Filtering*, North Atlantic Treaty Organization AGARD Report No. 139 (February 1970).

17.   H. W. Sorenson, *Parameter Estimation*, New York: Marcel Dekker, 1980.

18.   H. W. Sorenson, "Kalman Filtering Techniques," in *Advances in Control Systems* (Vol. 3), C. T. Leondes (Ed.), New York: Academic Press, 1966, pp. 219–289.

19.   M. Aoki, *State Space Modeling of Time Series*, Berlin: Springer-Verlag, 1987.

20.   R. F. Stengel, *Stochastic Optimal Control–Theory and Application*, New York: Wiley, 1986.

6

Applications and Additional Topics on Discrete Kalman Filtering

In the period immediately following Kalman's original work, many extensions and variations were developed that enhanced the usefulness of the technique in applied work. This chapter continues the subject of discrete Kalman filtering with some of the more important extensions and related topics. Also a limited number of applications are presented to illustrate the versatility of Kalman filtering. The treatment of applications here must be brief. More applications are given in Chapters 9 and 10. There also exist a wealth of application papers in various journals and conference proceedings over the past 30 years. In particular, many papers dealing with navigation and trajectory determination will be found in *Navigation* (Journal of the Institute of Navigation), *IEEE Transactions on Aerospace and Electronic Systems*, and *IEEE Transactions on Automatic Control*. Also, Sorenson (1) gives an especially valuable collection of applied papers in his IEEE Press book.

6.1
AUGMENTING THE STATE VECTOR

The Kalman filter is especially sensitive to format in that the process to be estimated and the measurement relationship must be of the exact form

$$\mathbf{x}_{k+1} = \boldsymbol{\phi}_k \mathbf{x}_k + \mathbf{w}_k \tag{6.1.1}$$

$$\mathbf{z}_k = \mathbf{H}_k \mathbf{x}_k + \mathbf{v}_k \tag{6.1.2}$$

where $\mathbf{w}_k$ and $\mathbf{v}_k$ must be white sequences. We are assured that $\mathbf{w}_k$ will be white if Eq. (6.1.1) is the sampled version of a continuous linear system driven by white noise. However, more often than not, the driving functions in the original problem are not white, and they, in turn, must be modeled with differential equations relating them to fictitious white noise processes. This creates additional state variables that must be appended to the original ones in the state model. An example will illustrate this. A second example will also be given that illustrates augmentation of the state vector to account for measurement noise that is not white.

EXAMPLE 6.1

This example is due to B. E. Bona and R. J. Smay (2), and it was one of the early on-line applications of Kalman filtering. In marine applications, the mission time is usually long, and the ship's inertial navigation system is expected to operate continuously for weeks with a minimum of attention. One of the principle sources of errors in inertial systems is gyro drift, and this causes the position error to grow with time. Thus, the system must be reset occasionally with position fixes from other sources such as LORAN or a satellite navigation system, or any of a multitude of navigation aids. In addition to resetting the position error, it is also desirable to correct the underlying sources giving rise to the error—the so-called gyro biases. These cannot be observed directly at sea, so one must be content only to observe them indirectly by observing their effect on the system position error. Thus, the main purpose of the Kalman filter in this application is to estimate these hidden gyro biases. Then they can be corrected and the system error growth can be reduced after the reset.

The applicable error propagation equations for a damped inertial navigation system in a slow-moving vehicle are*

$$\dot{\psi}_x - \Omega_z \psi_y = \varepsilon_x \tag{6.1.3}$$

$$\dot{\psi}_y + \Omega_z \psi_x - \Omega_x \psi_z = \varepsilon_y \tag{6.1.4}$$

$$\dot{\psi}_z + \Omega_x \psi_y = \varepsilon_z \tag{6.1.5}$$

where x, y, and z denote the platform (gyro and accelerometer instrument cluster) coordinate axes in the north, west, and up directions, and

ψ_x = inertial system's west position error (in terms of angular measure in radians)

ψ_y = inertial system's south position error (in terms of angular measure in radians)

ψ_z = [platform azimuth error] − [west position error] · [tan (latitude)]

---

*Equations (6.1.3) to (6.1.5) are certainly not obvious, and a considerable amount of background in inertial navigation theory is needed to understand the assumptions and approximations leading to this simple set of equations (3, 4). We do not attempt to derive the equations here. For purposes of understanding the Kalman filter, simply assume that these equations do, in fact, accurately describe the error propagation in this application and proceed on to the details of the Kalman filter. (This example is developed further in Problems 6.1, 6.2, and 6.3.)

Also

$$\Omega_x = x \text{ component of earth rate } \Omega \text{ [i.e., } \Omega_x = \Omega \cos(\text{Lat.})]$$
$$\Omega_z = z \text{ component of earth rate } \Omega \text{ [i.e., } \Omega_z = \Omega \sin(\text{Lat.})]$$

and

$$\varepsilon_x, \varepsilon_y, \varepsilon_z = \text{gyro drift rates for the } x, y, \text{ and } z \text{ axis gyros}$$

We assume that the ship's latitude is known approximately; therefore Ω_x and Ω_z are known and may be assumed to be constant over reasonable time intervals.

The three differential equations [(6.1.3) to (6.1.5)] represent a third-order linear system with ε_x, ε_y, and ε_z as the driving functions. It is not reasonable to assume these to be white noise in this situation. Quite to the contrary, instead of "jumping around wildly" like white noise, just the reverse is to be expected. Gyro drift is due to mechanical causes that produce minute random torques about the gyro's output axis, and these usually change slowly with time. In the referenced paper, Bona and Smay suggest that the respective gyro drift rates may be properly modeled as a sum of a random constant plus a Markov component with a large time constant. This model accommodates slowly varying changes in the "bias" as well as a constant offset. Thus, we let

$$\varepsilon_x = \varepsilon_{xm} + \varepsilon_{xc}$$

$$\varepsilon_y = \varepsilon_{ym} + \varepsilon_{yc}$$

$$\varepsilon_z = \varepsilon_{zm} + \varepsilon_{zc} \tag{6.1.6}$$

where subscripts m and c denote Markov and constant components, respectively. Now, from Section 5.2 we know that random constants and Markov processes can be modeled with simple first-order differential equations. Thus, we have for the Markov components

$$\dot{\varepsilon}_{xm} = -\beta_x \varepsilon_{xm} + u_x$$

$$\dot{\varepsilon}_{ym} = -\beta_y \varepsilon_{ym} + u_y$$

$$\dot{\varepsilon}_{zm} = -\beta_z \varepsilon_{zm} + u_z \tag{6.1.7}$$

where u_x, u_y, and u_z are independent white noise processes. For the constant components, we have

$$\dot{\varepsilon}_{xc} = 0$$

$$\dot{\varepsilon}_{yc} = 0$$

$$\dot{\varepsilon}_{zc} = 0 \tag{6.1.8}$$

We now have six additional first-order equations that must be appended to the original three equations. The addition of the new state variables is often called "augmenting the state vector." Regardless of terminology, the system is now ninth order. Let

$$\psi_x, \psi_y, \psi_z = \text{state variables } x_1, x_2, x_3$$

$$\varepsilon_{xm}, \varepsilon_{ym}, \varepsilon_{zm} = \text{state variables } x_4, x_5, x_6$$

$$\varepsilon_{xc}, \varepsilon_{yc}, \varepsilon_{zc} = \text{state variables } x_7, x_8, x_9$$

The complete state model is then

$$
\begin{bmatrix} \dot{x}_1 \\ \dot{x}_2 \\ \dot{x}_3 \\ \dot{x}_4 \\ \dot{x}_5 \\ \dot{x}_6 \\ \dot{x}_7 \\ \dot{x}_8 \\ \dot{x}_9 \end{bmatrix}
=
\begin{bmatrix}
\begin{array}{ccc|ccc|c}
0 & \Omega_z & 0 & & & & \\
-\Omega_z & 0 & \Omega_x & & \mathbf{I} & & \mathbf{I} \\
0 & -\Omega_x & 0 & & & & \\
\hline
 & & & -\beta_x & 0 & 0 & \\
 & \mathbf{0} & & 0 & -\beta_y & 0 & \mathbf{0} \\
 & & & 0 & 0 & -\beta_z & \\
\hline
 & & & & & & \\
 & \mathbf{0} & & & \mathbf{0} & & \mathbf{0} \\
 & & & & & &
\end{array}
\end{bmatrix}
\begin{bmatrix} x_1 \\ x_2 \\ x_3 \\ x_4 \\ x_5 \\ x_6 \\ x_7 \\ x_8 \\ x_9 \end{bmatrix}
+
\begin{bmatrix} 0 \\ 0 \\ 0 \\ u_x \\ u_y \\ u_z \\ 0 \\ 0 \\ 0 \end{bmatrix}
$$

(6.1.9)

Note that the forcing functions appearing in the continuous model are now white; hence, we are assured that the discrete model, that is, the solution of Eq. (6.1.9), will be of the proper form given by Eq. (6.1.1). Once the parameters in the continuous model are set, it is a routine matter to determine ϕ_k and $\mathbf{Q}_k$ for any time interval (t_k, t_{k+1}) using the methods given in Section 5.3. It is assumed that the continuous-model parameters are known from physical considerations of the particular equipment at hand.

The measurements in this application will be a sequence of position fixes obtained from sources completely independent of the inertial system that is being reset. The difference between the independently determined position and the inertial system's output will yield noisy measurements of the inertial *N-S* and *E-W* position errors; the noise in this case being the error associated with external reference providing the discrete position fix. Presumably, this error will be random and independent of the $\mathbf{w}_k$ sequence in the inertial error model. Furthermore, if external fixes are available only on an occasional basis, it is reasonable to assume that their errors will be uncorrelated. Thus, $\mathbf{v}_k$ in the measurement model satisfies the necessary requirements, that is, it must also be a white sequence. With each position fix, we get a 2-tuple. Thus, the measurement model is

$$\begin{bmatrix} z_1 \\ z_2 \end{bmatrix} = \underbrace{\begin{bmatrix} 1 & 0 & 0 & & \\ & & & \mathbf{0} & \mathbf{0} \\ 0 & 1 & 0 & & \end{bmatrix}}_{\mathbf{H}_k} \begin{bmatrix} x_1 \\ x_2 \\ x_3 \\ \hline x_4 \\ x_5 \\ x_6 \\ \hline x_7 \\ x_8 \\ x_9 \end{bmatrix} + \begin{bmatrix} v_1 \\ v_2 \end{bmatrix} \qquad (6.1.10)$$

The $\mathbf{H}_k$ matrix is indicated in Eq. (6.1.10), and $\mathbf{R}_k$ will be assumed to be a diagonal matrix whose elements are the variances of the E-W and N-S components of external fix error. Since these will depend on the type of reference being used, the on-line computer must be programmed to accommodate to whatever references might be used. For example, one would not expect to obtain the same quality position fix using a handheld sextant as would be obtained from a navigation satellite system or LORAN. The elements of $\mathbf{R}_k$ must accurately reflect the quality of the measurement in order for the Kalman filter to produce good estimates. Once the estimates are obtained, the position and gyro bias corrections may be made in accordance with these estimates, and the inertial system is ready to operate unaided until another position fix becomes available. The system will then be reset again and so forth.

Before leaving this example, it is especially important to note the versatility of the Kalman filter in accommodating measurements from a variety of sources and at arbitrary times. This is a dynamical problem. The mathematical connections among the nine quantities being estimated and the measurements are relatively complicated and are changing with time. In effect, the Kalman filter automatically accounts for these complex relationships via the key parameters ϕ_k, $\mathbf{Q}_k$, $\mathbf{H}_k$, and $\mathbf{R}_k$; then it makes the best use of whatever reference information that may be available, be it meager or plentiful, high in quality or low in quality! ■

EXAMPLE 6.2

A classical problem in Wiener filter theory is one of estimating a Gauss–Markov process (signal) in the presence of additive noise that is also Gauss–Markov. (A stationary version of this problem was presented in Problem 4.5.) Of course, the spectral characteristics of the two processes must be different if the problem is to be nontrivial. Let us assume that the autocorrelation functions of the signal and noise processes are

$$R_s(\tau) = \sigma_s^2 e^{-\beta_s|\tau|} \qquad (6.1.11)$$

$$R_n(\tau) = \sigma_n^2 e^{-\beta_n|\tau|} \qquad (6.1.12)$$

and that $s(t)$ and $n(t)$ are independent. The measurement situation is given by

$$z(t) = s(t) + n(t) \qquad (6.1.13)$$

A Kalman filter is to be designed (i.e., modeled) to process a sequence of discrete samples of $z(t)$. These will be denoted as $z_0, z_1, z_2, \ldots$, and the sample spacing is Δt.

Obviously, we can model the signal process with one state variable driven by white noise (see Example 5.2). The modeling difficulty here lies with the measurement noise, which is nonwhite. This violates one of the assumptions associated with Eqs. (6.1.1) and (6.1.2), namely that v_k must be a white sequence. The solution to the difficulty in this case is simply to model the noise with a single state variable (just as was done with the signal) and absorb it in with the process to be estimated. The state vector then becomes a 2-tuple, and the continuous state equations are

$$\begin{bmatrix} \dot{x}_1 \\ \dot{x}_2 \end{bmatrix} = \begin{bmatrix} -\beta_s & 0 \\ 0 & -\beta_n \end{bmatrix} \begin{bmatrix} x_1 \\ x_2 \end{bmatrix} + \begin{bmatrix} \sqrt{2\sigma_s^2 \beta_s}\, u_s \\ \sqrt{2\sigma_n^2 \beta_n}\, u_n \end{bmatrix} \qquad (6.1.14)$$

where u_s and u_n are independent white noises. We note that x_1 and x_2 are decoupled, so we can write the corresponding discrete equations by inspection:

$$\begin{bmatrix} x_1 \\ x_2 \end{bmatrix}_{k+1} = \begin{bmatrix} e^{-\beta_s \Delta t} & 0 \\ 0 & e^{-\beta_n \Delta t} \end{bmatrix} \begin{bmatrix} x_1 \\ x_2 \end{bmatrix}_k + \begin{bmatrix} w_1 \\ w_2 \end{bmatrix}_k \qquad (6.1.15)$$

The transition and $\mathbf{Q}_k$ matrices can now be written by using the results of Example 5.3.

$$\Phi_k = \begin{bmatrix} e^{-\beta_s \Delta t} & 0 \\ 0 & e^{-\beta_n \Delta t} \end{bmatrix} \qquad (6.1.16)$$

$$\mathbf{Q}_k = \begin{bmatrix} \sigma_s^2(1 - e^{-2\beta_s \Delta t}) & 0 \\ 0 & \sigma_n^2(1 - e^{-2\beta_n \Delta t}) \end{bmatrix} \qquad (6.1.17)$$

In our two-state model the measurement noise was absorbed into the state vector, so the measurement model is now

$$z_k = \begin{bmatrix} 1 & 1 \end{bmatrix} \begin{bmatrix} x_1 \\ x_2 \end{bmatrix}_k + [0] \qquad (6.1.18)$$

Thus, the $\mathbf{H}_k$ and $\mathbf{R}_k$ parameters are

$$\mathbf{H}_k = \begin{bmatrix} 1 & 1 \end{bmatrix}, \qquad \mathbf{R}_k = \begin{bmatrix} 0 \end{bmatrix} \tag{6.1.19}$$

Note that $\mathbf{R}_k$ is zero in this model. This is perfectly legitimate in the discrete Kalman filter, provided that $\mathbf{Q}_k$ is positive definite. This assures that $(\mathbf{H}_k \mathbf{P}_k^- \mathbf{H}_k^T + \mathbf{R}_k)$ will always be invertible in the gain computation step.

Our model will be complete with the specification of the initial conditions. If we assume that the signal and noise processes are in a stationary condition when sampling begins at $t = 0$, then we are forced to choose $\hat{\mathbf{x}}_0^-$ and $\mathbf{P}_0^-$ to be

$$\hat{\mathbf{x}}_0^- = \begin{bmatrix} 0 \\ 0 \end{bmatrix}, \qquad (s \text{ and } n \text{ are zero-mean processes}) \tag{6.1.20}$$

$$\mathbf{P}_0^- = \begin{bmatrix} \sigma_s^2 & 0 \\ 0 & \sigma_n^2 \end{bmatrix}, \qquad (s \text{ and } n \text{ are assumed to be independent}) \tag{6.1.21}$$

Our model is now complete with the specification of $\boldsymbol{\Phi}_k, \mathbf{Q}_k, \mathbf{H}_k, \mathbf{R}_k$ and the initial conditions. ∎

A comment about modeling is in order before going on to another subject. Only rarely in applied work does the analyst/designer have the luxury of having the physical problem presented to him or her in the exact form given by Eqs. (6.1.1) and (6.1.2). More often than not, assumptions have to be made about the various processes involved, and the original problem has to be twisted around into the appropriate form, just as was the case with the previous two examples. This "twisting around" to fit the right form is often referred to as modeling, and it is all important in Kalman filtering applications. Good modeling leads to good results; bad modeling leads to bad results. It is as simple as that. There are no set rules to follow in the modeling procedure, and it often requires some imagination. Perhaps the best way to become adept at modeling is to look at a wide variety of examples. Hopefully, the examples given in this chapter and those in Chapters 9 and 10, where nonlinearities are considered, will be helpful in this regard.

6.2
ALTERNATIVE FORM OF THE DISCRETE KALMAN FILTER

The Kalman filter equations given in Chapter 5 can be algebraically manipulated into a variety of forms (5, 6, 7). An alternative form that is especially useful will now be presented. We begin with the expression for updating the error covariance, Eq. (5.4.22), and we temporarily omit the subscripts to save writing.

$$\mathbf{P} = (\mathbf{I} - \mathbf{KH})\mathbf{P}^- \tag{6.2.1}$$

Recall that the Kalman gain is given by Eq. (5.4.17).

$$\mathbf{K} = \mathbf{P}^-\mathbf{H}^T(\mathbf{H}\mathbf{P}^-\mathbf{H}^T + \mathbf{R})^{-1} \tag{6.2.2}$$

Substituting Eq. (6.2.2) into (6.2.1) yields

$$\mathbf{P} = \mathbf{P}^- - \mathbf{P}^-\mathbf{H}^T(\mathbf{H}\mathbf{P}^-\mathbf{H}^T + \mathbf{R})^{-1}\mathbf{H}\mathbf{P}^- \tag{6.2.3}$$

We now wish to show that if the inverses of $\mathbf{P}$, $\mathbf{P}^-$, and $\mathbf{R}$ exist, $\mathbf{P}^{-1}$ can be written as

$$\mathbf{P}^{-1} = (\mathbf{P}^-)^{-1} + \mathbf{H}^T\mathbf{R}^{-1}\mathbf{H} \tag{6.2.4}$$

Justification of Eq. (6.2.4) is straightforward. We simply form the product of the right sides of Eqs. (6.2.3) and (6.2.4) and show that this reduces to the identity matrix. Proceeding as indicated,

$$\left[\mathbf{P}^- - \mathbf{P}^-\mathbf{H}^T(\mathbf{H}\mathbf{P}^-\mathbf{H}^T + \mathbf{R})^{-1}\mathbf{H}\mathbf{P}^-\right]\left[(\mathbf{P}^-)^{-1} + \mathbf{H}^T\mathbf{R}^{-1}\mathbf{H}\right]$$

$$= \mathbf{I} - \mathbf{P}^-\mathbf{H}^T\left[(\mathbf{H}\mathbf{P}^-\mathbf{H}^T + \mathbf{R})^{-1} - \mathbf{R}^{-1} + (\mathbf{H}\mathbf{P}^-\mathbf{H}^T + \mathbf{R})^{-1}\mathbf{H}\mathbf{P}^-\mathbf{H}^T\mathbf{R}^{-1}\right]\mathbf{H}$$

$$= \mathbf{I} - \mathbf{P}^-\mathbf{H}^T\left[(\mathbf{H}\mathbf{P}^-\mathbf{H}^T + \mathbf{R})^{-1}(\mathbf{I} + \mathbf{H}\mathbf{P}^-\mathbf{H}^T\mathbf{R}^{-1}) - \mathbf{R}^{-1}\right]\mathbf{H}$$

$$= \mathbf{I} - \mathbf{P}^-\mathbf{H}^T[\mathbf{R}^{-1} - \mathbf{R}^{-1}]\mathbf{H}$$

$$= \mathbf{I}$$

An alternative expression for the Kalman gain may also be derived. Beginning with Eq. (6.2.2) we have

$$\mathbf{K} = \mathbf{P}^-\mathbf{H}^T(\mathbf{H}\mathbf{P}^-\mathbf{H}^T + \mathbf{R})^{-1}$$

Insertion of $\mathbf{P}\mathbf{P}^{-1}$ and $\mathbf{R}^{-1}\mathbf{R}$ will not alter the gain. Thus, $\mathbf{K}$ can be written as

$$\mathbf{K} = \mathbf{P}\mathbf{P}^{-1}\mathbf{P}^-\mathbf{H}^T\mathbf{R}^{-1}\mathbf{R}(\mathbf{H}\mathbf{P}^-\mathbf{H}^T + \mathbf{R})^{-1}$$

$$= \mathbf{P}\mathbf{P}^{-1}\mathbf{P}^-\mathbf{H}^T\mathbf{R}^{-1}(\mathbf{H}\mathbf{P}^-\mathbf{H}^T\mathbf{R}^{-1} + \mathbf{I})^{-1}$$

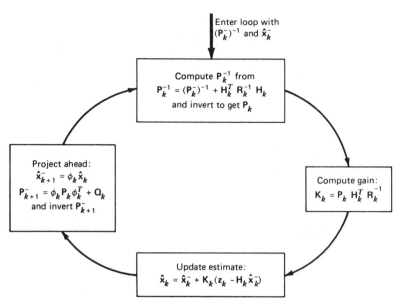

Figure 6.1. Alternative Kalman filter recursive loop.

We now use Eq. (6.2.4) for $\mathbf{P}^{-1}$ and obtain

$$
\begin{aligned}
\mathbf{K} &= \mathbf{P}\Big[(\mathbf{P}^-)^{-1} + \mathbf{H}^T\mathbf{R}^{-1}\mathbf{H}\Big]\mathbf{P}^-\mathbf{H}^T\mathbf{R}^{-1}(\mathbf{H}\mathbf{P}^-\mathbf{H}^T\mathbf{R}^{-1} + \mathbf{I})^{-1} \\
&= \mathbf{P}(\mathbf{I} + \mathbf{H}^T\mathbf{R}^{-1}\mathbf{H}\mathbf{P}^-)\mathbf{H}^T\mathbf{R}^{-1}(\mathbf{H}\mathbf{P}^-\mathbf{H}^T\mathbf{R}^{-1} + \mathbf{I})^{-1} \\
&= \mathbf{P}\mathbf{H}^T\mathbf{R}^{-1}(\mathbf{I} + \mathbf{H}\mathbf{P}^-\mathbf{H}^T\mathbf{R}^{-1})(\mathbf{H}\mathbf{P}^-\mathbf{H}^T\mathbf{R}^{-1} + \mathbf{I})^{-1} \\
&= \mathbf{P}\mathbf{H}^T\mathbf{R}^{-1} \qquad\qquad\qquad\qquad\qquad\qquad\qquad\qquad (6.2.5)
\end{aligned}
$$

The main results have now been derived, and Eqs. (6.2.4) and (6.2.5) may be rewritten with the subscripts reinserted:

$$
\mathbf{P}_k^{-1} = (\mathbf{P}_k^-)^{-1} + \mathbf{H}_k^T\mathbf{R}_k^{-1}\mathbf{H}_k \qquad\qquad (6.2.6)
$$

$$
\mathbf{K}_k = \mathbf{P}_k\mathbf{H}_k^T\mathbf{R}_k^{-1} \qquad\qquad (6.2.7)
$$

Note that the updated error covariance can be computed without first finding the gain. Also, the expression for gain now involves $\mathbf{P}_k$; therefore, if Eq. (6.2.7) is to be used, $\mathbf{K}_k$ must be computed *after* the $\mathbf{P}_k$ computation. Thus, the order in which the $\mathbf{P}_k$ and $\mathbf{K}_k$ computation appear in the recursive algorithm is reversed from that presented in Chapter 5. The alternative Kalman filter algorithm just derived is summarized in Fig. 6.1.

Note from Fig. 6.1 that two ($n \times n$) matrix inversions are required for each recursive loop. If the order of the state vector is large, this leads to obvious computational problems. Nevertheless, the alternative algorithm has some useful applications. One of these will now be presented as an example.

EXAMPLE 6.3

Suppose we wish to estimate a random constant based on a sequence of independent noisy measurements of the constant. We can think of the constant as being a deterministic random process that satisfies the differential equation

$$\dot{x} = 0 \tag{6.2.8}$$

Thus, ϕ_k and Q_k are 1 and 0, respectively. Let us also speculate that very little is known about the process initially. The constant is equally likely to be positive or negative, and its magnitude could be quite large. It might be thought of as a random variable with a flat probability density function extending from $-\infty$ to $+\infty$ or, more realistically, a normal zero-mean random variable with a very large variance. This being the case, the a priori estimate and associated error covariance should be

$$\hat{x}_0^- = 0 \tag{6.2.9}$$

$$P_0^- = \infty \tag{6.2.10}$$

However, this is not permitted in the usual Kalman filter algorithm (Fig. 5.9), because it leads to the indeterminant form ∞/∞ in the gain expression. The alternative algorithm will accommodate this situation, though, because $(P_0^-)^{-1}$ rather than P_0^- appears in the first step.

Time is of no consequence in this example because we are estimating a constant. So let us assume we have N independent noisy measurements of x, each made at $t = 0$ and having an error variance of σ^2. The measurement model is then

$$\begin{bmatrix} z_1 \\ z_2 \\ \vdots \\ z_N \end{bmatrix} = \begin{bmatrix} 1 \\ 1 \\ \vdots \\ 1 \end{bmatrix} [x] + \begin{bmatrix} v_1 \\ v_2 \\ \vdots \\ v_N \end{bmatrix} \tag{6.2.11}$$

and $\mathbf{R}_k$ is the $(N \times N)$ diagonal matrix

$$\mathbf{R}_k = \begin{bmatrix} \sigma^2 & 0 & & \cdots \\ 0 & \sigma^2 & & \\ \vdots & & \ddots & \\ & & & \sigma^2 \end{bmatrix} = \sigma^2 \mathbf{I} \tag{6.2.12}$$

Proceeding with the first step of the alternative algorithm yields

$$P_0^{-1} = (P_0^-)^{-1} + H_0^T R_0^{-1} H_0$$

$$= (\infty)^{-1} + [1 \quad 1 \quad \cdots \quad 1] \frac{1}{\sigma^2} I \begin{bmatrix} 1 \\ 1 \\ \vdots \\ 1 \end{bmatrix} \qquad (6.2.13)$$

or

$$P_0 = \frac{\sigma^2}{N} \qquad (6.2.14)$$

Next, the gain is computed as

$$K_0 = P_0 H_0^T R_0^{-1}$$

$$= \frac{\sigma^2}{N} [1 \quad 1 \quad \cdots \quad] \begin{bmatrix} \frac{1}{\sigma^2} & 0 & \cdots \\ 0 & \frac{1}{\sigma^2} & \\ \vdots & & \ddots \end{bmatrix}$$

$$= \begin{bmatrix} \dfrac{1}{N} & \dfrac{1}{N} & \cdots & \dfrac{1}{N} \end{bmatrix} \qquad (6.2.15)$$

Finally, the estimate is given by

$$\hat{x}_0 = \hat{x}_0^- + K_0(z - H_0 \hat{x}_0^-)$$

$$= K_0 z = \frac{z_1}{N} + \frac{z_2}{N} + \cdots + \frac{z_N}{N} \qquad (6.2.16)$$

The final result is no surprise; it is exactly the result one would expect from elementary statistics. The main point of this example is this: The alternative algorithm provides a means of starting the Kalman filter with "infinite uncertainty" if the physical situation under consideration so dictates. ■

6.3
SEQUENTIAL PROCESSING
OF THE MEASUREMENT DATA

We now have two different Kalman filter algorithms as summarized in Figs. 5.9 and 6.1. They are, of course, algebraically equivalent and produce identical estimates (with perfect arithmetic). The choice as to which should be used in a particular application is a matter of computational convenience. Both algorithms involve matrix inverse operations, and these may lead to difficulties. When using the alternative algorithm of Fig. 6.1, there is no reasonable way to avoid two ($n \times n$) matrix inversions with each recursive cycle. If the dimension of the state vector n is large, this is, at best, awkward computationally. On the other hand, the matrix inverse that appears in the regular algorithm given in Fig. 5.9 is the same order as the measurement vector. Since this is often less than the order of the state vector, it is usually the preferred algorithm. Furthermore, if the measurement errors at time t_k are uncorrelated, the inverse operation can be eliminated entirely by processing the data sequentially. This will now be shown.

We begin with the expression for the updated error covariance, Eq. (6.2.6):

$$\mathbf{P}_k^{-1} = (\mathbf{P}_k^-)^{-1} + \begin{bmatrix} \mathbf{H}_k^{aT} & \vdots & \mathbf{H}_k^{bT} & \vdots & \cdots \end{bmatrix} \begin{bmatrix} (\mathbf{R}_k^a)^{-1} & 0 & 0 \\ 0 & (\mathbf{R}_k^b)^{-1} & 0 \\ 0 & 0 & \ddots \end{bmatrix} \begin{bmatrix} \mathbf{H}_k^a \\ \mathbf{H}_k^b \\ \vdots \end{bmatrix}$$

$$(6.3.1)$$

The second term in Eq. (6.3.1) is intentionally written in partitioned form and $\mathbf{R}_k$ is assumed to be at least block diagonal. Physically, this means that the measurements available at t_k can be grouped together such that the measurements errors among the $a, b, \ldots$ blocks are uncorrelated. This is often the case when redundant measurements come from different instruments. We next expand the partitions of Eq. (6.3.1) to get

$$\mathbf{P}_k^{-1} = \underbrace{(\mathbf{P}_k^-)^{-1} + \mathbf{H}_k^{aT}(\mathbf{R}_k^a)^{-1}\mathbf{H}_k^a}_{\substack{\mathbf{P}_k^{-1} \text{ after assimilating} \\ \text{block } a \text{ measurements}}} + \mathbf{H}_k^{bT}(\mathbf{R}_k^b)^{-1}\mathbf{H}_k^b + \cdots \qquad (6.3.2)$$

$$\underbrace{\phantom{(\mathbf{P}_k^-)^{-1} + \mathbf{H}_k^{aT}(\mathbf{R}_k^a)^{-1}\mathbf{H}_k^a + \mathbf{H}_k^{bT}(\mathbf{R}_k^b)^{-1}\mathbf{H}_k^b}}_{\substack{\mathbf{P}_k^{-1} \text{ after assimilating both} \\ \text{block } a \text{ and } b \text{ measurements}}}$$

and so forth

Note that the sum of the first two terms is just the $\mathbf{P}_k^{-1}$ one would obtain after assimilating the "block a" measurement just as if no further measurements were available. The Kalman gain associated with this block of measurements may now be used to update the state estimate accordingly. Now think of making a trivial projection ahead through zero time. The a posteriori $\mathbf{P}$ then becomes the a priori $\mathbf{P}$ for the next step. When this is added to the b term of Eq. (6.3.2), we have the updated $\mathbf{P}_k^{-1}$ after assimilating the second block of data. This can now be repeated sequentially until all blocks are processed. The final estimate and associated error is then the same as would be obtained if all the measurements at t_k had been processed simultaneously. Thus, the designer has some flexibility in the design of the system software. The available measurements at any particular time may be processed either sequentially in blocks or all at once, as best suits the situation at hand.

Sequential processing is also useful from a system organization viewpoint. Often the system must have the flexibility to accommodate a variety of measurement combinations at each update point. By block processing, the system may be programmed to cycle through all possible measurement blocks sequentially, processing those that are available and skipping those that are not. Simultaneous processing requires a somewhat more complicated system organization whereby the system must be able to form appropriate $\mathbf{H}_k$ and $\mathbf{R}_k$ matrices for all possible combinations of measurements, and it must be prepared to do the corresponding matrix operations with various dimensionality.

There are bound to be some applications where the measurement errors are all mutually correlated. The $\mathbf{R}_k$ matrix is then "full." If this is the case, and sequential processing is desirable, linear combinations of the measurements may be formed in such a way as to form a new set of measurements whose errors are uncorrelated. One technique for accomplishing this is known as the Gram–Schmidt orthogonalization procedure (7). This procedure is straightforward and an exercise is included to demonstrate its application to the problem of decoupling the measurement errors (see Problem 6.4).

6.4
AIDED INERTIAL NAVIGATION SYSTEMS

Kalman filtering has been used extensively in navigation applications since the mid-1960s, and interest continues unabated with the most recent applications having to do with the navigation satellite system known as Global Positioning System (GPS) [(8, 9; also, see Chapter 10)]. Navigation problems seem to form a natural setting for Kalman filtering for the following reasons:

1. The dynamics are usually linear (or may be linearized with reasonable accuracy—see Chapter 9).

2. There is often a redundancy of measurement information from a variety of navigational sources. Also, we usually want to estimate more

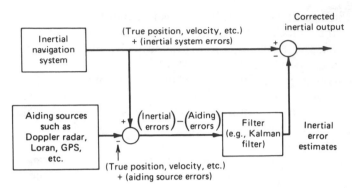

Figure 6.2. Aided inertial navigation system — feedforward configuration.

than one quantity, for example, both position and velocity. Thus the problem is of the multiple-input/multiple-output variety.

3. The navigation problem is basically an on-line problem. Measurements must be processed essentially in real time, and thus efficient processing is a necessity.

4. There is frequently a need to squeeze the best possible performance out of the system, and thus the data processing must be optimal (or at least nearly so).

One particular integrated navigation system problem that keeps recurring is that of integrating inertial system measurements with other navigation aids such as LORAN, GPS, or Doppler radar. It was mentioned in Section 6.1 that inertial navigation systems have drift characteristics that cause the system errors to grow with time. They also exhibit undamped oscillatory errors that are undesirable (3, 4). Thus, in applications where the mission time is relatively long, inertial systems are usually used in conjunction with other navigation instruments that perform an aiding role in the overall navigation system. Such integrated systems are sometimes called *aided* inertial systems, in contrast to *pure* inertial systems, which operate without the benefit of any supplementary navigation information.

Since aided inertial systems always involve redundant measurements, the question arises as to how all the redundant information should be blended together to yield the best estimates of position, velocity, and so on. The system configuration usually used is shown in Fig. 6.2.* This will be recognized as being similar to the complementary filter discussed in Section 4.6. Conceptually, we think of the inertial system as being the prime system that provides continuous outputs of position, velocity, and vehicle angular orientation relative to some

---

*The configuration shown in Fig. 6.2 is known as the feedforward configuration. It should be apparent that the estimates of the inertial errors could also be fed back and the corrections made internally in the inertial system. This is called the feedback configuration. Within the linearity assumption, both configurations yield the same results. However, in practice there are subtle differences that involve nonlinearities. This is discussed further in Chapter 9.

known reference such as an earth-fixed reference frame. The aiding sources are supplementary to the inertial system and their outputs may be available only intermittently, depending on the situation at hand.

The purpose of the filter is to estimate the inertial system errors. The estimates may then be used to correct the inertial system's output as indicated in Fig. 6.2. It is important to note that the filter, be it optimal or otherwise, operates only on the system errors. The total navigation quantities, such as position and velocity, pass through the system without distortion or delay. This is important in the navigation problem, because vehicle position and velocity are usually *not* properly modeled as random processes. If the filter indicated in Fig. 6.2 is to be a Kalman filter, then all the errors in the inertial and aiding sources must be modeled as random processes in vector form, and the filter must estimate all the state variables of the system. The filter estimates are then used to correct the inertial system.

As mentioned previously, the particular combination of inertial data integrated with GPS is of special interest in the period of the 1990s, when GPS is becoming operational. Nonlinear measurement relationships are involved in this system, however, so further discussion of this is deferred until Chapter 10 following the material on linearization in Chapter 9. The reader who is especially interested in the integrated navigation system problem (and is also short of time) might do well to skip Chapters 7 and 8 and proceed directly to Chapter 9 after finishing this chapter.

6.5
POWER SYSTEM
RELAYING APPLICATION

New applications of Kalman filtering keep appearing regularly, and many of these are now outside the original application area of navigation. One such application has to do with power system relaying (12, 13). When a fault (short) occurs on a three-phase transmission line, it is desirable to sense the problem promptly and take appropriate relaying action to protect the remainder of the system. The hierarchy of decisions that must be made as to which relays should trip (and where) is relatively complicated. It suffices to say here that it is desirable to determine the distance to the fault as soon as possible; and, in order to do this, the steady-state postfault currents and voltages must be estimated. Transient components are superimposed on the steady-state signals immediately after the fault, so that the transients become the corrupting noise in the problem. Normally, these transients are not considered as random noise. However, to model them otherwise complicates the problem immensely because of the many variables involved. So the basic problem is to estimate the steady-state components of the sending-end voltages and current in the presence of the transient (noise) components. It is assumed that digital samples of the various phase voltages and currents are available for processing at a reasonably fast rate, say 64 samples per cycle of the 60-Hz signal.

Girgis and Brown (13) made an extensive simulation study of the transients accompanying various types of faults for various lengths of line, and so forth. They concluded that the transients could be approximated as nonstationary random processes, and they developed models accordingly that would fit the required format of a Kalman filter. The simulation study indicated that the voltage transients consisted primarily of high-frequency components that decayed exponentially with time. Thus it seemed reasonable to model the time samples of the process as a white sequence with an exponentially decaying variance. The current transients, however, showed (on the average) a sizable long-time-constant exponential component in addition to the high-frequency components. (Power engineers sometimes refer to this as the "dc offset.") Thus the model chosen for the current noise was an exponential process with random initial amplitude plus a white sequence with exponentially decaying variance. The signal process to be estimated for both current and voltage was a sine wave with random amplitude and phase. This is readily modeled as a two-element vector, where the state variables are the coefficients of the sine and cosine components of the wave [see Section 5.2 and Problem 5.8(b)]. The final models for the currents and voltages may be summarized as follows:

Voltage Model (same for each phase)

1. State equations:

$$\begin{bmatrix} x1_{k+1} \\ x2_{k+1} \end{bmatrix} = \begin{bmatrix} 1 & 0 \\ 0 & 1 \end{bmatrix} \begin{bmatrix} x1_k \\ x2_k \end{bmatrix} + \begin{bmatrix} 0 \\ 0 \end{bmatrix} \qquad (6.5.1)$$

2. Measurement equation:

$$z_k = \begin{bmatrix} \cos \omega_0 k \, \Delta t & -\sin \omega_0 k \, \Delta t \end{bmatrix} \begin{bmatrix} x1_k \\ x2_k \end{bmatrix} + v_k \qquad (6.5.2)$$

3. Initial conditions:

$$\mathbf{P}_0^- = \begin{bmatrix} \sigma_v^2 & 0 \\ 0 & \sigma_v^2 \end{bmatrix} \qquad (6.5.3)$$

$$\hat{\mathbf{x}}_0^- = \text{Measured value of } \mathbf{x} \text{ just prior to the fault} \qquad (6.5.4)$$

The σ_v^2 parameter was determined by simulation. It is fixed and not determined on-line. On the other hand, the initial estimates of $\mathbf{x}$ are assumed to be determined on-line. There is no reason to waste the available measurement information just prior to the fault. It should be obvious from Eq. (6.5.1) that $\mathbf{Q}_k = 0$ for the voltage model. Also, as mentioned previously, $\mathbf{R}_k$ is assumed to decay exponentially with k, and the exponential parameters are predetermined by simulation or experimental data.

Current Model (same for all phases)

1. State equations:

$$
\begin{bmatrix} x1_{k+1} \\ x2_{k+1} \\ x3_{k+1} \end{bmatrix} = \begin{bmatrix} 1 & 0 & 0 \\ 0 & 1 & 0 \\ 0 & 0 & e^{-\beta \Delta t} \end{bmatrix} \begin{bmatrix} x1_k \\ x2_k \\ x3_k \end{bmatrix} + \begin{bmatrix} 0 \\ 0 \\ w_k \end{bmatrix} \qquad (6.5.5)
$$

2. Measurement equation:

$$
z_k = \begin{bmatrix} \cos \omega_0 k \, \Delta t & -\sin \omega_0 k \, \Delta t & 1 \end{bmatrix} \begin{bmatrix} x1_k \\ x2_k \\ x3_k \end{bmatrix} + v_k \qquad (6.5.6)
$$

3. Initial conditions:

$$
\mathbf{P}_0^- = \begin{bmatrix} \sigma_i^2 & 0 & 0 \\ 0 & \sigma_i^2 & 0 \\ 0 & 0 & \sigma_i^2 \end{bmatrix} \qquad (6.5.7)
$$

$$
\hat{\mathbf{x}}_0^- = \begin{bmatrix} x1^- \ (\text{meas.}) \\ x2^- \ (\text{meas.}) \\ 0 \end{bmatrix} \qquad (6.5.8)
$$

Just as in the voltage model, σ_i^2 is determined off-line and the first two elements of $\hat{\mathbf{x}}_0^-$ are obtained from measurements just prior to the fault. The third element of $\hat{\mathbf{x}}_0^-$ (i.e., the exponential component) is assumed to be zero. The measurement error variance R_k is assumed to decay exponentially just as in the voltage model. The $\mathbf{Q}_k$ parameter is not zero, though, because the exponential component was observed in the simulation studies to have a small residual noise associated with it. This is accounted for in the model with $\mathbf{w}_k$, and thus the "33" element of $\mathbf{Q}_k$ is nonzero. In effect, $x3$ is modeled as a nonstationary process with a large random initial value, and then it relaxes to a Markov process with a relatively small rms value in the steady-state condition. This is an unusual model, but perfectly legitimate. Again, this illustrates the versatility of the Kalman filter to adapt to a wide variety of situations.

Figure 6.3. Kalman filter estimation of the postfault voltage states.

Figures 6.3 and 6.4 show the voltage and current estimates for a particular simulation of a line-to-ground fault located 90 miles from the sending end. The details of the simulation are not important here, because the results are all relative. Recall that $x1$ and $x2$ are the coefficients of the sine and cosine components of the steady-state values, so they are constants. Note that the Kalman filter estimates converge reasonably well to the correct values after about 8 ms (half cycle at 60 Hz). Figure 6.5 shows the result of using the voltage and current estimates of this simulation to compute distances to the fault. A similar distance calculation was also made using current and voltages as determined by a discrete Fourier transform algorithm. Both the Fourier transform and Kalman filter results are shown in Fig. 6.5, and it is clear that the Kalman filter converges on the correct result faster than the other algorithm. This is as expected. The discrete Fourier transform approach does not account for the time-varying nature of the noise, nor does it allow for any a priori knowledge of

Figure 6.4. Kalman filter estimation of the postfault current states.

Figure 6.5. The computed distance to the fault using the Kalman filter algorithm and the discrete Fourier transform.

the parameters being estimated. Of course, both algorithms converge to the correct result eventually. However, time is of the essence! Why accept inferior performance when optimal performance is readily available.

Improvements have been made on this basic scheme, since it was first presented in 1981 by Girgis and Brown (13). The refinements involve adaptive Kalman filtering, though, so further discussion of this application will be deferred until Chapter 9.

6.6
DIVERGENCE PROBLEMS

Since the discrete Kalman filter is recursive, the looping can be continued indefinitely, in principle, at least. There are practical limits, though, and under certain conditions divergence problems can arise. We elaborate briefly on three common sources of difficulty.

Roundoff Errors

As with any numerical procedure, roundoff error can lead to problems as the number of steps becomes large. There is no one simple solution to this, and each case has to be examined on its own merits. Fortunately, if the system is observable and process noise drives each of the state variables, the Kalman filter has a degree of natural stability. In this case a stable, steady-state solution for the **P** matrix will normally exist, even if the process is nonstationary. If the **P** matrix is perturbed from its steady-state solution in such a way as not to lose positive definiteness, then it tends to return to the same steady-state solution. This is obviously helpful, provided **P** does not lose its positive definiteness. (See Section 7.6 for more on filter stability.)

Some techniques that have been found useful in preventing, or at least forestalling, roundoff error problems are:

1. Use high-precision arithmetic, especially in off-line analysis work.

2. If measurement data are sparse, beware of propagating the **P** matrix in many tiny steps between measurements. (This simplifies the transition matrix calculation but opens "Pandora's box" with regard to roundoff error.)

3. If possible, avoid deterministic (undriven) processes in the filter modeling. (*Example:* a random constant.) These usually lead to a situation where the **P** matrix approaches a semidefinite condition as the number of steps becomes large. A small error may then trip the **P** matrix into a non-positive-definite condition, and this can then lead to divergence. A good solution is to add (if necessary) small positive quantities to the major-diagonal terms of the **Q** matrix. This amounts to inserting a small amount of process noise to each of the states. This leads to a degree of suboptimality, but that is better than having the filter diverge!

4. Symmetrize the **P** and **P**$^-$ matrices with each recursive step. This is very important. We know that a covariance matrix must be symmetric, so any asymmetry must be due to roundoff error. The asymmetry can grow if left unchecked. [The symmetry problem is automatically solved if, in programming the recursive equations, one assumes symmetry and uses only the upper (or lower) triangular part of the covariance matrix in all required matrix-multiply operations.]

5. In some on-line applications where the filter is expected to operate for long periods without reinitialization, it may be desirable to implement an alternative algorithm known as *U-D* factorization. This algorithm is mathematically equivalent to the regular and alternative Kalman filter algorithms given in Sections 5.4 and 6.2, but factors of **P** are propagated with each step rather than **P** itself. The net effect is better numerical behavior (see Chapter 9).

Modeling Errors

Another type of divergence may arise because of inaccurate modeling of the process being estimated. This has nothing to do with numerical roundoff; it occurs simply because the designer (engineer) "told" the Kalman filter that the process behaved one way, whereas, in fact, it behaves another way. As a simple example, if you tell the filter that the process is a random constant (i.e., zero slope), and the actual process is a random ramp (nonzero slope), the filter will be continually trying to fit the wrong curve to the measurement data! This can also occur with nondeterministic as well as deterministic processes, as will now be demonstrated.

EXAMPLE 6.4

Consider a process that is actually random walk but is incorrectly modeled as a random constant. We have then (with numerical values inserted to correspond to a subsequent simulation):

(a) The "truth model":

$$\dot{x} = u(t), \qquad u(t) = \text{unity Gaussian white noise, and } \text{Var}[x(0)] = 1$$

$$z_k = x_k + v_k, \qquad \begin{array}{l} \text{measurement samples at } t = 0, 1, 2, \ldots \\ \text{and } \text{Var}(v_k) = .1. \end{array}$$

(b) Incorrect Kalman filter model:

$$x = \text{constant}, \qquad \text{where } x \sim N(0, 1)$$

$$z_k = x_k + v_k \qquad (\text{same as for truth model})$$

The Kalman filter parameters for the incorrect model (b) are: $\phi_k = 1$, $Q_k = 0$, $H_k = 1$, $R_k = .1$, $\hat{x}_0 = 0$, and $P_0^- = 1$. For the truth model the parameters are the same except that $Q_k = 1$, rather than zero.

The random walk process (a) was simulated using Gaussian random numbers with zero mean and unity variance. The resulting sample process for 35 sec is shown in Fig. 6.6. A measurements sequence z_k of this sample process was also generated using another set of $N(0, 1)$ random numbers for v_k. This measurement sequence was first processed using the incorrect model (i.e., $Q_k = 0$), and again with the correct model (i.e., $Q_k = 1$). The results are shown along with the sample process in Fig. 6.6. In this case, the measurement noise is relatively small ($\sigma \approx .3$), and we note that the estimates of the correctly modeled filter follow the random walk quite well. On the other hand, the incorrectly modeled filter does very poorly after the first few steps. This is due to the filter's gain becoming less and less with each succeeding step. At the 35th step the gain is almost two orders of magnitude less than at the beginning. Thus, the filter becomes very sluggish and will not follow the random walk. Had the simulation been allowed to go on further, it would have become even more sluggish. ∎

The moral to Example 6.4 is simply this. Any model that assumes the process, or any facet of the process, to be absolutely constant forever and ever is

Figure 6.6. Simulation results for random walk example.

a risky model. In the physical world, very few things remain absolutely constant. Instrument biases, even though called "biases," have a way of slowly changing with time. Thus, most instruments need occasional recalibration. The obvious remedy for this type of divergence problem is always to insert some process noise into each of the state variables. Do this even at the risk of some degree of suboptimality; it makes for a much safer filter than otherwise. It also helps with potential roundoff problems. (*Note:* Don't "blame" the filter for this kind of divergence problem. It is the fault of the designer/analyst, not the filter!)

In view of the preceding remarks, it is of interest to look again at the gyro-bias models used in Example 6.1. In the Bona–Smay example the gyro biases were modeled as the sum of Markov and constant components. Obviously the constant component is somewhat risky if the system is intended to operate for long periods of time. It is completely inflexible, and the filter will give measurement data that is months old just as much weight as current data in estimating the constant. This can be corrected simply by changing the q_{77}, q_{88}, and q_{99} terms in the **Q** matrix from zero to appropriate small positive values. In effect, this allows the "constants" the flexibility to random walk if the measurements so dictate. A random walk model is a safer model where the time span is large, and it is usually preferred over the truly constant model.

Choosing an appropriate process model is always an important consideration, and a certain amount of common sense judgment is called for in deciding
on a model that will fit the situation at hand reasonably well, but at the same
time will not be too complicated. There are no firm rules on this. You have to
consider each case on its own merits.

Observability Problem

There is a third kind of divergence problem that may occur when the system is
not observable. Physically, this means that there are one or more state variables
(or linear combinations thereof) that are hidden from the view of the observer
(i.e., the measurements). As a result, if the unobserved processes are unstable,
the corresponding estimation errors will be similarly unstable. This problem has
nothing to do with roundoff error or inaccurate system modeling. It is just a
simple fact of life that sometimes the measurement situation does not provide
enough information to estimate all the state variables of the system. In a sense,
this type of problem should not even be referred to as divergence, because the
filter is still doing the best estimation possible under adverse circumstances.

There are formal tests of observability that may be applied to systems of
low dimensionality. These tests are not always practical to apply, though, in
higher-order systems. Sometimes one is not even aware that a problem exists
until after extensive error analysis of the system (see Section 6.7). If unstable
estimation errors exist, this will be evidenced by one or more terms along the
major diagonal of **P** tending to increase without bound. If this is observed, and
proper precautions against roundoff error have been taken, the analyst knows
an observability problem exists. The only really good solution to this kind of
divergence is to improve the observability situation by adding appropriate
measurements to make the system completely observable.

6.7
OFF-LINE SYSTEM ERROR ANALYSIS

In addition to its value as an estimator, the Kalman filter also provides a
convenient means of system error analysis. In preliminary analysis and design,
the analyst often needs to assess various competing designs relative to their
effect on system accuracy. Recall that in the filter recursive equations, the error
covariance **P** is propagated along with the estimate $\hat{x}$. As a matter of fact, this
cannot be avoided because **P** is needed for the gain computation and subsequent updating of the estimate. The reverse is not true, though. The **P** matrix
can be propagated without forming the estimate. With a modest amount of
algebra, one can write $\mathbf{P}_{k+1}$ explicitly as a function of $\mathbf{P}_k$, and thus have a single
recursive equation for the error covariance matrix (see Problem 6.6). This leads
to no significant saving in the number of arithmetic steps needed in the loop,

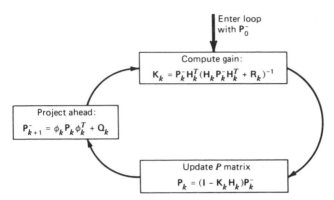

Figure 6.7.   Recursive loop for propagating optimal error covariance matrix.

though, so the analyst usually programs the filter loop shown in Fig. 5.9 with the estimate computations omitted. The abbreviated loop usually used for analysis of the optimal estimator is summarized in Fig. 6.7. It should be emphasized here that the **P** matrix does not depend on the actual measurement data received on any particular sample run. Thus, the Kalman filter in its pure form is not adaptive. The exception would be the case where the system has the capability of accepting or rejecting measurement data on-line based on some sort of validity check. Many systems are designed with this feature, and they might be said to be adaptive in this primitive sense.

When the optimal gain is used in error analysis, the results represent the ideal situation where the filter model and the actual process model are identical. For purposes of comparison, it is also desirable to be able to assess the error in suboptimal systems. This can be done by cycling suboptimal gains through a truth-model loop that uses the appropriate **P**-update equation. Recall that Eq. (5.4.11) is valid for any gain, optimal or suboptimal. The procedure for performing suboptimal error analysis is then summarized in Fig. 6.8. If you wish to make a comparison between optimal and suboptimal systems, the optimal loop of Fig. 6.7 must also be run in addition to the loops shown in Fig. 6.8. Usually the loops are all run in parallel, because some of the computations required are common to all loops. Be careful, though, to keep the gains and

Figure 6.8.   Block diagram for suboptimal error analysis.

P-matrix computations of each loop separated because they are all different (even though we have not attempted to distinguish among them in the notation in Fig. 6.7 and 6.8). It is a bit confusing, but there are three **P** matrices to keep straight:

1. **P** of the optimal estimator: This always yields the most optimistic results—that which might be expected if the filter model and the truth model match perfectly.
2. **P** of the truth model with suboptimal gains: This yields more realistic results. It represents the actual estimation errors to be expected when the filter model and the truth model are not the same.
3. **P** of the suboptimal filter that generates the suboptimal gain sequence. This matrix has no significance in terms of the errors of anything. It can be viewed just as a necessary adjunct for computing the suboptimal gain sequence.

Note that the gain and **P** matrices of the optimal and suboptimal models may not even have the same dimensions. Often the analyst wishes to assess the degradation in performance caused by deletion of some of the less important state variables in the state model. In this case, the suboptimal model is of lesser order than the truth model, and suboptimal gains for the deleted state variables are not generated by the suboptimal model. Thus, zero gains must be used for the deleted state variables in the truth-model calculations. However, a word of caution is in order. In deriving the recursive equations, it was tacitly assumed that the estimates and the true states are propagated ahead via the same transition matrix. This will not be true with the lesser-order model, because its ϕ_k will be of less order than ϕ_k for the truth model, which in turn governs the true $\mathbf{x}_k$. However, if the *partitioned part* of the truth ϕ_k corresponding to the reduced-order system is the same as that used in the suboptimal system, valid results may be obtained for the state variables of interest. One should be careful, though, and always check this for the particular problem at hand.

A simple example will now illustrate suboptimal error analysis.

EXAMPLE 6.5

We return to Example 6.4 in which a random-walk process was incorrectly modeled as a random constant. One sample run of a random-walk process was used to demonstrate the divergence phenomenon. However, this can hardly be called proof of divergence. We could, of course, make many more runs using new sets of random numbers, and then average the results to find the rms error. This would be doing the analysis the hard way, though. With suboptimal variance analysis, we can obtain the rms error with one run! All we need to do is consider the random-walk model as the truth model and the random-constant model as the suboptimal system, and then cycle the suboptimal gains through the truth model as indicated in Fig. 6.8. This was done for the first 15 steps of this example, and the resulting rms error along with suboptimal gains is shown in Fig. 6.9. The optimal rms error is also shown for comparison. It can be seen

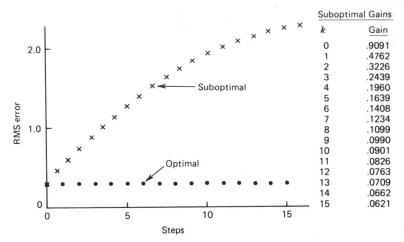

Figure 6.9.   Suboptimal analysis for random-walk example.

that divergence does occur in this situation. In this example, it can be easily verified that the suboptimal filter approaches a steady-state condition where the error variance increases by a fixed amount with each step. Thus, the rms error increases as the square root of the number of steps as indicated in Fig. 6.9.   ■

6.8
RELATIONSHIP TO DETERMINISTIC
LEAST SQUARES AND NOTE
ON ESTIMATING A CONSTANT

Both Kalman and Wiener filtering are sometimes referred to simply as least-squares filtering (14, 15, 16). It was mentioned in Chapter 4 that this is somewhat an oversimplification, because the criterion for optimization is minimum *mean*-square error and not the squared error in a deterministic sense. There is, however, a coincidental connection between Kalman/Wiener filtering and deterministic least squares, and this will now be demonstrated. The presentation here follows closely that of Sorenson (7).

Consider a set of m linear equations in $\mathbf{x}$ specified in matrix form by

$$\mathbf{Mx} = \mathbf{b} \qquad (6.8.1)$$

In Eq. (6.8.1) we think of $\mathbf{M}$ and $\mathbf{b}$ as being given, and $\mathbf{x}$ is $(n \times 1)$, $\mathbf{b}$ is $(m \times 1)$, and thus $\mathbf{M}$ is $(m \times n)$. Let us assume that $m > n$, and that $\mathbf{x}$ is overdetermined by the system of equations represented by Eq. (6.8.1). Thus no solution for $\mathbf{x}$ will satisfy all equations. This situation arises frequently in physical experiments where redundant noisy measurements are made of linear combinations of fixed parameters. In such cases it is logical to ask, "What solution will best fit all the equations?" The term *best* must, of course, be defined and it is frequently defined to be the particular $\mathbf{x}$, say $\mathbf{x}_{\text{opt}}$, that minimizes the sum of the squared residuals. That is, move $\mathbf{b}$ to the left side of Eq. (6.8.1) and substitute $\mathbf{x}_{\text{opt}}$ for $\mathbf{x}$.

This yields a residual vector $\boldsymbol{\varepsilon}$ given by

$$\mathbf{M}\mathbf{x}_{\text{opt}} - \mathbf{b} = \boldsymbol{\varepsilon} \tag{6.8.2}$$

and $\mathbf{x}_{\text{opt}}$ is chosen such that $\boldsymbol{\varepsilon}^T\boldsymbol{\varepsilon}$ is minimized. A perfect fit, of course, would make $\boldsymbol{\varepsilon}^T\boldsymbol{\varepsilon} = 0$.

We can generalize at this point and consider a weighted sum of squared residuals specified by

$$\begin{bmatrix} \text{Weighted sum of} \\ \text{squared residuals} \end{bmatrix} = (\mathbf{M}\mathbf{x}_{\text{opt}} - \mathbf{b})^T \mathbf{W}(\mathbf{M}\mathbf{x}_{\text{opt}} - \mathbf{b}) \tag{6.8.3}$$

We assume that the weighting matrix $\mathbf{W}$ is symmetric and positive definite and, hence, so is its inverse. If we wish equal weighting of the residuals, we simply let $\mathbf{W}$ be the identity matrix. The problem now is to find the particular $\mathbf{x}$ (i.e., $\mathbf{x}_{\text{opt}}$) that minimizes the weighted sum of the residuals. Toward this end, the expression given by Eq. (6.8.3) may be expanded and differentiated term by term and then set equal to zero.* This leads to

$$\frac{d}{d\mathbf{x}_{\text{opt}}}\left[\mathbf{x}_{\text{opt}}^T(\mathbf{M}^T\mathbf{W}\mathbf{M})\mathbf{x}_{\text{opt}} - \mathbf{b}^T\mathbf{W}\mathbf{M}\mathbf{x}_{\text{opt}} - \mathbf{x}_{\text{opt}}^T\mathbf{M}^T\mathbf{W}\mathbf{b} + \mathbf{b}^T\mathbf{b}\right]$$

$$= 2(\mathbf{M}^T\mathbf{W}\mathbf{M})\mathbf{x}_{\text{opt}} - (\mathbf{b}^T\mathbf{W}\mathbf{M})^T - \mathbf{M}^T\mathbf{W}\mathbf{b} = 0 \tag{6.8.4}$$

---

*The derivative of a scalar s with respect to a vector $\mathbf{x}$ is defined to be

$$\frac{ds}{d\mathbf{x}} = \begin{bmatrix} \dfrac{ds}{dx_1} \\ \dfrac{ds}{dx_2} \\ \vdots \\ \dfrac{ds}{dx_n} \end{bmatrix}$$

The two matrix differentiation formulas used to arrive at Eq. (6.8.4) are

$$\frac{d(\mathbf{x}^T\mathbf{A}\mathbf{x})}{d\mathbf{x}} = 2\mathbf{A}\mathbf{x} \quad \text{(for symmetric } \mathbf{A})$$

and

$$\frac{d(\mathbf{a}^T\mathbf{x})}{d\mathbf{x}} = \frac{d(\mathbf{x}^T\mathbf{a})}{d\mathbf{x}} = \mathbf{a}$$

Both of these formulas can be verified by writing out a few scalar terms of the matrix expressions and using ordinary differentiation methods.

Equation (6.8.4) may now be solved for $\mathbf{x}_{opt}$. The result is

$$\mathbf{x}_{opt} = \left[(\mathbf{M}^T\mathbf{W}\mathbf{M})^{-1}\mathbf{M}^T\mathbf{W}\right]\mathbf{b} \tag{6.8.5}$$

and this is the solution of the deterministic least-squares problem.

Next consider the Kalman filter solution for the same measurement situation. The vector $\mathbf{x}$ is assumed to be a random constant, so the differential equation for $\mathbf{x}$ is

$$\dot{\mathbf{x}} = 0 \tag{6.8.6}$$

The corresponding discrete model is then

$$\mathbf{x}_{k+1} = \mathbf{I} \cdot \mathbf{x}_k + 0 \tag{6.8.7}$$

The measurement equation is

$$\mathbf{z}_k = \mathbf{H}_k\mathbf{x}_k + \mathbf{v}_k \tag{6.8.8}$$

where $\mathbf{z}_k$ and $\mathbf{H}_k$ play the same roles as $\mathbf{b}$ and $\mathbf{M}$ in the deterministic problem. Since time is of no consequence, we assume that all measurements occur simultaneously. Furthermore, we assume that we have no a priori knowledge of $\mathbf{x}$, so the initial $\hat{\mathbf{x}}_0^-$ will be zero and its associated error covariance will be ∞. Therefore, using the alternative form of the Kalman filter (Section 6.2), we have

$$\mathbf{P}_0^{-1} = (\infty)^{-1} + \mathbf{H}_0^T\mathbf{R}_0^{-1}\mathbf{H}_0$$

$$= \mathbf{H}_0^T\mathbf{R}_0^{-1}\mathbf{H}_0 \tag{6.8.9}$$

The Kalman gain is then

$$\mathbf{K}_0 = \left(\mathbf{H}_0^T\mathbf{R}_0^{-1}\mathbf{H}_0\right)^{-1}\mathbf{H}_0^T\mathbf{R}_0^{-1}$$

and the Kalman filter estimate of $\mathbf{x}$ at $t = 0$ is

$$\hat{\mathbf{x}}_0 = \left[\left(\mathbf{H}_0^T\mathbf{R}_0^{-1}\mathbf{H}_0\right)^{-1}\mathbf{H}_0^T\mathbf{R}_0^{-1}\right]\mathbf{z}_0 \tag{6.8.10}$$

This is the same identical expression obtained for $\mathbf{x}_{opt}$ in the deterministic least-squares problem with $\mathbf{R}_0^{-1}$ playing the role of the weighting matrix $\mathbf{W}$.

Let us now recapitulate the conditions under which the Kalman filter estimate coincides with the deterministic least-squares estimate. First, the system state vector was assumed to be a random constant (the dynamics are thus trivial). Second, we assumed the measurement sequence was such as to yield an overdetermined system of linear equations [otherwise $(\mathbf{H}_0^T \mathbf{R}_0^{-1} \mathbf{H}_0)^{-1}$ will not exist]. And, finally, we assumed that we had no prior knowledge about the constant vector being estimated. This latter assumptions is unusual because in many situations we have at least some a priori knowledge of the process being estimated. One of the things that distinguishes the Kalman filter from other estimators is the convenient way in which it accounts for this prior knowledge via the initial conditions of the recursive process. (This was used to good advantage in the power system relaying application of Section 6.5.) Of course, if there is truly no prior knowledge to use, the Kalman filter advantage is lost (in this respect), and it degenerates to a least-squares fit under the conditions just stated.

The coincidence in the deterministic least-squares and Kalman filter estimates is really rather remarkable. Remember, one solution was obtained by posing a *deterministic* optimization problem; the other by posing a similar *stochastic* problem. There is no reason *offhand* to think these two approaches would lead to identical solutions. Yet they do under certain circumstances. The circumstances may be generalized somewhat from those of this example, but not to the complete extent of the general process model used in the Kalman filter. [See Sorenson (15) for more on this point.] Thus, this happy coincidence in the two solutions will not always exist.

There are subtleties that need to be recognized when using Kalman filtering to estimate a constant. These are related to deterministic least squares. A simple example will illustrate the significance of the Kalman filter estimate in the unknown-constant estimation problem.

EXAMPLE 6.6

Consider an elementary physics experiment that is intended to measure the gravity constant g. A mass is released at $t = 0$ in a vertical, evacuated column, and multiple-exposure photographs of the falling mass are taken at .05-sec intervals beginning at $t = .05$ sec. A sequence of N such exposures is taken, and then the position of the mass at each time is read from a scale in the photograph. There will be experimental errors for a number of reasons; let us assume that they are random (i.e., not systematic) and are such that the statistical uncertainties in all position readings are the same and that the standard deviation of these is σ.

Consider g to be an unknown constant, and suppose we say that we have no prior knowledge about its value. (We will elaborate on this assumption later in the example.) We wish to develop a Kalman filter for processing the noisy position measurements. We begin our model by letting g be the single state variable x. The discrete state equation is then

$$x_{k+1} = 1 \cdot x_k + 0 \qquad (6.8.11)$$

The measurement sequence z_k is related to x_k via the equations

$$z_1 = \left(\tfrac{1}{2}t_1^2\right)x_1 + v_1 \qquad \text{(The first measurement is at } t_1, \text{ not } t_0.)$$

$$z_2 = \left(\tfrac{1}{2}t_2^2\right)x_2 + v_2$$

$$\vdots$$

$$z_N = \left(\tfrac{1}{2}t_N^2\right)x_N + v_N \tag{6.8.12}$$

The filter parameters are then

$$\phi_k = 1, \qquad Q_k = 0$$

$$H_k = \tfrac{1}{2}t_k^2, \qquad R_k = \sigma^2 \tag{6.8.13}$$

The measurement sequence begins at $t = t_1 = .05$ sec, so we enter the recursive loop at $k = 1$ with the initial conditions

$$\hat{x}_1^- = 0$$

$$P_1^- = \infty \qquad \text{(no prior knowledge of } g) \tag{6.8.14}$$

The alternative algorithm is useful here to get the recursive process started. Therefore, we update P_1^- first using Eq. (6.2.6).

$$P_1^{-1} = \left(P_1^-\right)^{-1} + \left(\tfrac{1}{2}t_1^2\right)\left(\frac{1}{\sigma^2}\right)\left(\tfrac{1}{2}t_1^2\right)$$

Or

$$P_1 = \frac{4}{t_1^4}\sigma^2 \tag{6.8.15}$$

The Kalman gain is computed next using Eq. (6.2.7).

$$K_1 = P_1 H_1^T R_1^{-1} = \frac{2}{t_1^2} \tag{6.8.16}$$

Finally, the estimate is updated using the usual update equation:

$$\hat{x}_1 = 0 + \left(\frac{2}{t_1^2}\right)(z_1 - 0) = \left(\frac{2}{t_1^2}\right)z_1 \tag{6.8.17}$$

The a posteriori P is finite, so we can now project ahead to the next measurement and proceed with the recursive processing using either the regular algorithm (Chapter 5) or the alternative algorithm (Chapter 6). This is routine, so we will not pursue the algebra further here. It is easily demonstrated that in these circumstances the Kalman filter estimate obtained at each step is identical to that obtained from deterministic least squares and batch processing the data (see Problem 6.9).

This example is similar to Example 6.3 in many respects, but there are subtle differences that warrant further elaboration. In Example 6.3 the process envisioned was an ensemble of random constants (see also Example 2.5). Thus, when we think of this in terms of a statistical experiment, each time function in the ensemble is constant with time, but their amplitudes are different in that they are sample realizations of a zero-mean random variable with a large variance. Thus, the setting for Example 6.3 is truly a stochastic setting, and all the conclusions about minimizing the mean-square error, and so forth, are applicable. Now contrast this with the setting for the gravity-determination experiment of the present example. Not only is it unrealistic to say that we have no prior knowledge of g, but the unknown g is not really properly modeled as a random process. Presumably, if we were to repeat our elementary physics experiment over and over again at the same location, we would have the same g each time the experiment was performed. Thus, the stochastic setting here is different from the one for Example 6.3. The gravity-experiment measurement noise may be random, but the quantity being estimated (even if unknown) is not. Now, there is nothing wrong with applying the Kalman filter algorithm in this setting. After all, it can be viewed as just one of many arithmetic rules for processing numerical data. The fault, if there is fault, comes with the interpretation of the results. In Example 6.3 we have a right to expect the Kalman filter to minimize the mean-square error in a truly stochastic (ensemble averaging) sense. In the present example we cannot draw the same inference. We do know in the present example that the resulting estimate is identical with the ordinary deterministic least-squares estimate (see Problem 6.9), but beyond this we should not draw hasty conclusions about optimality. ∎

6.9 TELEPHONE LOAD FORECASTING EXAMPLE

It was mentioned previously that applications of Kalman filtering have been spreading into areas far afield from the original applications in the aerospace field. For example, the January 1982 issue of the *Bell System Technical Journal* was devoted entirely to the application of Kalman filtering to telephone load forecasting. C. R. Szelag (17) had a particularly interesting paper in this issue of the *BSTJ*, and we will examine briefly a simplified version of this application.

The load history for one of the telephone trunks considered by Szelag (17) is repeated here in Fig. 6.10. The "squares" in the figure indicate the actual

Figure 6.10.   Telephone trunk load.

load data experienced by the system during the 1975–79 period. (The circles may be ignored for the moment.) Clearly, the load exhibits a general yearly oscillatory trend superimposed on a linearly increasing trend. However, there is some noisiness in the trend because the oscillatory part is not exactly sinusoidal, nor is the linear part exactly linear, which indicates that this process might be modeled as the sum of oscillatory and linear components that are driven by white noise inputs. That is, we speculate that the following model might fit this situation reasonably well:

1.   Linear part:

$$\ddot{x} = u_1(t), \qquad u_1(t) \text{ is white noise} \qquad (6.9.1)$$

2.   Oscillatory part:

$$\ddot{y} + \omega^2 y = u_2(t), \qquad u_2(t) \text{ white and independent of } u_1(t) \quad (6.9.2)$$

Next, Eqs. (6.9.1) and (6.9.2) may be written in vector form by defining state variables as follows:

$$x_1 = x \qquad\qquad (6.9.3)$$

$$x_2 = \dot{x} \qquad\qquad (6.9.4)$$

$$x_3 = y \qquad\qquad (6.9.5)$$

$$x_4 = \dot{y} \qquad\qquad (6.9.6)$$

The state equations are then

$$
\begin{bmatrix} \dot{x}_1 \\ \dot{x}_2 \\ \dot{x}_3 \\ \dot{x}_4 \end{bmatrix} = \left[\begin{array}{cc|cc} 0 & 1 & & \\ 0 & 0 & & \mathbf{0} \\ \hline & & 0 & 1 \\ & \mathbf{0} & -\omega^2 & 0 \end{array} \right] \begin{bmatrix} x_1 \\ x_2 \\ x_3 \\ x_4 \end{bmatrix} + \begin{bmatrix} 0 \\ u_1(t) \\ 0 \\ u_2(t) \end{bmatrix}
\tag{6.9.7}
$$

Let us now assume that the measurements are discrete and are uniformly spaced with a sampling interval of T. The discrete state model is then

$$
\mathbf{x}_{k+1} = \left[\begin{array}{cc|cc} 1 & T & & \\ 0 & 1 & & \mathbf{0} \\ \hline & & \cos \omega T & \sin \omega T \\ & \mathbf{0} & -\sin \omega T & \cos \omega T \end{array} \right] \mathbf{x}_k + \mathbf{w}_k
\tag{6.9.8}
$$

The transition matrix $\boldsymbol{\phi}_k$ is now determined as shown in component form in Eq. (6.9.8). The $\mathbf{Q}_k$ matrix will depend on the amplitudes of the white-noise inputs $u_1(t)$ and $u_2(t)$. This parameter is usually the most elusive one of the model and often has to be determined by trial and error. Szelag (17) does not provide a numerical value for $\mathbf{Q}_k$ in this case, but he does indicate that it was determined empirically.

The measurement equation is especially simple in this example because the total process being measured is just the sum of the linear and oscillatory parts. Thus z_k is scalar and we have

$$
z_k = \begin{bmatrix} 1 & 0 & 1 & 0 \end{bmatrix} \mathbf{x}_k + v_k
\tag{6.9.9}
$$

Therefore $\mathbf{H}_k$ is as shown in component form in Eq. (6.9.9). We would expect the R_k parameter associated with the measurement noise v_k to be relatively small in this case. As with the case of the $\mathbf{Q}_k$ parameter, no numerical value for R_k was given in the reference paper, so we have no check on this assumption. It is reasonable to assume, though, that the trunk load measurement is relatively accurate.

Let us now assume the $\boldsymbol{\phi}_k$, $\mathbf{Q}_k$, $\mathbf{H}_k$, and R_k parameters of the model have been determined. In this application the problem is to predict the load at the next step. That is, we have the past history of load measurements up to and including time t_k, and we wish to estimate the load at t_{k+1}. Thus, it is the a priori estimate $\hat{\mathbf{x}}_{k+1}^-$ that is of primary interest in this application. (To be more specific, it is the sum of the first and third elements of the vector that is of interest.) The a priori or one-step prediction estimate is, of course, one of the normal computations in the usual Kalman filter algorithm; therefore, the recursive equations shown in Fig. 5.9 are directly applicable without any

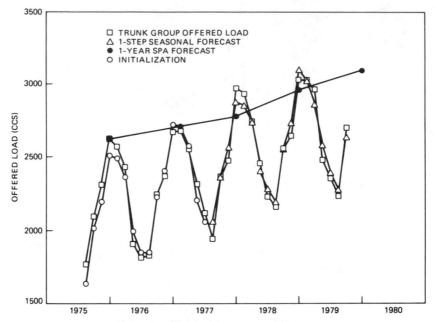

Figure 6.11. Telephone trunk load data along with one-step predictions.

modification. There is a start-up problem, though, when using the algorithm either in real time or in an empirical evaluation of the estimator's merits using real-life data from past experience. With respect to the latter, the author chose to use the first 16 data points to initialize the filter. This then established an a posteriori estimate and associated error covariance matrix at step 16. These were then projected ahead to step 17, and the regular filter loop was continued to the end of the data stream.

Figure 6.11 shows the results of a sample run for the load history shown previously in Fig. 6.10. As before, the squares indicate the actual load data. The triangles are the a priori estimates of interest, and note that they do not begin until step 17. (The circles relate to the initialization and may be ignored for our purposes.) It can be seen that the Kalman filter does a reasonably good job of one-step prediction on this run.

A few further comments about initialization and other matters are in order before leaving this example. First, Szelag (17) chose to use a batch-processing weighted-least-squares approach in initializing his Kalman filter. In view of our comments about deterministic least squares in Section 6.8, it would appear that this could just as well have been done recursively. To do this, we would simply assume an artificially large $\mathbf{P}^-$ matrix at the beginning of the data stream, and then the Kalman filter would *automatically* give the initial a priori estimate (e.g., zero) negligible weight. In this example the state vector being estimated is fourth order and the measurements are scalar; thus it would take the first four steps to bring the initial $\mathbf{P}$ matrix down to normal size, so to speak. From that point on the filter would behave more or less normally, and the **P** matrix should reach a near stationary condition by step 16. Assessment of the filter's accuracy as a one-step predictor could then begin with step 17.

We might also note that the model chosen here for tutorial purposes is relatively simple. Szelag notes in his paper that higher-order harmonics may be added to the model. Of course, this refinement is not without a price because each harmonic added to the model increases the dimensionality of the state vector by two units. Also, the input noise model could be refined by allowing white noise to enter the $\dot{x}_2$ and $\dot{x}_4$ equations directly, as well as the $\dot{x}_1$ and $\dot{x}_3$ equations; see Eq. (6.9.7). This does not appear to be worth the effort in this example, though, in view of the fact that $\mathbf{Q}_k$ is being determined empirically. In any event, this is a new and novel application, and it is interesting to see Kalman filtering spread to fields outside of aerospace engineering.

6.10
DETERMINISTIC INPUTS

In many situations the random processes under consideration are driven by deterministic as well as random inputs. That is, the process equation may be of the form

$$\dot{\mathbf{x}} = \mathbf{F}\mathbf{x} + \mathbf{G}\mathbf{u} + \mathbf{B}\mathbf{u}_d \qquad (6.10.1)$$

where $\mathbf{B}\mathbf{u}_d$ is the additional deterministic input. Since the system is linear we can use superposition and consider the random and deterministic responses separately. Thus the discrete Kalman filter equations are modified only slightly. The only change required is in the estimate projection equation. In this equation the contribution due to $\mathbf{B}\mathbf{u}_d$ must be properly accounted for. Using the same zero-mean argument as before, relative to the random response, we then have

$$\hat{\mathbf{x}}_{k+1}^- = \boldsymbol{\phi}_k \hat{\mathbf{x}}_k + 0 + \int_{t_k}^{t_{k+1}} \boldsymbol{\phi}(t_{k+1}, \tau) \mathbf{B}(\tau) \mathbf{u}_d(\tau) \, d\tau \qquad (6.10.2)$$

where the integral term is the contribution due to $\mathbf{B}\mathbf{u}_d$ in the interval (t_k, t_{k+1}). The associated equation for $\mathbf{P}_{k+1}^-$ is $(\boldsymbol{\phi}_k \mathbf{P}_k \boldsymbol{\phi}_k^T + \mathbf{Q}_k)$, as before, because the uncertainty in the deterministic term is zero. Also, the estimate update and associated covariance expressions (see Fig. 5.9) are unchanged, provided the deterministic contribution has been properly accounted for in computing the a priori estimate $\hat{\mathbf{x}}_k^-$.

Another way of accounting for the deterministic input is to treat the problem as a superposition of two entirely separate estimation problems, one deterministic and the other random. The deterministic one is trivial, of course, and the random one is not trivial. This complete separation approach is not necessary, though, provided one properly accounts for the deterministic contribution in the projection step.

6.11
PERSPECTIVE

We have now presented the basics of Kalman filtering and looked at a few examples of how the technique can be applied in physical situations. More is yet to come in the remaining chapters. However, this is a good place to pause for a moment and reflect on just what we have (and do not have) with this thing we call a Kalman filter.

1. The Kalman filter is intended to be used for estimating *random* processes. Any application in a nonrandom setting must be viewed with caution (see Example 6.6).

2. The Kalman filter is model dependent. This is to say that we assume that we know a priori the model parameters. These, in turn, come from the second-order "statistics" of the various processes involved in the application at hand. Therefore, in its most primitive form, the Kalman filter is not adaptive or self-learning.

3. The Kalman filter is a linear estimator. When all the processes involved are Gaussian, the filter is optimal in the minimum-mean-square-error sense within a class of *all* estimators, linear and nonlinear. [See Meditch (5) for a good discussion of optimality.]

4. Various Kalman filter recursive algorithms exist. The "usual" algorithm was given in Chapter 5, an alternative one was presented in Chapter 6, and a third one (*U-D* factorization) will be presented in Chapter 9. All of these yield identical results (assuming perfect arithmetic).

5. Under certain special circumstances, the Kalman filter yields the same result obtained from deterministic least squares (see Section 6.8).

6. Kalman filtering is especially useful as an analysis tool in offline error analysis studies. The filter error covariance equation can be propagated recursively without actual (or simulated) measurement data.

With these brief comments in mind, we are now ready to proceed to variations on the original discrete Kalman filter. It is worth mentioning that the discrete filter came first historically (1960). The continuous version and other variations followed the discrete filter.

PROBLEMS

6.1. The ψ equations that describe the error propagation in a slow-moving inertial navigation system were given in Section 6.1 [Eqs. (6.1.3) to (6.1.5)]. Let ψ_x, ψ_y, ψ_z be state vector elements x_1, x_2, x_3 and $\varepsilon_x, \varepsilon_y, \varepsilon_z$ be the system *inputs*.

First, write the equations in the standard state-space form:

$$\dot{\mathbf{x}} = \mathbf{Fx} + \mathbf{Bu} \qquad (\mathbf{u} \text{ is not white noise})$$

Next, find the eigenvalues of the $\mathbf{F}$ matrix. (*Note:* These describe the natural modes of the system.)

6.2. Consider a simplified version of Example 6.1 in which the gyro drifts are modeled simply as random constants. The system state vector is then a 6-tuple. Also, assume the sampling interval for the measurements Δt is constant and small relative to the natural period of the system (i.e., 1 day, see Problem 6.1). Find the $\boldsymbol{\phi}_k, \mathbf{Q}_k, \mathbf{H}_k, \mathbf{R}_k$ and the initial parameters $\mathbf{P}_0^-, \hat{\mathbf{x}}_0^-$ for this system. Assume the earth-rate components Ω_x and Ω_z are known constants. Also, assume the initial uncertainties in ψ_x, ψ_y, ψ_z are all the same with variance σ_ψ^2, and the corresponding initial variances for $\varepsilon_x, \varepsilon_y, \varepsilon_z$ are all σ_ε^2. All initial crosscorrelations are zero, and the measurement-error variances are σ_x^2 and σ_y^2.

6.3. Is the system of Problem 6.2 observable? Recall from linear system theory that the test for observability is to form the matrix

$$\mathbf{M} = \left[\mathbf{H}^T, \mathbf{F}^T \mathbf{H}^T, (\mathbf{F}^T)^2 \mathbf{H}^T, \ldots, (\mathbf{F}^T)^{n-1} \mathbf{H}^T \right]$$

and then test the rank of $\mathbf{M}$. If it is of full rank, the system is observable; if not, the system is not observable. (*Answer:* The system is not observable. This can also be seen by writing out the explicit solution of the state equations in terms of the initial conditions. Certain linear combinations of state variables always appear together and are thus inseparable.)

6.4. Consider the measurement to be a 2-tuple $[z_1, z_2]^T$, and assume that the measurement errors are correlated such that the $\mathbf{R}$ matrix is of the form

$$\mathbf{R} = \begin{bmatrix} r_{11} & r_{12} \\ r_{12} & r_{22} \end{bmatrix}$$

(a) Form a new measurement pair, z_1' and z_2', as a linear combination of the original pair such that the errors in the new pair are uncorrelated. (*Hint:* First, let $z_1' = z_1$ and then assume $z_2' = c_1 z_1 + c_2 z_2$ and choose the constants c_1 and c_2 such that the new measurement errors are uncorrelated.)

(b) Find the $\mathbf{R}$ matrix associated with the new $\mathbf{z}'$ measurement vector.

6.5. The current model of the relaying application presented in Section 6.5 contained a nonstationary state variable with exponential properties. Assume that the initial value of this state variable is a random variable and has a variance that is 100 times larger than the steady-state variance of the process. Sketch three typical sample functions of the process.

6.6. (a) Derive an explicit expression for $\mathbf{P}_{k+1}$ in terms of $\mathbf{P}_k$ and the $\phi, \mathbf{H}, \mathbf{Q}, \mathbf{R}$ parameters. (The a priori error covariance matrix $\mathbf{P}^-$ and gain $\mathbf{K}$ should *not* appear in your final expression.)

(b) Find a similar difference equation for $\mathbf{P}_{k+1}^-$ in terms of $\mathbf{P}_k^-$ and the system parameters.

6.7. Suboptimal filter analysis is often used in sensitivity analysis. This is where the analyst wishes to assess the sensitivity of the system error to changes in certain parameters. In the random-walk example of Section 6.7, the true value of Q was said to be 1.0, and the rms estimation error worked out to be about .3027.

(a) Now suppose the designer incorrectly models the filter to be used in this application using $Q = 1.1$ rather than the true value. The filter will then be suboptimal to some extent. Compute the filter rms error for this suboptimal situation.

(b) Suppose the designer errs in the other direction and chooses the filter Q to be 0.9 rather than the true value of 1.0. Find the filter rms error for this situation.

(c) The percent variation of Q on either side of truth in parts (a) and (b) is about 10 percent. What is the corresponding percent variation in the rms estimation error? Would you call this a high or low sensitivity situation?

(*Note:* Convergence to the steady-state condition is quite rapid in this case. Thus, the analysis can be carried out easily with a hand-held calculator.)

6.8. A block diagram showing how errors propagate in one level channel of an undamped, Schuler-tuned inertial navigation system is given in the accompanying figure. Some approximations have been made, but the diagram is reasonably accurate for a few cycles of the Schuler oscillation (about 84 min). The accelerometer and gyro-drift forcing functions are modeled as random-walk processes in this model. This allows for "biases" that can slowly walk off in a random manner. We have a 5-tuple state vector in this example, and the physical state variables are

$$x_1 = \text{platform tilt in rad}$$

$$x_2 = \text{velocity error in m/sec}$$

$$x_3 = \text{position error in m}$$

$$x_4 = \text{gyro drift rate in rad/sec}$$

$$x_5 = \text{accelerometer error in m/sec}^2$$

Assume that we have a noninertial measure of position that is noisy but independent of the inertial system's estimate of position. Is the system observ-

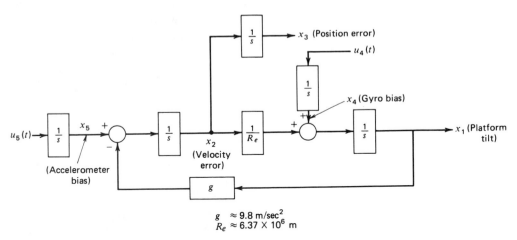

$$g \approx 9.8 \text{ m/sec}^2$$
$$R_e \approx 6.37 \times 10^6 \text{ m}$$

Problem 6.8

able on the basis of a sequence of independent position measurements alone? See Problem 6.3 for the test of observability. (This problem is continued in Chapter 7, Problem 7.12*.)

6.9. The recursive process in Example 6.6 was carried through only one step. Continue the process through a second recursive step, and obtain an explicit expression for $\hat{x}$ in terms of z_1 and z_2. Next, compute the ordinary least-squares estimate of g using just two measurements z_1 and z_2. Do this on a batch basis using the equation

$$\hat{g} = (\mathbf{H}^T \mathbf{H})^{-1} \mathbf{H}^T \mathbf{z}$$

where

$$\mathbf{H} = \begin{bmatrix} \frac{1}{2}t_1^2 \\ \frac{1}{2}t_2^2 \end{bmatrix}, \quad \text{and} \quad \mathbf{z} = \begin{bmatrix} z_1 \\ z_2 \end{bmatrix}$$

Compare the least-squares result with that obtained after carrying the Kalman filter recursive process through two steps.

6.10. In Example 6.6 the initial position and velocity for the falling mass were assumed to be known, that is, they were exactly zero; the gravity constant was presumed to be unknown and to be estimated. Bozic (18) presents an interesting variation on this problem where the situation is reversed; g is assumed to be known perfectly and the initial position and velocity are random variables with known Gaussian distribution. Even though the trajectory obeys a known deterministic equation of motion, the random initial conditions add sufficient uncertainty to the motion to make the trajectory a legitimate random process. Assume that the initial position and velocity are normal random variables described by $N(0, \sigma_p^2)$ and $N(0, \sigma_v^2)$. Let state variables x_1 and x_2 be position and velocity measured downward, and let the measurements take place at uniform intervals Δt beginning at $t = 0$. The measurement error variance is R. Work out the key parameters for the Kalman filter model for this situation. That is, find $\boldsymbol{\phi}_k, \mathbf{Q}_k, \mathbf{H}_k, \mathbf{R}_k$ and the initial $\hat{\mathbf{x}}_0^-$ and $\mathbf{P}_0^-$. (Note that a deterministic forcing function has to be accounted for in this example. The effect of this forcing function, though, will appear in projecting $\hat{\mathbf{x}}_k$ ahead to get $\hat{\mathbf{x}}_{k+1}^-$, but not in the model parameters.)

6.11*. The Gauss–Markov model has been used extensively in both Chapters 5 and 6. Another somewhat more complicated process model that requires two state variables is also used frequently in Kalman filter applications. This process has a damped harmonic autocorrelation function, and one special version of this type of process can be generated with the shaping filter shown in the accompanying figure. Note that three parameters describe this process completely: σ_y^2, the variance; ω_0, the undamped natural frequency; and ζ, the damping ratio.

Consider a Gaussian process y as described in the figure with the following parameters:

$$\sigma_y = 20 \text{ units}, \qquad \omega_0 = .02 \text{ rad/sec}, \qquad \zeta = .5$$

This will be referred to as the "signal" in our estimator. We wish now to look at a discrete Kalman filter where we have a direct one-to-one measurement of y corrupted by measurement noise. The measurement error in this case consists of an additive combination of a Gauss–Markov process and a Gaussian white sequence. The parameters for the measurement noises are as follows:

Markov component: $\qquad \sigma_m = 10 \text{ units}, \qquad \beta_m = .002 \text{ rad/sec}$

White component: $\qquad \sigma_{\text{white}} = 10 \text{ units}$

The sampling interval is 10 sec.

We wish to determine the steady-state rms estimator error for this situation. Using appropriate covariance analysis software, cycle the Kalman filter through the required number of steps to reach a steady-state condition. Do this two ways:

(a) First, initialize the filter $\mathbf{P}$ matrix by assuming that the initial state estimate is set equal to zero and that the system state is in a stationary condition when the filter begins to process measurements. (Note that the initial $\mathbf{P}_0^-$ can be found in this situation by evaluating $\mathbf{Q}$ empirically for very large Δt.)

(b) Next, arbitrarily initialize $\mathbf{P}_0^-$ to be zero (more precisely, the null matrix). Then run the error covariance program to the steady-state condition and compare the result with that obtained in part (a). This might be thought of as the lazy way to solve the problem (but effective!).

Unity white noise $u(t)$ → $\dfrac{2\omega_0 \sigma_y \sqrt{\zeta\omega_0}}{s^2 + 2\zeta\omega_0 s + \omega_0^2}$ → $y(t)$

Problem 6.11*

It should be noted that the 3-state model for this problem partitions into nontrivial upper 2×2 and lower 1×1 parts. Thus, if you wish to simplify the calculations, $\boldsymbol{\phi}_k$ and $\mathbf{Q}_k$ can be evaluated in terms of their partitioned parts separately; then the partitioned parts can be put back together for the final $\boldsymbol{\phi}_k$ and $\mathbf{Q}_k$ matrices to be used in the covariance analysis.

6.12*. The accompanying block diagram shows two cascaded integrators driven by white noise. The two state variables x_1 and x_2 can be thought of as position and velocity for convenience, and the forcing function is acceleration. Let us suppose that we have a noisy measurement of velocity, but there is no direct observation of position. Application of the observability test given in Problem 6.3 indicates that this system is not observable on the basis of a velocity measurement alone. Clearly, there will be no divergence of the estimation error in x_2, because we have a direct measurement of it. However, divergence of the estimation error of x_1 is not so obvious.

The question of divergence of the error in estimating x_1 is easily answered empirically by cycling through the Kalman filter error covariance equations until either (a) a stationary condition for p_{11} is reached, or (b) divergence becomes obvious by continued growth of p_{11} with each recursive step. Perform the suggested experiment using appropriate covariance analysis software. You will find the following numerical values suitable for this exercise:

$$\text{Power spectral density of } f(t) = .1 \; (\text{m/sec}^2)^2/(\text{rad/sec})$$

$$\text{Step size } \Delta t = 1 \text{ sec}$$

$$\text{Measurement error variance} = .01 \; (\text{m/sec})$$

Note that if divergence is found, it is not the "fault" of the filter. It simply reflects the incomplete measurement situation.

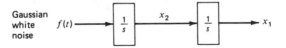

Gaussian white noise $f(t)$ → $\dfrac{1}{s}$ → x_2 → $\dfrac{1}{s}$ → x_1

Problem 6.12*

6.13*. The accompanying figure shows how errors propagate in one level channel of an inertial navigation system. In the interest of simplicity, the diagram shows only the effect of acceleration measurement errors (i.e., accelerometer instrument errors), and they have been modeled as white noise. It is easily verified that this system has its characteristic poles on the imaginary axis in the s-plane, so the system is unstable. Thus, one would expect the mean-square values of all the state variables to increase without bound when the system is driven with white noise.

Suppose now that we are able to make a measurement of velocity that is completely independent of the inertial system. The noninertial measurement could then be compared with the inertial one, and we would obtain a noisy measurement of the inertial system velocity error (i.e., state variable x_2). The question then arises: Would the system be completely observable on the basis of this external velocity observation? Clearly, x_2 is being observed, but observation of x_1 and x_3 is not so obvious. Using the observability test given in Problem 6.3, determine analytically whether or not the system is observable.

The answer to the observability question just posed is no. But that fact, in itself, does not tell us which state variables are being observed and which are not. Some further insight into this can be gained by considering the system to be excited by white noise (as shown in the figure), and then cycling through the Kalman filter error covariance equations until either a stationary condition is reached or divergence is obvious, relative to each of the state variables. Perform this experiment and form conclusions as to which state variables are being observed. The following numerical values for the filter parameters are suitable for this experiment:

Power spectral density of the accelerometer error $= 10^{-3}\,(\text{m/sec}^2)^2/(\text{rad/sec})$

External velocity measurement error variance $= .01\,(\text{m/sec})^2$

Step size $\Delta t = 120$ sec

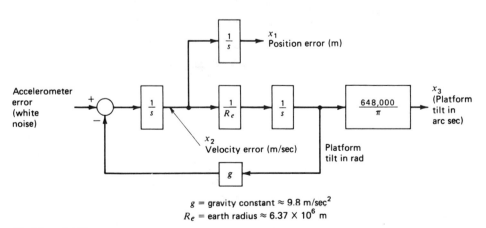

g = gravity constant ≈ 9.8 m/sec^2
R_e = earth radius $\approx 6.37 \times 10^6$ m

Problem 6.13*

Note that the natural period of this system is about 84 minutes, so you may have to let the covariance equations cycle through many steps to reach valid conclusions. Also note that the divergence here is not the "fault" of the filter. It is simply a matter of inadequate measurement information.

6.14*. In Section 6.5 a harmonic random process was modeled with two random constant state variables, and the oscillatory dynamics were accounted for in the measurement part of the model. We wish to consider the same model here and see what happens if we misjudge the natural frequency in the model. We will use the analysis methods discussed in Section 6.7. We will assume the truth model to be as follows:

Truth Model
Process model:

$$\begin{bmatrix} x_1 \\ x_2 \end{bmatrix}_{k+1} = \begin{bmatrix} 1 & 0 \\ 0 & 1 \end{bmatrix} \begin{bmatrix} x_1 \\ x_2 \end{bmatrix}_k + \begin{bmatrix} 0 \\ 0 \end{bmatrix}$$

Measurement model:

$$z_k = \begin{bmatrix} \cos \omega_0 k \, \Delta t & -\sin \omega_0 k \, \Delta t \end{bmatrix} \begin{bmatrix} x_1 \\ x_2 \end{bmatrix}_k + v_k$$

where
$\omega_0 = 2\pi(.2)$ rad/sec
$R_k = E[v_k^2] = 100$ units
$\Delta t = .1$ sec
$k = 0, 1, 2, \ldots, 100$

The initial conditions are

$$\hat{x}_0^- = \begin{bmatrix} 0 \\ 0 \end{bmatrix}, \qquad P_0^- = \begin{bmatrix} 10^4 & 0 \\ 0 & 10^4 \end{bmatrix} \text{ units}$$

(a) First, work out the error covariances for the optimal filter for the specified range of k. The p_{11} and p_{22} terms are both important in this model, because we need estimates of both x_1 and x_2 to determine the amplitude and phase of the random harmonic process. (We are doing only covariance analysis here, though, so the estimates of x_1 and x_2 will not be obtained in this problem.)

(b) Now suppose that the filter designer misjudges the natural frequency of the process and actually implements a frequency of $2\pi(.25)$ rad/sec in the on-line measurement equation rather than the true frequency. That is, the implemented filter has for its H matrix:

$$H(\text{implemented}) = \begin{bmatrix} \cos 2\pi(.25)k \, \Delta t & -\sin 2\pi(.25)k \, \Delta t \end{bmatrix}$$

Compute the sequence of suboptimal gains for this improperly modeled filter and save them in a file. (Note that the sequence of P matrices resulting from this operation has no physical meaning in terms of estimation errors.)

(c) Now cycle the suboptimal gain sequence from part (b) through the truth model and determine the actual error covariance sequence for the mismodeled filter. Compare the optimal and suboptimal error covariances and note any divergence that takes place. Plots will be helpful in making the comparison.

REFERENCES CITED IN CHAPTER

1. H. W. Sorenson (Ed.), *Kalman Filtering: Theory and Application*, New York: IEEE Press, 1985.

2. B. E. Bona and R. J. Smay, "Optimum Reset of Ship's Inertial Navigation System," *IEEE Trans. on Aerospace and Electronic Systems, AES-2:* No. 4, 409–414 (July 1966).

3. G. R. Pitman (Ed.), *Inertial Guidance*, New York: Wiley, 1962.

4. J. C. Pinson, "Inertial Guidance for Cruise Vehicles," in C. T. Leondes (Ed.), *Guidance and Control of Aerospace Vehicles*, New York: McGraw-Hill, 1963.

5. J. S. Meditch, *Stochastic Optimal Linear Estimation and Control*, New York: McGraw-Hill, 1969.

6. A. Gelb (Ed.), *Applied Optimal Estimation*, Cambridge, Mass.: MIT Press, 1974.

7. H. W. Sorenson, "Kalman Filtering Techniques," in C. T. Leondes (Ed.), *Advances in Control Systems*, Vol. 3, New York: Academic Press, 1966.

8. R. P. Denaro, "Navstar: The All-Purpose Satellite," *IEEE Spectrum*, Vol. 18, No. 5, 35–40 (May 1981).

9. B. W. Parkinson and S. W. Gilbert, "NAVSTAR: Global Positioning System—Ten Years Later," *Proc. of the IEEE, 71*, No. 10, 1177–1186 (October 1983).

10. R. G. Brown, "Integrated Navigation Systems and Kalman Filtering: A Perspective," *Navigation, Jour. of the Inst. of Navigation, 19:* No. 4, 335–362 (Winter 1972–73).

11. J. D. Salisbury, "Comments on Integrated Navigation Systems and Kalman Filtering: A Perspective," *Navigation, Jour. of the Inst. of Navigation, 20:* No. 2, 190 (Summer 1973).

12. A. A. Girgis, "Application of Kalman Filtering in Computer Relaying of Power Systems," Ph.D. dissertation, Iowa State University, Ames, 1981.

13. A. A. Girgis and R. G. Brown, "Application of Kalman Filtering in Computer Relaying," *IEEE Trans. on Power Apparatus and Systems, PAS-100:* No. 7, 3387–3397 (July 1981).

14. H. W. Bode and C. E. Shannon, " A Simplified Derivation of Linear Least Squares Smoothing and Prediction Theory," *Proc. I.R.E., 38*, 417–424 (April 1950).

15. H. W. Sorenson, "Least-Squares Estimation: From Gauss to Kalman," *IEEE Spectrum, 7,* 63–68 (July 1970).

16. T. Kailath, "A View of Three Decades of Linear Filtering Theory," *IEEE Trans. in Information Theory, IT-20:* No. 2, 146–181 (March 1974).

17. C. R. Szelag, "A Short-Term Forecasting Algorithm for Trunk Demand Servicing," *The Bell System Technical Journal, 61:* No. 1, 67–76 (January 1982).

18. S. M. Bozic, *Digital and Kalman Filtering*, London: E. Arnold, Publisher, 1979.

7

The Continuous Kalman Filter

About a year after his paper on discrete-data filtering, R. E. Kalman coauthored a second paper with R. S. Bucy on continuous filtering (1). This paper also proved to be a milestone in the area of optimal filtering. Our approach here will be somewhat different from theirs, in that we will derive the continuous filter equations as a limiting case of the discrete equations as the step size becomes small. Philosphically, it is of interest to note that we begin with the discrete equations and then go to the continuous equations. So often in numerical procedures, we begin with the continuous dynamical equations; these are then discretized and the discrete equations become approximations of the continuous dynamics. Not so with the Kalman filter! The discrete equations are exact and stand in their own right, provided, of course, that the difference equation model of the process is exact and not an approximation.

The continuous Kalman filter is probably not as important in applications as the discrete filter, especially in real-time systems. However, the continuous filter is important for both conceptual and theoretical reasons, so this chapter will be devoted primarily to continuous filtering. In addition, sections are included on correlated process and measurement noise and also on filter stability. These topics were deferred to this chapter because they are better developed after deriving the continuous filter equations rather than before.

7.1
TRANSITION FROM THE DISCRETE TO CONTINUOUS FILTER EQUATIONS

First, we assume the process and measurement models to be of the form:

$$\text{Process model: } \dot{\mathbf{x}} = \mathbf{F}\mathbf{x} + \mathbf{G}\mathbf{u} \tag{7.1.1}$$

$$\text{Measurement model: } \mathbf{z} = \mathbf{H}\mathbf{x} + \mathbf{v} \tag{7.1.2}$$

where

$$E\left[\mathbf{u}(t)\mathbf{u}^T(\tau)\right] = \mathbf{Q}\delta(t - \tau) \qquad (7.1.3)$$

$$E\left[\mathbf{v}(t)\mathbf{v}^T(\tau)\right] = \mathbf{R}\delta(t - \tau) \qquad (7.1.4)$$

$$E\left[\mathbf{u}(t)\mathbf{v}^T(\tau)\right] = 0 \qquad (7.1.5)$$

We note that in Eqs. (7.1.1) and (7.1.2), $\mathbf{F}$, $\mathbf{G}$, and $\mathbf{H}$ may be time varying. Also, by analogy with the discrete model, we assume that $\mathbf{u}(t)$ and $\mathbf{v}(t)$ are vector white noise processes with zero crosscorrelation. The covariance parameters $\mathbf{Q}$ and $\mathbf{R}$ play roles similar to $\mathbf{Q}_k$ and $\mathbf{R}_k$ in the discrete filter, but they do not have the same numerical values. The relationships between the corresponding discrete and continuous filter parameters will be derived presently.

Recall that for the discrete filter,

$$\mathbf{Q}_k = E\left[\mathbf{w}_k\mathbf{w}_k^T\right] \qquad (7.1.6)$$

$$\mathbf{R}_k = E\left[\mathbf{v}_k\mathbf{v}_k^T\right] \qquad (7.1.7)$$

To make the transition from the discrete to continuous case, we need the relations between $\mathbf{Q}_k$ and $\mathbf{R}_k$ and the corresponding $\mathbf{Q}$ and $\mathbf{R}$ for a small step size Δt. Looking at $\mathbf{Q}_k$ first, and referring to Eq. (5.3.6), we note that $\boldsymbol{\phi} \approx \mathbf{I}$ for small Δt and thus

$$\mathbf{Q}_k \approx \underset{\text{small } \Delta t}{\iint} \mathbf{G}(\xi)E\left[\mathbf{u}(\xi)\mathbf{u}^T(\eta)\right]\mathbf{G}^T(\eta)\,d\xi\,d\eta \qquad (7.1.8)$$

Next, substituting Eq. (7.1.3) into (7.1.8) and integrating over the small Δt interval yields

$$\mathbf{Q}_k = \mathbf{G}\mathbf{Q}\mathbf{G}^T\,\Delta t \qquad (7.1.9)$$

The derivation of the equation relating $\mathbf{R}_k$ and $\mathbf{R}$ is more subtle. In the continuous model $\mathbf{v}(t)$ is white, so simple sampling of $\mathbf{z}(t)$ leads to measurement noise with infinite variance. Hence, in the sampling process, we have to imagine averaging the continuous measurement over the Δt interval to get an equivalent discrete sample. This is justified because $\mathbf{x}$ is not white and may be approximated as a constant over the interval. Thus we have

$$\mathbf{z}_k = \frac{1}{\Delta t}\int_{t_{k-1}}^{t_k} \mathbf{z}(t)\,dt = \frac{1}{\Delta t}\int_{t_{k-1}}^{t_k} \left[\mathbf{H}\mathbf{x}(t) + \mathbf{v}(t)\right]dt$$

$$\approx \mathbf{H}\mathbf{x}_k + \frac{1}{\Delta t}\int_{t_{k-1}}^{t_k} \mathbf{v}(t)\,dt \qquad (7.1.10)$$

The discrete-to-continuous equivalence is then

$$\mathbf{v}_k = \frac{1}{\Delta t} \int_{\text{small } \Delta t} \mathbf{v}(t)\, dt \qquad (7.1.11)$$

From Eq. (7.1.7) we have

$$E\left[\mathbf{v}_k \mathbf{v}_k^T\right] = \mathbf{R}_k = \frac{1}{\Delta t^2} \iint_{\text{small } \Delta t} E\left[\mathbf{v}(u)\mathbf{v}^T(v)\right] du\, dv \qquad (7.1.12)$$

Substituting Eq. (7.1.4) into (7.1.12) and integrating yields the desired relationship

$$\mathbf{R}_k = \frac{\mathbf{R}}{\Delta t} \qquad (7.1.13)$$

At first glance, it may seem strange to have the discrete measurement error approach ∞ as $\Delta t \to 0$. However, this is offset by the sampling rate becoming infinite at the same time.

In making the transition from the discrete to continuous case, we first note from the error covariance projection equation (i.e., $\mathbf{P}_{k+1}^- = \boldsymbol{\phi}_k \mathbf{P}_k \boldsymbol{\phi}_k^T + \mathbf{Q}_k$) that $\mathbf{P}_{k+1}^- \to \mathbf{P}_k$ as $\Delta t \to 0$. Thus, we do not need to distinguish between a priori and a posteriori $\mathbf{P}$ matrices in the continuous filter. We proceed with the derivation of the continuous gain expression. Recall that the discrete Kalman gain is given by (see Fig. 5.9)

$$\mathbf{K}_k = \mathbf{P}_k^- \mathbf{H}_k^T \left(\mathbf{H}_k \mathbf{P}_k^- \mathbf{H}_k^T + \mathbf{R}_k\right)^{-1} \qquad (7.1.14)$$

Using Eq. (7.1.13) and noting that $\mathbf{R}/\Delta t \gg \mathbf{H}_k \mathbf{P}_k^- \mathbf{H}_k^T$ leads to

$$\mathbf{K}_k = \mathbf{P}_k^- \mathbf{H}_k^T (\mathbf{H}_k \mathbf{P}_k^- \mathbf{H}_k + \mathbf{R}/\Delta t)^{-1} \approx \mathbf{P}_k^- \mathbf{H}_k^T \mathbf{R}^{-1} \Delta t$$

We can now drop the subscripts and the super minus on the right side and we obtain

$$\mathbf{K}_k = (\mathbf{P}\mathbf{H}^T \mathbf{R}^{-1})\, \Delta t \qquad (7.1.15)$$

We *define* the continuous Kalman gain as the coefficient of Δt in Eq. (7.1.15), that is,

$$\mathbf{K} \triangleq \mathbf{P}\mathbf{H}^T \mathbf{R}^{-1} \qquad (7.1.16)$$

Next, we look at the error covariance equation. From the projection and update equations (Fig. 5.9), we have

$$\mathbf{P}_{k+1}^{-} = \boldsymbol{\phi}_k \mathbf{P}_k \boldsymbol{\phi}_k^T + \mathbf{Q}_k$$

$$= \boldsymbol{\phi}_k (\mathbf{I} - \mathbf{K}_k \mathbf{H}_k) \mathbf{P}_k^{-} \boldsymbol{\phi}_k^T + \mathbf{Q}_k$$

$$= \boldsymbol{\phi}_k \mathbf{P}_k^{-} \boldsymbol{\phi}_k^T - \boldsymbol{\phi}_k \mathbf{K}_k \mathbf{H}_k \mathbf{P}_k^{-} \boldsymbol{\phi}_k^T + \mathbf{Q}_k \qquad (7.1.17)$$

We now approximate $\boldsymbol{\phi}_k$ as $\mathbf{I} + \mathbf{F}\Delta t$ and note from Eq. (7.1.15) that $\mathbf{K}_k$ is of the order of Δt. After neglecting higher-order terms in Δt, Eq. (7.1.17) becomes

$$\mathbf{P}_{k+1}^{-} = \mathbf{P}_k^{-} + \mathbf{F}\mathbf{P}_k^{-}\,\Delta t + \mathbf{P}_k^{-}\mathbf{F}^T \Delta t - \mathbf{K}_k \mathbf{H}_k \mathbf{P}_k^{-} + \mathbf{Q}_k \qquad (7.1.18)$$

We next substitute the expressions for $\mathbf{K}_k$ and $\mathbf{Q}_k$, Eqs. (7.1.15) and (7.1.9), and form the finite difference expression

$$\frac{\mathbf{P}_{k+1}^{-} - \mathbf{P}_k^{-}}{\Delta t} = \mathbf{F}\mathbf{P}_k^{-} + \mathbf{P}_k^{-}\mathbf{F}^T - \mathbf{P}_k^{-}\mathbf{H}^T \mathbf{R}^{-1}\mathbf{H}_k \mathbf{P}_k^{-} + \mathbf{G}\mathbf{Q}\mathbf{G}^T \qquad (7.1.19)$$

Finally, passing to the limit as $\Delta t \to 0$ and dropping the subscripts and super minus leads to the matrix differential equation

$$\dot{\mathbf{P}} = \mathbf{F}\mathbf{P} + \mathbf{P}\mathbf{F}^T - \mathbf{P}\mathbf{H}^T \mathbf{R}^{-1}\mathbf{H}\mathbf{P} + \mathbf{G}\mathbf{Q}\mathbf{G}^T \qquad (7.1.20)$$

$$\mathbf{P}(0) = \mathbf{P}_0$$

Next, consider the state estimation equation. Recall the discrete equation is

$$\hat{\mathbf{x}}_k = \hat{\mathbf{x}}_k^{-} + \mathbf{K}_k (\mathbf{z}_k - \mathbf{H}_k \hat{\mathbf{x}}_k^{-}) \qquad (7.1.21)$$

We now note that $\hat{\mathbf{x}}_k^{-} = \boldsymbol{\phi}_{k-1}\hat{\mathbf{x}}_{k-1}$. Thus, Eq. (7.1.21) can be written as

$$\hat{\mathbf{x}}_k = \boldsymbol{\phi}_{k-1}\hat{\mathbf{x}}_{k-1} + \mathbf{K}_k (\mathbf{z}_k - \mathbf{H}_k \boldsymbol{\phi}_{k-1}\hat{\mathbf{x}}_{k-1}) \qquad (7.1.22)$$

Again, we approximate $\boldsymbol{\phi}$ as $\mathbf{I} + \mathbf{F}\Delta t$. Then, neglecting higher-order terms in Δt and noting that $\mathbf{K}_k = \mathbf{K}\Delta t$ leads to

$$\hat{\mathbf{x}}_k - \hat{\mathbf{x}}_{k-1} = \mathbf{F}\hat{\mathbf{x}}_{k-1}\,\Delta t + \mathbf{K}\Delta t (\mathbf{z}_k - \mathbf{H}_k \hat{\mathbf{x}}_{k-1}) \qquad (7.1.23)$$

Finally, dividing by Δt, passing to the limit, and dropping the subscripts yields

Figure 7.1. On-line block diagram for the continuous Kalman filter.

the differential equation

$$\dot{\hat{\mathbf{x}}} = \mathbf{F}\hat{\mathbf{x}} + \mathbf{K}(\mathbf{z} - \mathbf{H}\hat{\mathbf{x}}) \qquad (7.1.24)$$

Equations (7.1.16), (7.1.20), and (7.1.24) comprise the continuous Kalman filter equations and these are summarized in Fig. 7.1. If the filter were to be implemented on-line, note that certain equations would have to be solved in real time as indicated in Fig. 7.1. Theoretically, the differential equation for **P** could be solved off-line, and the gain profile could be stored for later use on-line. However, the main $\hat{\mathbf{x}}$ equation must be solved on-line, because $\mathbf{z}(t)$, that is, the noisy measurement, is the input to the differential equation.

The continuous filter equations as summarized in Fig. 7.1 are innocent looking because they are written in matrix form. They should be treated with respect, though. It does not take much imagination to see the degree of complexity that results when they are written out in scalar form. If the dimensionality is high, an analog implementation is completely unwieldy.

Note that the error covariance equation must be solved in order to find the gain, just as in the discrete case. In the continuous case, though, a differential rather than difference equation must be solved. Furthermore, the differential equation is nonlinear because of the $\mathbf{PH}^T\mathbf{R}^{-1}\mathbf{HP}$ term, which complicates matters. This will be explored further in the next section.

7.2
SOLUTION OF THE MATRIX RICCATI EQUATION

The error covariance equation

$$\dot{\mathbf{P}} = \mathbf{FP} + \mathbf{PF}^T - \mathbf{PH}^T\mathbf{R}^{-1}\mathbf{HP} + \mathbf{GQG}^T \qquad (7.2.1)$$

$$\mathbf{P}(0) = \mathbf{P}_0$$

is a special form of nonlinear differential equation known as the matrix Riccati equation. This equation has been studied extensively, and an analytical solution exists for the constant-parameter case. The general procedure is to transform the single nonlinear equation into a system of two simultaneous linear equations; and, of course, analytical solutions exist for linear differential equations with constant coefficients. Toward this end we assume that $\mathbf{P}$ can be written in product form as

$$\mathbf{P} = \mathbf{XZ}^{-1}, \qquad \mathbf{Z}(0) = \mathbf{I} \qquad (7.2.2)$$

or

$$\mathbf{PZ} = \mathbf{X} \qquad (7.2.3)$$

Differentiating both sides of Eq. (7.2.3) leads to

$$\dot{\mathbf{P}}\mathbf{Z} + \mathbf{P}\dot{\mathbf{Z}} = \dot{\mathbf{X}} \qquad (7.2.4)$$

Next, we substitute $\dot{\mathbf{P}}$ from Eq. (7.2.1) into Eq. (7.2.4) and obtain

$$(\mathbf{PF} + \mathbf{PF}^T - \mathbf{PH}^T\mathbf{R}^{-1}\mathbf{HP} + \mathbf{GQG}^T)\mathbf{Z} + \mathbf{P}\dot{\mathbf{Z}} = \dot{\mathbf{X}} \qquad (7.2.5)$$

Rearranging terms and noting that $\mathbf{PZ} = \mathbf{X}$ leads to

$$\mathbf{P}(\mathbf{F}^T\mathbf{Z} - \mathbf{H}^T\mathbf{R}^{-1}\mathbf{HX} + \dot{\mathbf{Z}}) + (\mathbf{FX} + \mathbf{GQG}^T\mathbf{Z} - \dot{\mathbf{X}}) = 0 \qquad (7.2.6)$$

Note that if both terms in parentheses in Eq. (7.2.6) are set equal to zero, equality is satisfied. Thus, we have the pair of linear differential equations

$$\dot{\mathbf{X}} = \mathbf{FX} + \mathbf{GQG}^T\mathbf{Z} \qquad (7.2.7)$$

$$\dot{\mathbf{Z}} = \mathbf{H}^T\mathbf{R}^{-1}\mathbf{HX} - \mathbf{F}^T\mathbf{Z} \qquad (7.2.8)$$

with initial conditions

$$\mathbf{X}(0) = \mathbf{P}_0$$

$$\mathbf{Z}(0) = \mathbf{I} \qquad (7.2.9)$$

These can now be solved by a variety of methods, including Laplace transforms. Once $\mathbf{P}$ is found, the gain $\mathbf{K}$ is obtained as $\mathbf{PH}^T\mathbf{R}^{-1}$, and the filter parameters are determined. An example illustrates the procedure.

EXAMPLE 7.1

We now return to the continuous filter problem considered previously in Example 4.5. The problem was solved there using Wiener methods, and we now wish to apply Kalman filtering methods. Let the signal and noise be statistically independent with autocorrelation functions

$$R_s(\tau) = e^{-|\tau|}, \quad \left(\text{or } S_s(s) = \frac{2}{-s^2 + 1} \right) \qquad (7.2.10)$$

$$R_n(\tau) = \delta(\tau), \quad (\text{or } S_n = 1) \qquad (7.2.11)$$

Since this is a one-state system, x is a scalar. Let x equal the signal. The additive measurement noise is white and thus no augmentation of the state vector is required. The process and measurement models are then

$$\dot{x} = -x + \sqrt{2}u, \quad u = \text{unity white noise} \qquad (7.2.12)$$

$$z = x + v, \quad v = \text{unity white noise} \qquad (7.2.13)$$

Thus, the system parameters are

$$F = -1, \quad G = \sqrt{2}, \quad Q = 1, \quad R = 1, \quad H = 1$$

The differential equations for X and Z are then

$$\dot{X} = -X + 2Z, \quad X(0) = P_0$$

$$\dot{Z} = X + Z, \quad Z(0) = 1 \qquad (7.2.14)$$

Equations (7.2.14) may be solved readily using Laplace-transform techniques.

The result is

$$X(t) = P_0 \cosh \sqrt{3}\, t + \frac{(2 - P_0)}{\sqrt{3}} \sinh \sqrt{3}\, t$$

$$Z(t) = \cosh \sqrt{3}\, t + \frac{(P_0 + 1)}{\sqrt{3}} \sinh \sqrt{3}\, t \qquad (7.2.15)$$

The solution for P may now be formed as $P = XZ^{-1}$:

$$P = \frac{P_0 \cosh \sqrt{3}\, t + \dfrac{(2 - P_0)}{\sqrt{3}} \sinh \sqrt{3}\, t}{\cosh \sqrt{3}\, t + \dfrac{(P_0 + 1)}{\sqrt{3}} \sinh \sqrt{3}\, t} \qquad (7.2.16)$$

Once P is found, the gain K is given by

$$K = PH^T R^{-1}$$

and the filter yielding $\hat{x}$ is determined.

Obviously there should be a correspondence between this solution and the one obtained by Wiener methods. The general connection between the two methods will be discussed in more detail in Section 7.7; here we simply consider a limiting case check as $t \to \infty$. This should yield the same steady-state (stationary) solution obtained previously with Wiener methods. Letting $t \to \infty$ in Eq. (7.2.16) and noting that $P_0 = 1$ yields

$$P \to \frac{1 \cdot e^{\sqrt{3}\, t} + \dfrac{2 - 1}{\sqrt{3}} e^{\sqrt{3}\, t}}{e^{\sqrt{3}\, t} + \dfrac{1 + 1}{\sqrt{3}} e^{\sqrt{3}\, t}} = \sqrt{3} - 1 \qquad (7.2.17)$$

The Kalman filter block diagram for this example is then as shown in Fig. 7.2.

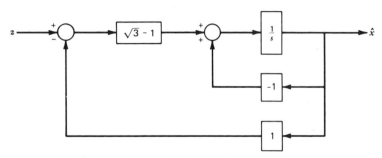

Figure 7.2. Stationary Kalman filter for Example 7.1.

This can be systematically reduced to yield the following overall transfer function relating $\hat{x}$ to z:

$$G(s) = \frac{\text{Laplace transform of } \hat{x}}{\text{Laplace transform of } z} = \frac{\sqrt{3} - 1}{s + \sqrt{3}} \qquad (7.2.18)$$

This is the same result obtained using Wiener methods. ■

7.3
CORRELATED MEASUREMENT
AND PROCESS NOISE

Thus far we have considered the process noise **u** and the measurement noise **v** to have zero crosscorrelation. This is often a reasonable assumption in physical problems, but not always. We will now see how the filter equations can be modified to account for crosscorrelation between the process and measurement noises. We will look at the continuous filter first, and then we will retrace the steps in the derivation of the discrete filter with the crosscorrelation properly accounted for.

Continuous Filter — Correlated Process
and Measurement Noise

We first define the process and measurement models. They are

$$\left. \begin{array}{l} \dot{\mathbf{x}} = \mathbf{Fx} + \mathbf{Gu} \\ \mathbf{z} = \mathbf{Hx} + \mathbf{v} \end{array} \right\} \quad \text{(as before)} \qquad \begin{array}{l} (7.3.1) \\ (7.3.2) \end{array}$$

where

$$\left. \begin{array}{l} E\left[\mathbf{u}(t)\mathbf{u}^T(\tau)\right] = \mathbf{Q}\delta(t - \tau) \\ E\left[\mathbf{v}(t)\mathbf{v}^T(\tau)\right] = \mathbf{R}\delta(t - \tau) \end{array} \right\} \quad \text{(as before)} \qquad (7.3.3)$$

and

$$E\left[\mathbf{u}(t)\mathbf{v}^T(\tau)\right] = \mathbf{C}\delta(t - \tau) \qquad \text{(rather than zero, as before)} \quad (7.3.4)$$

Our general approach is to form a new process model whose input noise has zero crosscorrelation with **v**. We first note that $\mathbf{z} - \mathbf{Hx} - \mathbf{v}$ is zero, and thus

$\mathbf{D}(\mathbf{z} - \mathbf{Hx} - \mathbf{v})$ may be added to the right side of Eq. (7.3.1). Thus we have

$$\dot{\mathbf{x}} = \mathbf{Fx} + \mathbf{Gu} + \mathbf{D}(\mathbf{z} - \mathbf{Hx} - \mathbf{v}) \qquad (7.3.5)$$

where the constant $\mathbf{D}$ will be chosen presently. However, first we rearrange the terms of Eq. (7.3.5) to obtain a new process model.

$$\text{New process model:} \qquad \dot{\mathbf{x}} = (\mathbf{F} - \mathbf{DH})\mathbf{x} + \mathbf{Dz} + (\mathbf{Gu} - \mathbf{Dv}) \qquad (7.3.6)$$

The random process $\mathbf{x}$ is the same as before, but Eq. (7.3.6) shows that we can think of $\mathbf{x}$ as a superposition of two responses. One part is due to $\mathbf{Dz}(t)$, which we will treat as if it were a known explicit input. (It is observed and available directly as a measurement.) The other part of the response is due to $(\mathbf{Gu} - \mathbf{Dv})$, and this is a white noise input that is not observed (directly, at least). For lack of better names, we will refer to these two component responses as the explicit and random parts, and they will be denoted as $\mathbf{x}_e$ and $\mathbf{x}_r$. The total response is then

$$\mathbf{x} = \mathbf{x}_e + \mathbf{x}_r \qquad (7.3.7)$$

Now, in the new process model, we wish to choose $\mathbf{D}$ such that $(\mathbf{Gu} - \mathbf{Dv})$ and $\mathbf{v}$ have zero crosscorrelation, that is,

$$E\left\{ \left[\mathbf{Gu}(t) - \mathbf{Dv}(t) \right] \mathbf{v}(\tau)^T \right\} = 0 \qquad (7.3.8)$$

or

$$E\left[\mathbf{Gu}(t)\mathbf{v}(\tau)^T \right] = E\left[\mathbf{Dv}(t)\mathbf{v}(\tau)^T \right] \qquad (7.3.9)$$

Next, substituting Eqs. (7.3.3) and (7.3.4) into Eq. (7.3.9) leads to

$$\mathbf{GC}\delta(t - \tau) = \mathbf{DR}\delta(t - \tau) \qquad (7.3.10)$$

We now see that if we choose $\mathbf{D}$ to be

$$\mathbf{D} = \mathbf{GCR}^{-1} \qquad (7.3.11)$$

the desired decorrelation is effected. The new process model now satisfies all the necessary conditions imposed in Section 7.1, so the error covariance expression may be written as

$$\dot{\mathbf{P}} = (\mathbf{F} - \mathbf{DH})\mathbf{P} + \mathbf{P}(\mathbf{F} - \mathbf{DH})^T - \mathbf{PH}^T\mathbf{R}^{-1}\mathbf{HP} + \mathbf{Q}' \qquad (7.3.12)$$

where $\mathbf{Q}'$ is defined by

$$E\left\{[\mathbf{Gu}(t) - \mathbf{Dv}(t)][\mathbf{Gu}(\tau) - \mathbf{DV}(\tau)]^T\right\} = \mathbf{Q}'\delta(t - \tau) \qquad (7.3.13)$$

Expanding Eq. (7.3.13) and using Eqs. (7.3.3), (7.3.4), and (7.3.11) leads to

$$\mathbf{Q}' = \mathbf{G}(\mathbf{Q} - \mathbf{CR}^{-1}\mathbf{C}^T)\mathbf{G}^T \qquad (7.3.14)$$

The equation for $\dot{\mathbf{P}}$ may now be rewritten as

$$\dot{\mathbf{P}} = (\mathbf{F} - \mathbf{DH})\mathbf{P} + \mathbf{P}(\mathbf{F} - \mathbf{DH})^T - \mathbf{PH}^T\mathbf{R}^{-1}\mathbf{HP} + \mathbf{G}(\mathbf{Q} - \mathbf{CR}^{-1}\mathbf{C}^T)\mathbf{G}^T$$
$$(7.3.15)$$

The expression for gain in the new model is then

$$\mathbf{K}' = \mathbf{PH}^T\mathbf{R}^{-1} \qquad (7.3.16)$$

and the motive for using the prime will be made clear shortly.

We now look at the equation for the estimate. Just as in the case of the process itself, we wish to think of the estimate as a superposition of an explicit part, which is an estimate of $\mathbf{x}_e$, and another part, which is an estimate of the random component $\mathbf{x}_r$. Thus we have for the two estimates

$$\dot{\hat{\mathbf{x}}}_e = (\mathbf{F} - \mathbf{DH})\hat{\mathbf{x}}_e + \mathbf{Dz}(t) \qquad (7.3.17)$$

$$\dot{\hat{\mathbf{x}}}_r = (\mathbf{F} - \mathbf{DH})\hat{\mathbf{x}}_r + \mathbf{K}'(\mathbf{z}_r - \mathbf{H}\hat{\mathbf{x}}_r) \qquad (7.3.18)$$

The error associated with $\hat{\mathbf{x}}_e$ is, of course, zero because $\mathbf{z}(t)$ is known, that is, in real time it is the total measurement and it is known exactly. The measurement of $\mathbf{x}_r$ has been denoted as $\mathbf{z}_r$, and this requires some explanation. From the basic measurement relationship, we have

$$\mathbf{z} = \mathbf{Hx} + \mathbf{v} = \mathbf{Hx}_e + \mathbf{Hx}_r + \mathbf{v}$$

or

$$(\mathbf{z} - \mathbf{Hx}_e) = \mathbf{Hx}_r + \mathbf{v} \qquad (7.3.19)$$

Thus, $(\mathbf{z} - \mathbf{Hx}_e)$ must be considered as the noisy measurement of $\mathbf{x}_r$. This is denoted as $\mathbf{z}_r$ in Eq. (7.3.18). Thus, making the substitution into Eq. (7.3.18) leads to

$$\dot{\hat{\mathbf{x}}}_r = (\mathbf{F} - \mathbf{DH})\hat{\mathbf{x}}_r + \mathbf{K}'(\mathbf{z} - \mathbf{H}\hat{\mathbf{x}}_e - \mathbf{H}\hat{\mathbf{x}}_r) \qquad (7.3.20)$$

Now, adding Eqs. (7.3.17) and (7.3.20), and noting that $\hat{\mathbf{x}} = \hat{\mathbf{x}}_e + \hat{\mathbf{x}}_r$, yields

$$\dot{\hat{\mathbf{x}}} = (\mathbf{F} - \mathbf{DH})\hat{\mathbf{x}} + (\mathbf{D} + \mathbf{K}')\mathbf{z} - \mathbf{K}'\mathbf{H}\hat{\mathbf{x}} = \mathbf{F}\hat{\mathbf{x}} + (\mathbf{D} + \mathbf{K}')(\mathbf{z} - \mathbf{H}\hat{\mathbf{x}}) \quad (7.3.21)$$

It can be seen from the form of Eq. (7.3.21) that $(\mathbf{D} + \mathbf{K}')$ plays the role of "gain" in the estimation equation for the total $\mathbf{x}$ quantity. The error associated with the $\hat{\mathbf{x}}_e$ component is zero, so the $\mathbf{P}$ equation, which was derived for the $\hat{\mathbf{x}}_r$ component, also applies to the total $\hat{\mathbf{x}}$. From Eqs. (7.3.11) and (7.3.16), we see that the gain $\mathbf{D} + \mathbf{K}'$ is just $(\mathbf{PH}^T + \mathbf{GC})\mathbf{R}^{-1}$. Thus, the final estimation equations for $\hat{\mathbf{x}}$ may be summarized as

1. Estimation equation:

$$\dot{\hat{\mathbf{x}}} = \mathbf{F}\hat{\mathbf{x}} + \mathbf{K}(\mathbf{z} - \mathbf{H}\hat{\mathbf{x}}), \qquad \hat{\mathbf{x}}(0) = \mathbf{x}_0 \quad (7.3.22)$$

2. Gain equation:

$$\mathbf{K} = (\mathbf{PH}^T + \mathbf{GC})\mathbf{R}^{-1} \quad (7.3.23)$$

3. Error covariance equation:

$$\dot{\mathbf{P}} = (\mathbf{F} - \mathbf{DH})\mathbf{P} + \mathbf{P}(\mathbf{F} - \mathbf{DH})^T - \mathbf{PH}^T\mathbf{R}^{-1}\mathbf{HP}$$

$$+ \mathbf{G}(\mathbf{Q} - \mathbf{CR}^{-1}\mathbf{C}^T)\mathbf{G}^T \quad (7.3.24)$$

$$\mathbf{P}(0) = \mathbf{P}_0$$

where

$$\mathbf{D} = \mathbf{GCR}^{-1} \quad (7.3.25)$$

We note that if $\mathbf{C}$ is zero, the above estimation equations reduce to the previous equations where we had zero crosscorrelation between $\mathbf{u}$ and $\mathbf{v}$. This is as expected. Also, note that in the process of deriving the filter equations for the correlated case, it was necessary to consider a superposition of explicit and random responses. Since this can always be done in a linear system, the addition of an explicit or deterministic driving function to the process equation presents no particular problem. We simply treat it separately and note that it contributes zero error to the estimate (see Problem 7.9). An example illustrating the use of the equations for correlated $\mathbf{u}$ and $\mathbf{v}$ will be presented in Section 7.4.

Discrete Filter — Correlated Process and Measurement Noise

Just as for the continuous filter, we first define the process and measurement models. They are as follows:

$$
\left.\begin{array}{l}
\mathbf{x}_{k+1} = \boldsymbol{\phi}_k \mathbf{x}_k + \mathbf{w}_k \\
\mathbf{z}_k = \mathbf{H}_k \mathbf{x}_k + \mathbf{v}_k
\end{array}\right\} \quad \text{(as before in Chapter 5)}
\qquad
\begin{array}{l}
(7.3.26) \\
(7.3.27)
\end{array}
$$

where

$$
E\left[\mathbf{w}_k \mathbf{w}_i^T\right] = \begin{cases} \mathbf{Q}_k, & i = k \\ 0, & i \neq k \end{cases} \quad \text{(as before in Chapter 5)} \quad (7.3.28)
$$

$$
E\left[\mathbf{v}_k \mathbf{v}_i^T\right] = \begin{cases} \mathbf{R}_k, & i = k \\ 0, & i \neq k \end{cases} \quad \text{(as before in Chapter 5)} \quad (7.3.29)
$$

and

$$
E\left[\mathbf{w}_{k-1} \mathbf{v}_k^T\right] = \mathbf{C}_k \quad \text{(new)} \quad (7.3.30)
$$

Before proceeding, an explanation is in order as to why we are concerned with the crosscorrelation of $\mathbf{v}_k$ with $\mathbf{w}_{k-1}$ rather than $\mathbf{w}_k$, which one might expect from just a casual look at the problem. Rewriting Eq. (7.3.26) with k retarded one step should help in this regard:

$$
\mathbf{x}_k = \boldsymbol{\phi}_{k-1} \mathbf{x}_{k-1} + \mathbf{w}_{k-1} \quad (7.3.31)
$$

Note that it is $\mathbf{w}_{k-1}$ (and not $\mathbf{w}_k$) that represents the cumulative effect of the white forcing function in the continuous model in the interval (t_{k-1}, t_k). Similarly, $\mathbf{v}_k$ represents the cumulative effect of the white measurement noise in the continuous model when averaged over the same interval (t_{k-1}, t_k) [see Eq. (7.1.11)]. Therefore, if we wish to have a correspondence between the continuous and discrete models for small Δt, it is the crosscorrelation between $\mathbf{v}_k$ and $\mathbf{w}_{k-1}$ that we need to include in the discrete model. This is, of course, largely a matter of notation, but an important one.

Now, we could go backward from the continuous equations just derived to get the corresponding discrete equations. However, it is somewhat easier in this case to start over and re-derive the discrete recursive equations with proper accounting for the crosscorrelation in the process. We begin with the general update equation.

$$
\hat{\mathbf{x}}_k = \hat{\mathbf{x}}_k^- + \mathbf{K}_k(\mathbf{z}_k - \mathbf{H}_k \hat{\mathbf{x}}_k^-) \quad (7.3.32)
$$

Next, we form the expression for the estimation error.

$$\mathbf{e}_k = \mathbf{x}_k - \hat{\mathbf{x}}_k$$

$$= \mathbf{x}_k - \left[\hat{\mathbf{x}}_k^- + \mathbf{K}_k(\mathbf{z}_k - \mathbf{H}_k\hat{\mathbf{x}}_k^-) \right]$$

$$= (\mathbf{I} - \mathbf{K}_k\mathbf{H}_k)\mathbf{e}_k^- - \mathbf{K}_k\mathbf{v}_k \qquad (7.3.33)$$

We anticipate now that $\mathbf{e}_k^-$ and $\mathbf{v}_k$ will be correlated, so we will work this out as a side problem:

$$E\left[\mathbf{e}_k^- \mathbf{v}_k^T \right] = E\left[(\mathbf{x}_k - \hat{\mathbf{x}}_k^-)\mathbf{v}_k^T \right]$$

$$= E\left[(\boldsymbol{\phi}_{k-1}\mathbf{x}_{k-1} + \mathbf{w}_{k-1} - \boldsymbol{\phi}_{k-1}\hat{\mathbf{x}}_{k-1})\mathbf{v}_k^T \right] \qquad (7.3.34)$$

Note that $\mathbf{v}_k$ will not be correlated with either $\mathbf{x}_{k-1}$ or $\hat{\mathbf{x}}_{k-1}$ because of its whiteness. Therefore, Eq. (7.3.34) reduces to

$$E\left[\mathbf{e}_k^- \mathbf{v}_k^T \right] = E\left[\mathbf{w}_{k-1}\mathbf{v}_k^T \right] = \mathbf{C}_k \qquad (7.3.35)$$

We now return to the main derivation. By using Eq. (7.3.33), we form the expression for the $\mathbf{P}_k$ matrix

$$\mathbf{P}_k = E\left[\mathbf{e}_k \mathbf{e}_k^T \right]$$

$$= E\left\{ \left[(\mathbf{I} - \mathbf{K}_k\mathbf{H}_k)\mathbf{e}_k^- - \mathbf{K}_k\mathbf{v}_k \right] \left[(\mathbf{I} - \mathbf{K}_k\mathbf{H}_k)\mathbf{e}_k^- - \mathbf{K}_k\mathbf{v}_k \right]^T \right\} \qquad (7.3.36)$$

Now, expanding Eq. (7.3.36) and taking advantage of Eq. (7.3.35) leads to

$$\mathbf{P}_k = (\mathbf{I} - \mathbf{K}_k\mathbf{H}_k)\mathbf{P}_k^-(\mathbf{I} - \mathbf{K}_k\mathbf{H}_k)^T + \mathbf{K}_k\mathbf{R}_k\mathbf{K}_k^T$$

$$- (\mathbf{I} - \mathbf{K}_k\mathbf{H}_k)\mathbf{C}_k\mathbf{K}_k^T - \mathbf{K}_k\mathbf{C}_k^T(\mathbf{I} - \mathbf{K}_k\mathbf{H}_k)^T \qquad (7.3.37)$$

This is a perfectly general expression for the error covariance and is valid for any gain $\mathbf{K}_k$. The last two terms in Eq. (7.3.37) are "new" and involve the crosscorrelation parameter $\mathbf{C}_k$.

We now follow the same procedure used in Section 5.4 to find the optimal gain. We differentiate Trace $\mathbf{P}_k$ with respect to $\mathbf{K}_k$ and set the result equal to zero. The necessary matrix differentiation formulas are given in Section 5.4, and the resulting optimal gain is

$$\mathbf{K}_k = \left(\mathbf{P}_k^- \mathbf{H}_k^T + \mathbf{C}_k \right) \left[\mathbf{H}_k\mathbf{P}_k^- \mathbf{H}_k^T + \mathbf{R}_k + \mathbf{H}_k\mathbf{C}_k + \mathbf{C}_k^T\mathbf{H}_k^T \right]^{-1} \qquad (7.3.38)$$

Note that this expression is similar to the gain formula of Chapter 5 except for the additional terms involving $\mathbf{C}_k$. Let $\mathbf{C}_k$ go to zero, and Eq. (7.3.38) reduces to the same gain as in the zero crosscorrelation model, which is as it should be.

We can now substitute the optimal gain expression, Eq. (7.3.38), into the general $\mathbf{P}_k$ equation, Eq. (7.3.37), to get the a posteriori $\mathbf{P}_k$ equation. After some algebraic manipulation, this leads to either of the two forms:

$$\mathbf{P}_k = \mathbf{P}_k^- - \mathbf{K}_k\left[\mathbf{H}_k\mathbf{P}_k^-\mathbf{H}_k^T + \mathbf{R}_k + \mathbf{H}_k\mathbf{C}_k + \mathbf{C}_k^T\mathbf{H}_k^T\right]\mathbf{K}_k^T \qquad (7.3.39)$$

or

$$\mathbf{P}_k = (\mathbf{I} - \mathbf{K}_k\mathbf{H}_k)\mathbf{P}_k^- - \mathbf{K}_k\mathbf{C}_k^T \qquad (7.3.40)$$

The projection equations are not affected by the crosscorrelation between $\mathbf{w}_{k-1}$ and $\mathbf{v}_k$ because of the whiteness property of each. Therefore, the projection equations are (repeated here for completeness)

$$\hat{\mathbf{x}}_{k+1}^- = \boldsymbol{\phi}_k\hat{\mathbf{x}}_k \qquad (7.3.41)$$

$$\mathbf{P}_{k+1}^- = \boldsymbol{\phi}_k\mathbf{P}_k\boldsymbol{\phi}_k^T + \mathbf{Q}_k \qquad (7.3.42)$$

Equations (7.3.32), (7.3.38), (7.3.40), (7.3.41), and (7.3.42) now comprise the complete set of recursive equations for the correlated process and measurement noise case. We could go one step further at this point and show that the discrete recursive equations will, in fact, go over into the continuous-model differential equations as $\Delta t \to 0$. This is fairly routine, though, so it is left as an exercise (see Problem 7.10).

7.4
COLORED MEASUREMENT NOISE

In the equations for the continuous Kalman filter, the inverse of the $\mathbf{R}$ matrix appears in both the gain and error covariance expressions. If the measurement noise is colored, rather than white, it must be modeled with additional state variables. This leaves zero for the white component, and thus $\mathbf{R}^{-1}$ will not exist. This leads to obvious difficulty. Of course, one remedy would be to add a small white component artificially, and then proceed on with the filter design using the usual equations. However, this is begging the issue. There are legitimate situations where one simply does not want to model any part of the measurement error as white noise.

A general solution to the colored measurement noise problem was first presented by Bryson and Johansen (2). Our approach here will be slightly different and, we hope, more intuitively satisfying. If $\mathbf{R}^{-1}$ does not exist, we have a situation where certain linear combinations of the state variables are being measured perfectly. Obviously, if a particular state variable is known perfectly, it can be removed from the estimation problem. There is certainly no need for filtering something that is already free of corrupting noise. Therefore, our general approach will be to remove those quantities that are known perfectly from those that arc not, and then solve the remaining estimation problem. This usually necessitates a linear transformation of the original state vector in order to decouple the known states from the others. This adds to the algebra, but it certainly presents no theoretical problem because the state estimate transforms with exactly the same transformation as the state itself.

EXAMPLE 7.2

Consider the problem of separating an additive combination of two independent Markov processes with identical spectral functions. We know the solution calls for a trivial filter with a gain function of $\frac{1}{2}$. Let the autocorrelation functions of the signal and noise be

$$R_s(\tau) = e^{-|\tau|}, \qquad R_n(\tau) = e^{-|\tau|} \tag{7.4.1}$$

The measurement z is the additive combination

$$z = s + n \tag{7.4.2}$$

Now the obvious way to model the continuous Kalman filter is to let s be x_1 and n be x_2, and then we have

$$\begin{bmatrix} \dot{x}_1 \\ \dot{x}_2 \end{bmatrix} = \begin{bmatrix} -1 & 0 \\ 0 & -1 \end{bmatrix} \begin{bmatrix} x_1 \\ x_2 \end{bmatrix} + \begin{bmatrix} \sqrt{2} & 0 \\ 0 & \sqrt{2} \end{bmatrix} \begin{bmatrix} u_1 \\ u_2 \end{bmatrix} \tag{7.4.3}$$

$$[z] = [1 \quad 1]\begin{bmatrix} x_1 \\ x_2 \end{bmatrix} + [0] \tag{7.4.4}$$

where u_1 and u_2 are independent, unity white noise driving functions. Notice that v is zero and thus R^{-1} does not exist. Since we have a perfect measurement of the linear combination $x_1 + x_2$, let us transform to new states, y_1 and y_2, according to

$$\begin{bmatrix} y_1 \\ y_2 \end{bmatrix} = \begin{bmatrix} 1 & 0 \\ 1 & 1 \end{bmatrix} \begin{bmatrix} x_1 \\ x_2 \end{bmatrix} \tag{7.4.5}$$

or

$$\mathbf{y} = \mathbf{\Lambda x}$$

Making this transformation leads to the new state and measurement models

$$\begin{bmatrix} \dot{y}_1 \\ \dot{y}_2 \end{bmatrix} = \begin{bmatrix} -1 & 0 \\ 0 & -1 \end{bmatrix}\begin{bmatrix} y_1 \\ y_2 \end{bmatrix} + \begin{bmatrix} \sqrt{2} & 0 \\ \sqrt{2} & \sqrt{2} \end{bmatrix}\begin{bmatrix} u_1 \\ u_2 \end{bmatrix} \qquad (7.4.6)$$

$$[z] = [0 \quad 1]\begin{bmatrix} y_1 \\ y_2 \end{bmatrix} + [0] \qquad (7.4.7)$$

Note that z is now a perfect estimate of y_2 alone, so that we need to worry only about estimating y_1. Thus, the order of the problem is reduced from 2 to 1.

Consider next the y_1 state equation

$$\dot{y}_1 = -y_1 + \sqrt{2}\,u_1 \qquad (7.4.8)$$

This is in the correct form as it is. [Usually y_2 would also appear, and if so, it would be treated as a *known* driving function because it is equal to $z(t)$.]

Next, we need a measurement relationship of the proper form, that is, there must be nontrivial additive white noise and some linear connection to y_1 (and y_1 alone). In this case, since z has zero additive white noise, we can consider its derivative $\dot{z}$ as also being available as a measurement.

$$\dot{z} = \dot{y}_2 = -y_2 + \sqrt{2}\,u_1 + \sqrt{2}\,u_2 \qquad (7.4.9)$$

Observe that $\dot{z}$ contains additive white noise as required. However, it also involves y_2. In order to eliminate y_2, we note

$$z = y_2 \qquad (7.4.10)$$

Now, add z and $\dot{z}$ to eliminate y_2. This leads to

$$\dot{z} + z = 0 \cdot y_1 + \sqrt{2}\,(u_1 + u_2) \qquad (7.4.11)$$

We now have a linear connection between $\dot{z} + z$ and y_1 with additive white noise, which is in the correct form. In the new problem of estimating y_1, we then have

$$F = -1$$
$$G = \sqrt{2}$$
$$u = u_1$$
$$Q = 1$$
$$H = 0$$
$$v = \sqrt{2}\,(u_1 + u_2) \qquad (7.4.12)$$

and $\dot{z} + z$ plays the role of the "measurement." (After all, if z is known, so is $\dot{z}$.)

Note that u and v are correlated, so we must use the correlated form of Kalman filter equations given in Section 7.3. Using the notation of Section 7.3, we then have the additional parameters

$$C = \sqrt{2}$$

$$R = 4$$

$$G(Q - CR^{-1}C^T)G^T = 1$$

$$D = GCR^{-1} = \tfrac{1}{2} \tag{7.4.13}$$

Therefore, the optimal filter is given by the differential equation

$$\dot{\hat{y}}_1 = -\hat{y}_1 + \tfrac{1}{2}[(\dot{z} + z) - 0 \cdot \hat{y}_1] \tag{7.4.14}$$

This will be recognized as the equivalent transfer function relationship

$$\frac{\hat{y}_1(s)}{z(s)} = \frac{\tfrac{1}{2}(s + 1)}{(s + 1)} = \frac{1}{2} \tag{7.4.15}$$

This checks with Wiener filter theory, that is,

$$\hat{y}_1 = \tfrac{1}{2}z \tag{7.4.16}$$

We now have optimal estimates of y_1 and y_2. All we have to do is transform back to the x domain to get optimal estimates of x_1 and x_2. This yields

$$\begin{bmatrix} \hat{x}_1 \\ \hat{x}_2 \end{bmatrix} = \begin{bmatrix} 1 & 0 \\ 1 & 1 \end{bmatrix}^{-1} \begin{bmatrix} \hat{y}_1 \\ \hat{y}_2 \end{bmatrix} = \begin{bmatrix} 1 & 0 \\ -1 & 1 \end{bmatrix} \begin{bmatrix} \tfrac{1}{2}z \\ z \end{bmatrix} = \begin{bmatrix} \tfrac{1}{2}z \\ \tfrac{1}{2}z \end{bmatrix} \tag{7.4.17}$$

Thus, the optimal estimates of both x_1 and x_2 check with the results of the Wiener theory. ∎

The procedure for dealing with colored measurement noise may now be summarized as follows:

1. Decouple the quantities that are known perfectly from those that are not. Use a linear transformation and, as a matter of convenience, define the new state variables such that the perfectly known variables

are at the "bottom" of the new state vector. The bottom elements should then have a one-to-one correspondence to the perfect measurements.

2. Consider a reduced state estimation problem where the state vector contains only the "top" elements of the new state vector. These are not known perfectly from the measurement information, so a nontrivial estimation problem exists relative to these variables. Note that the perfectly known bottom elements that appear on the right side of the state equation may be replaced with measurements z_1, z_2, etc. They may then be grouped with the driving functions because they are known quantities. The reduced state equations must fit the general form

$$\dot{\mathbf{x}}_T = \mathbf{F}_T \mathbf{x}_T + (\text{known inputs}) + (\text{white noise inputs}) \quad (7.4.18)$$

where the subscript T is used to indicate top elements of the transformed state vector.

3. Rearrange the measurement equation to achieve the desired form

$$\mathbf{z}_T = \mathbf{H}_T \mathbf{x}_T + (\text{white noise}) \quad (7.4.19)$$

Note that a nontrivial white noise component must be associated with each measurement element. Also, the number of elements in $\mathbf{z}_T$ must be the same as for the original $\mathbf{z}$ vector in order that no measurement information is ignored in estimating $\mathbf{x}_T$. The white noise components are brought in by repeated differentiation of the perfect z elements until a white noise term appears. The new $\mathbf{z}_T$ vector is then formed from appropriate linear combinations of the elements of $\mathbf{z}$ and their derivatives. Sometimes considerable algebraic manipulation is needed to twist the equations into the form demanded by Eq. (7.4.19), but this can always be done.

4. Solve the reduced state estimation problem. Usually the filter equations for correlated $\mathbf{u}$ and $\mathbf{v}$, as given in Section 7.3, will have to be used. The result will be a differential equation for the reduced state vector. This equation will normally contain derivatives of some of the measurements, as well as the measurements, as driving functions. The derivatives usually present no problems and may be left there as driving functions. If, however, one wishes to remove them, there are standard techniques in linear system theory for doing so [e.g., see Ogata (3) or Chen (4)]. The reduced state equation is usually solved (conceptually, at least) for either the transient or steady-state solution, whichever is desired. The end result is the best estimate of the reduced state vector (i.e., the imperfectly known elements) in the transformed domain.

5. Finally, append the perfectly known elements to the bottom of those estimated in step 4, and transform the total state estimate back to the original state space. This gives the optimal estimate for the original problem.

7.5
SUBOPTIMAL ERROR ANALYSIS

Just as in the discrete case, an error covariance equation can be derived that is applicable to suboptimal as well as optimal gain. The derivation is similar to that given in Section 7.1, and the relationships between $\mathbf{Q}$ and $\mathbf{Q}_k$ and $\mathbf{R}$ and $\mathbf{R}_k$ are given there. We begin with the projection equation for the discrete filter

$$\mathbf{P}_{k+1}^- = \boldsymbol{\phi}_k \mathbf{P}_k \boldsymbol{\phi}_k^T + \mathbf{Q}_k \tag{7.5.1}$$

We now replace $\mathbf{P}_k$ with the general $\mathbf{P}$-update expression given by Eq. (5.4.11). This yields

$$\mathbf{P}_{k+1}^- = \boldsymbol{\phi}_k \left[(\mathbf{I} - \mathbf{K}_k \mathbf{H}_k) \mathbf{P}_k^- (\mathbf{I} - \mathbf{K}_k \mathbf{H}_k)^T + \mathbf{K}_k \mathbf{R}_k \mathbf{K}_k^T \right] \boldsymbol{\phi}_k^T + \mathbf{Q}_k \tag{7.5.2}$$

Next, the following substitutions are made in Eq. (7.5.2):

$$\boldsymbol{\phi}_k = \mathbf{I} + \mathbf{F} \Delta t$$

$$\mathbf{K}_k = \mathbf{K} \, \Delta t$$

$$\mathbf{Q}_k = \mathbf{G} \mathbf{Q} \mathbf{G}^T \Delta t$$

$$\mathbf{R}_k = \frac{\mathbf{R}}{\Delta t} \tag{7.5.3}$$

Then, after neglecting higher-order terms in Δt, the expression for $\mathbf{P}_{k+1}^-$ becomes

$$\mathbf{P}_{k+1}^- = \mathbf{P}_k^- - \mathbf{K} \mathbf{H}_k \mathbf{P}_k^- \, \Delta t - \mathbf{P}_k^- \mathbf{H}_k^T \mathbf{K}^T \, \Delta t + \mathbf{K} \mathbf{R} \mathbf{K}^T \, \Delta t + \mathbf{F} \mathbf{P}_k^- \, \Delta t$$

$$+ \, \mathbf{P}_k^- \mathbf{F}^T \Delta t + \mathbf{G} \mathbf{Q} \mathbf{G}^T \, \Delta t \tag{7.5.4}$$

Finally, forming the difference $\mathbf{P}_{k+1}^- - \mathbf{P}_k^-$, dividing by Δt, and passing to the limit leads to

$$\dot{\mathbf{P}} = \mathbf{F} \mathbf{P} + \mathbf{P} \mathbf{F}^T - \mathbf{K} \mathbf{H} \mathbf{P} - \mathbf{P} \mathbf{H}^T \mathbf{K}^T + \mathbf{K} \mathbf{R} \mathbf{K}^T + \mathbf{G} \mathbf{Q} \mathbf{G}^T \tag{7.5.5}$$

This equation may now be used with any gain $\mathbf{K}$ and is useful in suboptimal error analysis. A word of caution is in order, though. Note that "gain" means the $\mathbf{K}$ coefficient in the differential equation

$$\dot{\hat{\mathbf{x}}} = \mathbf{F} \hat{\mathbf{x}} + \mathbf{K}(\mathbf{z} - \mathbf{H} \hat{\mathbf{x}}) \tag{7.5.6}$$

and the equations describing the suboptimal filter under consideration must be put into the proper format before Eq. (7.5.5) can be applied [see Problem 7.3 for an example of the use of Eq. (7.5.5)].

7.6
FILTER STABILITY
IN STEADY-STATE CONDITION

In many applications the Kalman filter will reach a steady-state condition (or at least quasi-steady-state) within a reasonable time after initialization. When this happens, the gain becomes constant. In the continuous case the filter then looks much the same as any other fixed-parameter filter except, perhaps, it may be of the multiple input–output variety rather than just a single input–output filter. Similarly, in the steady state the discrete Kalman filter looks much like any other digital filter, provided that the sampling rate is uniform. Conceptually, the filter is simply a linear operator that processes a set of inputs (the measurements) and transforms them to a corresponding set of outputs (the state estimates). We know from classical filter theory that the characteristic poles of the filter (roots of the transfer function denominator) tell us much about the stability of the filter, so we will now extend this idea to the Kalman filter (in the steady state, of course).

Continuous Filter Stability

We will go directly to the differential equation relating the input $\mathbf{z}(t)$ to the output $\hat{\mathbf{x}}(t)$. Equation (7.1.24) is repeated for convenience:

$$\dot{\hat{\mathbf{x}}} = \mathbf{F}\hat{\mathbf{x}} + \mathbf{K}(\mathbf{z} - \mathbf{H}\hat{\mathbf{x}}) \qquad (7.6.1)$$

Now assume that $\mathbf{F}$, $\mathbf{K}$, and $\mathbf{H}$ are constant and take the Laplace transform of both sides of Eq. (7.6.1). (We will tacitly assume that the initial conditions are zero.) After rearranging terms, this leads to

$$[s\mathbf{I} - (\mathbf{F} - \mathbf{KH})]\hat{\mathbf{X}}(s) = \mathbf{KZ}(s) \qquad (7.6.2)$$

It can be seen that the bracketed term on the left side of Eq. (7.6.2) is the characteristic matrix of the system, and its determinant will be the characteristic polynomial of the system, that is,

$$\text{Characteristic polynomial} = |s\mathbf{I} - (\mathbf{F} - \mathbf{KH})| \qquad (7.6.3)$$

The roots of this polynomial are then the characteristic poles of the system. (Or, if you prefer, the eigenvalues of $(\mathbf{F} - \mathbf{KH})$ are the characteristic eigenvalues of the system.)

A simple example will illustrate the use of Eq. (7.6.3). Refer to Example 7.1 and note that the steady-state gain is

$$K = PH^T R^{-1} = (\sqrt{3} - 1)(1)(1) = \sqrt{3} - 1$$

Also, in this example F and H are

$$F = -1, \qquad H = 1$$

Therefore, the characteristic polynomial is

$$s - \left[-1 - (\sqrt{3} - 1)(1) \right] = s + \sqrt{3}$$

The characteristic pole is seen to be at $-\sqrt{3}$ in the s-plane, indicating that the filter is stable. This is, of course, the same result that was arrived at by block diagram reduction in Example 7.1. However, it is important to note that we did this without any such block diagram manipulation. Remember that state-space block diagrams like the one shown in Fig. 7.1 are vector *time-domain* diagrams, and extracting the transfer function in the s-domain can be very complicated (if not virtually impossible) in higher-order systems. On the other hand, it is routine to find the roots of a polynomial (or eigenvalues of a square matrix) via computer methods, even in high-order systems.

Discrete Filter Stability

Stability analysis for the discrete Kalman filter is similar to that of the continuous filter, except that we use z-transforms rather than Laplace transforms. We begin with the basic filter update equation, Eq. (5.4.8), which is repeated here for convenience:

$$\hat{\mathbf{x}}_k = \hat{\mathbf{x}}_k^- + \mathbf{K}_k(\mathbf{z}_k - \mathbf{H}_k \hat{\mathbf{x}}_k^-) \qquad (7.6.4)$$

to get the difference equation for $\hat{\mathbf{x}}_k$ we replace $\hat{\mathbf{x}}_k^-$ with $\boldsymbol{\phi}_{k-1}\hat{\mathbf{x}}_{k-1}$ in Eq. (7.6.4). The result is

$$\hat{\mathbf{x}}_k = (\boldsymbol{\phi}_{k-1} - \mathbf{K}_k \mathbf{H}_k \boldsymbol{\phi}_{k-1})\mathbf{x}_{k-1} + \mathbf{K}_k \mathbf{z}_k \qquad (7.6.5)$$

We now take the z-transform of both sides of Eq. (7.6.5) and note that retarding $\hat{\mathbf{x}}_k$ one step in the time domain is the equivalent of multiplying by z^{-1} in the z-domain. The result is

$$\hat{\mathbf{X}}(z) = (\boldsymbol{\phi} - \mathbf{K}_k \mathbf{H}\boldsymbol{\phi}) z^{-1}\hat{\mathbf{X}}(z) + \mathbf{K}_k \mathbf{Z}(z)$$

Or, after rearranging terms, we have

$$[z\mathbf{I} - (\boldsymbol{\phi} - \mathbf{K}_k\mathbf{H}\boldsymbol{\phi})]\hat{\mathbf{X}}(z) = z\mathbf{K}_k\mathbf{Z}(z) \qquad (7.6.6)$$

Note here that the subscripts have been dropped because of the stationary assumption, except for $\mathbf{K}_k$ where the subscript has been retained as a reminder that this is the discrete filter gain, and not the continuous gain $\mathbf{K}$. Also, italic z is the z-transform variable, whereas boldface $\mathbf{z}$ and $\mathbf{Z}$ refer to the vector measurements as usual.

Note the similarity between Eq. (7.6.6) and Eq. (7.6.2). Clearly, the characteristic polynomial is

$$\text{Characteristic polynomial} = |z\mathbf{I} - (\boldsymbol{\phi} - \mathbf{K}_k\mathbf{H}\boldsymbol{\phi})| \qquad (7.6.7)$$

and the roots of this polynomial tell the story about the filter stability. As before, the roots of the characteristic polynomial are the same as the eigenvalues of $(\boldsymbol{\phi} - \mathbf{K}_k\mathbf{H}\boldsymbol{\phi})$. A simple example will illustrate the usefulness of Eq. (7.6.7).

EXAMPLE 7.3

Let us return to the random-walk problem of Example 6.4 and investigate the stability of the filter in the steady-state condition. The discrete model in this example is

$$x_{k+1} = x_k + w_k \qquad (7.6.8)$$

$$z_k = x_k + v_k \qquad (7.6.9)$$

and the discrete filter parameters are

$$\phi_k = 1, \quad H_k = 1, \quad Q_k = 1, \quad R_k = .1, \quad P_0^- = 1, \quad \hat{x}_0^- = 0$$

In this example the gain reaches steady state in just a few steps, and it is easily verified that its steady-state value is

$$K_k \approx .916$$

We can now form the characteristic polynomial from Eq. (7.6.7):

$$\text{Characteristic polynomial} = z - [1 - (.916)(1)(1)]$$

$$= z - .084 \qquad (7.6.10)$$

The characteristic root is at .084, which is well within the unit circle in the z-plane. Thus we see that the filter is highly stable in this case.

Note that even though the input in this case is nonstationary, the filter itself is intrinsically stable. Furthermore, the filter pole location tells us that any small perturbation from the steady-state condition (e.g., due to roundoff error) will damp out quickly. Any such perturbations will be attenuated by a factor of .084 with each step in this case, so their effect "vaporizes" rapidly. This same kind of reasoning can be extended to the vector case, provided that the **P** matrix is kept symmetric in the recursive process and that it is never allowed to lose its positive definiteness. Thus, we see that we can gain considerable insight into the filter operation just by looking at its characteristic poles in the steady-state condition, provided, of course, that a steady-state condition exists. ∎

7.7
RELATIONSHIP BETWEEN WIENER
AND KALMAN FILTERS

It is appropriate here to pause and reflect on the connection between Wiener and Kalman filter theories. Both the Wiener and Kalman filters are minimum mean-square-error estimators, both require the same a priori information about the processes being estimated, and both yield identical estimates. We saw in Chapter 4 that the Wiener approach leads to an integral equation with the filter weighting function as the unknown. After solution of the integral equation, the weighting function then describes the relationship between input and output. On the other hand, the end result of continuous Kalman filter theory is a differential equation relating input and output. There must be equivalence, and books on linear system theory [e.g., Chen (4)] may be consulted for the details of how to get back and forth between the two descriptions. There are, however, certain subtleties about this particular problem that warrant some further amplification.

First, in the Wiener theory we related the filter response to the input via the superposition (convolution) integral, which was written in the form

$$\hat{x}(t) = \int_0^t g(\tau, t) z(t - \tau) \, d\tau \qquad (7.7.1)$$

This form was convenient because the first argument of g has the significance of the "age" variable, and g tells us how the past values of the input are weighted to yield the present value of the output (5). The second argument of g (i.e., t) simply appears as a parameter that may be considered fixed in the optimization process. Frequently, though, books on linear system theory [e.g., see Chen (4)] write the superposition integral in the form

$$\hat{x}(t) = \int_0^t h(t, \tau) z(\tau) \, d\tau \qquad (7.7.2)$$

(This was mentioned before in Chapter 4.) When $\hat{x}$ is written in the form of Eq. (7.7.2), $h(t, \tau)$ has a physical interpretation as the system response at time t to a unit impulse applied at time τ. By making a simple change of dummy variables in Eq. (7.7.1), we can obtain the relationship between g and h:

$$g(t - \tau, t) = h(t, \tau) \qquad (7.7.3)$$

This is not a point of confusion in constant-parameter systems. It is, however, in time-variable systems, and we have to be watchful of this little detail in converting from a weighting-function description to a state-realization (differential equation) description.

In addition, the Wiener filter is always a single-output estimator. That is, we may have multiple inputs, but we always choose a single scalar process (usually called "signal") as the quantity being estimated. For example, consider the problem where we have one Markov process that we call signal s, and the additive noise consists of another Markov process n_1 plus a white component n_2. In Wiener theory, n_1 gets lumped with n_2 as the corrupting noise and the estimator yields just an estimate of s. On the other hand, with Kalman filtering, one models both s and n_1 as state variables, and the filter estimates both s and n_1 simultaneously (even though $\hat{n}_1$ may not be of interest). In effect, the Kalman filter does the work of two Wiener filters. Furthermore, the Kalman filter automatically provides information about the quality of the estimates (i.e., their mean-square errors) while doing the estimation. The Wiener filter does not provide this information, and one has to go to considerable extra effort to get mean-square error information. Thus, we have in the Kalman filter a group of estimators, all packaged into a single matrix algorithm. The price, of course, is extra computation effort. We often find in Kalman filtering that we are forced to carry along considerable excess baggage in order to obtain estimates of a few select quantities of interest.

Another subtlety that appears when comparing Wiener and Kalman methods has to do with the initial conditions. It may appear at first glance that we are allowed to choose $\hat{\mathbf{x}}_0$ and $\mathbf{P}_0$ as we wish in a Kalman filter, whereas no such explicit choice exists in the corresponding Wiener formulation. This apparent discrepancy exists because the Wiener filter output was written as a superposition integral, which tacitly implies zero initial conditions for the filter. This is justified, provided the input processes have zero mean and we have no prior information about the processes other than what was already assumed in modeling the various spectral functions. To get correspondence in the Kalman filter we must choose $\hat{\mathbf{x}}_0$ to be zero; and, having done this, $\mathbf{P}_0$ must be the covariance of the process itself. Thus, we really do not have as much legitimate choice in the initial $\hat{\mathbf{x}}_0$ and $\mathbf{P}_0$ as it seems at first glance; at least this is true if we want the result to be optimal. The correspondence between the Kalman and Wiener solutions, properly initialized, is easily verified for simple cases, and Problems 7.4 and 7.5 are intended to demonstrate this.

In summary, it is not proper to say that a Kalman filter is either better or worse than a Wiener filter, because both yield identical results once they are realized. However, in the realization, either for analysis purposes or on-line, the Kalman approach clearly has two distinct advantages:

1. With one common matrix formulation, we can accommodate a large class of estimation problems with relatively complicated process and measurement relationships.

2. The recursive feature of the Kalman filter makes it readily adaptable to computer solutions. This is certainly of considerable practical importance. Solutions of relatively complex estimation problems are often quite feasible using Kalman filtering methods, whereas the same problems may be completely intractable using Wiener methods.

PROBLEMS

7.1.   Consider a scalar process $x(t)$ that may be thought of as the result of integrating Gaussian white noise with a spectral amplitude of 16 units. Let us say the integrator is "zeroed" at $t = 0$, and thus we know $x(0) = 0$. (This is a Wiener process.) Let us further say that we have a continuous noisy measurement of x where the measurement noise is white, Gaussian, independent of x, and has a spectral amplitude of 4 units.

(a) Find the optimal Kalman filter for this situation. (Your solution may be left in the form of a differential equation, but all parameters of the differential equation are to be written out explicitly.)

(b) Find the optimal filter transfer function and the corresponding rms error for the steady-state condition.

7.2.   In the two-input Wiener problem described in Problem 4.17, there were two independent noisy measurements of the signal:

$$z_1 = s(t) + n_1(t)$$

$$z_2 = s(t) + n_2(t)$$

The autocorrelation functions for the signal and noises were given as

$$R_s(\tau) = e^{-|\tau|}$$

$$R_{n_1}(\tau) = e^{-2|\tau|}$$

$$R_{n_2}(\tau) = \delta(\tau)$$

and all three processes were independent of each other. Solve this same problem again for the steady-state transfer functions for the z_1 and z_2 inputs using Kalman filtering methods. (*Hint:* One of the measurement noises is colored, so the procedure given in Section 7.4 should be helpful.)

7.3. The optimal estimator for a Markov signal plus additive white noise was given in Example 7.1. The gain worked out to be time variable, but it did approach a constant value of $(\sqrt{3} - 1)$ as $t \to \infty$. Investigate the effect of using this constant gain for all $t \geq 0$ relative to the filter rms error. Sketch plots of both the optimal and suboptimal rms errors for purposes of comparison.

7.4. Consider a stationary Gaussian Markov signal $x(t)$ whose autocorrelation function is

$$R_x(\tau) = 4e^{-|\tau|}$$

We make two noisy measurements of the signal, one at $t = 0$ and the other at $t = 1$, and we denote these as z_0 and z_1. Each discrete measurement is known to have an error variance of unity, and the errors are statistically independent. We have no prior knowledge of the signal other than its autocorrelation function as stated above.

(a) Using the methods of Section 4.8 (i.e., a weighting function approach), write an explicit expression for the optimal estimate of $x(1)$ in terms of z_0 and z_1.

(b) Repeat part (a) using discrete Kalman filtering and compare the result with that obtained in part (a). (*Note:* The Wiener and Kalman estimates should be identical.)

(c) Let the initial condition for a discrete Kalman filter be

$$\hat{x}_0^- = 0 \quad \text{and} \quad P_0^- = \infty$$

Then work out the explicit expression for $\hat{x}(1)$ and note that it is different from those obtained in parts (a) and (b).

7.5. The same estimation problem was used in both Examples 4.5 and 7.1. Show that the two estimators yield identical results for all $t > 0$ as well as for the steady-state condition. [*Hint:* In the Wiener solution, $\hat{x}(t)$ can be written in terms of a convolution integral where the weighting function is known explicitly. First, convert the weighting function to impulsive-response form and then substitute the integral expression for $\hat{x}(t)$ into the differential equation describing the Kalman estimator and show that it satisfies the differential equation.]

7.6. A certain noisy measurement is known to be of the form

$$z(t) = a_0 + n(t), \qquad t \geq 0$$

where a_0 is an unknown random constant and $n(t)$ is white Gaussian noise with a spectral amplitude of A. Initially, at $t = 0$, all that is known about a_0 is that it has a zero-mean normal distribution with a variance σ^2. We wish to estimate a_0 on a "running time" basis beginning at $t = 0$. Find the appropriate continuous Kalman filter for estimating a_0 and sketch a block diagram for the filter. (*Note:* This is the same estimation problem as that of Problem 4.12. However, when using Kalman filtering, we do not need to consider a random constant as a limiting case of a Markov process.)

7.7. The resulting optimal filter for Problem 7.6 is described by a differential equation relating the input $z(t)$ to the output $\hat{x}(t)$. When the same estimation problem is solved with Wiener methods, the result is a weighting function relating $z(t)$ and $\hat{x}(t)$. Show that the two results are equivalent.

7.8. Consider the two-state Gaussian random process shown in the block diagram below. Clearly, the measurement is of state variable 1, and it may be assumed that the measurement begins at $t = 0$. The initial condition on the first integrator (state variable 2) is zero, whereas the second integrator has an initial condition that is a zero-mean normal random variable with $\sigma^2 = 1$. Write the differential equation for the **P** matrix for the continuous Kalman filter for this situation. Be sure to specify the appropriate initial conditions for **P**, and also be sure to specify all elements of the matrix parameters (such as **F**, **R**, and **Q**) of the differential equation. You do not need to solve the differential equation.

Problem 7.8

7.9. Modification of the discrete Kalman filter equations to include a deterministic input was discussed in Section 6.10. Show that the corresponding modification in the continuous case is accomplished with the addition of $\mathbf{Bu}_d$ as a driving function in the $\hat{\mathbf{x}}$ differential equation. Specifically, assume the process equation is of the form

$$\dot{\mathbf{x}} = \mathbf{Fx} + \mathbf{Gu} + \mathbf{Bu}_d \qquad (P7.9.1)$$

where $\mathbf{Bu}_d$ is the deterministic input. Then show that the corresponding differential equation for the estimate is

$$\dot{\hat{\mathbf{x}}} = \mathbf{F}\hat{\mathbf{x}} + \mathbf{K}(\mathbf{z} - \mathbf{H}\hat{\mathbf{x}}) + \mathbf{Bu}_d \qquad (P7.9.2)$$

7.10.   The discrete filter equations that account for the correlation between $\mathbf{w}_{k-1}$ and $\mathbf{v}_k$ should, for small step size Δt, go over into the continuous filter equations derived in Section 7.3 by other means. In order to show this, we must first derive the appropriate connection between $\mathbf{C}$ (continuous model) and $\mathbf{C}_k$ (discrete model). We do this by noting that

$$\mathbf{w}_{k-1} = \int_{t_{k-1}}^{t_k} \boldsymbol{\phi}(t_k, \xi) \mathbf{G}(\xi) \mathbf{u}(\xi) \, d\xi$$

$$\mathbf{v}_k = \frac{1}{\Delta t} \int_{t_{k-1}}^{t_k} \mathbf{v}(\eta) \, d\eta, \qquad \text{(small } \Delta t)$$

The $\mathbf{C}_k$ parameter is then (for small Δt)

$$\mathbf{C}_k = E\left[\mathbf{w}_{k-1}\mathbf{v}_k^T\right]$$

$$= \frac{1}{\Delta t} \int_{t_{k-1}}^{t_k} \int_{t_{k-1}}^{t_k} \boldsymbol{\phi}(t_k, \xi)\mathbf{G}(\xi) \underbrace{E\left[\mathbf{u}(\xi)\mathbf{v}^T(\eta)\right]}_{\mathbf{C}\delta(\xi - \eta)} \, d\xi \, d\eta$$

$$= \frac{1}{\Delta t} \mathbf{G}\mathbf{C}\,\Delta t$$

Therefore, the desired relationship is

$$\mathbf{C}_k = \mathbf{G}\mathbf{C}$$

This is then added to the other necessary correspondences for small Δt:

$$\mathbf{H}_k = \mathbf{H}$$

$$\mathbf{R}_k = \frac{\mathbf{R}}{\Delta t}$$

$$\boldsymbol{\phi}_k = \mathbf{I} + \mathbf{F}\,\Delta t$$

$$\mathbf{Q}_k = \mathbf{G}\mathbf{Q}\mathbf{G}^T\,\Delta t$$

Now, following the same method used in Section 7.1, show that the discrete filter equations for the correlated case go over into the continuous filter differential equations derived in Section 7.3.

7.11*. The accompanying figure shows two integrators in cascade that are driven by white noise with a spectral amplitude of 100 units. The state variables have been chosen as the two integrator outputs. Suppose that discrete measurements of x_1 are available at intervals of 0.1 sec, and the associated discrete measurement error variance is 25 units. We will assume that the process is initiated at $t = 0$, and the initial conditions for both x_1 and x_2 are known to be zero.

(a) It is easily verified that this system is observable using the observability test given in Problem 6.3. Thus, even though the process is nonstationary, we might expect the filter gain and error covariance to approach a steady-state condition. Verify this empirically by cycling the discrete filter error covariance equations through a sufficient number of steps to reach the steady-state condition. Note the final gain and error covariance matrices. Also note the number of recursive steps required to get within about 1 percent of the final value.

(b) The values of the characteristic poles that control the time required to reach steady state are not obvious in this situation. However, they can be calculated in a number of ways. First, using the software accompanying this text, find the characteristic poles of the filter in the steady-state condition. [The computer algorithm used in the software follows the method given in Ref. (6), pp. 256–265.] Is the time required to reach steady state as determined in part (a) consistent with the pole locations as found in this part?

(c) It was shown in Section 7.6 that the filter characteristic poles can be found by determining the eigenvalues of $(\phi - KH\phi)$. Of course, the steady-state gain must be found first before this method can be used. Using the steady-state gain as determined empirically in part (a) and the methods of Section 7.6, find the characteristic poles of the filter. Compare these with the results of part (b).

(d) It was mentioned in Section 6.6 that the Kalman filter will recover from roundoff errors if all the state variables are driven and the symmetry and positive-definiteness of the **P** matrix are maintained. This can easily be demonstrated (not proved) with the example at hand. Consider the filter steady-state condition as determined in part (a). Now increase the p_{11} and p_{22} terms by about 10 percent and cycle the filter through an additional 50 steps. Observe the filter **P** matrix as it "relaxes" back to the same steady-state condition that was observed in part (a).

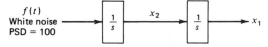

Problem 7.11*

7.12*.   If you worked Problem 6.8, you should have found that the five-state system of that problem was not observable. Now consider a simplified version of that system where we say that the gyro bias is the predominant forcing function, and the accelerometer bias is negligible. This eliminates u_5 and state variable x_5, and the combination can be replaced with a relatively small white noise forcing function at the summing point. Call this u_2; it will account for the accelerometer instrument error. After eliminating x_5, the model reduces to a four-state model. Choose the following spectral amplitudes for the accelerometer and gyro white noise forcing functions:

$$S_a = 1.2 \times 10^{-4} \text{ m}^2/\text{sec}^3 \quad \text{(for } u_2\text{)}$$

$$S_g = .65 \times 10^{-14} \text{ rad}^2/\text{sec}^3 \quad \text{(for } u_4\text{)}$$

As before, we will assume that we have available only a sequence of discrete noisy measurements of x_3. Assume that the step size Δt is 120 sec, which is small relative to the Schuler period (84 min). Thus, the small Δt approximation can be used in both the ϕ_k and Q_k calculations. We will also assume that the measurement error variance R_k is 40,000 m^2.

   (a) Is the simplified four-state system observable? (Use the observability test given in Problem 6.3 for this part.)
   (b) Is the resulting four-state Kalman filter stable? If so, find the steady-state characteristic roots of the filter.

7.13*.   The matrix Riccati equation, Eq. (7.2.1), describes the estimation error in a continuous Kalman filter. Note that setting $R^{-1} = 0$ corresponds to no measurement. In this case the best estimate of the state is zero, and $P(t)$ is just the covariance of the x process beginning at $t = 0$. If we let the initial value of $P(t)$ be zero, we then obtain the differential equation for the Q_k matrix that was discussed in Chapter 5. That is, with the step size Δt as the argument, we have

$$\mathring{Q}_k = FQ_k + Q_kF^T + GQG^T, \; Q_k(0) = 0 \qquad \text{(P7.13.1)}$$

   (a) Show that Eq. (P7.13.1) yields the same result as Eq. (5.3.6) for small step size Δt.
   (b) For the scalar Markov process discussed in Example 5.7, Q_k was found to be approximately .03921. Show that both the Kalm 'N Smooth software provided with the text and the solution of Eq. (P7.13.1) yield this same answer.

REFERENCES CITED IN CHAPTER 7

1. R. E. Kalman and R. S. Bucy, "New Results in Linear Filtering and Prediction," *Trans. of the ASME, Jour. of Basic Engr.*, *83:* 95–108 (1961).
2. A. E. Bryson, Jr., and D. E. Johansen, "Linear Filtering for Time-Varying Systems Using Measurements Containing Colored Noise," *IEEE Trans. on Automatic Control, AC-10:* No. 1, 4–10 (Jan. 1965).
3. K. Ogata, *State Space Analysis of Control Systems*, Englewood Cliffs, N.J.: Prentice-Hall, 1967.
4. C. T. Chen, *Linear System Theory and Design*, New York: Holt, Rinehart, and Winston, 1984.
5. R. G. Brown and J. W. Nilsson, *Introduction to Linear Systems Analysis*, New York: Wiley, 1962.
6. G. F. Franklin and J. D. Powell, *Digital Control of Dynamic Systems*, Reading, Mass.: Addison-Wesley, 1980.

Additional References on Continuous Kalman Filtering

7. A. Gelb (Ed.), *Applied Optimal Estimation*, Cambridge, Mass.: MIT Press, 1974.
8. A. P. Sage and J. L. Melsa, *Estimation Theory with Applications to Communications and Control*, New York: McGraw-Hill, 1971.
9. J. S. Meditch, *Stochastic Optimal Linear Estimation and Control*, New York: McGraw-Hill, 1969.
10. A. H. Jazwinski, *Stochastic Processes and Filtering Theory*, New York: Academic Press, 1970.
11. P. S. Maybeck, *Stochastic Models, Estimation and Control* (Vol. 1), New York: Academic Press, 1979.
12. S. M. Bozic, *Digital and Kalman Filtering*, London: E. Arnold, Publisher, 1979.

8

Discrete Smoothing and Prediction

The emphasis in the preceding chapters on Kalman filtering has been on the *filter* problem, that is, estimation of the process at the same time as the current measurement. We now wish to generalize the theory to include prediction and smoothing just as was done with the Wiener theory. It works out that discrete prediction is relatively easy, so it will be considered first. The smoothing problem is somewhat more difficult; but it is still manageable, and recursive solutions are available. We will see that the Kalman theory accommodates a larger class of smoothing-type problems than the Wiener theory, and thus it has more potential in applied work, especially in off-line data processing. We now proceed to the discrete prediction problem.

8.1
PREDICTION

Recall that in the discrete filter equations there is a projection step where the a posteriori estimate $\hat{\mathbf{x}}_k$ is projected ahead to yield the a priori estimate $\hat{\mathbf{x}}_{k+1}^-$. This is, of course, the best estimate of $\mathbf{x}$ at t_{k+1}, given all the measurement data up through $\mathbf{z}_k$. Thus, this is one-step prediction. To predict N steps ahead we simply use the same projection equations (and justification) with an appropriate change in the ϕ and $\mathbf{Q}$ matrices. Thus

$$\hat{\mathbf{x}}(k + N|k) = \phi_{k+N,k}\hat{\mathbf{x}}(k|k) \tag{8.1.1}$$

$$\mathbf{P}(k + N|k) = \phi_{k+N,k}\mathbf{P}(k|k)\phi_{k+N,k}^T + \mathbf{Q}_{k+N,k} \tag{8.1.2}$$

where

$\hat{\mathbf{x}}(k + N|k)$ = estimate of **x** at t_{k+N} given measurement data through $\mathbf{z}_k$

$\mathbf{P}(k + N|k)$ = error covariance associated with $\hat{\mathbf{x}}(k + N|k)$

$\boldsymbol{\phi}_{k+N,k}$ = state transition matrix for the time interval from t_k to t_{k+N}

$\mathbf{Q}_{k+N,k}$ = covariance of the driven response for the time interval from t_k to t_{k+N}

$\hat{\mathbf{x}}(k|k), \mathbf{P}(k|k)$ = filter estimate and its error covariance at t_k

A word of explanation about notation is in order here. In both smoothing and prediction we need to keep track of the estimate and measurement times separately. This will be done with the parenthetical notation indicated above; for example, we read $\hat{\mathbf{x}}(k + N|k)$ as "estimate of **x** at time t_{k+N} given all the measurement data up through time t_k." Our previous single-subscript notation then translates into:

$$\hat{\mathbf{x}}_k^- = \hat{\mathbf{x}}(k|k - 1)$$

$$\mathbf{P}_k^- = \mathbf{P}(k|k - 1)$$

$$\hat{\mathbf{x}}_k = \hat{\mathbf{x}}(k|k)$$

$$\mathbf{P}_k = \mathbf{P}(k|k)$$

The beauty of the explicit parenthetical notation is that prediction, filtering, and smoothing can all be accommodated within the same notational structure. The price, of course, is more writing.

Note that the expressions for $\hat{\mathbf{x}}(k + N|k)$ and $\mathbf{P}(k + N|k)$ involve the filter quantities $\hat{\mathbf{x}}(k|k)$ and $\mathbf{P}(k|k)$. Thus, the recursive filter equations must be solved as usual. The prediction of **x** at t_{k+N} is an extra computation. This is summarized in Fig. 8.1. The equation for the predictor error covariance $\mathbf{P}(k + N|k)$ is

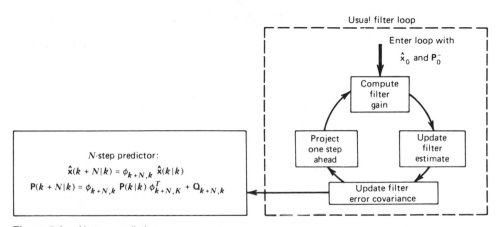

Figure 8.1. *N*-step prediction.

included in the predictor box in Fig. 8.1. This error covariance is not needed in the recursive filter loop, so this computation may be omitted if $\mathbf{P}(k + N|k)$ is not if interest.

8.2
CLASSIFICATION OF SMOOTHING PROBLEMS

Smoothing seems to be inherently more difficult than either filtering or prediction. In the Wiener theory of Chapter 4, α is negative in Eq. (4.3.22) and this causes a cusp to appear in the positive-time part of the shifted time function. (At least this is true for rational spectral functions—see Problem 8.1.) This cusp, in turn, leads to a considerably more complicated expression for the Fourier transform than occurs in the corresponding filter or prediction problems. Similarly, we will see that the recursive algorithms for smoothing are considerably more complicated than those for filtering and prediction.

Meditch (1) classifies smoothing into three categories:

1. *Fixed-interval smoothing.* Here the time interval of measurements (i.e., the data span) is fixed, and we seek optimal estimates at some, or perhaps all, interior points. This is the typical problem encountered when processing noisy measurement data off-line.

2. *Fixed-point smoothing.* In this case, we seek an estimate at a single fixed point in time, and we think of the measurements continuing on indefinitely ahead of the point estimation. An example of this would be the estimation of initial conditions based on noisy trajectory observations after $t = 0$. In fixed-point smoothing there is no loss of generality in letting the fixed point be at the beginning of the data stream, because all prior measurements can be processed with the filter algorithm.

3. *Fixed-lag smoothing.* In this problem, we again envision the measurement information proceeding on indefinitely with the running time variable t, and we seek an optimal estimate of the process at a fixed length of time back in the past. Clearly, the Wiener problem with α negative is fixed-lag smoothing. It is of interest to note that the Wiener formulation will not accommodate either fixed-interval or fixed-point smoothing without using multiple sweeps through the same data with different values of α. This would be a most awkward way to process measurement data.

Obviously, the three smoothing problems are related, and it is not especially difficult to devise correct, but clumsy, solutions. Thus, the central problem is one of finding reasonably efficient recursive algorithms for each of the types of smoothing. This has been studied extensively since the early 1960s, and we will present only those solutions that are generally considered to be best from a

Figure 8.2. Procedure for fixed-interval smoothing.

computational viewpoint. We will not attempt to derive all the algorithms. The derivations are adequately documented in the references cited. (Also, see Problem 8.2.) We now proceed to look at algorithms for the three specified categories of smoothing.

8.3
DISCRETE FIXED-INTERVAL SMOOTHING

The algorithm to be presented here is due to Rauch, Tung, and Striebel (2, 3), and its derivation is given in Meditch (1) as well as the referenced papers. Also, a new simplified derivation is presented in Problem 8.2. In the interest of brevity, the algorithm will be subsequently referred to as the RTS algorithm. Consider a fixed-length interval containing $N + 1$ measurements. These will be indexed in ascending order $z_0, z_1, \ldots, z_N$. The assumptions relative to the process and measurement models are the same as for the filter problem. The computational procedure for the RTS algorithm consists of a forward recursive sweep followed by a backward sweep. This is illustrated in Fig. 8.2. We enter the algorithm as usual at $k = 0$ with the initial conditions $\hat{\mathbf{x}}_0^-$ and $\mathbf{P}_0^-$. We then sweep forward using the conventional filter algorithm. With each step of the forward sweep, we must save the computed a priori and a posteriori estimates and their associated $\mathbf{P}$ matrices. These are needed for the backward sweep. After completing the forward sweep, we begin the backward sweep with "initial" conditions $\hat{\mathbf{x}}(N|N)$ and $\mathbf{P}(N|N)$ obtained as the final computation in the forward sweep. With each step of the backward sweep the old filter estimate is updated to yield an improved smoothed estimate, which is based on all the measurement data. The recursive equations for the backward sweep are

$$\hat{\mathbf{x}}(k|N) = \hat{\mathbf{x}}(k|k) + \mathbf{A}(k)[\hat{\mathbf{x}}(k+1|N) - \hat{\mathbf{x}}(k+1|k)] \qquad (8.3.1)$$

where the smoothing gain $\mathbf{A}(k)$ is given by

$$\mathbf{A}(k) = \mathbf{P}(k|k)\boldsymbol{\phi}^T(k + 1, k)\mathbf{P}^{-1}(k + 1|k) \qquad (8.3.2)$$

and

$$k = N - 1, N - 2, \ldots, 0$$

The error covariance matrix for the smoothed estimates is given by the recursive equation

$$\mathbf{P}(k|N) = \mathbf{P}(k|k) + \mathbf{A}(k)[\mathbf{P}(k + 1|N) - \mathbf{P}(k + 1|k)]\mathbf{A}^T(k) \quad (8.3.3)$$

It is of interest to note that the *smoothing* error covariance matrix is not needed for the computation of the estimates in the backward sweep. This is in contrast to the situation in the filter (forward) sweep where the **P**-matrix sequence is needed for the gain and associated estimate computations. An example illustrating the use of the RTS algorithm is now in order.

EXAMPLE 8.1

Let us consider the same Gauss–Markov process used previously in Example 5.7, Section 5.5. Recall that the process has an autocorrelation function

$$R_x(\tau) = 1 \cdot e^{-|\tau|}$$

and we have a sequence of noisy measurements of this process taken every .02 sec. The measurement sequence begins at $t = 0$ and ends at $t = 1$ sec giving a total of 51 discrete measurements. The measurement errors are white and have unity variance. A sample of this situation was simulated using Gaussian random numbers, and the filtering results were described previously, in Section 5.5. We now continue this same example with a backward sweep and obtain smoothed estimates.

A partial listing of the results from the filtering solution is repeated in Table 8.1 for reference purposes. It can be seen from the error covariances that the filter has reached a steady-state condition by the end of the forward sweep.

Table 8.1 Filtering Results

k	$\hat{x}(k|k)$	$P(k|k)$	$\hat{x}(k|k - 1)$	$P(k|k - 1)$	$z(k)$
⋮	⋮	⋮	⋮	⋮	⋮
46	− .630	.1653	− .595	.1980	− .811
47	− .372	.1653	− .618	.1980	.871
48	− .521	.1653	− .364	.1980	− 1.310
49	− .436	.1653	− .510	.1980	− .064
50	− .545	.1653	− .428	.1980	− 1.138

We enter the backward sweep at the end point where $k = 50$. Here we have

$$\hat{x}(50|50) = -.545 \qquad (8.3.4)$$

$$P(50|50) = .1653 \qquad (8.3.5)$$

Since the filter solution at this point is conditioned on *all* the measurement data, it is also the smoothed estimate at $k = 50$. We are now ready to compute the smoothed estimate one step back at $k = 49$. From Eqs. (8.3.1) and (8.3.2) we have

$$\hat{x}(49|50) = \hat{x}(49|49) + A(49)[\hat{x}(50|50) - \hat{x}(50|49)] \qquad (8.3.6)$$

where

$$A(49) = P(49|49)\phi P^{-1}(50|49) \qquad (8.3.7)$$

Using the filtering results from Table 8.1, and noting that the transition matrix is $e^{-.02} \approx .98020$ leads to

$$A(49) = .8183$$

$$\hat{x}(49|50) = -.5317$$

This may now be repeated at $k = 48, 47, \ldots$, etc., with the results

$$\hat{x}(48|50) = -.5388$$

$$\hat{x}(47|50) = -.5150$$

$$\hat{x}(46|50) = -.5457$$

$$\vdots$$

etc.

The smoothing results for the entire 50 steps are summarized in Fig. 8.3 along the filter results for comparison. It is evident from the plots that the smoothed plot is, in fact, smoother than the filter plot. This is as expected, because the smoother uses both past and future data in the makeup of its estimate. The smoothed estimate is not conspicuously better than the filter estimate for this simulation. However, we should not expect to be able to draw firm conclusions from such a small sample.

Figure 8.3. Smoothed and filter estimate plots for Example 8.1.

It is also of interest in this example to compare the error variances of the filter and smoothed estimates. These are shown in Fig. 8.4. Both the filter and smoother converge to a steady-state condition after about 15 steps. The steady-state values are

$$P_{\text{filter}} \approx .1653 \qquad (\sigma_{\text{filter}} = .406)$$

$$P_{\text{smoother}} \approx .099 \qquad (\sigma_{\text{smoother}} = .315)$$

Note that when the comparison is made on the basis of rms value, there is not a dramatic difference in the errors. In addition, the smoothed error curve is minimum in the middle and then gets larger as either end point is approached. This indicates that the best situation for estimation occurs in the interior region

Figure 8.4. Error variances for filter and smoothed estimates of Example 8.1.

Table 8.2 Fixed-Interval Smoothing Solution

k	$\hat{x}(k\mid 50)$	$P(k\mid 50)$
0	-1.601	.1653
1	-1.569	.1434
2	-1.567	.1287
3	-1.568	.1189
4	-1.565	.1123
5	-1.556	.1079
$\vdots$	$\vdots$	$\vdots$
25	$-.146$	.0990
$\vdots$	$\vdots$	$\vdots$
45	$-.567$	.1079
46	$-.546$	.1123
47	$-.515$	.1189
48	$-.539$	.1287
49	$-.532$	.1434
50	$-.545$	.1653

where there is an abundance of measurement data in either direction from the point of estimation. This is exactly what we would expect intuitively.

Before leaving this example, the steady-state behavior in the middle of the variance plot of Fig. 8.4 should be noted. This frequently happens when the data span is large and the process is stationary and observable. In the steady-state region, the filter and smoother gains and error covariance matrices are constant, and considerable computational simplification occurs, both in terms of the number of arithmetic operations and the amount of storage required for the algorithm. Thus, sometimes a problem that appears to be quite formidable at first glance works out to be feasible because a large amount of the measurement data can be processed in the steady-state region.

For future reference purposes, a partial listing of the fixed-interval smoothing solution is given in Table 8.2.

8.4
DISCRETE FIXED-POINT SMOOTHING

The fixed-point smoothing algorithm to be presented here is taken directly from Meditch (1). Alternative algorithms that, under certain circumstances, may be better computationally are also given by Meditch. In order to keep the discussion here as simple as possible, we present just one algorithm, which was chosen because of its simplicity and similarity to the fixed-interval algorithm. The algorithm is

$$\hat{\mathbf{x}}(k|j) = \hat{\mathbf{x}}(k|j-1) + \mathbf{B}(j)[\hat{\mathbf{x}}(j|j) - \hat{\mathbf{x}}(j|j-1)] \qquad (8.4.1)$$

where

$$k \text{ is fixed} \qquad (\text{usually } k = 0)$$

$$j = k + 1, k + 2, \text{ etc.}$$

and

$$\mathbf{B}(j) = \prod_{i=k}^{j-1} \mathbf{A}(i) \qquad (8.4.2)$$

$$\mathbf{A}(i) = \mathbf{P}(i|i)\boldsymbol{\phi}^T(i+1|i)\mathbf{P}^{-1}(i+1|i) \qquad (8.4.3)$$

(Note that the equation for **A** is the same as for the fixed-interval algorithm.) The error covariance associated with the smoothed estimate is given by

$$\mathbf{P}(k|j) = \mathbf{P}(k|j-1) + \mathbf{B}(j)[\mathbf{P}(j|j) - \mathbf{P}(j|j-1)]\mathbf{B}^T(j) \qquad (8.4.4)$$

Note that the filter estimates and their error covariances are needed in this algorithm, just as in the fixed-interval case. The computational procedure is summarized in Fig. 8.5. We enter the algorithm at the beginning of the data stream with the usual a priori $\hat{\mathbf{x}}_0^-$ and $\mathbf{P}_0^-$. (We have let $k = 0$.) These initial parameters may come from prior knowledge of the process in the event there are no prior measurements, or $\hat{\mathbf{x}}_0^-$ and $\mathbf{P}_0^-$ may come from processing previous data via the filter algorithm. In either event, the first step is to update the

Figure 8.5.   Procedure for fixed-point smoothing.

a priori $\hat{\mathbf{x}}_0^-$ and $\mathbf{P}_0^-$ with the measurement $\mathbf{z}_0$ and obtain $\hat{\mathbf{x}}(0|0)$ and $\mathbf{P}(0|0)$. This is done with the usual filter equations and, of course, the result can be interpreted as either a filter estimate or the zero-stage smoothed estimate—they are one and the same. We next let $j = 1$, and we are ready to solve the one-stage smoothing problem using Eqs. (8.4.1) to (8.4.4). This may then be continued indefinitely, theoretically, at least. Of course, as with any recursive algorithm, roundoff error may eventually lead to difficulties. We now look at an example of fixed-point smoothing.

EXAMPLE 8.2

To illustrate the application of the fixed-point algorithm, we use the same process and simulated data used in Example 8.1. We let the fixed point be at $k = 0$ and the data for the first 20 steps are given in Table 8.3. Since the filtering results are needed for the solution of the smoothing problem, they are also listed in the table.

The first step is to compute $\hat{x}(0|0)$ and $P(0|0)$. This was done in the previous filter solution with the result

$$\hat{x}(0|0) = -1.610$$

$$P(0|0) = .5$$

Table 8.3 Data for Fixed-Point Example

j	$\hat{x}(j|j)$	$\hat{x}(j|j-1)$	$A(j)$	$B(j)$	$\hat{x}(0|j)$	$P(j|j)$	$P(j|j-1)$	$P(0|j)$
0	-1.610		.9432		-1.610	.5		.5
1	-1.317	-1.578	.9114	.9432	-1.364	.3419	.5196	.3419
2	-1.350	-1.291	.8860	.8596	-1.414	.2689	.3677	.2689
3	-1.406	-1.323	.8661	.7616	-1.477	.2293	.2975	.2293
4	-1.454	-1.378	.8513	.6597	-1.528	.2060	.2595	.2060
5	-1.815	-1.425	.8411	.5616	-1.747	.1917	.2372	.1917
6	-1.911	-1.779	.8336	.4723	-1.809	.1826	.2234	.1826
7	-1.879	-1.873	.8287	.3937	-1.812	.1767	.2147	.1767
8	-1.592	-1.842	.8255	.3263	-1.730	.1729	.2090	.1729
9	-1.365	-1.560	.8227	.2693	-1.677	.1703	.2053	.1703
10	-1.238	-1.338	.8214	.2216	-1.655	.1686	.2029	.1686
11	-1.338	-1.214	.8201	.1820	-1.678	.1675	.2012	.1675
12	-1.436	-1.307	.8199	.1493	-1.697	.1668	.2002	.1668
13	-1.346	-1.408	.8191	.1224	-1.689	.1663	.1994	.1663
14	-.848	-1.320	.8189	.1002	-1.642	.1660	.1990	.1660
15	-.821	-.831	.8186	.0821	-1.641	.1657	.1987	.1657
16	-.540	-.804	.8186	.0672	-1.623	.1656	.1984	.1656
17	-.290	-.529	.8185	.0550	-1.610	.1655	.1983	.1655
18	-.204	-.284	.8184	.0450	-1.607	.1654	.1982	.1654
19	-.077	-.200	.8184	.0368	-1.602	.1654	.1981	.1654
20	-.134	-.076		.0302	-1.604	.1653	.1981	.1653

We are now ready to find $\hat{x}(0|1)$ by letting $k = 0$ and $j = 1$ in Eqs. (8.4.1), (8.4.2), and (8.4.3). The results are

$$B(1) = A(0) = P(0|0)\phi P^{-1}(1|0)$$

$$= .9432$$

$$\hat{x}(0|1) = \hat{x}(0|0) + B(1)[\hat{x}(1|1) - \hat{x}(1|0)]$$

$$= -1.364$$

We now have the needed information to go on to the next step and compute $\hat{x}(0|2)$, and so on. Note that it is not necessary to compute the smoothed error covariance in order to find the smoothed estimates. It was computed for checking purposes, though, and is listed in Table 8.3. The solution has nearly converged after 20 steps, and the results are essentially the same as those obtained from the fixed-interval solution, that is,

$$\hat{x}(0|20) \approx \hat{x}(0|50) \approx -1.60$$

$$P(0|20) \approx P(0|50) \approx .1653$$

This indicates that the correlation between the process and the measurement 20 steps away from the point of estimation is so weak that measurements beyond 20 steps are of little value in forming the smoothed estimate.   ■

8.5
FIXED-LAG SMOOTHING

The fixed-lag smoothing problem was originally solved by Wiener in the 1940s. However, he considered only the stationary case, that is, the smoother was assumed to have the entire past history of the input available for weighting in its determination of the estimate. Of the three smoothing categories mentioned in Section 8.2, the fixed-lag problem is the most complicated. One reason for this is the start-up problem. For example, let us say we wish to estimate our process at a point three steps back from the current point of measurement. If we enter the problem with the first measurement at $t = 0$, we have only the one measurement on which to base the estimate, and there is a gap in the measurement sequence between the current time and point of estimation. This gap becomes filled as time progresses, but we have to be watchful of this little detail for the first few steps.

Meditch (1) gives an algorithm for fixed-lag smoothing that is considerably more complicated than the algorithms for the other two smoothing categories. We take a somewhat simpler approach here. Knowing the RTS algorithm for fixed-interval smoothing, we can always do fixed-lag smoothing by first filtering

Table 8.4 Forward and Backward Sweeps (Current Measurement Is at $k = 0$)

	Forward Sweep				Backward Sweep		
k	$\hat{x}(k\|k-1)$	$P(k\|k-1)$	$\hat{x}(k\|k)$	$P(k\|k)$	k	$A(k)$	$\hat{x}(k\|0)$
-3	0	1	0	1	0	—	-1.610
-2	0	1	0	1	-1	.9802	-1.578
-1	0	1	0	1	-2	.9802	-1.547
0	0	1	-1.610	.5	-3	.9802	-1.516

up to the current measurement and then sweeping back a fixed number of steps with the RTS algorithm. If the number of backward steps is small, this is a simple and effective way of doing fixed-lag smoothing. If, however, the number of backward steps is large, this method becomes cumbersome and one should look for more efficient algorithms. To illustrate the sweeping back on a running time basis and the start-up problem, we return to the Gauss–Markov example previously considered in Examples 8.1 and 8.2.

EXAMPLE 8.3

The needed filtering data for our Gauss–Markov example is given in Example 8.2. Let us say that we wish to estimate the process three steps back from the current point of measurement, and that the first measurement occurs at $k = 0$. Index k is the discrete running time variable. Our first estimate is then to be $\hat{x}(-3|0)$. Anticipating that we will use the Rauch backward sweep from $k = 0$ to $k = -3$, we see that we will need all the filter estimates for this interval. Now, even though we have no measurements prior to $k = 0$, we can still form a trivial forward sweep beginning at $k = -3$. Recall that filter updating without the benefit of a measurement simply amounts to setting the a posteriori $\hat{x}$ and P equal to the a priori $\hat{x}^-$ and P^-. The result of this trivial sweep is shown on the left in Table 8.4. We now have all the filtering data needed for the backward sweep. The desired estimate $\hat{x}(-3|0)$ is obtained from Eqs. (8.3.1) and (8.3.2), and the results are given on the right side of Table 8.4. Presumably, the only estimate of interest here is the last entry in the table, $\hat{x}(-3|0) = -1.516$.

Next, suppose we get a measurement at $k = 1$, and we wish the smoothed estimate three steps back at $k = -2$. The forward sweep that began at $k = -3$ may now be extended one step, and the results are shown at the left in Table

Table 8.5 Forward and Backward Sweeps (Current Measurement Is at $k = 1$)

	Forward Sweep				Backward Sweep		
k	$\hat{x}(k\|k-1)$	$P(k\|k-1)$	$\hat{x}(k\|k)$	$P(k\|k)$	k	$A(k)$	$\hat{x}(k\|1)$
-2	0	1	0	1	1	—	-1.317
-1	0	1	0	1	0	.9432	-1.364
0	0	1	-1.610	.5	-1	.9802	-1.337
1	-1.578	.5196	-1.317	.3419	-2	.9802	-1.310

Table 8.6 Forward and Backward Sweeps (Current Measurement Is at $k = 2$)

	Forward Sweep				Backward Sweep		
k	$\hat{x}(k\|k-1)$	$P(k\|k-1)$	$\hat{x}(k\|k)$	$P(k\|k)$	k	$A(k)$	$\hat{x}(k\|2)$
-1	0	1	0	1	2	—	-1.350
0	0	1	-1.610	.5	1	.9114	-1.371
1	-1.578	.5196	-1.317	.3419	0	.9432	-1.415
2	-1.291	.3677	-1.350	.2689	-1	.9802	-1.387

Table 8.7 Forward and Backward Sweeps (Current Measurement Is at $k = 3$)

	Forward Sweep				Backward Sweep		
k	$\hat{x}(k\|k-1)$	$P(k\|k-1)$	$\hat{x}(k\|k)$	$P(k\|k)$	k	$A(k)$	$\hat{x}(k\|3)$
0	0	1	-1.610	.5	3	—	-1.406
1	-1.578	.5196	-1.317	.3419	2	.8860	-1.423
2	-1.291	.3677	-1.350	.2689	1	.9114	-1.438
3	-1.323	.2975	-1.406	.2293	0	.9432	-1.478

8.5. Note that the only new entries are in the bottom row. The results of the backward sweep are also given in Table 8.5, and all the entries here are new; that is, the backward sweep must be completely redone with each new measurement. Note that each time we increment the filter forward one step, the oldest estimate and its error covariance may be discarded, because we need only the four most recent filter results for the backward sweep. Thus, we have sort of a "sliding window" of data that we must store as time increments forward.

The procedure just described may now be repeated indefinitely. For reference purposes, the solutions for the next two steps are given in Tables 8.6 and 8.7. The procedure has been demonstrated with tables at each step for the purpose of clarity, but it should be recognized that much of the data in the tables is either repetitious or not necessary for succeeding steps. Thus, the actual programming of the running forward-plus-backward algorithm is not as complicated as would appear from the tabular listings. With each new measurement, it simply involves moving one step forward and M steps back, where M is the desired fixed lag. ■

8.6 FORWARD–BACKWARD FILTER APPROACH TO SMOOTHING

In 1969, D. C. Fraser and J. E. Potter presented a novel solution to the smoothing problem (4). With reference to the fixed-interval problem, their approach was to *filter* the measurement data from both ends to the interior

Figure 8.6. Forward–backward filtering. Smoothed estimate $\hat{\mathbf{x}}(t'|T)$ is a linear combination of $\hat{\mathbf{x}}(t')$ and $\hat{\mathbf{x}}_b(t')$.

point of interest, and then the two filter estimates were combined to obtain the optimal smoothed estimate. This is illustrated in Fig. 8.6. The equations for the continuous problem will be developed first following the derivation given in Gelb (5). We will then deviate from both Gelb and the Fraser–Potter paper and present a new simplified derivation for the discrete case.

The Continuous Problem

Suppose we have two independent, unbiased estimates of some parameter x. Call these $\hat{x}_1$ and $\hat{x}_2$ and let the corresponding estimation errors have variances σ_1^2 and σ_2^2. We wish to form a new estimate $\hat{x}$ as a linear combination of $\hat{x}_1$ and $\hat{x}_2$, where $\hat{x}$ is to have minimum mean-square error. Thus, we write $\hat{x}$ as

$$\hat{x} = k_1\hat{x}_1 + k_2\hat{x}_2 \tag{8.6.1}$$

If the new estimate is to be unbiased,

$$k_1 + k_2 = 1 \tag{8.6.2}$$

Thus, Eq. (8.6.1) may be rewritten as

$$\hat{x} = k_1\hat{x}_1 + (1 - k_1)\hat{x}_2 \tag{8.6.3}$$

The mean-square error for $\hat{x}$ is then

$$E\left[(x - \hat{x})^2\right] = E\left\{[x - k_1\hat{x}_1 - (1 - k_1)\hat{x}_2]^2\right\} \tag{8.6.4}$$

or

$$E[e^2] = E\left\{[k_1(e_1 - e_2) + e_2]^2\right\} \tag{8.6.5}$$

where e, e_1, and e_2 are the errors in $\hat{x}$, $\hat{x}_1$, and $\hat{x}_2$, respectively. Equation (8.6.5) may now be differentiated with respect to k_1 and set equal to zero to find the optimal k_1. Performing this operation and noting that e_1 and e_2 are independent yields

$$k_1 = \frac{E[e_2^2]}{E[e_1^2] + E[e_2^2]} = \frac{\sigma_2^2}{\sigma_1^2 + \sigma_2^2} \qquad (8.6.6)$$

and

$$k_2 = \frac{E[e_1^2]}{E[e_1^2] + E[e_2^2]} = \frac{\sigma_1^2}{\sigma_1^2 + \sigma_2^2} \qquad (8.6.7)$$

The corresponding variance of the optimal estimate $\hat{x}$ is obtained from Eq. (8.6.5) and is

$$\sigma^2 = \frac{\sigma_1^2 \sigma_2^2}{\sigma_1^2 + \sigma_2^2} \qquad (8.6.8)$$

Notice that the weighting factors k_1 and k_2 used for blending $\hat{x}_1$ and $\hat{x}_2$ are inversely proportional to the variances of the individual estimates. (Electrical engineers usually notice the obvious analogy between this and the division of electric current between two parallel resistors. Variance in the statistics problem plays the same role as resistance in the circuit problem.) We note in passing that Eqs. (8.6.6), (8.6.7), and (8.6.8) may also be written in terms of reciprocals as

$$\sigma^2 = \left[(\sigma_1^2)^{-1} + (\sigma_2^2)^{-1} \right]^{-1} \qquad (8.6.9)$$

$$k_1 = \sigma^2 (\sigma_1^2)^{-1} \qquad (8.6.10)$$

$$k_2 = \sigma^2 (\sigma_2^2)^{-1} \qquad (8.6.11)$$

We now return to the forward–backward filter problem depicted in Fig. 8.6. The forward filter is the usual continuous Kalman filter, as discussed in Chapter 7. Thus, the process model and filter equations are
  Forward process model:

$$\left. \begin{array}{l} \dot{\mathbf{x}} = \mathbf{Fx} + \mathbf{Gu} \\ \mathbf{z} = \mathbf{Hx} + \mathbf{v} \end{array} \right\} \begin{array}{l} \mathbf{u} \text{ and } \mathbf{v} \text{ are white noise processes} \\ \text{with zero cross correlation} \end{array} \qquad \begin{array}{l} (8.6.12) \\ (8.6.13) \end{array}$$

  Forward filter equations:

$$\dot{\mathbf{x}} = \mathbf{F}\hat{\mathbf{x}} + \mathbf{PH}^T \mathbf{R}^{-1}(\mathbf{z} - \mathbf{H}\hat{\mathbf{x}}) \qquad (8.6.14)$$

$$\dot{\mathbf{P}} = \mathbf{FP} + \mathbf{PF}^T - \mathbf{PH}^T \mathbf{R}^{-1} \mathbf{HP} + \mathbf{GQG}^T \qquad (8.6.15)$$

Initial conditions:

$$\hat{\mathbf{x}}(0) = \mathbf{0}, \qquad \mathbf{P}(0) = E\left[\mathbf{x}(0)\mathbf{x}^T(0)\right] \tag{8.6.16}$$

The resulting forward estimate and its associated error covariance will be denoted without subscripts as $\hat{\mathbf{x}}(t')$ and $\mathbf{P}(t')$.

For the backward filter, it is convenient to define a new running time variable τ that proceeds backward in time. The backward process model is then obtained by replacing the time derivative in Eq. (8.6.12) with $-d/d\tau$. This leads to the process equation.

Backward process model:

$$\frac{d\mathbf{x}}{d\tau} = -\mathbf{F}\mathbf{x} - \mathbf{G}\mathbf{u} \tag{8.6.17}$$

The corresponding filter equations are then obtained by replacing $\mathbf{F}$ and $\mathbf{G}$ in Eqs. (8.6.14) and (8.6.15) with $-\mathbf{F}$ and $-\mathbf{G}$.

Backward filter equations:

$$\frac{d\hat{\mathbf{x}}_b}{d\tau} = -\mathbf{F}\hat{\mathbf{x}}_b + \mathbf{P}_b\mathbf{H}^T\mathbf{R}^{-1}(\mathbf{z} - \mathbf{H}\hat{\mathbf{x}}_b) \tag{8.6.18}$$

$$\frac{d\mathbf{P}_b}{d\tau} = -\mathbf{F}\mathbf{P}_b - \mathbf{P}_b\mathbf{F}^T - \mathbf{P}_b\mathbf{H}^T\mathbf{R}^{-1}\mathbf{H}\mathbf{P}_b + \mathbf{G}\mathbf{Q}\mathbf{G}^T \tag{8.6.19}$$

The boundary conditions for the backward filter equations are awkward, to say the least. Note that we have already used our a priori knowledge about the process in the forward filter. Thus, this information must not be used again in the backward filter. (If it were used in both filters, the two estimates, $\hat{\mathbf{x}}$ and $\hat{\mathbf{x}}_b$, would not be independent when they meet at the interior point t'.) This forces us to demand that $\mathbf{P}_b(0) = \infty$. Note that $\tau = 0$ corresponds to $t = T$. The corresponding initial estimate of $\hat{\mathbf{x}}_b$ is arbitrary and immaterial, because the initial $\hat{\mathbf{x}}_b$ is given zero weight due to the infinite initial $\mathbf{P}_b$. Both Gelb and Fraser–Potter give a procedure for avoiding the "infinite-P" problem. Their procedure involves propagating $\mathbf{P}_b^{-1}$ rather than $\mathbf{P}_b$. The boundary conditions, while awkward in the continuous problem, do not present similar problems in the discrete counterpart, so we will be content here to beg the question in the continuous problem and simply say that the initial conditions are awkward. (See Problem 8.6 for a further discussion of this.)

Assume now that forward and backward filters have been solved for $\hat{\mathbf{x}}, \mathbf{P}$ and $\hat{\mathbf{x}}_b, \mathbf{P}_b$ at the meeting point t'. The appropriate blend of $\hat{\mathbf{x}}$ and $\hat{\mathbf{x}}_b$ and its error covariance are then obtained from Eqs. (8.6.1) and (8.6.9) to (8.6.11) by simply replacing scalar variances with covariance matrices. The final result is then

$$\mathbf{P}(t'|T) = \left(\mathbf{P}^{-1} + \mathbf{P}_b^{-1}\right)^{-1} \tag{8.6.20}$$

$$\hat{\mathbf{x}}(t'|T) = \mathbf{P}(t'|T)\left[\mathbf{P}^{-1}\hat{\mathbf{x}} + \mathbf{P}_b^{-1}\hat{\mathbf{x}}_b\right] \tag{8.6.21}$$

The Discrete Problem

When experimental measurements are recorded for later processing off-line, they are usually sampled and recorded in digital form. Thus, the discrete smoothing problem is often of more practical interest than its continuous counterpart. It can be seen that the Fraser–Potter forward–backward filter approach has considerable potential for computational savings, as compared with the RTS algorithm. First, by filtering from both ends to the middle, there is no need to store any of the intermediate calculations along the way. The only ones needed for calculating the smoothed estimate are those obtained at the meeting point. Also, filtering does not involve inverting a $\mathbf{P}$ matrix with each step, as is required on the backward sweep of the RTS algorithm. Since the $\mathbf{P}$ matrix often has a much higher dimension than $\mathbf{R}$, having to form $\mathbf{P}^{-1}$ with each step is a time-consuming computation. Thus, a discrete version of Fraser–Potter approach is attractive and will now be pursued further.

The discrete forward filter has been discussed in some detail in Chapters 5 and 6, and there is no change in either the model or the recursive equations relative to the forward part of the estimator. As before, the a priori knowledge about the process will be incorporated into the forward filter and not the backward one. It is important to note, though, that the discrete forward filter equations are *not* approximations of the continuous filter equations. They are exact in their own right, provided that $\boldsymbol{\phi}_k, \mathbf{Q}_k, \mathbf{H}_k$, and $\mathbf{R}_k$ are exact. We now seek a similar exact set of equations for the backward filter.

Once we have obtained an a priori estimate and its $\mathbf{P}$ matrix at any point in the backward process, the updating is the same for the forward filter. This should be apparent from the derivation given in Chapter 5, because there is nothing in the derivation that depends on how we got the a priori estimate and its associated error covariance. Thus, the only different aspect in the backward procedure has to do with the projection of $\mathbf{x}_b$ and $\mathbf{P}_b$ from one step to the next. The projection equations can be obtained from the continuous backward filter equations, Eqs. (8.6.18) and (8.6.19). We do this by thinking of having arrived at some point i in the backward process, and then we proceed through the $(i, i + 1)$ interval without benefit of measurements. Saying that we have no measurements is the same as letting $\mathbf{R}^{-1} \to 0$, which indicates worthless measurements. If we let $\mathbf{R}^{-1} = 0$ in Eqs. (8.6.18) and (8.6.19), they reduce to

$$\frac{d\hat{\mathbf{x}}_b}{d\tau} = -\mathbf{F}\hat{\mathbf{x}}_b \qquad (8.6.22)$$

$$\frac{d\mathbf{P}_b}{d\tau} = -\mathbf{F}\mathbf{P}_b - \mathbf{P}_b\mathbf{F}^T + \mathbf{G}\mathbf{Q}\mathbf{G}^T \qquad (8.6.23)$$

These equations must now be solved for the $(i, i + 1)$ interval, subject to the initial conditions $\hat{\mathbf{x}}_b(\tau_i)$ and $\mathbf{P}_b(\tau_i)$. (We assume here that i increments in unit steps in the same direction as τ.) The solution of Eq. (8.6.22) is obtained from linear system theory, just as in the case of the forward filter. In the interest of simplicity, we assume $\mathbf{F}$, $\mathbf{G}$, and $\mathbf{Q}$ are constant over the interval. Then

$$\hat{\mathbf{x}}_b^*(\tau_{i+1}) = e^{-\mathbf{F}\Delta\tau}\hat{\mathbf{x}}_b(\tau_i) \qquad (8.6.24)$$

or

$$\hat{\mathbf{x}}_{b(i+1)}^* = \boldsymbol{\phi}^{-1}(\Delta\tau)\hat{\mathbf{x}}_{bi} \qquad (8.6.25)$$

where the asterisk * indicates "a priori" and $\boldsymbol{\phi}^{-1}(\Delta\tau)$ is the inverse of the forward transition matrix for the interval. This is exactly what we would expect intuitively.

Since the backward projection of $\mathbf{P}_b$ is not as intuitively obvious as that for $\hat{\mathbf{x}}_b$, we will rely on the formal solution of Eq. (8.6.23) and not try to justify the answer intuitively. We will show that if $\mathbf{P}_b$ is projected according to

$$\mathbf{P}_{b(i+1)}^* = \boldsymbol{\phi}^{-1}(\Delta\tau)[\mathbf{P}_{bi} + \mathbf{Q}_i]\boldsymbol{\phi}^{-1}(\Delta\tau)^T \qquad (8.6.26)$$

the differential equation is satisfied. Let the right end of the projection interval be fixed at $\tau = \tau_0$. The left end is then the negatively running time variable τ, and $\tau > \tau_0$. We speculate that

$$\mathbf{P}_b(\tau) = \boldsymbol{\phi}^{-1}(\tau - \tau_0)[\mathbf{P}_b(\tau_0) + \mathbf{Q}_i(\tau)]\boldsymbol{\phi}^{-1^T}(\tau - \tau_0) \qquad (8.6.27)$$

will satisfy Eq. (8.6.23). Note that the discrete $\mathbf{Q}_i$ is written as a function of τ, because it varies as the interval increases. Differentiation of the assumed solution, Eq. (8.6.27), leads to

$$\frac{d\mathbf{P}_b}{d\tau} = \frac{d\boldsymbol{\phi}^{-1}}{d\tau}[\mathbf{P}_b(\tau_0) + \mathbf{Q}_i]\boldsymbol{\phi}^{-1^T} + \boldsymbol{\phi}^{-1}\frac{d\mathbf{Q}_i}{d\tau}\boldsymbol{\phi}^{-1^T} + \boldsymbol{\phi}^{-1}[\mathbf{P}_b(\tau_0) + \mathbf{Q}_i]\frac{d\boldsymbol{\phi}^{-1^T}}{d\tau}$$

$$(8.6.28)$$

where the parenthetical dependence has been omitted in places to save writing. We first note from linear system theory that

$$\frac{d\boldsymbol{\phi}^{-1}}{d\tau} = -\mathbf{F}\boldsymbol{\phi}^{-1} \tag{8.6.29}$$

Thus, the first and third terms of Eq. (8.6.28) combine to yield $(-\mathbf{F}\mathbf{P}_b - \mathbf{P}_b\mathbf{F}^T)$, which is identical to the first two terms on the right side of Eq. (8.6.23). Looking next at the $\boldsymbol{\phi}^{-1}(d\mathbf{Q}_i/d\tau)\boldsymbol{\phi}^{-1^T}$ term in Eq. (8.6.28), we see that we need to form the derivative of $\mathbf{Q}_i$. Equation (5.3.6) is useful here. Let $t_k = T - \tau$, $t_{k+1} = T - \tau_0$, and note that $E[\mathbf{u}(\xi)\mathbf{u}^T(\eta)] = \mathbf{Q}\,\delta(\xi - \eta)$. After performing the inner integration we have

$$\mathbf{Q}_i = \int_{T-\tau}^{T-\tau_0} \boldsymbol{\phi}(T - \tau_0, \eta)\mathbf{G}\mathbf{Q}\mathbf{G}^T\boldsymbol{\phi}^T(T - \tau_{0,}\,\eta)\,d\eta \tag{8.6.30}$$

Differentiating this with respect to τ leads to

$$\frac{d\mathbf{Q}_i}{d\tau} = \boldsymbol{\phi}(\tau - \tau_0)\mathbf{G}\mathbf{Q}\mathbf{G}^T\boldsymbol{\phi}^T(\tau - \tau_0) \tag{8.6.31}$$

This may now be pre- and postmultiplied by $\boldsymbol{\phi}^{-1}(\tau - \tau_0)$ and $\boldsymbol{\phi}^{-1^T}(\tau - \tau_0)$, as indicated in Eq. (8.6.28), and the result is obviously $\mathbf{G}\mathbf{Q}\mathbf{G}^T$, which equals the third term on the right side of Eq. (8.6.23). Thus, the discrete covariance projection given by Eq. (8.6.26) is seen to satisfy the continuous differential equation for $\mathbf{P}_b$ with $\mathbf{R}^{-1}$ set equal to zero. It is important to note that the $\boldsymbol{\phi}$ and $\mathbf{Q}_i$ in the projection equations are computed from the *forward* process equations. That is, they are the same parameters that would have been used in the forward filter for the interval, had the forward recursive steps been continued up through that interval of time.

The boundary conditions for the backward filter are awkward, but not impossible. One way of implementing the infinite initial $\mathbf{P}$ matrix in off-line processing is to make all terms along its major diagonal very large, say about 10 orders of magnitude larger than their initial counterparts in the forward filter. The off-diagonal terms may then be set equal to zero. For all practical purposes, this completely deweights the prior information about the process relative to the weight that it is given in the forward filter. This is not a very elegant approach, but it is an effective, practical solution in many applications. In simple models where the order of the state vector is small, the boundary-condition problem may be handled analytically by using the alternative Kalman filter algorithm for the first step in the backward filter. This is illustrated in the one-state example presented at the end of this section, and it is also discussed further in problem 8.6. If all else fails, we can resort to propagating $\mathbf{P}$ inverse as suggested in the Fraser–Potter paper cited earlier.

We assume now that the forward filter has been stopped ahead recursively at the estimation point, say k, and the end result is an a posteriori estimate $\mathbf{x}_k$ and an associated $\mathbf{P}_k$. The backward filter steps backward from the end point N, and it stops at $k + 1$ where it assimilates the $\mathbf{z}_{k+1}$ measurement. It then projects this estimate one more step to obtain an a priori estimate $\mathbf{x}_{bk}^*$ and its associated $\mathbf{P}_{bk}^*$. It does not assimilate the $\mathbf{z}_k$ measurement, because this has already been used in the forward filter. Finally, the forward and backward estimates are blended together in accordance with the equation

$$\hat{\mathbf{x}}(k|N) = \mathbf{P}(k|N)\left[\mathbf{P}_k^{-1}\hat{\mathbf{x}}_k + \mathbf{P}_{bk}^{*-1}\hat{\mathbf{x}}_{bk}^*\right] \tag{8.6.32}$$

where

$$\mathbf{P}(k|N) = \left[\mathbf{P}_k^{-1} + \mathbf{P}_{bk}^{*-1}\right]^{-1} \tag{8.6.33}$$

An example will now illustrate the procedure.

EXAMPLE 8.4

Again, we consider the same Gauss–Markov process used for the previous examples of this chapter. Let us say we want to find the smoothed estimate at $k = 48$, that is, $\hat{x}(48|50)$. We first forward-filter up through z_{48}. This has already been done in previous examples and the results are given in Table 8.1. The pertinent forward-filter results are

$$\hat{x}_{48} = -.521$$

$$P_{48} = .165$$

We now need to generate a similar estimate at $k = 48$ with the backward filter. We begin at end point where $k = 50$. The initial conditions for the backward filter are

$$\hat{x}_{b0}^* = 0, \qquad P_{b0}^* = \infty \tag{8.6.34}$$

where the asterisk indicates "a priori" in the backward filter, and the subscript indicates the backward index i, rather than k. We update the initial estimate with the alternative Kalman filter algorithm in order to avoid the P-equal-infinity problem. Thus we have at $k = 50$ $(i = 0)$

$$P_{b0}^{-1} = P_{b0}^{*-1} + H_0^T R_0^{-1} H_0$$

$$= 0 + 1 \tag{8.6.35}$$

or

$$P_{b0} = 1 \tag{8.6.36}$$

The gain is then

$$K_{b0} = P_{b0} H_0^T R_0^{-1} = 1 \qquad (8.6.37)$$

and the updated estimate is

$$\hat{x}_{b0} = \hat{x}_{b0}^* + K_{b0}[z_0 - H_0 \hat{x}_{b0}^*] = z_0 = -1.138$$

$$(\text{from } k = 50 \text{ entry of Table 8.1}) \qquad (8.6.38)$$

Note that we could have taken a more heuristic approach here and simply said: "Initially, we knew nothing about the backward process at the end point, and hence we must accept the measurement at $k = 50$ at face value. The resulting estimation error is then just the measurement error." (This philosophy can also be extended to the vector case, as shown in Problem 8.6.)

Continuing the backward filter, we next project x_b and P_b to $k = 49$ ($i = 1$) using Eqs. (8.6.25) and (8.6.26). The results are

$$\hat{x}_{b1}^* = e^{.02} \hat{x}_{b0} = e^{.02}(-1.138) = -1.161 \qquad (8.6.39)$$

$$P_{b1}^* = e^{.02}(P_{b0} + Q_i)e^{.02} = (e^{.02})^2[1 + .0392] = 1.0816 \qquad (8.6.40)$$

Next, we compute the gain and update the a priori estimate using the regular Kalman filter algorithm:

$$K_{b1} = \frac{P_{b1}^*}{P_{b1}^* + 1} = .5196 \qquad (8.6.41)$$

$$x_{b1} = \hat{x}_{b1}^* + K_{b1}(z_1 - \hat{x}_{b1}^*) = -.591 \qquad (8.6.42)$$

(Note from Table 8.1 that the measurement at $k = 49$ is $-.064$.) The a posteriori error covariance matrix is then

$$P_{b1} = (1 - K_{b1})P_{b1}^* = .5196 \qquad (8.6.43)$$

Now we project $\hat{x}_{b1}$ and P_{b1} to $k = 48$ ($i = 2$).

$$\hat{x}_{b2}^* = e^{.02} \hat{x}_{b1} = -.603 \qquad (8.6.44)$$

$$P_{b2}^* = e^{.02}(P_{b1} + Q_i)e^{.02} = .5816 \qquad (8.6.45)$$

The backward filter is stopped at this point, because the measurement at $k = 48$ has already been assimilated in the forward filter.

We are now ready to blend together the forward and backward estimates in accordance with Eqs. (8.6.32) and (8.6.33). The results are

$$P(48|50) = \left[(.1653)^{-1} + (.5816)^{-1} \right]^{-1} = .1287$$

and

$$\hat{x}(48|50) = P(48|50) \left[P_{48}^{-1}\hat{x}_{48} + P_{b2}^{-1}\hat{x}_{b2}^* \right] = -.539$$

Notice that these results are the same as those given in Table 8.2, which were obtained using the RTS algorithm. ∎

PROBLEMS

8.1. Consider the same signal and noise situation used in Example 4.3. We found there that the causal Wiener filter was a simple first-order low-pass filter. Using Eq. (4.3.22) find the stationary, fixed-lag smoothing solution for $\alpha = -1$. Recall that the signal and noise are independent processes with autocorrelation functions

$$R_s(\tau) = e^{-|\tau|}$$

$$R_n(\tau) = \delta(\tau)$$

The solution may be given as a transfer function, that is, find $G(s)$ rather than $g(t)$. Note that the smoothing solution is considerably more complicated than the corresponding filter solution.

8.2. (a) Consider the one-step back, fixed-interval smoothing problem. Its solution may be obtained from the usual Kalman filter equations by batching together the last two measurements z_{N-1} and z_N and considering them as one measurement occurring at $N - 1$. Note that z_N can be linearly related to x_{N-1} as follows:

$$z_N = H_N x_N + v_N$$

$$= H_N(\phi_{N-1} x_{N-1} + w_{N-1}) + v_N$$

$$= (H_N \phi_{N-1}) x_{N-1} + (H_N w_{N-1} + v_N)$$

The batched measurement relationship at $N - 1$ is then

$$\left[\frac{z_{N-1}}{z_N} \right] = \left[\frac{H_{N-1}}{H_N \phi_{N-1}} \right] x_{N-1} + \left[\frac{v_{n-1}}{H_N w_{N-1} + v_N} \right]$$

Note that the upper and lower components of the batched measurement can be processed sequentially because their errors are uncorrelated. Assume now that we have an a priori estimate $\hat{\mathbf{x}}(N - 1|N - 2)$ and its error covariance matrix $\mathbf{P}(N - 1|N - 2)$. Proceed to update the estimate by assimilating the two components of the measurement *separately* in two steps, and show the final result is the same as that obtained using the RTS algorithm. (*Hint*: The first sequential step yields $\hat{\mathbf{x}}(N - 1|N - 1)$. Next, show that the gain for the second sequential step is $\mathbf{P}(N - 1|N - 1)\boldsymbol{\phi}_{N-1}^{T}\mathbf{P}(N|N - 1)^{-1}\mathbf{K}_{N}$ where $\mathbf{K}_{N}$ is the usual Kalman filter gain for the Nth stage. Finally, replace $\hat{\mathbf{x}}(N|N)$ in the RTS formula with $[\hat{\mathbf{x}}(N|N - 1) + \mathbf{K}_{N}(\mathbf{z}_{N} - \mathbf{H}_{N}\hat{\mathbf{x}}(N|N - 1))]$ and show the equivalence.)

(b) The exercise of part (a) can be generalized to justify the RTS algorithm for any interior point within the fixed interval from 0 to N. To do this, let the interior point be denoted k and batch together all subsequent measurements $\mathbf{z}_{k+1}, \mathbf{z}_{k+2}, \ldots, \mathbf{z}_{N}$. Call the batched measurement $\mathbf{y}_{k+1}$, that is,

$$\mathbf{y}_{k+1} = \begin{bmatrix} \mathbf{z}_{k+1} \\ \hline \mathbf{z}_{k+2} \\ \hline \cdot \\ \cdot \\ \cdot \\ \mathbf{z}_{N} \end{bmatrix}$$

We can now form a linear connection between $\mathbf{x}_{k+1}$ and $\mathbf{y}_{k+1}$ as

$$\mathbf{y}_{k+1} = \mathbf{M}_{k+1}\mathbf{x}_{k+1} + \mathbf{v}_{k+1}'$$

The batched measurement $\mathbf{y}_{k+1}$ now plays the same role as $\mathbf{z}_{N}$ in part (a). (We do not actually have to write out $\mathbf{M}_{k+1}$ and $\mathbf{v}_{k+1}'$ explicitly. We simply need to know that such a relationship exists.) We can now consider the interior-point smoothing problem in terms of an equivalent filter problem that terminates at $k + 1$. This is the same as the problem considered in (a) except for notation. Proceed through the steps of exercise (a) again with appropriate changes in notation and show that the generalization is valid.

8.3. Consider a stationary, Gauss–Markov process with an autocorrelation function

$$R_{x}(\tau) = e^{-|\tau|}$$

Assume that we have two noisy measurements of this process that were made at $t = 0$ and $t = 1$. Call these z_{0} and z_{1}. The measurement errors associated with z_{0} and z_{1} are white and have a variance of unity. We wish to find the optimal estimate of x at $t = 0$, given z_0 and z_1, that is, we desire $\hat{x}(0|1)$. Write an expression for $\hat{x}(0|1)$ explicitly in terms of the measurements z_0 and z_1 using:

(a) The RTS algorithm.
(b) The Fraser–Potter forward–backward filter method.
(c) The weighting function method (see Section 4.8).
(d) The fixed-point algorithm of Section 8.4.

8.4. Show that the continuous version of the RTS algorithm is as follows:

$$\dot{\hat{\mathbf{x}}}(t|T) = \mathbf{F}\hat{\mathbf{x}}(t|T) + \mathbf{G}\mathbf{Q}\mathbf{G}^T\mathbf{P}^{-1}(t|t)\left[\hat{\mathbf{x}}(t|T) - \hat{\mathbf{x}}(t|t)\right]$$

$$\dot{\mathbf{P}}(t|T) = \left[\mathbf{F} + \mathbf{G}\mathbf{Q}\mathbf{G}^T\mathbf{P}^{-1}(t|t)\right]\mathbf{P}(t|T)$$

$$+ \mathbf{P}(t|T)\left[\mathbf{F} + \mathbf{G}\mathbf{Q}\mathbf{G}^T\mathbf{P}^{-1}(t|t)\right]^T - \mathbf{G}\mathbf{Q}\mathbf{G}^T$$

Boundary conditions: $\hat{\mathbf{x}}(T|T)$ and $\mathbf{P}(T|T)$ obtained from filter solution. (*Hint*: Begin with the discrete RTS algorithm and then let the step size approach zero, just as was done in deriving the filter equations in Chapter 7. Recall that $\mathbf{Q}_k$ in the discrete model approaches $\mathbf{G}\mathbf{Q}\mathbf{G}^T\Delta t$ for small Δt. Also note that $[\mathbf{I} - \mathbf{A}(k)]$ is of the order of Δt in the smoothing algorithm; thus, the gain $\mathbf{A}(k)$ approaches $\mathbf{I}$ in the limit as $\Delta t \to 0$.)

8.5. Consider a Wiener process and measurement situation as follows:

$$\dot{x} = u(t), \qquad x(0) = 0$$

$$z = x + v(t)$$

where $u(t)$ and $v(t)$ are independent Gaussian white noise processes with spectral amplitudes q and r, respectively.

(a) Assume a fixed interval T that is sufficiently large to allow the filter solution to reach a steady-state condition on the forward sweep. This then becomes the boundary condition for the backward sweep. Using the continuous RTS algorithm given in Problem 8.4, show that $P(t|T)$ for the terminal region of the $[0, T]$ interval is approximated by

$$P(t|T) \approx \frac{\alpha}{2}[1 + e^{-2\beta(T-t)}]$$

where $\alpha = \sqrt{rq}$ and $\beta = \sqrt{q/r}$.

(b) Note that the solution of part (a) approaches $\alpha/2$ as $t \to 0$. This is obviously not compatible with the known a priori boundary condition for a Wiener process; that is, $P(0) = 0$. Explain this discrepancy.

8.6. Consider the two-state system

$$\begin{bmatrix} \dot{x}_1 \\ \dot{x}_2 \end{bmatrix} = \begin{bmatrix} -1 & 0 \\ 0 & -2 \end{bmatrix} \begin{bmatrix} x_1 \\ x_2 \end{bmatrix} + \begin{bmatrix} u_1 \\ u_2 \end{bmatrix}$$

where u_1 and u_2 are independent Gaussian white-noise inputs with unity-amplitude spectral functions. We have a sequence of scalar measurements of this process $z_0, z_1, \ldots, z_N$ that are related to the process by the equation

$$z_k = \begin{bmatrix} 1 & 1 \end{bmatrix} \mathbf{x}_k + v_k, \qquad k = 0, 1, \ldots, N$$

where v_k has unity variance. That is, we are allowed to observe only the sum of the state variables at each step and not their individual values. Suppose we want an estimate of the process at some interior point, and we wish to get it using the Fraser–Potter forward–backward-filter method. In this case, the single measurement z_N does not provide enough information to yield a finite error-covariance estimate of x at $t = N$. Thus, we cannot start the backward filter quite as easily as was done in Example 8.4. In this problem, let the Δt interval be unity and show that the backward filter may be started at $t = N - 1$ by batching together z_N and z_{N-1} into an equivalent vector measurement at $t = N - 1$. The same technique used in Problem 8.2 will be helpful here. In effect, you need to show that the error covariance after assimilating z_N and z_{N-1} is finite and nonsingular, and that the estimate is the same as would be obtained by deterministic methods. (The extension of this technique to higher-order systems is fairly obvious, provided the system is observable. We simply batch together an appropriate number of measurements at the end of the interval and then solve for the system state, just as if the measurements were noise-free. This then becomes the initial state estimate for the backward filter, and we start the backward filtering an appropriate number of steps back from the end point.)

8.7. Table 8.3 summarizes the results of the fixed-point smoothing simulation of Example 8.2. Notice that the filter error variances listed under the column $P(j|j)$ are identical with the smoothing error variances $P(0|j)$. Give an intuitive explanation of this coincidence.

8.8*. The estimates obtained from fixed-interval smoothing should always be statistically better (or at least as good) than the corresponding filtered estimates. This is not always obvious from looking at just one sample realization of a process. Consider the 50-step process described in Example 8.1, and run four sample realizations of the x_k and z_k processes using the Kalm 'N Smooth software provided with the text. Then run the corresponding filtered and smoothed estimates for each of the Monte Carlo realizations. For each realization plot the true x_k, the filtered estimate, and the smoothed estimate together on one graph and draw your own conclusion as to which of the estimates is best (in a statistical sense, of course).

REFERENCES CITED IN CHAPTER 8

1. S. Meditch, *Stochastic Optimal Linear Estimation and Control*, New York: McGraw-Hill, 1969.
2. H. E. Rauch, "Solutions to the Linear Smoothing Problem," *IEEE Trans. on Auto. Control*, AC-8: 371 (1963).
3. H. E. Rauch, F. Tung, and C. T. Striebel, "Maximum Likelihood Estimates of Linear Dynamic Systems," *AIAA J., 3:* 1445 (1965).
4. D. C. Fraser and J. E. Potter, "The Optimum Linear Smoother as a Combination of Two Optimum Linear Filters," *IEEE Trans. on Auto Control, AC-14:* No. 4, 387 (August 1969).
5. A. Gelb (Ed.), *Applied Optimal Estimation*, Cambridge, Mass.: MIT Press, 1974.

Additional References on Smoothing

6. A. P. Sage and J. L. Melsa, *Estimation Theory with Applications to Communications and Control*, New York: McGraw-Hill, 1971.
7. B. D. O. Anderson and J. B. Moore, *Optimal Filtering*, Englewood Cliffs, N.J.: Prentice-Hall, 1979.

9

Linearization and Additional Topics on Applied Kalman Filtering

Kalman's papers of the early 1960s (1, 2) were recognized almost immediately as new and important contributions to least-squares filtering. As a result, there was a renewal of research interest in this area, and a flurry of papers expanding on Kalman's original work followed during the next decade or so. Kailath (3) gives an especially comprehensive bibliography of papers for this period. Research work in this area still continues (although perhaps at a somewhat reduced rate), and new applications and extensions continue to appear regularly in the technical literature (4). A few of the more significant extensions and related topics have been selected for comment here. The list is by no means comprehensive. However, there is a hierarchy of importance, and it is the authors' recommendation that the reader begin with the first topic on linearization. This is, by far, the most important section in this chapter. The other sections may be studied in any desired order as time permits.

9.1
LINEARIZATION

Some of the most successful applications of Kalman filtering have been in situations with nonlinear dynamics and/or nonlinear measurement relationships. We now examine two basic ways of linearizing the problem. One is to linearize about some nominal trajectory in state space that does not depend on the measurement data. The resulting filter is usually referred to as simply a

356

linearized Kalman filter. The other method is to linearize about a trajectory that is continually updated with the state estimates resulting from the measurements. When this is done, the filter is called an *extended Kalman filter*. A brief discussion of each will now be presented.

Linearized Kalman Filter

We begin by assuming the process to be estimated and the associated measurement relationship may be written in the form

$$\dot{\mathbf{x}} = \mathbf{f}(\mathbf{x}, \mathbf{u}_d, t) + \mathbf{u}(t) \tag{9.1.1}$$

$$\mathbf{z} = \mathbf{h}(\mathbf{x}, t) + \mathbf{v}(t) \tag{9.1.2}$$

where $\mathbf{f}$ and $\mathbf{h}$ are known functions, $\mathbf{u}_d$ is a deterministic forcing function, and $\mathbf{u}$ and $\mathbf{v}$ are white noise processes with zero crosscorrelation as before. Note that nonlinearity may enter into the problem either in the dynamics of the process or in the measurement relationship. Also, note that the forms of Eqs. (9.1.1) and (9.1.2) are somewhat restrictive in that $\mathbf{u}$ and $\mathbf{v}$ are assumed to be separate additive terms and are not included with the $\mathbf{f}$ and $\mathbf{h}$ terms. However, to do otherwise complicates the problem considerably, and thus we will stay with these restrictive forms.

Let us now assume that an approximate trajectory $\mathbf{x}^*(t)$ may be determined by some means. This will be referred to as the nominal or reference trajectory, and it is illustrated along with the actual trajectory in Fig. 9.1. The actual trajectory $\mathbf{x}(t)$, may then be written as

$$\mathbf{x}(t) = \mathbf{x}^*(t) + \Delta\mathbf{x}(t) \tag{9.1.3}$$

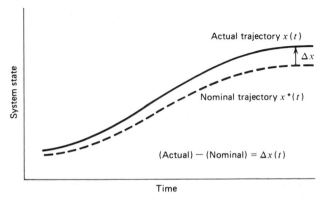

Figure 9.1. Nominal and actual trajectories for a linearized Kalman filter.

Equations (9.1.1) and (9.1.2) then become

$$\dot{\mathbf{x}}^* + \Delta\dot{\mathbf{x}} = \mathbf{f}(\mathbf{x}^* + \Delta\mathbf{x}, \mathbf{u}_d, t) + \mathbf{u}(t) \tag{9.1.4}$$

$$\mathbf{z} = \mathbf{h}(\mathbf{x}^* + \Delta\mathbf{x}, t) + \mathbf{v}(t) \tag{9.1.5}$$

We now assume $\Delta\mathbf{x}$ is small and approximate the $\mathbf{f}$ and $\mathbf{h}$ functions with Taylor's series expansions, retaining only first-order terms. The result is

$$\dot{\mathbf{x}}^* + \Delta\dot{\mathbf{x}} \approx \mathbf{f}(\mathbf{x}^*, \mathbf{u}_d, t) + \left[\frac{\partial\mathbf{f}}{\partial\mathbf{x}}\right]_{\mathbf{x}=\mathbf{x}^*} \cdot \Delta\mathbf{x} + \mathbf{u}(t) \tag{9.1.6}$$

$$\mathbf{z} \approx \mathbf{h}(\mathbf{x}^*, t) + \left[\frac{\partial\mathbf{h}}{\partial\mathbf{x}}\right]_{\mathbf{x}=\mathbf{x}^*} \cdot \Delta\mathbf{x} + \mathbf{v}(t) \tag{9.1.7}$$

where

$$\frac{\partial\mathbf{f}}{\partial\mathbf{x}} = \begin{bmatrix} \frac{\partial f_1}{\partial x_1} \frac{\partial f_1}{\partial x_2} \cdots \\ \frac{\partial f_2}{\partial x_1} \frac{\partial f_2}{\partial x_2} \cdots \\ \vdots \end{bmatrix}; \quad \frac{\partial\mathbf{h}}{\partial\mathbf{x}} = \begin{bmatrix} \frac{\partial h_1}{\partial x_1} \frac{\partial h_1}{\partial x_2} \cdots \\ \frac{\partial h_2}{\partial x_1} \frac{\partial h_2}{\partial x_2} \cdots \\ \vdots \end{bmatrix} \tag{9.1.8}$$

It is customary to choose the nominal trajectory $\mathbf{x}^*(t)$ to satisfy the deterministic differential equation.

$$\dot{\mathbf{x}}^* = \mathbf{f}(\mathbf{x}^*, \mathbf{u}_d, t) \tag{9.1.9}$$

Substituting this into (9.1.6) then leads to the linearized model

$$\Delta\dot{\mathbf{x}} = \left[\frac{\partial\mathbf{f}}{\partial\mathbf{x}}\right]_{\mathbf{x}=\mathbf{x}^*} \cdot \Delta\mathbf{x} + \mathbf{u}(t), \quad \text{(linearized dynamics)} \tag{9.1.10}$$

$$[\mathbf{z} - \mathbf{h}(\mathbf{x}^*, t)] = \left[\frac{\partial\mathbf{h}}{\partial\mathbf{x}}\right]_{\mathbf{x}=\mathbf{x}^*} \cdot \Delta\mathbf{x} + \mathbf{v}(t), \quad \text{(linearized measurement equation)}$$

$$\tag{9.1.11}$$

Note that the "measurement" in the linear model is the actual measurement less that predicted by the nominal trajectory in the absence of noise. Also the equivalent $\mathbf{F}$ and $\mathbf{H}$ matrices are obtained by evaluating the partial derivative matrices [Eqs. (9.1.8)] along the *nominal* trajectory. We will now look at two

Figure 9.2. Geometry for DME example.

examples that illustrate the linearization procedure. In the first example the nonlinearity appears only in the measurement relationship, so it is relatively simple. In the second, nonlinearity occurs in both the measurement and process dynamics, so it is somewhat more involved than the first.

EXAMPLE 9.1 _____

In many electronic navigation systems the basic observable is a noisy measurement of range (distance) from the vehicle to a known location. One such system that has enjoyed wide use in aviation is Distance Measuring Equipment (DME). We do not need to go into detail here as to how the equipment works. It suffices to say that the airborne equipment transmits a pulse that is returned by the ground station, and then the aircraft equipment interprets the transit time in terms of distance. In our example we will simplify the geometric situation by assuming that the aircraft and the two DME stations are all in a horizontal plane as shown in Fig. 9.2 (slant range ≈ horizontal range). The coordinates of the two DME stations are assumed to be known, and the aircraft coordinates are unknown and to be estimated.

   We will look at the aircraft dynamics first. This, in turn, will determine the process state model. To keep things as simple as possible, we will assume a nominal straight-and-level flight condition with constant velocity. The true trajectory will be assumed to be the nominal one plus small perturbations due to random horizontal accelerations, which will be assumed to be white. This leads to random walk in velocity and integrated random walk in position. This is probably unrealistic for long time spans because of the control applied by the pilot (or autopilot). However, this would be a reasonable model for short intervals of time. The basic differential equations of motion in the x and y directions are then

$$\ddot{x} = 0 + u_x$$

$$\ddot{y} = \underbrace{0}_{\substack{\text{Deterministic} \\ \text{forcing} \\ \text{function}}} + \underbrace{u_y}_{\substack{\text{Random} \\ \text{forcing} \\ \text{function}}} \tag{9.1.12}$$

The dynamical equations are seen to be linear in this case, so the differential equations for the incremental quantities are the same as for the total x and y, that is,

$$\Delta \ddot{x} = u_x$$

$$\Delta \ddot{y} = u_y \tag{9.1.13}$$

We now define filter state variables in terms of the incremental positions and velocities:

$$x_1 = \Delta x, \qquad x_2 = \Delta \dot{x}$$

$$x_3 = \Delta y, \qquad x_4 = \Delta \dot{y} \tag{9.1.14}$$

The state equations are then

$$\begin{bmatrix} \dot{x}_1 \\ \dot{x}_2 \\ \dot{x}_3 \\ \dot{x}_4 \end{bmatrix} = \begin{bmatrix} 0 & 1 & 0 & 0 \\ 0 & 0 & 0 & 0 \\ 0 & 0 & 0 & 1 \\ 0 & 0 & 0 & 0 \end{bmatrix} \begin{bmatrix} x_1 \\ x_2 \\ x_3 \\ x_4 \end{bmatrix} + \begin{bmatrix} 0 \\ u_x \\ 0 \\ u_y \end{bmatrix} \tag{9.1.15}$$

The state variables are driven by the white noise processes u_x and u_y, so we are assured that the corresponding discrete equations will be in the appropriate form for a Kalman filter.

We now turn to the measurement relationships. We will assume that we have two simultaneous range measurements, one to DME_1 and the other to DME_2. The two measurement equations in terms of the total x and y are then

$$z_1 = \sqrt{(x - a_1)^2 + (y - b_1)^2} + v_1$$

$$z_2 = \sqrt{(x - a_2)^2 + (y - b_2)^2} + v_2 \tag{9.1.16}$$

where v_1 and v_2 are additive white measurement noises. We see immediately that the connection between the observables (z_1 and z_2) and the quantities to be estimated (x and y) is nonlinear. Thus, linearization about the nominal trajectory is in order. We assume that an approximate nominal position is known at the time of the measurement, and that the locations of the two DME stations are known exactly. We now need to form the $\partial\mathbf{h}/\partial\mathbf{x}$ matrix as specified by Eq. (9.1.8). (We note a small notational problem here. The variables $x_1, x_2, x_3,$ and x_4 are used in Eq. (9.1.8) to indicate total state variables, and then the same symbols are used again to indicate incremental state variables as defined by Eqs. (9.1.14). However, the meanings of the symbols are never mixed in any one set of equations, so this should not lead to confusion.) We now note that the x and y position variables are the first and third elements of the state vector. Thus, evaluation of the partial derivatives indicated in Eq. (9.1.8) leads to

$$\frac{\partial\mathbf{h}}{\partial\mathbf{x}} = \begin{bmatrix} \dfrac{(x_1-a_1)}{\sqrt{(x_1-a_1)^2+(x_3-b_1)^2}} & 0 & \dfrac{(x_3-b_1)}{\sqrt{(x_1-a_1)^2+(x_3-b_1)^2}} & 0 \\ \dfrac{(x_1-a_2)}{\sqrt{(x_1-a_2)^2+(x_3-b_2)^2}} & 0 & \dfrac{(x_3-b_2)}{\sqrt{(x_1-a_2)^2+(x_3-b_2)^2}} & 0 \end{bmatrix}$$

(9.1.17)

or

$$\frac{\partial\mathbf{h}}{\partial\mathbf{x}} = \begin{bmatrix} -\cos\theta_1 & 0 & -\sin\theta_1 & 0 \\ -\cos\theta_2 & 0 & -\sin\theta_2 & 0 \end{bmatrix}$$

(9.1.18)

Finally, we note that Eq. (9.1.18) can be generalized even further, since the sine and cosine terms are actually direction cosines between the x and y axes and the respective lines of sight to the two DME stations. Therefore, we will write the linearized $\mathbf{H}$ matrix in its final form as

$$\mathbf{H} = \frac{\partial\mathbf{h}}{\partial\mathbf{x}}\bigg|_{\mathbf{x}=\mathbf{x}^*} = \begin{bmatrix} -\cos\theta_{x1} & 0 & -\cos\theta_{y1} & 0 \\ -\cos\theta_{x2} & 0 & -\cos\theta_{y2} & 0 \end{bmatrix}$$

(9.1.19)

where the subscripts on θ indicate the respective axes and lines of sight to the DME stations. Note that $\mathbf{H}$ is evaluated at a point on the *nominal* trajectory. (The true trajectory is not known to the filter.) The nominal aircraft position will change with each step of the recursive process, so the terms of $\mathbf{H}$ are time-variable and must be recomputed with each recursive step. Also, recall from Eq. (9.1.11) that the measurement presented to the linearized filter is the total $\mathbf{z}$ minus the predicted $\mathbf{z}$ based on the nominal position $\mathbf{x}^*$.

Strictly speaking, the linearized filter is always estimating incremental quantities, and then the total quantity is reconstructed by adding the incremental estimate to the nominal part. However, we will see later that when it comes to the actual mechanics of handling the arithmetic on the computer, we can avoid working with incremental quantities if we choose to do so. Further discussion of this will be deferred until Section 9.2 in the part on aided INS systems. We will now proceed to a second linearization example, where the process dynamics as well as the measurement relationship has to be linearized.

∎

EXAMPLE 9.2

This example is taken from Sorenson (6) and is a classic example of linearization of a nonlinear problem. Consider a near-earth space vehicle in a nearly circular orbit. It is desired to estimate the vehicle's position and velocity on the basis of a sequence of angular measurements made with a horizon sensor. With reference to Fig. 9.3, the horizon sensor is capable of measuring:

1. The angle γ between the earth's horizon and the local vertical.

2. The angle α between the local vertical and a known reference line (say to a celestial object).

In the interest of simplicity, we assume all motion and measurements to be within a plane as shown in Fig. 9.3. Thus the motion of the vehicle can be described with the usual polar coordinates r and θ.

The equations of motion for the space vehicle may be obtained from either Newtonian or Lagrangian mechanics. They are (see Section 2-10, Ref. 7):

$$\ddot{r} - r\dot{\theta}^2 + \frac{K}{r^2} = u_r(t) \tag{9.1.20}$$

$$r\ddot{\theta} + 2\dot{r}\dot{\theta} = u_\theta(t) \tag{9.1.21}$$

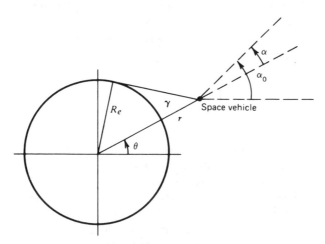

Figure 9.3. Coordinates for space vehicle example.

where K is a constant proportional to the universal gravitational constant, and u_r and u_θ are small random forcing functions in the r and θ directions (due mainly to gravitational anomalies unaccounted for in the K/r^2 term). It can be seen that the constant K must be equal to gR_e^2 if the gravitational forcing function is to match the earth's gravity constant g at the surface. The random forcing functions u_r and u_θ will be assumed to be white. We will look at the linearized process dynamics first and then consider the nonlinear measurement situation later.

The equations of motion, Eqs. (9.1.20) and (9.1.21), are clearly nonlinear, so we must linearize the dynamics if we are to apply Kalman filter methods. We have assumed that random forcing functions u_r and u_θ are small, so the corresponding perturbations from a circular orbit will also be small. By direct substitution into Eqs. (9.1.20) and (9.1.21), it can be verified that

$$r^* = R_0 \qquad \text{(a constant radius)} \tag{9.1.22}$$

$$\theta^* = \omega_0 t \qquad \left(\omega_0 = \sqrt{\frac{K}{R_0^3}}\right) \tag{9.1.23}$$

will satisfy the differential equations. Thus, this will be the reference trajectory that we linearize about.

We note that we have two second-order differential equations describing the dynamics. Therefore, we must have four state variables in our state model. We choose the usual phase variables as state variables as follows:

$$x_1 = r, \qquad x_2 = \dot{r}$$

$$x_3 = \theta, \qquad x_4 = \dot{\theta} \tag{9.1.24}$$

The nonlinear state equations are then

$$\dot{x}_1 = x_2$$

$$\dot{x}_2 = x_1 x_4^2 - \frac{K}{x_1^2} + u_r(t)$$

$$\dot{x}_3 = x_4$$

$$\dot{x}_4 = -\frac{2x_2 x_4}{x_1} + \frac{u_\theta(t)}{x_1} \tag{9.1.25}$$

We must now form the $\partial \mathbf{f}/\partial \mathbf{x}$ matrix indicated in Eq. (9.1.8) to get the linearized $\mathbf{F}$ matrix.

$$
\frac{\partial \mathbf{f}}{\partial \mathbf{x}} = \begin{bmatrix} 0 & 1 & 0 & 0 \\ \left(x_4^2 + \dfrac{2K}{x_1^3}\right) & 0 & 0 & 2x_1 x_4 \\ 0 & 0 & 0 & 1 \\ \dfrac{2x_2 x_4}{x_1^2} & \dfrac{-2x_4}{x_1} & 0 & \dfrac{-2x_2}{x_1} \end{bmatrix}
$$

$$
= \begin{bmatrix} 0 & 1 & 0 & 0 \\ \left(\dot{\theta}^2 + \dfrac{2K}{r^3}\right) & 0 & 0 & 2r\dot{\theta} \\ 0 & 0 & 0 & 1 \\ \dfrac{2\dot{r}\dot{\theta}}{r^2} & \dfrac{-2\dot{\theta}}{r} & 0 & \dfrac{-2\dot{r}}{r} \end{bmatrix} \qquad (9.1.26)
$$

Next, we evaluate $\partial \mathbf{f}/\partial \mathbf{x}$ along the reference trajectory.

$$
\frac{\partial \mathbf{f}}{\partial \mathbf{x}} \Bigg|_{\substack{r=R_0 \\ \theta=\omega_0 t}} = \begin{bmatrix} 0 & 1 & 0 & 0 \\ 3\omega_0^2 & 0 & 0 & 2R_0\omega_0 \\ 0 & 0 & 0 & 1 \\ 0 & \dfrac{-2\omega_0}{R_0} & 0 & 0 \end{bmatrix} \qquad (9.1.27)
$$

Equation (9.1.27) then defines the $\mathbf{F}$ matrix that characterizes the linearized dynamics. Note that in the linear equations, Δr, $\dot{\Delta r}$, $\Delta \theta$, and $\dot{\Delta \theta}$ become the four state variables.

We now turn to the measurement model. The idealized (no noise) relationships are given by

$$
\begin{bmatrix} z_1 \\ z_2 \end{bmatrix} = \begin{bmatrix} \gamma \\ \alpha \end{bmatrix} = \begin{bmatrix} \sin^{-1}\left(\dfrac{R_e}{r}\right) \\ \alpha_0 - \theta \end{bmatrix} \qquad (9.1.28)
$$

We next replace r with x_1 and θ with x_3, and then perform the partial derivatives indicated by Eq. (9.1.8). The result is

$$
\left[\frac{\partial \mathbf{h}}{\partial \mathbf{x}}\right] = \begin{bmatrix} -\dfrac{R_e}{\sqrt{r^2 - R_e^2}} & 0 & 0 & 0 \\ 0 & 0 & -1 & 0 \end{bmatrix} \qquad (9.1.29)
$$

Finally, we evaluate $\partial \mathbf{h}/\partial \mathbf{x}$ along the reference trajectory

$$
\left. \frac{\partial \mathbf{h}}{\partial \mathbf{x}} \right|_{\substack{r=R_0 \\ \theta=\omega_0 t}} = \begin{bmatrix} -\dfrac{R_e}{\sqrt{R_0^2 - R_e^2}} & 0 & 0 & 0 \\ 0 & & 0 & -1 & 0 \end{bmatrix} \tag{9.1.30}
$$

This then becomes the linearized $\mathbf{H}$ matrix of the Kalman filter. The linearized model is now complete with the determination of the $\mathbf{F}$ and $\mathbf{H}$ matrices. Before leaving this example, though, it is worth noting that the forcing function $u_\theta(t)$ must be scaled by $1/R_0$ in the linear model because of the $1/x_1$ factor in Eq. (9.1.25). ∎

The Extended Kalman Filter

The extended Kalman filter is similar to a linearized Kalman filter except that the linearization takes place about the filter's estimated trajectory, as shown in Fig. 9.4, rather than a precomputed nominal trajectory. That is, the partial derivatives of Eq. (9.1.8) are evaluated along a trajectory that has been updated with the filter's estimates; these, in turn, depend on the measurements, so the filter gain sequence will depend on the sample measurement sequence realized on a particular run of the experiment. Thus, the gain sequence is not predetermined by the process model assumptions as in the usual Kalman filter.

A general analysis of the extended Kalman filter is difficult because of the feedback of the measurement sequence into the process model. However, qualitatively it would seem to make sense to update the trajectory that is used

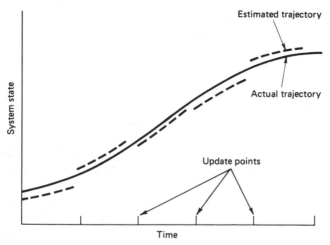

Figure 9.4. Reference and actual trajectories for an extended Kalman filter.

for the linearization—after all, why use the old trajectory when a better one is available? The flaw in this argument is this: the "better" trajectory is only better in a *statistical* sense. There is a chance (and maybe a good one) that the updated trajectory will be poorer than the nominal one. In that event, the estimates may be poorer; this, in turn, leads to further error in the trajectory, which causes further errors in the estimates, and so forth and so forth, leading to eventual divergence of the filter. The net result is that the extended Kalman filter is a somewhat riskier filter than the regular linearized filter, especially in situations where the initial uncertainty and measurement errors are large. It may be better *on the average* than the linearized filter, but it is also more likely to diverge in unusual situations.

Both the regular linearized Kalman filter and the extended Kalman filter have been used in a variety of applications. Each has its advantages and disadvantages, and no general statement can be made as to which is best because it depends on the particular situation at hand. The aided inertial navigation application, which was introduced in Section 6.4, serves as a good example to illustrate the difference between these two forms of linearization. Therefore, the discussion of aided inertial systems will be continued from a nonlinear viewpoint in the next section.

9.2
AIDED INERTIAL NAVIGATION SYSTEMS
— THE NONLINEAR VIEWPOINT

The idea of aiding an inertial navigation system with measurements from other sensors was introduced in Section 6.4. The complementary filter viewpoint was presented there as a matter of expediency, because linearization had not been discussed as yet at that point in the text. We are now in a position to look at this same system integration problem as a linearized nonlinear filter problem.

The Feedforward Configuration

Figure 9.5 will be recognized as being almost the same as the block diagram used in Section 6.4 (Fig. 6.2), except that we have added some symbols to the diagram and have reversed signs in four places. (The sign changes were necessitated by the definition of Δx in Eq. (9.1.3).) In Fig. 9.5 we think of the unaided inertial system as providing a reference trajectory in state-space. This reference trajectory is not precomputable (as it might be in some applications), but it unfolds in real time in accordance with whatever path the vehicle might take. Therefore, it is perfectly proper to use it as the nominal $x^*(t)$. The reference trajectory will not be perfect because of inertial instrument errors; hence, the reference consists of truth minus inertial system errors. Clearly, if we can estimate the inertial system errors, we can add them to the uncorrected inertial system output and

Figure 9.5.   Feedforward configuration for an aided inertial navigation system (an ordinary linearized Kalman filter).

achieve a corrected inertial system output. It should be apparent that the system integration scheme shown in Fig. 9.5 is, in fact, a regular linearized Kalman filter in every sense of the word. The process dynamics in the inertial system and the measurement relationships may be nonlinear. So be it. The nonlinearities are washed out in the differencing operation $[\mathbf{z}_k - \mathbf{h}_k(\mathbf{x}^*)]$, and the filter subproblem becomes linear, provided, of course, that the deviation of the reference trajectory from truth remains small throughout the time span of interest.

Some further comments about the block diagram of Fig 9.5 are in order before proceeding. First, it is tacitly assumed that in forming the difference $[\mathbf{z} - \mathbf{h}(\mathbf{x}^*)]$, we difference *like* quantities. For example, we do not try to compare inertially derived position directly with velocity as seen by the doppler radar. Rather, we compare inertial *velocity* with doppler radar *velocity*; and, furthermore, we compare the two in the same frame of reference. This often means that inertially derived quantities must be converted to another coordinate system before the appropriate comparison can be made. It should also be noted that each of the aiding sources does not have to measure all of the same dynamical quantities that are normally computed in the inertial system. If the "aiding" sources are really to aid, there needs to be at least one connection to the inertial quantities, but that is all. One of the reasons for choosing the inertial system as the reference system is that it is completely self-contained, and it normally provides all of the dynamical variables of interest, such as position, velocity, and vehicle attitude. Usually, no one of the aiding sources (regardless of how accurate) provides all of these quantities and, therefore, it is convenient conceptually to lump all of the aiding sources together in the "aiding" category. This particular system integration methodology has been used successfully with a wide variety of mixes of aiding sensors dating back to the early 1960s (8, 9, 10), and it continues to be used today.

Note that the Kalman filter operates only on system *errors* when using the methodology shown in Fig. 9.5. The inertial system errors are of prime interest, so we need a state-space model describing these errors, and it must be linear with white noise processes as the driving functions. Various models have been used, and the model to be used in any particular application will depend on the type of inertial system at hand and the degree of complexity that the systems engineer is willing to live with in the design. This gets rather involved, so we will not pursue it further here. (An example of an inertial error model is given in Chapter 10 in the section dealing with an integrated INS/GPS system.) In addition to modeling the inertial system errors, one must also augment the system state vector with any nonwhite aiding-source errors. Even though they may not be of primary interest in the navigation problem, they must, nevertheless, be carried along in the estimator if one is to have a properly modeled Kalman filter. The aiding-source error estimates are not, of course, used in the feedforward correction of the inertial system, but they are properly accounted for when forming the measurement residual.

One last comment should be made about the system methodology of Fig. 9.5 before proceeding to the extended filter. It is especially important to note that the inertial system measurements are not used as "measurements" in the Kalman filter. Rather, they are used exclusively to provide the reference trajectory. The inertial acceleration measurements are not filtered in any way; they are simply processed in a dynamically exact manner to yield position and velocity in the desired coordinate system. A measure of vehicle attitude with respect to some frame of reference is also a natural consequence of the inertial system. These quantities are not filtered. The filtering is done on the system *errors* and not on the total quantities. Thus, with this kind of system integration, the dynamic response in high-*g* maneuvers is exact with no time lag. This is one of the main advantages of this type of system integration; namely, the system *errors* get filtered, but without loss of high-fidelity response in the presence of high dynamics.

The Feedback Configuration

The block diagram of Fig. 9.6 shows a variation on the system integration scheme shown in Fig. 9.5. In the scheme of Fig. 9.6 the inertial system errors are fed back, and the corrections are made internally in the inertial system. Thus, we no longer have raw uncorrected outputs coming out of the inertial system. This mechanization is usually referred to as the feedback configuration, in contrast to the feedforward configuration. The key difference between the two schemes lies in the computation of $\mathbf{h}(\mathbf{x}^*)$, the predictive measurement. In the feedforward scheme, this is computed on the basis of the nominal trajectory, which may slowly diverge from truth as time evolves. However, in the feedback mechanization, $\mathbf{h}(\mathbf{x}^*)$ is computed on the basis of a trajectory that gets updated with each step in accordance with the filter's best estimate of the inertial system's errors. In short, this leads to an extended Kalman filter mode of operation, in contrast to an ordinary linearized Kalman filter.

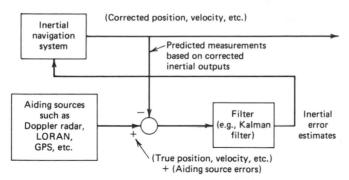

Figure 9.6. Feedback configuration for an aided inertial navigation system (an extended Kalman filter).

One should be careful not to be too literal in the interpretation of the block diagrams of Figs. 9.5 and 9.6. They are intended to be conceptual, not literal. For example, the "internal" corrections shown in Fig. 9.6 must be done in a computer somewhere in the system. In some systems this might be a central flight management computer that also does many other computational chores including, possibly, a separate feedforward solution. In other systems the computational effort may be distributed among various "boxes" in the system. Never mind which box does the computation; regardless, the key item is $h(x^*)$. If $h(x^*)$ is computed before the corrections are made to the inertial outputs, the filter is an ordinary linearized Kalman filter; if $h(x^*)$ is computed after the corrections are made, the filter is an extended Kalman filter. Both modes of operation have been used successfully, and each has its place. Both modes of operation have the same performance, within the linearity assumption. Clearly, though, the extended Kalman filter is to be preferred in applications where the mission time is long, as would be true of a ship at sea for many weeks or months. Otherwise, the reference trajectory would diverge from truth beyond acceptable limits. On the other hand, in a ballistic space vehicle launch, the mission time is short and, hopefully, the actual trajectory will match the preprogrammed one closely. In this case an ordinary linearized Kalman filter might well be preferred.

It should be apparent that the system integration philosophy shown in Fig. 9.5 (or Fig. 9.6) is not the only way of integrating inertial measurements with noninertial data. One has only to peruse the conference proceedings of recent navigation meetings to find dozens of different integration schemes, each with a little different twist because of some special circumstance. The scheme presented here is optimal though (within the constraint of dynamic exactness in the use of the inertial measurements), and it represents the best way of managing the system integration theoretically. However, the systems engineer often does not have the luxury of designing a truly optimal system, and he or she must settle for less because of equipment constraints, cost, and so forth. Even so, the optimal methodology presented here is still valuable for analysis purposes, because it provides a lower bound on system errors for a given mix of sensors. That is, the optimal system serves as a "yardstick" that can be used for comparison purposes in evaluating the degree of suboptimality of various candidate integration schemes.

Keeping Track of Total Estimates in an Extended Kalman Filter

It should be remembered that the basic state variables in a linearized Kalman filter are incremental quantities, and not the total quantities such as position, velocity, and so forth. However, in an extended Kalman filter it is usually more convenient to keep track of the total estimates rather than the incremental ones, so we will now proceed to show how this is done and why it is valid to do so.

We begin with the basic linearized measurement equation, Eq. (9.1.11)

$$\mathbf{z} - \mathbf{h}(\mathbf{x}^*) = \mathbf{H}\Delta\mathbf{x} + \mathbf{v} \tag{9.2.1}$$

Note that when working with incremental state variables, the measurement presented to the Kalman filter is $[\mathbf{z} - \mathbf{h}(\mathbf{x}^*)]$ rather than the total measurement $\mathbf{z}$. Next, consider the incremental estimate update equation at time t_k.

$$\Delta\hat{\mathbf{x}}_k = \Delta\hat{\mathbf{x}}_k^- + \mathbf{K}_k\left[\underbrace{\mathbf{z}_k - \mathbf{h}(\mathbf{x}_k^*)}_{\text{Inc. meas.}} - \mathbf{H}_k\Delta\hat{\mathbf{x}}_k^-\right] \tag{9.2.2}$$

Now, in forming the measurement residual in Eq. (9.2.2), suppose we associate the $\mathbf{h}(\mathbf{x}_k^*)$ term with $\mathbf{H}_k\Delta\hat{\mathbf{x}}_k^-$ rather than $\mathbf{z}_k$. The measurement residual can then be written as

$$\text{Measurement residual} = \left(\mathbf{z}_k - \hat{\mathbf{z}}_k^-\right) \tag{9.2.3}$$

because the predictive estimate of the measurement is just the sum of $\mathbf{h}(\mathbf{x}_k^*)$ and $\mathbf{H}_k\Delta\hat{\mathbf{x}}_k^-$. Note that the measurement residual as given by Eq. (9.2.3) is formed exactly as would be done in an extended Kalman filter, that is, it is the noisy measurement minus the predictive measurement based on the corrected trajectory rather than the nominal one.

We now return to the update equation, Eq. (9.2.2), and add $\mathbf{x}_k^*$ to both sides of the equation:

$$\underbrace{\mathbf{x}_k^* + \Delta\hat{\mathbf{x}}_k}_{\hat{\mathbf{x}}_k} = \underbrace{\mathbf{x}_k^* + \Delta\hat{\mathbf{x}}_k^-}_{\hat{\mathbf{x}}_k^-} + \mathbf{K}_k(\mathbf{z}_k - \hat{\mathbf{z}}_k^-) \tag{9.2.4}$$

$$\hat{\mathbf{x}}_k = \hat{\mathbf{x}}_k^- + \mathbf{K}_k(\mathbf{z}_k - \hat{\mathbf{z}}_k^-) \tag{9.2.5}$$

Equation (9.2.5) is, of course, the familiar linear estimate update equation written in terms of *total* rather than incremental quantities. It simply says that we correct the a priori estimate by adding the measurement residual appropriately weighted by the Kalman gain $\mathbf{K}_k$. Note that after the correction is made in the extended Kalman filter, the incremental $\Delta\hat{\mathbf{x}}_k$ is reduced to zero. Its projection to the next step is then trivial. The only nontrivial projection is to project $\hat{\mathbf{x}}_k$ (which has become the nominal $\mathbf{x}$ at t_k) to $\hat{\mathbf{x}}_{k+1}^-$. This must be done through the nonlinear dynamics as dictated by Eq. (9.1.1). That is,

$$\hat{\mathbf{x}}_{k+1}^- = \left\{ \begin{array}{l} \text{Solution of the nonlinear differential equation} \\ \dot{\mathbf{x}} = \mathbf{f}(\mathbf{x}, \mathbf{u}_d, t) \text{ at } t = t_{k+1}, \text{ subject to the} \\ \text{initial condition } \mathbf{x} = \hat{\mathbf{x}}_k \text{ at } t_k \end{array} \right\} \qquad (9.2.6)$$

Note that the additive white noise forcing function $\mathbf{u}(t)$ is zero in the projection step, but the deterministic $\mathbf{u}_d$ is included in the $\mathbf{f}$ function. Once $\hat{\mathbf{x}}_{k+1}^-$ is determined, the predictive measurement $\hat{\mathbf{z}}_{k+1}^-$ can be formed as $\mathbf{h}(\hat{\mathbf{x}}_{k+1}^-)$, and the measurement residual at t_{k+1} is formed as the difference $(\mathbf{z}_{k+1} - \hat{\mathbf{z}}_{k+1}^-)$. The filter is then ready to go through another recursive loop.

For completeness, we repeat the familiar error covariance update and projection equations,

$$\mathbf{P}_k = (\mathbf{I} - \mathbf{K}_k\mathbf{H}_k)\mathbf{P}_k^- \qquad (9.2.7)$$

$$\mathbf{P}_{k+1}^- = \boldsymbol{\phi}_k\mathbf{P}_k\boldsymbol{\phi}_k^T + \mathbf{Q}_k \qquad (9.2.8)$$

where $\boldsymbol{\phi}_k$, $\mathbf{H}_k$, and $\mathbf{Q}_k$ come from the linearized model. Equations (9.2.7) and (9.2.8) and the gain equation (which is the same as in the linear Kalman filter) should serve as a reminder that the extended Kalman filter is still working in the world of linear dynamics, even though it keeps track of total estimates rather than incremental ones.

We can now relate the preceeding comments about keeping track of total estimates to the aided inertial navigation application. In the feedforward configuration of Fig. 9.5, the filter outputs are literally the incremental quantities, that is, the inertial system errors (and perhaps some of the aiding source errors). The total estimates are then reconstructed as an additive combination of nominal and incremental quantities. Thus, the preceeding remarks about keeping track of total estimates in the filter do not apply to the feedforward configuration. In this configuration the filter actually keeps track of incremental estimates explicitly. On the other hand, the feedback arrangement of Fig. 9.6 is an extended Kalman filter, and it normally keeps track of total estimates in this mode of operation. To do otherwise would be most awkward.

9.3
ADAPTIVE (SELF-LEARNING)
KALMAN FILTER

In the usual Kalman filter we assume that all the process parameters, that is, ϕ_k, $\mathbf{H}_k$, $\mathbf{R}_k$, and $\mathbf{Q}_k$, are known. They may vary with time (index k) but, if so, the nature of the variation is assumed to be known. In physical problems this is often a rash assumption. There may be large uncertainty in some parameters because of inadequate prior test data about the process. Or some parameter might be expected to change slowly with time, but the exact nature of the change is not predictable. In such cases, it is highly desirable to design the filter to be self-learning, so that it can adapt itself to the situation at hand, whatever that might be. This problem has received considerable attention since Kalman's original papers of the early 1960s. However, it is not an easy problem with one simple solution. This is evidenced by the fact that 30 years later we still see occasional papers on the subject in current control system journals.

We will concentrate our attention here on an adaptive filter scheme that was first presented by D. T. Magill (11). Also a more intuitive scheme is discussed in Problem 9.3. We will see presently that Magill's adaptive filter is not just one filter but, instead, is a whole bank of Kalman filters running in parallel. At the time that this scheme was first suggested in 1965, it was considered to be impractical for implementation on-line. However, the spectacular advances in computer technology over the past few decades have made Magill's parallel-filter scheme quite feasible in a number of applications (12, 13, 14, 15). Because of the parallel bank of filters, this scheme also goes under the name Multiple Model Estimation Algorithm (MMEA). We will now proceed to the derivation that leads to the bank of parallel filters.

We begin with the simple statement that the desired estimator is to be the conditional mean given by

$$\hat{\mathbf{x}}_k = \int_{\mathbf{x}} \mathbf{x}\, p(\mathbf{x}|\mathbf{z}_k^*)\, d\mathbf{x} \qquad (9.3.1)$$

where $\mathbf{z}_k^*$ denotes all the measurements up to and including time t_k (i.e., $\mathbf{z}_1, \mathbf{z}_2, \ldots, \mathbf{z}_k$), and $p(\mathbf{x}|\mathbf{z}_k^*)$ is the probability density function of $\mathbf{x}_k$ with the conditioning shown in parentheses.* The indicated integration is over the entire $\mathbf{x}$ space. If the $\mathbf{x}$ and $\mathbf{z}$ processes are Gaussian, we are assured that the estimate given by Eq. (9.3.1) will be optimal by almost any reasonable criterion of optimality, least-mean-square or otherwise (16). We also wish to assume that some parameter of the process, say α, is unknown to the observer, and that this

---

*Throughout this section we will use a looser notation than that used in Chapter 1 in that p will be used for both probability density and discrete probability. In this way we avoid the multitudinous subscripts that would otherwise be required for conditioned multivariate random variables. However, this means that the student must use a little imagination and interpret the symbol p properly within the context of its use in any particular derivation.

parameter is a random variable (not necessarily Gaussian). Thus, on any particular sample run it will be an unknown constant, but with a known statistical distribution. Hence, rather than beginning with $p(\mathbf{x}|\mathbf{z}_k^*)$, we really need to begin with the joint density $p(\mathbf{x}, \alpha|\mathbf{z}_k^*)$ and sum out out on α to get $p(\mathbf{x}|\mathbf{z}_k^*)$. Thus, we will rewrite Eq. (9.3.1) in the form

$$\hat{\mathbf{x}}_k = \int_\mathbf{x} \mathbf{x} \int_\alpha p((\mathbf{x}, \alpha)|\mathbf{z}_k^*)\, d\alpha\, d\mathbf{x} \qquad (9.3.2)$$

But the joint density in Eq. (9.3.2) can be written as

$$p(\mathbf{x}, \alpha|\mathbf{z}_k^*) = p(\mathbf{x}|\alpha, \mathbf{z}_k^*)p(\alpha|\mathbf{z}_k^*) \qquad (9.3.3)$$

Substituting Eq. (9.3.3) into (9.3.2) and interchanging the order of integration leads to

$$\hat{\mathbf{x}}_k = \int_\alpha p(\alpha|\mathbf{z}_k^*)\int_\mathbf{x} \mathbf{x}\, p(\mathbf{x}|\alpha, \mathbf{z}_k^*)\, d\mathbf{x}\, d\alpha \qquad (9.3.4)$$

The inner integral will be recognized as just the usual Kalman filter estimate for a given α. This is denoted as $\hat{\mathbf{x}}_k(\alpha)$ where α shown in parentheses is intended as a reminder that there is α dependence. Equation (9.3.4) may now be rewritten as

$$\hat{\mathbf{x}}_k = \int_\alpha \hat{\mathbf{x}}_k(\alpha)p(\alpha|\mathbf{z}_k^*)\, d\alpha \qquad (9.3.5)$$

Or the discrete random variable equivalent to Eq. (9.3.5) would be

$$\hat{\mathbf{x}}_k = \sum_{i=1}^{L} \hat{\mathbf{x}}_k(\alpha_i)p(\alpha_i|\mathbf{z}_k^*) \qquad (9.3.6)$$

where $p(\alpha_i|\mathbf{z}_k^*)$ is the discrete probability for α_i, conditioned on the measurement sequence $\mathbf{z}_k^*$. We will concentrate on the discrete form from this point on in our discussion.

Equation (9.3.6) simply says that the optimal estimate is a weighted sum of Kalman filter estimates with each Kalman filter operating with a separate assumed value of α. This is shown in Fig. 9.7. The problem now reduces to one of determining the weight factors $p(\alpha_1|\mathbf{z}_k^*)$, $p(\alpha_2|\mathbf{z}_k^*)$, etc. These, of course, change with each recursive step as the measurement process evolves in time. Presumably, as more and more measurements become available, we learn more about the state of the process and the unknown parameter α. (Note that it is constant for any particular sample run of the process.)

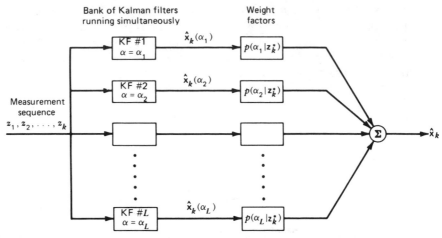

Figure 9.7.   Weighted sum of Kalman filter estimates.

We now turn to the matter of finding the weight factors indicated in Fig. 9.7. Toward this end we use Bayes' rule:

$$p(\alpha_i|\mathbf{z}_k^*) = \frac{p(\mathbf{z}_k^*|\alpha_i)p(\alpha_i)}{p(\mathbf{z}_k^*)} \tag{9.3.7}$$

But

$$p(\mathbf{z}_k^*) = \sum_{j=1}^{L} p(\mathbf{z}_k^*, \alpha_j)$$

$$= \sum_{j=1}^{L} p(\mathbf{z}_k^*|\alpha_j)p(\alpha_j) \tag{9.3.8}$$

Equation (9.3.8) may now be substituted into Eq. (9.3.7) with the result

$$p(\alpha_i|\mathbf{z}_k^*) = \left[\frac{p(\mathbf{z}_k^*|\alpha_i)p(\alpha_i)}{\displaystyle\sum_{j=1}^{L} p(\mathbf{z}_k^*|\alpha_j)p(\alpha_j)} \right], \qquad i = 1, 2, \ldots, L \tag{9.3.9}$$

The distribution $p(\alpha_i)$ is presumed to be known, so it remains to determine $p(\mathbf{z}_k^*|\alpha_i)$ in Eq. (9.3.9). Toward this end we will write $p(\mathbf{z}_k^*|\alpha_i)$ as a product of conditional density functions. Temporarily omitting the α_i conditioning (just to save writing) we have

$$p(\mathbf{z}_k^*) = p(\mathbf{z}_k, \mathbf{z}_{k-1}, \ldots \mathbf{z}_0)$$

$$= p(\mathbf{z}_k, \mathbf{z}_{k-1}, \ldots \mathbf{z}_1 | \mathbf{z}_0) p(\mathbf{z}_0)$$

$$= p(\mathbf{z}_k, \mathbf{z}_{k-1}, \ldots \mathbf{z}_2 | \mathbf{z}_1, \mathbf{z}_0) p(\mathbf{z}_1 | \mathbf{z}_0) p(\mathbf{z}_0)$$

$$\vdots$$

$$= p(\mathbf{z}_k | \mathbf{z}_{k-1}, \mathbf{z}_{k-2}, \ldots \mathbf{z}_0) p(\mathbf{z}_{k-1} | \mathbf{z}_{k-2}, \mathbf{z}_{k-3}, \ldots \mathbf{z}_0) \ldots p(\mathbf{z}_1 | \mathbf{z}_0) p(\mathbf{z}_0).$$

$$k = 1, 2, \ldots \quad (9.3.10)$$

We now note that the first term in the product string of Eq. (9.3.10) is just $p(\hat{\mathbf{z}}_k^-)$, and that the remaining product is just $p(\mathbf{z}_{k-1}^*)$. Thus we can rewrite Eq. (9.3.10) in the form

$$p(\mathbf{z}_k^*) = p(\hat{\mathbf{z}}_k^-) p(\mathbf{z}_{k-1}^*) \quad (9.3.11)$$

We now make the Gaussian assumption for the $\mathbf{x}$ and $\mathbf{z}$ processes (but not for α). Also, to simplify matters, we will assume $\mathbf{z}_k^*$ to be a sequence of scalar measurements $z_0, z_1, \ldots z_k$. Equation (9.3.11) then becomes

$$p(\mathbf{z}_k^*) = \frac{1}{(2\pi)^{1/2} (\mathbf{H}_k \mathbf{P}_k^- \mathbf{H}_k^T + R_k)^{1/2}} \exp\left[-\frac{1}{2} \frac{(z_k - \mathbf{H}_k \hat{\mathbf{x}}_k^-)^2}{(\mathbf{H}_k \mathbf{P}_k^- \mathbf{H}_k^T + R_k)} \right] p(\mathbf{z}_{k-1}^*),$$

$$k = 1, 2, \ldots \quad (9.3.12)$$

Bear in mind that $p(\mathbf{z}_k^*)$ will, in general, be different for each α_i. For example, if the unknown parameter is R_k, each filter in the bank of filters will be modeled around a different value for R_k.

It should be helpful at this point to go through an example step by step (in words, at least) to see how the parallel bank of filters works.

1. We begin with the prior distribution of α and set the filter weight factors accordingly. Frequently, we have very little prior knowledge of the unknown parameter α. In this case we would assume a uniform probability distribution and set all the weight factors equal initially. This does not have to be the case in general, though.

2. The initial prior estimates $\hat{\mathbf{x}}_0^-$ for each filter are set in accordance with whatever prior information is available. Usually, if $\mathbf{x}$ is a zero-mean process, $\hat{\mathbf{x}}_0^-$ is simply set equal to zero for each filter. We will assume that this is true in this example.

3. Usually, the initial estimate uncertainty does not depend on α, so the initial $\mathbf{P}_0^-$ for each filter will just be the covariance matrix of the $\mathbf{x}$ process.

4. Initially then, before any measurements are received, the prior estimates from each of the filters are weighted by the initial weight factors and summed to give the optimal prior estimate from the bank of filters. In the present example this is trivial, because the initial estimates for each filter were set to zero.

5. At $k = 0$ the bank of filters receives the first measurement z_0, and the unconditional $p(z_0)$ must be computed for each permissible α_i. We note that $z = \mathbf{H}\mathbf{x} + v$, so $p(z_0)$ may be written as

$$p(z_0) = \frac{1}{(2\pi)^{1/2}\left(\mathbf{H}_0\mathbf{C}_x\mathbf{H}_0^T + R_0\right)^{1/2}} \exp\left[-\frac{1}{2}\frac{z_0^2}{\left(\mathbf{H}_0\mathbf{C}_x\mathbf{H}_0^T + R_0\right)}\right]$$

$$(9.3.13)$$

where $\mathbf{C}_x$ is the covariance matrix of $\mathbf{x}$.

We note again that one or more of the parameters in Eq. (9.3.13) may have α dependence; thus, in general, $p(z_0)$ will be different for each α_i.

6. Once $p(z_0)$ for each α_i has been determined, Eq. (9.3.9) may be used to find $p(\alpha_i|z_0)$. These are the weight factors to be used in summing the updated estimates $\hat{\mathbf{x}}_0$ that come out of each of the filters in the bank of filters. This then yields the optimal adaptive estimate, given the measurement z_0, and we are ready to project on to the next step.

7. Each of the individual Kalman filter estimates and their error covariances are projected ahead to $k = 1$ in the usual manner. The adaptive filter must now compute $p(\mathbf{z}_1^*)$ for each α_i, and it uses the recursive formula, Eq. (9.3.12), in doing so. Therefore, for $p(\mathbf{z}_1^*)$ we have

$$p(\mathbf{z}_1^*) = \frac{1}{(2\pi)^{1/2}\left(\mathbf{H}_1\mathbf{P}_1^-\mathbf{H}_1^T + R_1\right)^{1/2}} \exp\left[-\frac{1}{2}\frac{(z_1 - \mathbf{H}_1\hat{\mathbf{x}}_1^-)^2}{\left(\mathbf{H}_1\mathbf{P}_1^-\mathbf{H}_1^T + R_1\right)}\right]p(\mathbf{z}_0^*)$$

$$(9.3.14)$$

Note that $p(\mathbf{z}_0^*)$ was computed in the previous step, and the prior $\hat{\mathbf{x}}_1^-$ and $\mathbf{P}_1^-$ for each α_i are obtained from the projection step.

8. Now, the $p(\mathbf{z}_1^*)$ determined in step 7 can be used in Bayes' formula, Eq. (9.3.9), and the weight factors $p(\alpha_i|\mathbf{z}_1^*)$ for $k = 1$ are thus determined. It should be clear now that this recursive procedure can be carried on ad infinitum.

We are now in a position to reflect on the whole adaptive filter in perspective. At each recursive step the adaptive filter does three things: (1) each filter in the bank of filters computes its own estimate, which is hypothesized on its own model; (2) the system computes the a posteriori probabilities for each of the hypotheses; and (3) the scheme forms the adaptive optimal estimate of **x** as a weighted sum of the estimates produced by each of the individual Kalman filters. As the measurements evolve with time, the adaptive scheme learns which of the filters is the correct one, and its weight factor approaches unity while the others are going to zero. The bank of filters accomplishes this, in effect, by looking at sums of the weighted squared measurement residuals. The filter with the smallest residuals "wins," so to speak.

The Magill scheme just described is not without some problems and limitations, but it is still important because it is optimum (within the various assumptions that were made), and it serves as a point of departure for other less rigorous schemes. One of the problems of this technique has to do with numerical behavior as the number of steps becomes large. Clearly, as the sum of the squared measurement residuals becomes large, there is a possibility of computer underflow. Also, note that the unknown parameter was assumed to be constant with time. Thus, there is no way this kind of adaptive filter can readjust if the parameter actually varies slowly with time. This adaptive scheme, in its purest form, never forgets; it tends to give early measurements just as much weight as later ones in its determination of the a posteriori probabilities. Some ad hoc procedure, such as periodic reinitialization, has to be used if the scheme is to adapt to a slowly varying parameter situation.

By its very nature, the Magill adaptive filter is a transient scheme, and it converges to the correct α_i in an optimal manner. This is one of the scheme's strong points, and it gives rise to a class of applications that are usually not thought of as adaptive filter problems. These are applications where we are more interested in the a posteriori probabilities than in the optimal estimate of the vector **x** process. The Magill scheme is an excellent multiple-hypothesis testor in the presence of Gauss–Markov noise. In this setting the main objective of the bank of filters is to determine which hypothesized model in the bank of filters is the correct one, and there have been a number of applications of this type reported in the recent literature (13, 14, 15). We will discuss one of these briefly at the end of this section. Before doing so, though, we wish to emphasize the assumptions that were used in deriving Magill's parallel filter scheme. These assumptions are most important in understanding the limitations of the scheme. The key assumptions are as follows:

1. We assume Gaussian statistics for the **x** and **z** processes, but not for α.

2. The unknown parameter α is assumed to be a discrete random variable with a known distribution. Thus, for any sample run, α will be constant with time. If, in fact, α is an unknown *deterministic* parameter (in contrast to being random), then we must be cautious about our conclusions relative of optimality. (See Example 6.6, Chapter 6, for further discussion of this.)

3. It is tacitly assumed that the process and measurement models are of the proper form for a linear Kalman filter for each allowable α_i. (Note

that finite-order state models will not exist for all random processes—band-limited white noise being a good example.)

We will now look at an application of the Magill adaptive filter where both the hypothesis and the optimal estimate are of interest.

EXAMPLE 9.3

We will now return to the power system relaying application that was discussed in Section 6.5. Recall that the problem there was one of estimating post-fault currents and voltages on a transmission line. The objective of the Kalman filter was to estimate the steady-state components based on a short observation span during the transient period immediately after the fault occurrence. We will now look at fault classification, which is a different facet of the overall power system relaying problem.

Classification of a fault on a three-phase transmission line within a few milliseconds after the fault is not as simple as it may seem at first glance. This is because a fault on any one line induces an imbalance that reflects into the unfaulted phases as well as the faulted one. However, the statistics of the voltage noises after the fault are significantly different for the faulted and unfaulted phases. For example, Girgis and Brown (13) found by extensive simulation that the measurement R_k parameters for the two cases could be approximated by

$$R_k(\text{faulted phase}) = 0.6e^{-k\Delta t/T_1}$$

$$R_k(\text{unfaulted phase}) = 0.1e^{-k\Delta t/T_1}$$

where R_k is in units of (per unit voltage)^2. Clearly, this is a significant difference. The measurement-noise differences, along with differences in initial conditions, can then be used to discriminate between the faulted and unfaulted conditions. A block diagram for accomplishing this with a Magill adaptive filter is shown in Fig. 9.8. In the scheme described by Girgis and Brown (13) there would be three such adaptive filters, one for each of the three phases. The results coming out of each of the filters are then combined to determine the type of fault (e.g., line a to ground, line b to ground, and so forth) and the optimal voltage estimates.

Most of the details of the referenced scheme will be omitted here because they are adequately covered in the reference. We will simply say that in all of the simulations conducted, the adaptive scheme worked quite well in identifying the faulted phase (or phases); and in all instances a firm identification was made within about 2 msec. Figure 9.9 shows the probability calculations for a typical run with a three-phase fault applied to the line. Only the probabilities associated with the "correct" Kalman filters are shown in the figure. The corresponding probabilities for the no-fault filters would just be the complements of the ones shown. Note that the plots for each of the phases are not the same, even

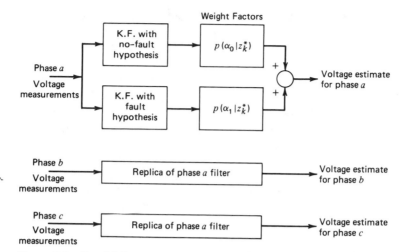

Figure 9.8. Parallel filter scheme for three-phase fault classification.

Figure 9.9. Probability profiles for the "correct" filter for each of the three phases.

though the fault is symmetric in this case. This is to be expected, though, because the instantaneous voltages on the three lines at the time of the fault are not the same. In effect, some of the resulting voltage profiles match the assumed fault model better than others; thus, the better the match, the faster the convergence. However, in spite of the differences in the probability responses for the three phases, all of them converge to the correct decision within 2 msec. This yields very fast fault classification.

Not only does the adaptive filter shown in Fig. 9.8 make a fast decision as to fault or no-fault, but it also estimates the post-fault voltage in the process. Of course, the estimate produced by the "winning" filter is the one of primary interest. After the decision is made as to the correct hypothesis, the other estimate can be ignored. The advantage of this type of operation over the nonadaptive scheme presented in Section 6.5 lies in the individualized modeling that is permitted in the adaptive scheme. There, each of the two filter models can be tailored to fit its own hypothesis, whereas in a nonadaptive scheme the model must be an all-purpose model that fits, to a degree at least, a wider class of situations. ■

9.4
CONVERSION OF CONTINUOUS MEASUREMENTS TO DISCRETE FORM — THE DELAYED STATE FILTER ALGORITHM

The continuous Kalman filter discussed in Chapter 7 is of considerable theoretical importance, but it would be a rare situation where one would actually implement the filter in analog form. Usually, in the end, the analyst/designer must make the transition to the discrete form, even though some (or perhaps all) of the measurements are continuous. Fortunately, the discrete filter equations are exact in their own right and are not approximations to the continuous case. Thus, there is no error in making the transition from the continuous to the discrete case relative to the recursive equations. However, in the transition the continuous measurements must be converted to equivalent sampled data (in some sense of equivalence), and there are subtleties associated with this conversion that warrant further comment.

To illustrate the problems that can arise in the continuous-to-discrete conversion, consider a scalar Gauss–Markov process whose spectral content is of the order of a few hertz. Let the additive noise be Gaussian and flat out to a few kilohertz. In order not to lose any signal information, it is tempting to sample the measurement as fast as possible, perhaps in the kilohertz range. However, this might impose unreasonable demands on the system computer. Certainly, we do not want to force the filter to cycle through millions of recursive steps when thousands would do just as well! On the other extreme, if we sample too slowly, say at less than the signal Nyquist rate, we should expect to lose signal information; and, as a result, the filter's performance will be poorer than need be. Thus, the sampling should be treated with some care. One technique that has been used successfully is to sample the measurement at a fairly high rate (if the frequency content of the noise so dictates), and then preaverage the samples over the Kalman-filter update interval to form a discrete measurement for the filter. In the example at hand, it might be reasonable to sample the continuous measurement at a rate of 1000 Hz and then preaverage 100-sample blocks to form "compressed" samples to present to the Kalman filter every .1 sec. The error variance associated with the compressed samples would then be less than that of the raw samples by a factor of 100, provided the raw sample errors were independent. Of course, the R_k parameter of the Kalman filter would be the variance associated with the compressed samples rather than that of the raw samples.

The preaverage technique just described works fine in many applications and nothing more sophisticated is required. However, there are cases where, for one reason or another, the Kalman filter must be relatively slow, and the resulting prefiltering interval is longer than desired. This presents problems for two reasons. First, some measurement information will be lost due to the low sampling rate. The loss may not be serious, though, if the sampling rate is not too low relative to the signal frequency content. A more serious problem may lie in the modeling inaccuracy that comes about from the assumption that the signal is constant over the averaging interval, when, in fact, it is not. In averaging, we integrate (or sum in the discrete case) over the update interval, and hence the compressed measurement presented to the filter is, in reality, of the form

$$\mathbf{z}_k = \int_{t_{k-1}}^{t_k} \left[\mathbf{Hx}(t) + \mathbf{v}(t) \right] dt \qquad (9.4.1)$$

If $\mathbf{x}(t)$ varies appreciably over the interval, the resultant $\mathbf{z}_k$ may be more representative of the value of $\mathbf{x}(t)$ at the midpoint than at the endpoint t_k, which is, of course, the assumption in the Kalman filter model. When this situation arises, there is a method of modifying the recursive equations to accommodate exactly to measurements of the form given by Eq. (9.4.1). This leads to the concept of the delayed-state filter algorithm, which we introduce by means of a simple example.

Assume for the moment that position and velocity are two of the state variables of the system. Suppose further that we have a continuous measurement of velocity that we wish to convert to discrete form (not an uncommon problem). Assume that we integrate (average) the noisy velocity measurement over the update interval. We then have

$$z_k = \int_{t_{k-1}}^{t_k} (\text{velocity} + \text{noise}) \, dt = (\text{position at } t_k)$$

$$- (\text{position at } t_{k-1}) + (\text{residual error}) \qquad (9.4.2)$$

Thus we find that z_k has a linear connection to both the present and previous states, rather than just the present state as assumed in the usual Kalman filter model. In more general terms, the measurement situation in this case is of the form

$$\mathbf{z}_k = \mathbf{H}_k \mathbf{x}_k + \mathbf{J}_k \mathbf{x}_{k-1} + \mathbf{v}_k \tag{9.4.3}$$

We now seek an appropriate set of recursive equations for the measurement model given by Eq. (9.4.3). As before, we assume $\mathbf{v}_k$ to be a white sequence with a covariance matrix $\mathbf{R}_k$. The desired solution is worked out in references 17 and 18 by combining the present and previous state vectors into a composite state vector with twice the dimension of the original ones. Here, we can accomplish the same result by a more direct path by taking advantage of the material in Chapter 5.

We begin with the estimate update equation

$$\hat{\mathbf{x}}_k = \hat{\mathbf{x}}_k^- + \mathbf{K}_k\big(\mathbf{z}_k - \mathbf{H}_k \hat{\mathbf{x}}_k^- - \mathbf{J}_k \hat{\mathbf{x}}_{k-1}\big) \tag{9.4.4}$$

Note that our best estimate of $\mathbf{z}_k$ prior to receiving the measurement is $\{\mathbf{H}_k \hat{\mathbf{x}}_k^- + \mathbf{J}_k \hat{\mathbf{x}}_{k-1}\}$, rather than just $\mathbf{H}_k \hat{\mathbf{x}}_k^-$ as in the usual Kalman filter. Next, we form the error and error covariance expressions just as was done in Chapter 5 [see Eqs. (5.4.9) to (5.4.11)].

$$\mathbf{e}_k = \mathbf{x}_k - \hat{\mathbf{x}}_k$$

$$= \mathbf{e}_k^- - \mathbf{K}_k(\mathbf{H}_k \mathbf{e}_k^- + \mathbf{J}_k \mathbf{e}_{k-1} + \mathbf{v}_k) \tag{9.4.5}$$

And

$$\mathbf{P}_k = E\big(\mathbf{e}_k \mathbf{e}_k^T\big)$$

$$= E\big\{\big[\mathbf{e}_k^- - \mathbf{K}_k(\mathbf{H}_k \mathbf{e}_k^- + \mathbf{J}_k \mathbf{e}_{k-1} + \mathbf{v}_k)\big]$$

$$\times \big[\mathbf{e}_k^- - \mathbf{K}_k(\mathbf{H}_k \mathbf{e}_k^- + \mathbf{J}_k \mathbf{e}_{k-1} + \mathbf{v}_k)\big]^T\big\} \tag{9.4.6}$$

We now note that

$$\mathbf{e}_k^- = \mathbf{x}_k - \hat{\mathbf{x}}_k^-$$

$$= \boldsymbol{\phi}_{k-1} \mathbf{x}_{k-1} + \mathbf{w}_{k-1} - \boldsymbol{\phi}_{k-1} \hat{\mathbf{x}}_{k-1}$$

$$= \boldsymbol{\phi}_{k-1} \mathbf{e}_{k-1} + \mathbf{w}_{k-1} \tag{9.4.7}$$

We can also take advantage of the knowledge that $\mathbf{v}_k$ is uncorrelated with $\mathbf{e}_k^-$

and $\mathbf{e}_{k-1}$, and $\mathbf{w}_{k-1}$ is uncorrelated with $\mathbf{e}_{k-1}$. Then, expanding Eq. (9.4.6) and performing the indicated expectations leads to (after a modest amount of algebra)

$$\mathbf{P}_k = \mathbf{P}_k^- - \mathbf{K}_k\left(\mathbf{H}_k\mathbf{P}_k^- + \mathbf{J}_k\mathbf{P}_{k-1}\boldsymbol{\phi}_{k-1}^T\right) - \left(\mathbf{P}_k^-\mathbf{H}_k^T + \boldsymbol{\phi}_{k-1}\mathbf{P}_{k-1}\mathbf{J}_k^T\right)\mathbf{K}_k^T$$

$$+ \mathbf{K}_k\left(\mathbf{H}_k\mathbf{P}_k^-\mathbf{H}_k^T + \mathbf{R}_k + \mathbf{J}_k\mathbf{P}_{k-1}\mathbf{J}_k^T + \mathbf{H}_k\boldsymbol{\phi}_{k-1}\mathbf{P}_{k-1}\mathbf{J}_k^T\right.$$

$$\left. + \mathbf{J}_k\mathbf{P}_{k-1}\boldsymbol{\phi}_{k-1}^T\mathbf{H}_k^T\right)\mathbf{K}_k^T \quad (9.4.8)$$

Equation (9.4.8) may now be compared with Eq. (5.4.15). Clearly, the forms are identical with the following replacements:

$$\mathbf{H}\mathbf{P}^- \rightarrow \mathbf{H}_k\mathbf{P}_k^- + \mathbf{J}_k\mathbf{P}_{k-1}\boldsymbol{\phi}_{k-1}^T \qquad (9.4.9)$$

$$(\mathbf{H}\mathbf{P}^-\mathbf{H}^T + \mathbf{R}) \rightarrow \mathbf{H}_k\mathbf{P}_k^-\mathbf{H}_k^T + \mathbf{R}_k + \mathbf{J}_k\mathbf{P}_{k-1}\mathbf{J}_k^T + \mathbf{H}_k\boldsymbol{\phi}_{k-1}\mathbf{P}_{k-1}\mathbf{J}_k^T$$

$$+ \mathbf{J}_k\mathbf{P}_{k-1}\boldsymbol{\phi}_{k-1}^T\mathbf{H}_k^T \qquad (9.4.10)$$

The same optimization arguments used in Chapter 5 may be used here, and hence the gain expression for the delayed-state algorithm is obtained simply by making the appropriate replacements in the usual Kalman gain formula. The result is

$$\mathbf{K}_k = \left(\mathbf{P}_k^-\mathbf{H}_k^T + \boldsymbol{\phi}_{k-1}\mathbf{P}_{k-1}\mathbf{J}_k^T\right)\left[\mathbf{H}_k\mathbf{P}_k^-\mathbf{H}_k^T + \mathbf{R}_k + \mathbf{J}_k\mathbf{P}_{k-1}\mathbf{J}_k^T\right.$$

$$\left. + \mathbf{H}_k\boldsymbol{\phi}_{k-1}\mathbf{P}_{k-1}\mathbf{J}_k^T + \mathbf{J}_k\mathbf{P}_{k-1}\boldsymbol{\phi}_{k-1}^T\mathbf{H}_k^T\right]^{-1}$$

$$(9.4.11)$$

The error covariance update expression may be written in the number of different forms, just as in the usual Kalman filter. Perhaps the simplest form for the delay-state algorithm is

$$\mathbf{P}_k = \mathbf{P}_k^- - \mathbf{K}_k\mathbf{L}_k\mathbf{K}_k^T \qquad (9.4.12)$$

where $\mathbf{L}_k$ is the bracketed term of Eq. (9.4.11) (not inverted). Since both $\mathbf{K}_k$ and $\mathbf{L}_k$ are computed in the gain computation step, Eq. (9.4.12) is easy to evaluate. The projection equations are the same as for the usual filter and need not be repeated.

Examples illustrating the use of the delayed state algorithm in aided inertial navigation systems are given in references 17, 18, and 19. In these examples position and velocity were "natural" system state variables. Thus measurements that were related to the integral of velocity fit the format of Eq.

(9.4.3) directly without any modification of the process model. This is not always the case, though, as will be seen presently. The referenced navigation examples are interesting practical applications, but they are somewhat lengthy; therefore, we will look at a simpler example here. We consider a scalar Markov process with preaveraged samples as the measurement sequence.

EXAMPLE 9.4

Consider a scalar Gauss–Markov process $x(t)$ with an autocorrelation function

$$R_x(\tau) = e^{-|\tau|} \tag{9.4.13}$$

From previous work in Chapter 5 (Example 5.2), we know that this process satisfies a first-order differential equation with white noise as the driving function. Suppose now that we have a continuous noisy measurement of $x(t)$ with the additive noise being Gaussian and white with a spectral amplitude of .02. The continuous Kalman filter model is then (see Chapter 7)

$$\dot{x} = -x + \sqrt{2}\,u(t), \qquad (Q = 1) \tag{9.4.14}$$

$$z = x + v(t), \qquad (R = .02) \tag{9.4.15}$$

Suppose further that we wish to implement the filter in discrete form (rather than continuous), and that the filter recursive cycle must be relatively slow in order not to impose undue hardship on the system computer (along with its other work load). Since the process has a time constant of unity, it would seem reasonable to average the measurement data over intervals of one-tenth of the time constant and then present the discrete filter with a sequence of preaveraged data at update intervals of .1 sec. In this case we know there will be some loss in measurement information due to the preaveraging; the question is, how much loss? For comparison purposes we examine three possibilities:

1. The continuous Kalman filter. This baseline system represents the best we could hope to do with an infinite sampling and update rate.
2. Optimal operation on the preaveraged data using the delayed-state Kalman filter algorithm. (There will be some degradation here as compared with the continuous case, but at least the sequence of compressed measurements will be processed optimally.)
3. The usual discrete Kalman filter, which processes the preaveraged data just as if they were truly instantaneous samples of $x(t)$ at the update times $t_1, t_2, \ldots, t_k$. (Here we have to contend with a faulty measurement model as well as preaveraged data.)

The Continuous Kalman Filter This example is similar to the one worked out in Example 7.1. Since the R parameter is the only element that is different in these two examples, the pair of linear equations to be solved here are

$$\dot{X} = -X + 2Z, \qquad X(0) = 1 \tag{9.4.16}$$

$$\dot{Z} = 50X + Z, \qquad Z(0) = 1 \tag{9.4.17}$$

These equations may be solved using Laplace transform methods. The result is

$$P(t) = \frac{X(t)}{Z(t)} \approx \frac{.450248e^{-\sqrt{101}\,t} + .549752e^{\sqrt{101}\,t}}{-2.037345e^{-\sqrt{101}\,t} + 3.037345e^{\sqrt{101}\,t}} \qquad (9.4.18)$$

In our comparison with the discrete models, we will be particularly interested in the error variance after a single step of .1 sec. Evaluation of Eq. (9.4.18) at $t = .1$ yields

$$P(.1) \approx .220696 \qquad (9.4.19)$$

Delayed-State Kalman Filter Model   The compressed measurement in this case is related to the integral of $x(t)$ because of the averaging over the Δt interval:

$$z_k = \frac{1}{\Delta t} \int_{t_{k-1}}^{t_k} [x(t) + v(t)]\, dt, \qquad \Delta t = .1 \text{ and}$$

$$k = 1, 2, \ldots n \qquad (9.4.20)$$

Since this does not fit the required format of the usual discrete Kalman filter, we modify the state model to make it amenable to solution with the delayed-state algorithm. This is done by defining the integral of $x(t)$ to be a new state variable. The augmented system will then have two state variables. They are defined as follows:

$$x_1(t) = \int_0^t x_2(\tau)\, d\tau, \qquad \text{(a new state variable)} \qquad (9.4.21)$$

$$x_2(t) = x(t), \qquad \text{(the original scalar Markov process)} \quad (9.4.22)$$

The details of the discrete model for this two-state process are given in Example 5.3. For a step size of .1 sec., the key parameters are

$$\phi_k \approx \begin{bmatrix} 1 & .0951626 \\ 0 & .9048374 \end{bmatrix} \qquad (9.4.23)$$

$$Q_k \approx \begin{bmatrix} .00061892 & .0090559 \\ .0090559 & .1812692 \end{bmatrix} \qquad (9.2.24)$$

The discrete measurement model is obtained by combining Eqs. (9.4.20), (9.4.21), and (9.4.22) for a step size of .1 sec. This yields

$$z_k = 10 \int_{t_{k-1}}^{t_k} x_2(\tau)\, d\tau$$

$$= 10x_1(t_k) - 10x_1(t_{k-1}) + 10 \int_{t_{k-1}}^{t_k} v(t)\, dt \qquad (9.4.25)$$

Clearly, this is of the form given by Eq. (9.4.3) where

$$\mathbf{H}_k = [10 \quad 0], \qquad \mathbf{J}_k = [-10 \quad 0] \tag{9.4.26}$$

and, for $\Delta t = .1$,

$$v_k = 10 \int_{t_{k-1}}^{t_{k-1}+.1} v(t)\, dt \tag{9.4.27}$$

Recall that

$$E[v(t)v(\tau)] = R\delta(t-\tau) = .02\delta(t-\tau) \tag{9.4.28}$$

We can now calculate the corresponding R_k in the discrete model.

$$R_k = E[v_k^2] = (10)^2 \int\int_{\Delta t \text{ interval}} E[v(t)v(\tau)]\, dt\, d\tau$$

$$= (10)^2(.02)(.1) = .2 \tag{9.4.29}$$

The discrete delayed-state model is now complete.

It is somewhat laborious to carry the delayed-state solution through a large number of steps numerically. We can, however, proceed through one step with only modest effort, and this will provide a limited comparison with the other models being considered. Thus consider the initial step from 0 to .1 sec. The first preaveraged measurement is not available until $t = .1$, and so we simply project the initial error covariance matrix ahead in accordance with

$$\mathbf{P}_1^- = \boldsymbol{\phi}_k \mathbf{P}_0 \boldsymbol{\phi}_k^T + \mathbf{Q}_k \tag{9.4.30}$$

where $\boldsymbol{\phi}_k$ and $\mathbf{Q}_k$ are given by Eqs. (9.4.23) and (9.4.24), and the initial $\mathbf{P}_0$ is

$$\mathbf{P}_0 = \begin{bmatrix} 0 & 0 \\ 0 & 1 \end{bmatrix} \tag{9.4.31}$$

(Note that x_1 is zero at $t = 0$ because of the definition of x_1, and it is known perfectly.) Numerical evaluation of Eq. (9.4.30) yields

$$\mathbf{P}_1^- \approx \begin{bmatrix} .0096748 & .095163 \\ .095163 & 1.0 \end{bmatrix} \tag{9.4.32}$$

It is now a routine matter to substitute into Eqs. (9.4.11) and (9.4.12) to find the gain and updated $\mathbf{P}_1$ matrix. The results are

$$\mathbf{K}_1 \approx \begin{bmatrix} \dfrac{.096748}{.96748 + .2} \\[2mm] \dfrac{.95163}{.96748 + .2} \end{bmatrix} \tag{9.4.33}$$

and

$$\mathbf{P}_1 \approx \begin{bmatrix} \overline{} & \overline{.22432} \\ \overline{} & \end{bmatrix} \tag{9.4.34}$$

Only the "22" term of the $\mathbf{P}_1$ matrix is shown, because it is the only variance of interest. Note that it is only slightly greater than the optimal value of .22069 obtained with the continuous filter.

Usual Kalman Filter with Only One State Variable We will now treat the averaged measurements just as if they were truly instantaneous measurements occurring at the sample points $t = .1, .2, \ldots$, etc. (See the sample function shown in Fig. 5.12 to get some idea as to the grossness of this approximation.) The additive noise variance R_k is .2 just as in the delayed-state model. Even though we know the measurement model is not exactly correct, it is still of interest to compute the resulting gain and associated P at $t = .1$. The process model parameters for this case are

$$\phi_k = e^{-.1} \approx .9048374$$

$$Q_k = 1 - e^{-.2} \approx .1812692$$

The a priori P at $t = .1$ is then

$$P_1^- = \phi_k P_0 \phi_k^T + Q_k = 1.0 \tag{9.4.35}$$

(which was more or less obvious at the outset). The gain and a posteriori P_1 are obtained from the usual Kalman filter equations with H_k set equal to 1. Their numerical values work out to be

$$K_1 = \frac{P_1^-}{P_1^- + R_1} = \frac{1.0}{1.0 + .2} \approx .83333 \tag{9.4.36}$$

$$P_1 = (1 - KH)P^- \approx .16667 \tag{9.4.37}$$

It is of interest to note that the gain of .83333 is not radically different from the correct gain of .815115 found in the delayed-state model. However, the computed P_1 of .16667 differs considerably from the optimal variance of .22432 computed previously. Thus, the oversimplified one-state model yields an unduly optimistic figure of merit for the estimation error.

Note that in this case we have intentionally refrained from referring to P as the error variance. It is not! It is, in fact, an erroneous figure of merit because of the rather gross approximation made in the measurement model. However, the fact that the gain worked out to be about "right" makes one think that the filter (with its admittedly erroneous model) might still be a reasonably good estimator. The delayed-state model provides a convenient means of assessing the degree of suboptimality in this case. All we need to do is let $\mathbf{K}_k$ in Eq. (9.4.8) be

$$\mathbf{K}_1 = \begin{bmatrix} 0 \\ .83333 \end{bmatrix} \tag{9.4.38}$$

This is the effective suboptimal gain produced by the one-state model. Equation (9.4.8) then yields the actual covariance realized by using the suboptimal gain in the "truth model." The result is

$$P_1(\text{actual}) = \begin{bmatrix} - & - \\ - & .22470 \end{bmatrix} \tag{9.4.39}$$

Surprisingly, the resultant value of .22470 is only slightly larger than the optimal value of .22432 (which, in turn, was only slightly larger than .22070 for the continuous filter).

There are two main results of this brief analysis:

1. The P_k computation of the simplified one-state model yields an overly optimistic estimate of the filter's accuracy.
2. However, in spite of the modeling errors, the one-state filter's performance is nearly as good as the optimal two-state filter, which uses the delayed-state measurement model. Roughly speaking, the P computation is bad, but the estimate is still fairly good.

The preceding conclusions are based on the results of just one recursive step. In order to be confident of these results, we should cycle through the recursive equations until the steady-state situation is reached. This is routine, but it involves considerable effort, so we will leave the example at this point. ■

It should be clear from this example that the delayed-state model is useful as an analysis tool in assessing the degree of suboptimality caused by compressing the measurement data, even if the delayed-state algorithm is not actually implemented on-line. Without this kind of suboptimality analysis, the effect of preaveraging the continuous measurement data has to be guesswork. With the analysis, one can compute quantitative measures of the degree of suboptimality introduced by averaging over any specified interval.

9.5
U-D FACTORIZATION

Computational problems associated with Kalman filtering were mentioned briefly in Chapter 6. Such problems are probably not as worrisome today as they were in the early 1960s because of the spectacular progress in computer technology.

Also, problems of divergence are better understood now than they were in the early days of Kalman filtering. Even so, there are occasional applications where roundoff errors can be a problem, and one must take all possible precautions against divergence. A class of Kalman filter algorithms known as square-root filtering have been developed, and they have somewhat better numerical behavior than the "usual" algorithm given in Chapter 5. (At least, with low-precision arithmetic this is so.) The basic idea is to propagate something analogous to $\sqrt{\mathbf{P}}$ (standard deviation) rather than $\mathbf{P}$ (variance). One of the critical situations is where the elements of $\mathbf{P}$ go through an extremely wide dynamical range in the course of the filter operation. For example, if the dynamical range of $\mathbf{P}$ is 20 orders of magnitude, the corresponding range of $\sqrt{\mathbf{P}}$ will be 10 orders of magnitude; and it does not take much imagination to see that we would be better off, numerically, manipulating $\sqrt{\mathbf{P}}$ rather than $\mathbf{P}$.

Both Bierman (20) and Maybeck (21) give good accounts of the development of square-root filtering, which apparently dates back to a 1964 paper by J. E. Potter [see Battin (22)]. We will be content here to look at just the **U-D** factorization algorithm, which is due to Bierman (20). It is now the favored algorithm in applications where numerical stability is of special concern. We will see presently that, technically, it is not really a square-root algorithm, because a square-rooting operation never appears anywhere in the algorithm. Nevertheless, it is usually considered in the square-root class because of the factorization of the **P** matrix. Our treatment of **U-D** factorization will be brief. The reader can consult either Bierman (20) or Maybeck (21) for more details. Maybeck gives an especially readable treatment of the subject.

We begin by defining what is meant by **U-D** factorization. Suppose we have a symmetric, positive-definite **P** matrix. We assert that it can always be decomposed into the factored form

$$\mathbf{P} = \mathbf{U}\mathbf{D}\mathbf{U}^T \tag{9.5.1}$$

where **U** is upper triangular (ones along the major diagonal, nontrival elements in the upper triangular part, and zeros in the lower triangular part), and **D** is diagonal. It might seem at first glance that carrying out the mechanics of the factorization would be difficult. However, this is not so. A simple (3×3) example will illustrate how easy it is to do the decomposition. We repeat Eq. (9.5.1) for a symmetric (3×3) **P** matrix.

$$\mathbf{P} = \begin{bmatrix} p_{11} & p_{12} & p_{13} \\ p_{12} & p_{22} & p_{23} \\ p_{13} & p_{23} & p_{33} \end{bmatrix} = \begin{bmatrix} 1 & u_{12} & u_{13} \\ 0 & 1 & u_{23} \\ 0 & 0 & 1 \end{bmatrix} \begin{bmatrix} d_{11} & 0 & 0 \\ 0 & d_{22} & 0 \\ 0 & 0 & d_{33} \end{bmatrix} \begin{bmatrix} 1 & 0 & 0 \\ u_{12} & 1 & 0 \\ u_{13} & u_{23} & 1 \end{bmatrix}$$

$$= \begin{bmatrix} d_{11} + d_{22}u_{12}^2 + d_{33}u_{13}^2 & d_{22}u_{12} + d_{33}u_{13}u_{23} & d_{33}u_{13} \\ d_{22}u_{12} + d_{33}u_{13}u_{23} & d_{22} + d_{33}u_{23}^2 & d_{33}u_{23} \\ d_{33}u_{13} & d_{33}u_{23} & d_{33} \end{bmatrix} \tag{9.5.2}$$

The decomposition problem is to determine d_{11}, d_{22}, d_{33}, u_{12}, u_{13}, and u_{23}, hopefully without having to solve simultaneous equations. Note that if we begin at the lower right corner, we can immediately write d_{33} as

$$d_{33} = p_{33} \tag{9.5.3}$$

Now, by inspection of the third column we can see that once we have determined d_{33}, we can write u_{13} and u_{23} as

$$u_{13} = \frac{p_{13}}{d_{33}} \quad \left(\text{or } \frac{p_{13}}{p_{33}} \right) \tag{9.5.4}$$

$$u_{23} = \frac{p_{23}}{d_{33}} \quad \left(\text{or } \frac{p_{23}}{p_{33}} \right) \tag{9.5.5}$$

Next, we continue by going to the 22 term and determine d_{22} from the equation

$$d_{22} = p_{22} - d_{33}u_{23}^2 \tag{9.5.6}$$

Finally, in a similar manner we go on to the 12 term and determine u_{12}, and then to the 11 term to find d_{11}, and all of this without solving any simultaneous equations or performing any square-root operations. It suffices to say here that this procedure can be automated in the form of an algorithm for computer use, and it is relatively easy to decompose $\mathbf{P}$ into its $\mathbf{U}$ and $\mathbf{D}$ factors. We will now go through the recursive equations, step by step, in terms of the $\mathbf{U}$-$\mathbf{D}$ factors.

We will begin with a priori $\mathbf{P}^-$ and will omit the k subscript just to save writing. We first factor $\mathbf{P}^-$ into its $\mathbf{U}$ and $\mathbf{D}$ parts.

$$\mathbf{P}^- = \mathbf{U}^- \mathbf{D}^- \mathbf{U}^{-T} \tag{9.5.7}$$

The Kalman gain is computed in the usual way, either as $\mathbf{P}^- \mathbf{H}^T (\mathbf{H}\mathbf{P}^- \mathbf{H}^T + \mathbf{R})^{-1}$ if $\mathbf{P}^-$ is used, or by the same formula with $\mathbf{P}^-$ replaced by $\mathbf{U}^- \mathbf{D}^- \mathbf{U}^{-T}$ if $\mathbf{P}^-$ is not available in reconstructed form from the previous step. After the gain is computed the estimate is updated by using the usual equation:

$$\hat{\mathbf{x}} = \hat{\mathbf{x}}^- + \mathbf{K}(\mathbf{z} + \mathbf{H}\hat{\mathbf{x}}^-) \tag{9.5.8}$$

Next, we need to update $\mathbf{P}^-$, and this is where we use $\mathbf{U}$-$\mathbf{D}$ factorization. We will begin with the $\mathbf{P}^-$ update form given by Eq. (5.4.20), which we repeat here for convenience.

$$\mathbf{P} = \mathbf{P}^- - \mathbf{P}^- \mathbf{H}^T (\mathbf{H}\mathbf{P}^- \mathbf{H}^T + \mathbf{R})^{-1} \mathbf{H}\mathbf{P}^- \tag{9.5.9}$$

We will assume that the measurement is scalar. (Note that we can always process the measurements sequentially in scalar form by forming appropriate linear combinations of the elements of **z**, if necessary, and thus diagonalize the **R** matrix. See Problem 6.4.) The $(\mathbf{HP}^-\mathbf{H}^T + \mathbf{R})$ term will then be scalar, and we will denote it as α, that is

$$\alpha \triangleq \mathbf{HP}^-\mathbf{H}^T + \mathbf{R} \tag{9.5.10}$$

Next, we will rewrite **P** and $\mathbf{P}^-$ in Eq. (9.5.9) in terms of their **U-D** factors. Let $\mathbf{P} = \mathbf{U}^+\mathbf{D}^+\mathbf{U}^{+T}$. Then

$$\mathbf{U}^+\mathbf{D}^+\mathbf{U}^{+T} = \mathbf{U}^-\mathbf{D}^-\mathbf{U}^{-T} - \frac{1}{\alpha}\mathbf{U}^-\mathbf{D}^-\mathbf{U}^{-T}\mathbf{H}^T\mathbf{H}\mathbf{U}^-\mathbf{D}^-\mathbf{U}^{-T}$$

$$= \mathbf{U}^-\left[\mathbf{D}^- - \frac{1}{\alpha}(\mathbf{D}^-\mathbf{U}^{-T}\mathbf{H}^T)(\mathbf{D}^-\mathbf{U}^{-T}\mathbf{H}^T)^T\right]\mathbf{U}^{-T} \tag{9.5.11}$$

Now note that the bracketed term in Eq. (9.5.11) is symmetric, so it can be factored into **U-D** factors. Let

$$\left[\mathbf{D}^- - \frac{1}{\alpha}(\mathbf{D}^-\mathbf{U}^{-T}\mathbf{H}^T)(\mathbf{D}^-\mathbf{U}^{-T}\mathbf{H}^T)\right] = \overline{\mathbf{U}}\,\overline{\mathbf{D}}\,\overline{\mathbf{U}}^T \tag{9.5.12}$$

Then, Eq. (9.5.11) can be rewritten as

$$\mathbf{U}^+\mathbf{D}^+\mathbf{U}^{+T} = \mathbf{U}^-\overline{\mathbf{U}}\,\overline{\mathbf{D}}\,\overline{\mathbf{U}}^T\mathbf{U}^{-T}$$

$$= (\mathbf{U}^-\overline{\mathbf{U}})\overline{\mathbf{D}}(\mathbf{U}^-\overline{\mathbf{U}})^T \tag{9.5.13}$$

Now note that $(\mathbf{U}^-\overline{\mathbf{U}})$ is upper triangular and $\overline{\mathbf{D}}$ is diagonal. Therefore, Eq. (9.5.13) is in **U-D** form and thus

$$\mathbf{U}^+ = \mathbf{U}^-\overline{\mathbf{U}} \tag{9.5.14}$$

$$\mathbf{D}^+ = \overline{\mathbf{D}} \tag{9.5.15}$$

Equations (9.5.14) and (9.5.15) then provide a means of updating the error covariance, but in terms of its **U-D** factors rather than in terms of $\mathbf{P}^-$.

It now remains to project the estimate and $\mathbf{U}^+$ and $\mathbf{D}^+$ ahead to the next step. The estimate is projected ahead as usual (with the subscripts reinserted).

$$\hat{\mathbf{x}}_{k+1}^- = \boldsymbol{\phi}_k\hat{\mathbf{x}}_k \tag{9.5.16}$$

The $\mathbf{U}^+$ and $\mathbf{D}^+$ factors can be projected ahead in a number of ways. The easiest way is to write the usual $\mathbf{P}$ projection equation as

$$\mathbf{P}_{k+1}^- = \boldsymbol{\phi}_k \mathbf{P}_k \boldsymbol{\phi}_k^T + \mathbf{Q}_k$$

$$= \boldsymbol{\phi}_k \mathbf{U}_k^+ \mathbf{D}_k^+ \mathbf{U}_k^{+T} \boldsymbol{\phi}_k^T + \mathbf{Q}_k$$

$$= (\boldsymbol{\phi}_k \mathbf{U}_k^+) \mathbf{D}_k^+ (\boldsymbol{\phi}_k \mathbf{U}_k^+)^T + \mathbf{Q}_k \qquad (9.5.17)$$

This computation is made easily, and it is especially efficient if $\boldsymbol{\phi}_k$ is upper triangular, or nearly so. (Integrators in cascade lead to an upper triangular transition matrix.) Also, the computation specified by Eq. (9.5.17) provides $\mathbf{P}_{k+1}^-$, which can then be used directly in the gain step at t_{k+1}. However, this method of projection begs the issue somewhat, because $\mathbf{U}^+$ and $\mathbf{D}^+$ have not been projected ahead individually. Rather, $\mathbf{P}_{k+1}^-$ has been reconstructed in the projection process. This technique has been used in equipment, though, and it is justified in applications where $\mathbf{Q}_k$ is "comfortably" positive-definite. This is to say, equivalently, that there is a substantial amount of process noise being added to each of the state variables in the projection step. This, in turn, adds to the estimation uncertainty in all the state variables. In such cases, the $\mathbf{P}$-update step will be the critical one, and this is where the $\mathbf{U}$-$\mathbf{D}$ factorization does the most good, so to speak.

If, for some reason, it is necessary to project $\mathbf{U}^+$ and $\mathbf{D}^+$ ahead individually and gain the corresponding numerical benefits, there are ways of doing so. Maybeck (21) gives one method that is straightforward, but it is somewhat more complicated and involves more computational effort than reconstructing $\mathbf{P}_{k+1}^-$ as given by Eq. (9.5.17). We will not pursue the matter further here other than to say that either Maybeck (21) or Bierman (20) may be consulted for methods of handling the projection step in terms of the $\mathbf{U}$-$\mathbf{D}$ factors.

We have one final comment before leaving the subject of $\mathbf{U}$-$\mathbf{D}$ factorization. Do not be unduly alarmed by the preceding remarks about numerical divergence. In many applications where the system is completely observable and there is adequate process noise feeding into all the system states, there is no divergence problem. This assumes, of course, that the programmer has taken the usual precaution of symmetrizing the error covariance matrix after both the projection and update steps. In most instances the usual update equation

$$\mathbf{P} = (\mathbf{I} - \mathbf{KH})\mathbf{P}^- \qquad (9.5.18)$$

is perfectly adequate and is, of course, easy to program. Therefore, do not be afraid of numerical stability in Kalman filtering. In rare situations it can be a problem but, even in these cases, there are methods of coping with it.

9.6
SCHMIDT – KALMAN FILTER. REDUCING
THE ORDER OF THE STATE VECTOR

In many real-life applications the size of the state vector can be a problem, and it becomes necessary to eliminate some of the elements to make the filter computationally feasible in real time. This is true for integrated navigation system applications, more often than not. Recall that when colored measurement noises are present, they must be absorbed as elements of the state vector to fit the required format for a Kalman filter (see Section 6.1). Estimates of these noise components are usually not of primary interest, so they immediately become candidates for elimination if the need arises. This cannot be done and maintain true optimality, but there is a way of at least partially compensating for the missing state variables in the resulting reduced-order filter. The technique that we will look at here was first suggested by S. F. Schmidt (23), and the resulting filter algorithm is often referred to as the Schmidt–Kalman filter. We will concentrate our attention here only on the case where the state variables to be eliminated are colored measurement noises.

We will begin the development of the Schmidt–Kalman filter by writing the total state vector for the truth model in partitioned form.

$$
\mathbf{X}_k = \left[-\frac{\mathbf{x}}{\mathbf{y}} - \right]_k =
\begin{bmatrix}
x_1 \\
x_2 \\
\vdots \\
x_n \\
\hline
y_1 \\
y_2 \\
\vdots \\
y_r
\end{bmatrix}_k
\left.\begin{array}{l}
\\ \\ \\ \\
\end{array}\right\} \text{Portion of the state vector to be retained and estimated}
\quad\quad
\left.\begin{array}{l}
\\ \\ \\
\end{array}\right\} \text{Part due to colored measurement noises which is to be eliminated}
\tag{9.6.1}
$$

Next, we write the process and measurement equations for the truth model in partitioned form.

Process model:

$$
\left[\begin{array}{c} \mathbf{x} \\ \hline \mathbf{y} \end{array} \right]_{k+1} =
\left[\begin{array}{c|c} \boldsymbol{\phi}_x & \mathbf{0} \\ \hline \mathbf{0} & \boldsymbol{\phi}_y \end{array} \right]_k
\left[\begin{array}{c} \mathbf{x} \\ \hline \mathbf{y} \end{array} \right]_k +
\left[\begin{array}{c} \mathbf{w}_x \\ \hline \mathbf{w}_y \end{array} \right]_k
\tag{9.6.2}
$$

Measurement model:

$$
\mathbf{z}_k = [\, \mathbf{H} \mid \mathbf{J} \,]_k \left[-\frac{\mathbf{x}}{\mathbf{y}} - \right]_k + \mathbf{v}_k
\tag{9.6.3}
$$

The error covariance associated with **X** will also be written in partitioned form as

$$[\mathbf{P}_X]_k = \left[\begin{array}{c|c} \mathbf{P}_x & \mathbf{P}_{xy} \\ \hline \mathbf{P}_{yx} & \mathbf{P}_y \end{array}\right]_k \tag{9.6.4}$$

Note from Eq. (9.6.2) that **x** and **y** are completely decoupled. Also, as usual we will assume that $(\mathbf{w}_x)_k$, $(\mathbf{w}_y)_k$, and $\mathbf{v}_k$ are white sequences with zero crosscorrelation and with covariances $(\mathbf{Q}_x)_k$, $(\mathbf{Q}_y)_k$, and $\mathbf{R}_k$.

We now need to write the expression for the optimal gain at step k in partitioned form, and at this point we will temporarily drop the k subscripts just to save writing. We use the usual optimal gain formula:

$$\left[\frac{\mathbf{K}_x}{\mathbf{K}_y}\right] = \left[\begin{array}{c|c} \mathbf{P}_x^- & \mathbf{P}_{xy}^- \\ \hline \mathbf{P}_{xy}^- & \mathbf{P}_y^- \end{array}\right]\left[\frac{\mathbf{H}^T}{\mathbf{J}^T}\right]\left[[\,\mathbf{H} \mid \mathbf{J}\,]\left[\begin{array}{c|c} \mathbf{P}_x^- & \mathbf{P}_{xy}^- \\ \hline \mathbf{P}_{yx}^- & \mathbf{P}_y^- \end{array}\right]\left[\frac{\mathbf{H}^T}{\mathbf{J}^T}\right] + \mathbf{R}\right]^{-1} \tag{9.6.5}$$

It is a routine matter now to expand out Eq. (9.6.5) and obtain explicit expressions for $\mathbf{K}_x$ and $\mathbf{K}_y$. However, we anticipate that $\mathbf{K}_y$ will not be needed in the reduced-order filter because the **y** variables will not be present. Hence, we will arbitrarily assume that the estimates of the **y** components are always zero (and invisible) in the reduced-order filter. Furthermore, we will use $\mathbf{K}_x$ in Eq. (9.6.5) as the gain for the remaining **x** state variables. This is, of course, the optimal gain, subject to the optimality of $\mathbf{P}_x^-$, $\mathbf{P}_{xy}^-$, and $\mathbf{P}_y^-$, which came from the previous step. Now, setting the **y** estimates equal to zero can be effected in the full-order filter by arbitrarily letting $\mathbf{K}_y$ be zero at each step, rather than the optimum value as determined by Eq. (9.6.5). If we do this, we then have a suboptimal full-order filter, and we can use suboptimal error analysis to evaluate the error covariance (see Section 6.7). In brief, this involves substituting the suboptimal gain into the general expression for the error covariance, Eq. (5.4.11) (repeated here in the notation of Chapter 5):

$$\mathbf{P} = (\mathbf{I} - \mathbf{KH})\mathbf{P}^-(\mathbf{I} - \mathbf{KH})^T + \mathbf{KRK}^T \tag{9.6.6}$$

Or, for the problem at hand, we have for the full-order suboptimal error covariance:

$$\left[\begin{array}{c|c} \mathbf{P}_x & \mathbf{P}_{xy} \\ \hline \mathbf{P}_{yx} & \mathbf{P}_y \end{array}\right] = \left[\left[\begin{array}{c|c} \mathbf{I} & \mathbf{0} \\ \hline \mathbf{0} & \mathbf{I} \end{array}\right] - \left[\frac{\mathbf{K}_x}{\mathbf{0}}\right][\,\mathbf{H} \mid \mathbf{J}\,]\right]$$

$$\times \left[\begin{array}{c|c} \mathbf{P}_x^- & \mathbf{P}_{xy}^- \\ \hline \mathbf{P}_{yx}^- & \mathbf{P}_y^- \end{array}\right]\left[\left[\begin{array}{c|c} \mathbf{I} & \mathbf{0} \\ \hline \mathbf{0} & \mathbf{I} \end{array}\right] - \left[\frac{\mathbf{K}_x}{\mathbf{0}}\right][\,\mathbf{H} \mid \mathbf{J}\,]\right]^T$$

$$+ \left[\frac{\mathbf{K}_x}{\mathbf{0}}\right]\mathbf{R}\left[\,\mathbf{K}_x^T \mid \mathbf{0}\,\right] \tag{9.6.7}$$

This equation can now be expanded out, and explicit expressions for updating $\mathbf{P}_x^-$, $\mathbf{P}_{xy}^-$, and $\mathbf{P}_y^-$ can be obtained.

We will summarize the recursive equations for the reduced-order (Schmidt–Kalman) filter.

1. The gain expression [from Eq. (9.6.5)]:

$$\mathbf{K}_x = \left(\mathbf{P}_x^- \mathbf{H}^T + \mathbf{P}_{xy}^- \mathbf{J}^T\right)\boldsymbol{\alpha}^{-1} \tag{9.6.8}$$

where

$$\boldsymbol{\alpha} = \mathbf{H}\mathbf{P}_x^- \mathbf{H}^T + \mathbf{H}\mathbf{P}_{xy}^- \mathbf{J}^T + \mathbf{J}\mathbf{P}_{yx}^- \mathbf{H}^T + \mathbf{J}\mathbf{P}_y^- \mathbf{J}^T + \mathbf{R} \tag{9.6.9}$$

2. Update $\hat{\mathbf{x}}^-$ (the usual equation because $\hat{\mathbf{y}}$ is zero):

$$\hat{\mathbf{x}} = \hat{\mathbf{x}}^- + \mathbf{K}_x(\mathbf{z} - \mathbf{H}\hat{\mathbf{x}}^-) \tag{9.6.10}$$

3. Covariance updates [from Eq. (9.6.7)]:

$$\mathbf{P}_x = (\mathbf{I} - \mathbf{K}_x \mathbf{H})\mathbf{P}_x^- - \mathbf{K}_x \mathbf{J}\mathbf{P}_{yx}^-$$

$$\mathbf{P}_{xy} = (\mathbf{I} - \mathbf{K}_x \mathbf{H})\mathbf{P}_{xy}^- - \mathbf{K}_x \mathbf{J}\mathbf{P}_y^-$$

$$\mathbf{P}_{yx} = \mathbf{P}_{xy}^T$$

$$\mathbf{P}_y = \mathbf{P}_y^- \tag{9.6.11}$$

4. Projection equations (with the k subscripts reinserted):
   (a) The state estimate (recall that x and y are decoupled):

$$\hat{\mathbf{x}}_{k+1}^- = (\boldsymbol{\phi}_x)_k \hat{\mathbf{x}}_k \tag{9.6.12}$$

   (b) The error covariances (from the usual $\boldsymbol{\phi}\mathbf{P}\boldsymbol{\phi}^T + \mathbf{Q}$ expression written in partitioned form):

$$(\mathbf{P}_x^-)_{k+1} = (\boldsymbol{\phi}_x)_k (\mathbf{P}_x)_k (\boldsymbol{\phi}_x)_k^T + (\mathbf{Q}_x)_k$$

$$(\mathbf{P}_{xy}^-)_{k+1} = (\boldsymbol{\phi}_x)_k (\mathbf{P}_{xy})_k (\boldsymbol{\phi}_y)_k^T$$

$$(\mathbf{P}_{yx}^-)_{k+1} = (\mathbf{P}_{xy}^-)_{k+1}^T$$

$$(\mathbf{P}_y^-)_{k+1} = (\boldsymbol{\phi}_y)_k (\mathbf{P}_y)_k (\boldsymbol{\phi}_y)_k^T + (\mathbf{Q}_y)_k \tag{9.6.13}$$

It should be apparent now that the recursive equations for the Schmidt–Kalman filter are more complicated than for an "ordinary" Kalman filter. Hence, one might logically ask: Where is the saving? The answer lies in the simplification that comes from setting $\mathbf{K}_y = 0$ and not having to carry $\mathbf{y}$ along in the recursive equations. The gain and update equations, bad as they look, would be even more complicated were $\mathbf{K}_y$ not equal to zero. The projection equations are really not simplified by setting $\mathbf{K}_y$ equal to zero. The simplification there comes from the block diagonal character of the transition matrix and not from anything that was done in the gain computation. A simple example will now be presented that compares the effectiveness of the Schmidt–Kalman filter with other ways of reducing the state vector.

EXAMPLE 9.5

Consider a simple situation where we wish to estimate a random-walk process whose initial value is a random variable described as $N(0, 1)$. The measurement noise in this situation will be assumed to be the sum of a unity variance Markov component plus a white component with a variance of 2 units. Thus, the state vector for the truth model will be a 2-tuple defined as follows:

$$\mathbf{X} = \begin{bmatrix} x \\ y \end{bmatrix}; \quad x \text{ is random walk}, \quad y \text{ is Markov} \tag{9.6.14}$$

The process equation for the truth model will then be

$$\begin{bmatrix} x \\ y \end{bmatrix}_{k+1} = \begin{bmatrix} 1 & 0 \\ 0 & e^{-\beta \Delta t} \end{bmatrix} \begin{bmatrix} x \\ y \end{bmatrix}_k + \begin{bmatrix} w_x \\ w_y \end{bmatrix}_k \tag{9.6.15}$$

In order to make hand calculations easy, we will choose the step size Δt and β such that $e^{-\beta \Delta t} = \frac{1}{2}$. We will also let the variance of the random walk in the Δt interval be $\frac{1}{4}$. Therefore,

$$\phi_x = 1, \quad \phi_y = \tfrac{1}{2}$$

$$Q_x = \tfrac{1}{4}, \quad Q_y = (1 - e^{-2\beta \Delta T}) = \tfrac{3}{4} \tag{9.6.16}$$

The measurement model is

$$z_k = \begin{bmatrix} 1 & 1 \end{bmatrix} \begin{bmatrix} x \\ y \end{bmatrix}_k + v_k \tag{9.6.17}$$

We will assume the variance of v_k to be 2 units; hence, the parameters for the measurement model are

$$H = 1, \quad J = 1, \quad R = 2 \tag{9.6.18}$$

The initial estimate at $k = 0$ and its associated error covariance are

$$\hat{\mathbf{X}}_0^- = \begin{bmatrix} 0 \\ 0 \end{bmatrix}, \qquad \mathbf{P}_0^- = \begin{bmatrix} 1 & 0 \\ 0 & 1 \end{bmatrix} \qquad (9.6.19)$$

We will assume that the first measurement occurs at $k = 0$, and we will cycle through two steps of the full-order optimal Kalman filter. The optimal filter will be the baseline system that will be used for comparison with various suboptimal estimators. The key result for the optimal estimator are as follows.

Optimal Estimator

Step (k)	Optimal Gain	Error Variance for $\hat{x}$
0	$\begin{bmatrix} 1/4 \\ 1/4 \end{bmatrix}$	$3/4 = .75$
1	$\begin{bmatrix} 14/59 \\ 13/59 \end{bmatrix}$	$187/236 \approx .7924$

Next, we will cycle through two steps of the reduced-order Schmidt–Kalman filter using Eqs. (9.6.8), (9.6.9), (9.6.11), and (9.6.13). The results are:

Kalman–Schmidt Filter

Step (k)	Gain for x	Error Variance for $\hat{x}$
0	$1/4$	$3/4 = .75$
1	$7/30$	$191/240 \approx .7958$

Note that the gain at $k = 1$ is slightly less than the optimal gain, and the corresponding estimation error variance is slightly larger than the optimal variance.

Now, another way of accounting for the colored measurement noise without including it as a state variable is to simply absorb it in with the white noise component. This is not rigorous, but it has been done on many occasions. This is sometimes referred to as "bumping up" the R matrix. If we use this approach for accounting for the Markov component, we simply "bump up" R from 2 to 3 units and calculate the gains by using the regular Kalman filter equations. These must then by cycled through the truth model (with $\mathbf{K}_y = 0$) to obtain the true error covariance sequence (see Section 6.7). The results are as follows.

Bumped-Up-R Model

Step (k)	Gain for x	Error Variance for $\hat{x}$
0	$1/4$	$.75$
1	$1/4$	$102/128 \approx .7969$

Note that the estimation error variance at $k = 1$ is slightly higher than that of the Schmidt–Kalman filter.

Figure 9.10. Comparison of four estimators.

Finally, in our comparison of various estimators, we can look at the simplest method of all for getting rid of the unwanted Markov state. This is to ignore it completely—pretend that it does not exist. This model (with $R = 2$) is analyzed in much the same manner as the bumped-up-R model, and the results are as follows.

Model with Markov State Ignored

Step (k)	Gain for x	Error Variance for $\hat{x}$
0	1/3	7/9
1	11/35	.8514

The estimation error variances for the four different models are summarized in Fig. 9.10. We should be careful about drawing general conclusions from a simple numerical example, but we can make the following statements with little risk: As might be expected, the optimal filter is best; the Schmidt–Kalman filter is next best; the filter with R bumped up is a close third; and the filter with the Markov noise completely ignored is a poor fourth. One message is clear from this example. Do not ignore colored noise components when they are of the same order of magnitude as the white noise. There is no need to do so. Some degree of suboptimality results in accounting for the discarded states with a Schmidt–Kalman filter (or by simply bumping up R), but this is much better than ignoring them completely. Our example here could be carried out through more steps until steady state is reached, but we will not do so here. This is left as an exercise (see Problem 9.5*). ∎

9.7
STOCHASTIC LINEAR
REGULATOR PROBLEM
AND THE SEPARATION THEOREM

The linear regulator problem is a classical problem of optimal control theory, and it can be posed (and solved) in either continuous or discrete form (16, 24 to 28). In the interest of brevity we will confine our remarks to the continuous version of the problem.

The *deterministic* linear regulator problem may be stated as follows: Given a linear dynamical system of the form

$$\dot{\mathbf{x}} = \mathbf{F}\mathbf{x} + \mathbf{B}\mathbf{u}_d \qquad (9.7.1)$$

what $\mathbf{u}_d(t)$ will minimize the quadratic performance index

$$J = \mathbf{x}^T(t_f)\mathbf{S}\mathbf{x}(t_f) + \int_{t_0}^{t_f}\left[\mathbf{x}^T(t)\mathbf{W}_x\mathbf{x}(t) + \mathbf{u}_d^T(t)\mathbf{W}_u\mathbf{u}_d(t)\right] dt \qquad (9.7.2)$$

where $\mathbf{S}$, $\mathbf{W}_x$ and $\mathbf{W}_u$ are symmetric positive definite weighting matrices that are chosen to fit the situation at hand? The intent of the regulator is to reduce the initial state of the system to zero (or nearly so) quickly and without undue control effort. The weighting factor $\mathbf{S}$ penalizes the system for not reaching zero at the specified terminal time t_f. The weighting factors $\mathbf{W}_x$ and $\mathbf{W}_u$ apply penalties for the trajectories of the state $\mathbf{x}(t)$ and the control $\mathbf{u}_d(t)$ over the time span $[t_0, t_f]$. It can be seen that if $\mathbf{W}_x$ is large and $\mathbf{W}_u$ small, the optimal system will apply a large control effort and force the system toward zero rapidly in order to avoid a large penalty due to large $\mathbf{W}_x$. On the other hand, if $\mathbf{W}_x$ is small and $\mathbf{W}_u$ large, the system will be frugal with its control effort, and it will approach zero slowly. Clearly, a wide variety of situations can be accommodated within the structure of this formulation. We will not derive the solution of the linear regulator problem here. This is adequately covered in the mentioned references (and many others, as well). We simply state that the optimal control law is a linear feedback law that specifies $\mathbf{u}_d(t)$ to be

$$\mathbf{u}_d(t) = -\mathbf{K}_1(t)\mathbf{x}(t) \qquad (9.7.3)$$

where the feedback gain $\mathbf{K}_1(t)$ is computed from the system parameters. (It is not a function of $\mathbf{x}$.) We need not be concerned with the detailed solution for $\mathbf{K}_1(t)$ (except to note parenthetically that there is a close duality between this and the optimal estimator problem.) It is important to note, though, that it is assumed that the state vector $\mathbf{x}$ is available for feedback purposes as indicated in Fig. 9.11. However, in many physical situations, we do not have the privilege of observing all the elements of $\mathbf{x}$ directly; quite to the contrary, we are usually

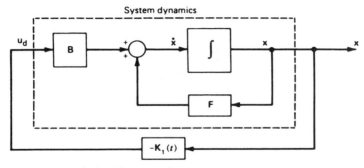

Figure 9.11. Optimal linear regulator.

allowed to observe **x** only through some output relationship, say

$$\mathbf{y} = \mathbf{Hx} \tag{9.7.4}$$

Now, based on **y**, we must somehow reconstruct **x**. If the observation is essentially error free, we can use observer theory in the reconstruction (28 to 30). However, if the observation of **x** is corrupted with noise, as is often the case, then the reconstruction of **x** becomes an estimation problem. This leads to the stochastic linear regulator problem,* which will now be formulated.

Let the system dynamics be specified by the linear equation

$$\dot{\mathbf{x}} = \mathbf{Fx} + \mathbf{Bu}_d + \mathbf{Gu} \tag{9.7.5}$$

where **F**, **B**, and $\mathbf{u}_d$ are as before, and **u** is an additional Gaussian white noise forcing function that is characterized by

$$E\left[\mathbf{u}(t)\mathbf{u}^T(\tau)\right] = \mathbf{Q}\delta(t - \tau) \tag{9.7.6}$$

The system state vector is assumed to be observed via the relationship

$$\mathbf{z} = \mathbf{Hx} + \mathbf{v} \tag{9.7.7}$$

where **v** is Gaussian white noise and is characterized by

$$E\left[\mathbf{v}(t)\mathbf{v}^T(\tau)\right] = \mathbf{R}\delta(t - \tau) \tag{9.7.8}$$

The two white noises **u** and **v** will be assumed to be independent, and, in order to avoid any questions about singular conditions, we assume that the system is controllable with respect to the control input $\mathbf{u}_d$ and is observable with respect to **z**. The optimization problem is to minimize the following cost function:

$$J = E\left\{\mathbf{x}^T(t_f)\mathbf{Sx}(t_f) + \int_{t_0}^{t_f}\left[\mathbf{x}^T(t)\mathbf{W}_x\mathbf{x}(t) + \mathbf{u}_d^T(t)\mathbf{W}_u\mathbf{u}_d(t)\right] dt\right\} \tag{9.7.9}$$

As before, we will not dwell on the details of the solution; we will simply state the results (16, 26, 27, 28). Design of the optimal stochastic linear regulator may be separated into two steps:

1. First, design the minimum mean-square error estimator for **x**, treating $\mathbf{u}_d$ just as if it were a known deterministic input. The optimal estimator is, of course, a Kalman filter with parameters **F**, $\mathbf{GQG}^T$, **H**, and **R**. Note that the process has a known input as well as a random input;

---

* This problem is also called the linear quadratic Gaussian (LQG) problem.

Figure 9.12. Optimal stochastic linear regulator.

therefore, the differential equation for $\hat{\mathbf{x}}$ is

$$\dot{\hat{\mathbf{x}}} = \mathbf{F}\hat{\mathbf{x}} + \mathbf{B}\mathbf{u}_d + \mathbf{K}(\mathbf{z} - \mathbf{H}\hat{\mathbf{x}}) \qquad (9.7.10)$$

where $\mathbf{K}$ is the continuous Kalman gain (see Problem 7.9).

2. Next, solve the deterministic linear regulator problem for the optimal feedback gain $\mathbf{K}_1(t)$ just as if a perfect measurement of $\mathbf{x}$ were available and $\mathbf{u}(t)$ were absent. Then let control input $\mathbf{u}_d$ be

$$\mathbf{u}_d = -\mathbf{K}_1(t)\hat{\mathbf{x}}(t) \qquad (9.7.11)$$

and the cost function is minimized. The resulting optimal system is summarized in Fig. 9.12.

The two-step solution just described is known as the *separation theorem* or *separation principle*. It is a most remarkable result in that we would normally expect that the feedback loop would horribly complicate matters by mixing together the control and estimation problems. However, it does not; the two problems separate nicely! The superficial reason for this is the duality between the optimal control and optimal filter problems. The underlying reason for the duality in the first place, though, is not all that obvious, so we will simply say that this is a fortuitous circumstance.

It is not our intent here to teach the subject of optimal control. The subject is well developed, and many fine books have been written in this area. We simply want to point out with this limited example that estimation theory (and Kalman filtering in particular) plays an important role in optimal control theory, and it behooves the control systems engineer/scientist to understand the rudiments of estimation theory.

PROBLEMS

9.1 Consider a simple one-dimensional trajectory determination problem as follows. A small object is launched vertically in the earth's atmosphere. The initial thrust exists for a very short time, and the object "free falls" ballistically for essentially all of its straight-up, straight-down trajectory. Let y be measured in the up direction, and assume that the nominal trajectory is governed by the following dynamical equation:

$$m\ddot{y} = -mg - D\dot{y}|\dot{y}|$$

where

$m = .05$ kg (mass of object)
$g = 9.087$ m/sec^2 (acceleration of gravity)
$D = 1.4 \times 10^{-4}$ n/(m/sec)^2 (drag coefficient)

The drag coefficient will be assumed to be constant for this relatively short trajectory, and note that drag force is proportional to (velocity)^2, which makes the differential equation nonlinear. Let the initial conditions for the nominal trajectory be as follows:

$$y(0) = 0$$

$$\dot{y}(0) = 85 \text{ m/sec}$$

The body is tracked and noisy position measurements are obtained at intervals of .1 sec. The measurement error variance is .25 m^2. The actual trajectory will differ from the nominal one primarily because of uncertainty in the initial velocity. Assume that the initial position is known perfectly but that initial velocity is best modeled as a normal random variable described by N (85 m/sec, 1 m^2/sec^2). Work out the linearized discrete Kalman filter model for the up portion of the trajectory. (*Hint:* An analytical solution for the nominal trajectory may be obtained by considering the differential equation as a first-order equation in velocity. Note $|\dot{y}| = \dot{y}$ during the up portion of the trajectory. Since variables are separable in the velocity equation, it can be integrated. The velocity can then be integrated to obtain position.)

9.2 (a) At the kth step of the usual nonadaptive Kalman filter, the measurement residual is $(\mathbf{z}_k - \mathbf{H}_k \hat{\mathbf{x}}_k^-)$. Let z_k be scalar and show that the expectation of the squared residual is minimized if $\hat{\mathbf{x}}_k^-$ is the optimal a priori estimate of $\mathbf{x}_k$, that is, the one normally computed in the projection step of the Kalman filter loop. (*Hint:* Use the measurement relationship $z_k = \mathbf{H}_k \mathbf{x}_k + v_k$ and note that v_k and the a priori estimation error have zero crosscorrelation. Also note that the a priori estimate $\hat{\mathbf{x}}_k^-$, optimal or otherwise, can only depend on the measurement sequence up through z_{k-1} and not z_k.)

(b) Show that the time sequence of residuals $(z_k - \mathbf{H}_k \hat{\mathbf{x}}_k^-)$, $(z_{k+1} - \mathbf{H}_{k+1} \hat{\mathbf{x}}_{k+1}^-)$,... is a white sequence if the filter is optimal.

9.3 In Problem 9.2 it was stated that the sequence of measurement residuals in a Kalman filter is a white sequence. This occurs only when the filter gain is adjusted to the optimal value. Thus, we can think of the filter gain as a "tuning" parameter that we tune to make the measurement-residual sequence white. This immediately suggests an intuitive sort of adaptive filter in which the filter monitors the measurement-residual sequence, and if it detects nonwhiteness, it readjusts the gain to force the sequence to be white. This will be demonstrated with a simple continuous Kalman filter example.

Let the process and measurement models be

$$\dot{x} = u(t), \qquad u(t) \text{ is white with spectral amplitude } q$$

$$z = x + v(t), \qquad v(t) \text{ is white with spectral amplitude } r$$

As usual, we will assume $u(t)$ and $v(t)$ to be independent. Their spectral amplitudes q and r are presumed to be unknown and, furthermore, they might

Problem 9.3

vary slowly with time. The continuous Kalman filter solution for this setting is given by (see Chapter 7)

$$\dot{\hat{x}} = K(z - \hat{x})$$

The solution is also shown in block diagram form in the accompanying figure. It is not difficult to imagine the filter performing spectral analysis on the measurement residual at the summing point in much the same manner as with any modern digital spectrum analyzer. This could be done repeatedly throughout the course of the filter operation. The result would be an experimentally derived power spectral density function (or autocorrelation function) that would be representative of the measurement-residual process over the time span of the sample.

Assume that an autocorrelation function has been determined as just described, and assume that q and r do not change appreciably during the observation span. Find q, r and the filter gain in terms of the parameters that describe the autocorrelation function. Note that once q and r are determined, the filter gain can be readjusted to the optimal condition. [If you have difficulty with this problem, see Gelb (31) pp. 319–320.]

9.4*. This problem is a variation on the DME example given in Section 9.1 (Example 9.1) with a simplification in the model of the aircraft dynamics. Suppose that the two DME stations are located on the x-axis as shown in the accompanying figure, and further suppose that the aircraft follows an approximate path from south to north as shown. The aircraft has a nominal velocity of 100 m/sec in a northernly direction, but there is random motion superimposed on this in both the x and y directions. The flight duration (for our purposes) is 200 sec, and the initial coordinates at $t = 0$ are properly described as normal random variables as follows:

$$x_0 \sim N(0, 2000 \text{ m}^2)$$

$$y_0 \sim N(-10000 \text{ m}, 2000 \text{ m}^2)$$

The aircraft kinematics are described by the following equations:

$$x_{k+1} = x_k + w_{1k}, \qquad k = 0, 1, 2, \ldots, 200$$

$$y_{k+1} = y_k + w_{2k} + 100 \, \Delta t, \qquad k = 0, 1, 2, \ldots, 200$$

where w_{1k} and w_{2k} are independent white Gaussian sequences described by

$$w_{1k} \sim N(0, 400 \text{ m}^2)$$

$$w_{2k} \sim N(0, 400 \text{ m}^2)$$

The sampling interval Δt is 1 sec. The aircraft motion will be recognized as simple random-walk in the x-coordinate, and random-walk superimposed on linear motion for the y-coordinate.

The aircraft obtains simultaneous discrete range measurements at 1 sec intervals on both DME stations, and the measurement errors consist of a superposition of Markov and white components. Thus we have (for a typical range measurement)

$$\begin{bmatrix} \text{Total} \\ \text{noisy} \\ \text{measurement} \end{bmatrix} = \begin{bmatrix} \text{True} \\ \text{total} \\ \text{range} \end{bmatrix} + \begin{bmatrix} \text{Markov} \\ \text{error} \\ \text{component} \end{bmatrix} + \begin{bmatrix} \text{White} \\ \text{error} \\ \text{component} \end{bmatrix}$$

We have two DME stations, so the measurement vector is a 2-tuple at each sample point beginning at $k = 0$ and continuing until $k = 200$. The Markov errors for each of the DME stations are independent Gaussian random processes, which are described by the autocorrelation function

$$R_m(\tau) = 900 e^{-.01|\tau|} \text{ m}^2$$

The white measurement errors for both stations have a variance of 225 m^2.

(a) Work out the linearized Kalman filter parameters for this situation. The linearization is to be done about the nominal linear-motion trajectory exactly along the y-axis. That is, the resultant filter is to be an "ordinary" linearized Kalman filter and not an extended Kalman filter.

(b) After the key filter parameters have been worked out, run a covariance analysis for $k = 0, 1, 2, \ldots, 200$, and plot the estimation error variances for both the x and y position coordinates.

(c) You should see a pronounced peak in the y-coordinate error curve as the aircraft goes through (or near) the origin. This simply reflects a bad geometric situation when the two DME stations are 180 degrees apart relative to the aircraft. One might think that the estimation error variance should go to infinity at exactly $k = 100$. Explain qualitatively why this is not true.

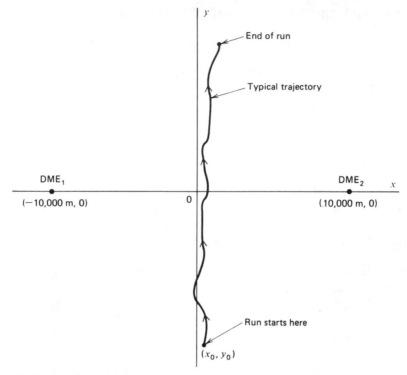

Problem 9.4*

9.5*. The optimal two-state system of Example 9.5 is completely observable, so one would expect the error covariance to reach a steady-state condition after a reasonable number of steps. We might guess that the same would be true for each of the one-state suboptimal estimators considered in the example, but there is some risk in this guess because of the mismatch between truth and the one-state models that were used.

    (a) Run the optimal estimator through 50 steps and verify that a steady-state condition is reached.

    (b) Consider next the suboptimal estimator where the Markov component of the measurement noise is completely ignored. Cycle the suboptimal error covariance through 50 steps and compare the results with those of part (a).

REFERENCES CITED IN CHAPTER 9

1. R. E. Kalman, "A New Approach to Linear Filtering and Prediction Problems," *Trans, of the ASME, Jour. of Basic Engr.*, 35–45 (March 1960).

2. R. E. Kalman and R. S. Bucy, "New Results in Linear Filtering and Prediction," *Trans. of the ASME, Jour. of Basic Engr., 83:* 95–108 (1961).

3. T. Kailath, "A View of Three Decades of Linear Filtering Theory," *IEEE Trans. on Information Theory, IT-20:* No. 2, 146–181 (March 1974).

4. H. W. Sorenson (Ed.), *Kalman Filtering: Theory and Applications*, New York: IEEE Press, 1985.

5. M. Kayton and W. R. Fried (Eds.), *Avionics Navigation Systems*, New York: Wiley, 1969.

6. H. W. Sorenson, "Kalman Filtering Techniques," in C. T. Leondes (Ed.), *Advances in Control Systems*, Vol. 3, New York: Academic Press, 1966, pp. 219–289.

7. R. G. Brown and J. W. Nilsson, *Introduction to Linear Systems Analysis*, New York: Wiley, 1962.

8. R. G. Brown and D. T. Friest, "Optimization of a Hybrid Inertial Solar-Tracker Navigation System," *1964 IEEE International Convention Record*, Part 7: 121–135.

9. C. T. Leondes (Ed.), *Theory and Applications of Kalman Filtering*, North Atlantic Treaty Organization AGARD Report No. 139 (February 1970).

10. R. G. Brown, "Integrated Navigation Systems and Kalman Filtering: A Perspective," *Navigation, Jour. of the Inst. of Navigation, 19:* No. 4, 355–362 (Winter 1972–73).

11. D. T. Magill, "Optimal Adaptive Estimation of Sampled Stochastic Processes," *IEEE Trans. on Automatic Control, AC-10:* No. 4, 434–439 (October 1965).

12. G. L. Mealy and W. Tang, "Application of Multiple Model Estimation to a Recursive Terrain Height Correlation System," *IEEE Trans. on Automatic Control, AC-28*, pp. 323–331 (March 1983).

13. A. A. Girgis and R. G. Brown, "Adaptive Kalman Filtering in Computer Relaying: Fault Classification Using Voltage Models," *IEEE Trans. on Power Apparatus and Systems, PAS-104:* No. 5, pp. 1168–1177 (May 1985).

14. R. G. Brown, "A New Look at the Magill Adaptive Filter as a Practical Means of Multiple Hypothesis Testing," *IEEE Trans. on Circuits and Systems, CAS-30,* pp. 765–768 (October 1983).

15. R. G. Brown and P. Y. C. Hwang, "A Kalman Filter Approach to Precision Geodesy," *Navigation, Jour. of the Institute of Navigation, 30:* No. 4, 338–349 (Winter 1983–84).

16. J. S. Meditch, *Stochastic Optimal Linear Estimation and Control*, New York: McGraw-Hill, 1969.

17. R. G. Brown and G. L. Hartman, "Kalman Filter with Delayed States as Observables," *Proc. of the National Electronics Conference*, Chicago, III., 1968.

18. R. G. Brown, "Analysis of an Integrated Inertial/Doppler-Satellite Navigation System: Part I, Theory and Mathematical Model," Tech. Report No. ERI 62600, Engr. Research Inst., Iowa State University, Ames, 1969.

19. R. G. Brown and L. L. Hagerman, "An Optimum Inertial Doppler-Satellite Navigation System," *Navigation, Jour. of the Inst. of Navigation, 16:* No. 3, 260–269 (Fall 1969).

20. G. J. Bierman, *Factorization Methods for Discrete Sequential Estimation*, New York: Academic Press, 1977.

21. P. S. Maybeck, *Stochastic Models, Estimation and Control*, Vol. 1, New York: Academic Press, 1979.

22. R. H. Battin, *Astronautical Guidance*, New York: McGraw-Hill, 1964, pp. 338–340.

23. S. F. Schmidt, "Application of State-Space Methods to Navigation Problems," in C. T. Leondes (Ed.), *Advances in Control Systems*, Vol. 3, New York: Academic Press, 1966.

24. M. Athans and P. L. Falb, *Optimal Control*, New York: McGraw-Hill, 1966.

25. D. E. Kirk, *Optimal Control Theory*, Englewood Cliffs, N.J.: Prentice-Hall, 1970.

26. A. E. Bryson, Jr., and Y. Ho, *Applied Optimal Control*, rev. ed., New York: Halsted Press, Div. of John Wiley & Sons, 1975.

27. A. P. Sage and C. C. White, *Optimum Systems Control*, 2nd ed., Englewood Cliffs, N.J.: Prentice-Hall, 1977.

28. G. F. Franklin and J. D. Powell, *Digital Control of Dynamic Systems*, Reading, Mass.: Addison-Wesley, 1980.

29. C. T. Chen, *Linear System Theory and Design*, New York: Holt, Rinehart and Winston, 1984.

30. T. Kailath, *Linear Systems*, Englewood Cliffs, N.J.: Prentice-Hall, 1980.

31. A. Gelb (Ed.), *Applied Optimal Estimation*, Cambridge, Mass.: MIT Press, 1974.

10

The Global Positioning System: A Case Study

The Global Positioning System (GPS) is a new satellite navigation system that is currently being deployed by the United States (1, 2). It is scheduled to become fully operational in the mid-1990s; and because of its worldwide coverage, it promises to be of great benefit to all nations of the world. GPS is, by far, the most ambitious navigation project ever undertaken by the United States, or by any nation for that matter, and its applications go beyond the usual positioning of aircraft and ships. Other applications include precise surveying, accurate land vehicle tracking, near-earth space navigation, and precise time dissemination on a worldwide basis. The central problem for the GPS receiver is the precise estimation of position, velocity, and time based on noisy observations of the satellite signals. It should come as no surprise then that this is an ideal setting for Kalman filtering. In fact, Kalman filtering has become a household word in the GPS industry. Our discussion of the subject here is intended to be tutorial and must be brief. Thus, we will confine our attention to receiver applications only and will leave all of the other interesting facets of Kalman filtering that are applied to GPS for extracurricular reading.

10.1
DESCRIPTION OF GPS

GPS is a satellite-based navigation system that allows a user with the proper equipment access to useful and accurate positioning information anywhere on the globe. Position and time determination is accomplished by the reception of GPS signals to obtain ranging information as well as messages transmitted by the satellites. The system of satellites that makes up the space segment of GPS will consist of 21 satellites or more in six 12-hour orbits when fully deployed. This ensures a user located anywhere on the globe a visibility of four satellites or more at any time. From an observer's viewpoint, the satellite geometry in the visible sky is always changing because the satellites are not in geosynchronous orbits. The maintenance of updated information embedded in the transmitted message is performed by ground monitoring stations that are collectively known as the Control Segment. The Control Segment periodically updates the information that is disseminated by all the satellites. This includes satellite ephemerides and health status, as well as a constellation almanac.

GPS signals are transmitted on two coherent carrier frequencies, L1 (1575.42 MHz) and L2 (1227.60 MHz), which are modulated by various spread-spectrum signals. The major carrier, L1, is biphase-modulated by two types of pseudorandom noise codes (see Section 2.14 of Chapter 2), one at 1.023 MHz called the C/A-code, and the other at 10.23 MHz called the P-code. The P-code or its military-classified counterpart, the Y-code, is intended only for authorized access because its one-chip wavelength of 30 m provides the most accurate positioning possible. The C/A-code, with its 300-m one-chip wavelength, is used in all cases for initial acquisition and code-signal alignment purposes. All users have access to this less-accurate C/A-code for positioning. The second carrier signal L2 contains only P-code (or Y-code) modulation and is intended to give authorized users the additional capability of actually measuring the ionospheric delays using the two frequencies, the delays being frequency-dependent. In official parlance, Y-code access is reserved for what is called the Precise Positioning Service (PPS) mode of operation, while everything else is classified as the Standard Positioning Service (SPS).

A 50-bit per second navigation message is also combined with the pseudorandom noise codes. This navigation message, 1500 bits in length and repeated every 30 sec, carries many kinds of information with varying degrees of functional and operational importance to the user. Foremost in significance is the satellite ephemerides, a collection of data making up 60 percent of the message which, when decoded, uniquely describes the position and trajectory of the satellite that transmitted it. The remaining 40 percent of space allotted to the message is common to all satellites and carries general almanac information. This information runs the gamut from providing approximate satellite positions for visibility checks and signal acquisition purposes to satellite health status and current operational modes.

To complete this brief description of the GPS operation, we now look at the basic function of position determination by an observer who is properly equipped to receive and decode GPS signals. In free space, there are three dimensions of position that need to be solved. Also, an autonomous user is not expected to be precisely synchronized to the satellite system time initially. In all, the standard GPS positioning problem poses four variables that can be solved from the following system of equations, representing measurements from four different satellites.

$$\psi_1 = \sqrt{(X_1 - x)^2 + (Y_1 - y)^2 + (Z_1 - x)^2} + c\,\Delta t$$
$$\psi_2 = \sqrt{(X_2 - x)^2 + (Y_2 - y)^2 + (Z_2 - z)^2} + c\,\Delta t$$
$$\psi_3 = \sqrt{(X_3 - x)^2 + (Y_3 - y)^2 + (Z_3 - z)^2} + c\,\Delta t$$
$$\psi_4 = \sqrt{(X_4 - x)^2 + (Y_4 - y)^2 + (Z_4 - z)^2} + c\,\Delta t \qquad (10.1.1)$$

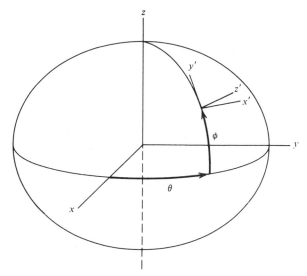

Figure 10.1. The WGS-84 coordinate reference frame (x, y, z) used by GPS and a locally level coordinate reference frame (x', y', z').

where

$$\psi_1, \psi_2, \psi_3, \psi_4 = \text{noiseless pseudorange}$$
$$[X_i, Y_i, Z_i]^T = \text{Cartesian position coordinates of satellite } i$$
$$[x, y, z]^T = \text{Cartesian position coordinates of observer}$$
$$\Delta t = \text{receiver offset from the satellite system time}$$
$$c = \text{speed of light}$$

The observer position $[x, y, z]^T$ is "slaved" to the coordinate frame of reference used by the satellite system. For GPS, this reference is a geodetic datum called WGS-84 (for World Geodetic System of 1984), which is Earth-Centered Earth-Fixed (3). The datum also defines the ellipsoid that crudely approximates the surface of the Earth (see Fig. 10.1). Although the satellite positions are reported in WGS-84 coordinates, it is sometimes useful to deal with a locally level frame of reference, where the x'-y' plane is tangential to the surface of the Earth ellipsoid. As depicted in Fig. 10.1, we shall define such a locally level reference frame by having the x'-axis pointing East, the y'-axis North, and the z'-axis normal to the plane, and equivalently, to the surface of the Earth at the observer's location.

It suffices here to say that the coordinate transformations to convert between the WGS-84 coordinates and any other derived reference frame, including the locally level one given here, is usually quite straightforward.

10.2
THE OBSERVABLES

Useful information can be derived from measurements made on the pseudorandom code and the carrier signal. The block diagram of a generic signal tracking

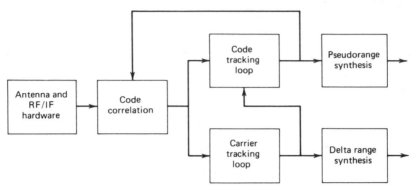

Figure 10.2. Generic signal tracking scheme for a GPS receiver.

scheme for a GPS receiver is shown in Fig. 10.2. In it, there are separate tracking loops for the code and the carrier, with the latter cross-feeding to partially aid the code tracking (4). The loop filters are generally simple low-pass filters, the bandwidths of which determine the noisiness of the measurement data that are synthesized.

The observable known as *pseudorange* is a timing measurement of the propagation delay that is due to the geometric range from the transmitting satellite to the receiver and also the receiver clock offset from satellite time—hence, the term *pseudorange* and not just *range*. At the point of reception, the measurement is made in the receiver by determining the amount of shift in the pseudorandom code position since the time of transmission. By using some coherent form of signal tracking, the binary pseudorandom code can be monitored by aligning the received signal with a replica of the known code generated by the receiver. Hence, the precision of the crosscorrelation process in determining the pseudorandom code position establishes the accuracy of the pseudorange measurement, which by state-of-the-art standards is considered to be about 1 m for P-code and a few meters for C/A-code under nominal signal reception strengths (4). In the context of Kalman filtering, these numbers represent the standard deviation of the white measurement noise. The pseudorange measurement can be summarily represented by the following equation:

$$\rho = \psi + \beta_\rho + v_\rho \tag{10.2.1}$$

where
- ψ = noiseless pseudorange consisting of geometric range and range error due to receiver timing error
- β_ρ = time-correlated errors associated with pseudorange
- v_ρ = pseudorange measurement noise

The term β_ρ represents other significant error sources, which may range from estimation errors in the reported satellite position and signal delays due to ionospheric and tropospheric refraction, to the intentional errors invoked under an accuracy-degradation scheme called Selective Availability. Sometimes collectively called *unmodeled biases*, these errors are generally difficult to estimate because of their poor observability characteristics, and discussion on their modeling is deferred to Section 10.3. However, because they are dependent on the spatial relationship between the observer and the satellites, two observers in proximity to one another and sighting the same satellites will encounter these errors with a high degree of correlation between them. To take advantage of this, if one of these observers is at a known reference location, these so-called unmodeled biases can be measured as a lumped error and the information, when shared, can be used by the other observer to correct for the error. This mode of operation is known as differential positioning, and corrections of this kind are very effective in enhancing position accuracy for the participants of such a positioning network. Today differential GPS plays an important role in the development of high-accuracy positioning systems (5, 6).

In addition to code tracking, most GPS navigation receivers possess the capability of tracking the carrier signal as well. The wavelength of the L1 carrier is little more than 19 cm, thus allowing very precise measurements of the phase of the carrier to be made. This type of measurement is made after the code modulation has been stripped off. The amount of noise in the carrier phase data depends largely on the parameters of the tracking loops used by the receiver. It may be less than 1 percent of the wavelength for a stationary receiver, but to maintain carrier tracking in high dynamics, the noise in the data for most navigation-type receivers may be as high as 2 percent (4, 7).

In most conventional GPS receivers, the rate of change in the carrier phase over a brief interval is used to represent the measured Doppler frequency. As an approximation to the true Doppler frequency, this observable, called *delta range*, is generally used to provide accurate velocity information. Although the Doppler frequency due to satellite motion as seen by a stationary observer is constantly changing, it is nevertheless accurately predictable. Hence, when the measured Doppler is compared against the predicted value, the difference reflects the velocity of the observer as well as the frequency error of the receiver clock. The use of the delta range measurement to approximate the Doppler frequency, of course, makes the assumption that the velocity is constant throughout the integration interval used to form the delta range. This assumption holds for the most part even if the velocity is not constant, provided that the integration interval remains short. Most receiver designs utilize integration intervals that are no more than a fraction of a second long. The accuracy of the delta range measurement in terms of velocity under nominal signal reception strengths is about 0.008 m/sec (4).

More recently, applications have surfaced whereby the Doppler data are exploited for information related to relative position rather than just velocity (8, 9). The difference between the integrated Doppler (sometimes also called *continuous carrier phase*) and the delta range is graphically compared in Fig. 10.3. The applications that utilize integrated Doppler measurements (discussed later in Sections 10.6 and 10.7) involve long integration intervals that demand uninterrupted and flawless tracking of the carrier. Under real operating conditions, such demands are seldom easily satisfied. But owing to the great benefits of accuracy that can be reaped, the efforts to overcome these limitations

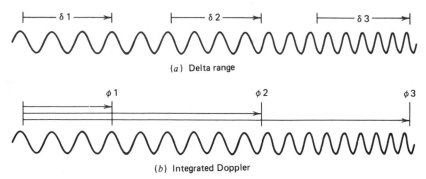

Figure 10.3.   Measurements formed from carrier phase data.

continue with measured success. The carrier phase measurement can be represented by the equation:

$$\phi = \psi + N_\phi + \beta_\phi + v_\phi \tag{10.2.2}$$

where

ψ = noiseless pseudorange consisting of geometric range and range error due to receiver timing error

N_ϕ = range uncertainty, sometimes called integer cycle ambiguity (see Section 10.8)

β_ϕ = time-correlated errors associated with carrier phase

v_ϕ = carrier phase measurement noise

When pseudorange and delta range (or integrated Doppler) data are used in a combined setting, it may be presumed that the measurement noises for both types of data can be regarded as independent of each other. In the receiver, different processes are involved in obtaining the two types of measurements (see Fig. 10.2). The cross-feed from the carrier to the code tracking loops can largely be ignored because the carrier data are virtually noiseless when compared with the pseudorange measurements made in the code loop.

10.3
DYNAMICAL PROCESSES AND KALMAN FILTER MODELS FOR A STAND-ALONE GPS RECEIVER

Dynamic State Process

In the GPS positioning problem, we need to solve for at least three degrees of freedom for position and one degree of freedom for time. The random process describing the time error is related to the characteristics of the receiver clock

Figure 10.4. Random-walk model for a stationary observer.

(oscillator), whereas the random process governing the position error states is dependent on the observer's dynamics. The former is independent of the latter, except when accelerative forces are large enough to physically stress the oscillator. It is very common for crystal oscillators to have some error in frequency, and it is the constancy of this quantity that defines their stability. Since a nearly constant frequency error gives rise to a predictable drift in the timing error, it is useful to have two error states instead of just one: a range bias and a range drift error. Further discussion of oscillators is reserved for Section 10.4.

In its most basic form to estimate position and time, the linearized Kalman filter state vector should consist of three position error states and two clock states. The clock states are usually defined in units of range and range rate for compatibility with the position states. This five-state filter is ideal for a stationary observer (i.e., constant position) where the random bias model might be appropriate for the position states, even though the random-walk model is usually preferred. Recall from Chapter 2 that with $\mathbf{Q} = \mathbf{0}$ for a random bias, no process noise is added with each iteration of the filter, thereby causing the error covariance matrix $\mathbf{P}$ to eventually converge to zero. We have seen, in Section 6.6, how such a situation invites the prospect of divergence as a result of numerical problems when one deals with infinitesimally small numbers, especially after a prolonged duration of processing. With the random-walk process, as shown in Fig. 10.4, $\mathbf{Q} \neq \mathbf{0}$. Note here that this dynamical model is also suitable for a near-stationary observer, such as a buoy drifting at sea. This will be known as the Position model. Note further that this is a linearized Kalman filter model, with the linearization taking place about a nominal fixed point.

When the observer is not stationary but moving with nearly constant velocity, we have a case resembling that of the oscillator with a stable frequency error. Here, the random-walk model is not ideally suitable because it makes the assumption that the driving white noise input to the integrator is unbiased when in fact the velocity-induced bias would be nontrivial. A better model, in this instance, would be one corresponding to the double-integrator transfer function diagram of Fig. 10.5, which we shall refer to as the Position–Velocity model. Here, the velocity is not white noise but a random-walk process. This choice is

Figure 10.5. Integrated random-walk model for a dynamic observer.

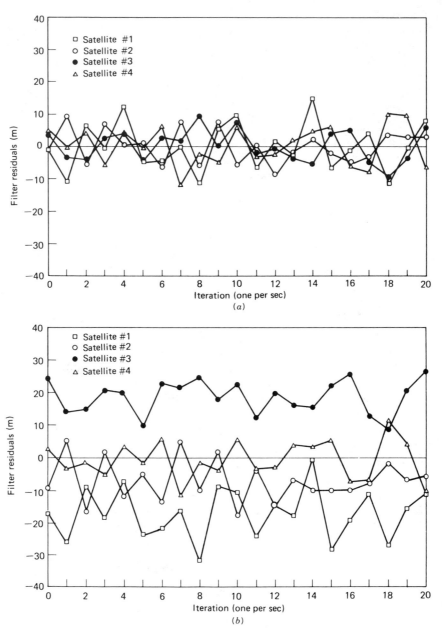

Figure 10.6. Kalman filter residuals for a dynamical situation using the (a) Position–Velocity model and (b) Position model.

substantiated by a comparison of the filter residuals generated by Kalman filters using both models (see Fig. 10.6).

Examination of the residual samples from a Kalman filter is a good way of determining the accuracy or fitness of the assumed model to the actual physical situation. The residuals for the Position–Velocity (PV) model appear unbiased and uncorrelated in time for the data associated with all four satellites. Conversely, the second set of residuals exhibit a deficiency in the Position model. Note that the PV model is also a linearized model, but in this case the filter must be an extended Kalman filter in order that the "nominal" position and velocity (about which the linearization takes place) remain reasonably close to the true values.

EXAMPLE 10.1

In the PV model, each spatial dimension will have two degrees of freedom, one of position and the other of velocity. Therefore, for the GPS problem in which there are three spatial dimensions and one time dimension, the state vector now becomes an 8-tuple. The PV dynamic process can be described by the following vector differential equation:

$$
\begin{bmatrix} \dot{x}_1 \\ \dot{x}_2 \\ \dot{x}_3 \\ \dot{x}_4 \\ \dot{x}_5 \\ \dot{x}_6 \\ \dot{x}_7 \\ \dot{x}_8 \end{bmatrix} =
\begin{bmatrix}
0 & 1 & 0 & 0 & 0 & 0 & 0 & 0 \\
0 & 0 & 0 & 0 & 0 & 0 & 0 & 0 \\
0 & 0 & 0 & 1 & 0 & 0 & 0 & 0 \\
0 & 0 & 0 & 0 & 0 & 0 & 0 & 0 \\
0 & 0 & 0 & 0 & 0 & 1 & 0 & 0 \\
0 & 0 & 0 & 0 & 0 & 0 & 0 & 0 \\
0 & 0 & 0 & 0 & 0 & 0 & 0 & 1 \\
0 & 0 & 0 & 0 & 0 & 0 & 0 & 0
\end{bmatrix}
\begin{bmatrix} x_1 \\ x_2 \\ x_3 \\ x_4 \\ x_5 \\ x_6 \\ x_7 \\ x_8 \end{bmatrix} +
\begin{bmatrix} 0 \\ u_2 \\ 0 \\ u_4 \\ 0 \\ u_6 \\ u_7 \\ u_8 \end{bmatrix} \qquad (10.3.1)
$$

where

x_1 = East position error
x_2 = East velocity
x_3 = North position error
x_4 = North velocity
x_5 = altitude error
x_6 = altitude rate
x_7 = range (clock) bias state
x_8 = range (clock) drift state

and the spectral amplitudes associated with the white noise driving functions are S_p for u_2, u_4, and u_6, S_f for u_7, and S_g for u_8.

From Eq. (10.3.1), the State Transition Matrix can be derived as

$$
\phi =
\begin{bmatrix}
1 & \Delta t & 0 & 0 & 0 & 0 & 0 & 0 \\
0 & 1 & 0 & 0 & 0 & 0 & 0 & 0 \\
0 & 0 & 1 & \Delta t & 0 & 0 & 0 & 0 \\
0 & 0 & 0 & 1 & 0 & 0 & 0 & 0 \\
0 & 0 & 0 & 0 & 1 & \Delta t & 0 & 0 \\
0 & 0 & 0 & 0 & 0 & 1 & 0 & 0 \\
0 & 0 & 0 & 0 & 0 & 0 & 1 & \Delta t \\
0 & 0 & 0 & 0 & 0 & 0 & 0 & 1
\end{bmatrix} \qquad (10.3.2)
$$

To obtain the process noise covariance matrix $\mathbf{Q}$, we resort to the methods given in Chapter 3. Note from Eq. (10.3.1) that we can treat the three Position–Velocity state variable pairs independently:

$$E\left[x_i^2(\Delta t)\right] = \int_0^{\Delta t}\int_0^{\Delta t} u \cdot v \cdot S_p \cdot \delta(u - v)\, du\, dv = \frac{S_p\, \Delta t^3}{3} \quad (10.3.3)$$

$$E\left[x_{i+1}^2(\Delta t)\right] = \int_0^{\Delta t}\int_0^{\Delta t} 1 \cdot 1 \cdot S_p \cdot \delta(u - v)\, du\, dv = S_p\, \Delta t \quad (10.3.4)$$

$$E\left[x_i(\Delta t)x_{i+1}(\Delta t)\right] = \int_0^{\Delta t}\int_0^{\Delta t} 1 \cdot v \cdot S_p \cdot \delta(u - v)\, du\, dv = \frac{S_p\, \Delta t^2}{2} \quad (10.3.5)$$

for $i = 1$, 3, and 5.

The equations involving the clock states x_7 and x_8 are developed in the next section as Eqs. (10.4.1)–(10.4.3). With these, the process noise covariance matrix is as follows:

$$\mathbf{Q} = \begin{bmatrix} \frac{S_p\, \Delta t^3}{3} & \frac{S_p\, \Delta t^2}{2} & 0 & 0 & 0 & 0 & 0 & 0 \\[2ex] \frac{S_p\, \Delta t^2}{2} & S_p\, \Delta t & 0 & 0 & 0 & 0 & 0 & 0 \\[2ex] 0 & 0 & \frac{S_p\, \Delta t^3}{3} & \frac{S_p\, \Delta t^2}{2} & 0 & 0 & 0 & 0 \\[2ex] 0 & 0 & \frac{S_p\, \Delta t^2}{2} & S_p\, \Delta t & 0 & 0 & 0 & 0 \\[2ex] 0 & 0 & 0 & 0 & \frac{S_p\, \Delta t^3}{3} & \frac{S_p\, \Delta t^2}{2} & 0 & 0 \\[2ex] 0 & 0 & 0 & 0 & \frac{S_p\, \Delta t^2}{2} & S_p\, \Delta t & 0 & 0 \\[2ex] 0 & 0 & 0 & 0 & 0 & 0 & S_f\, \Delta t + \frac{S_g\, \Delta t^3}{3} & \frac{S_g\, \Delta t^2}{2} \\[2ex] 0 & 0 & 0 & 0 & 0 & 0 & \frac{S_g\, \Delta t^2}{2} & S_g\, \Delta t \end{bmatrix}$$

$$(10.3.6)$$

where S_p, S_f, and S_g are white noise spectral amplitudes defined in Eq. (10.3.1). ∎

The determination of the spectral amplitude S_p for the position random process is at best a "guesstimate" based roughly on expected vehicle dynamics.

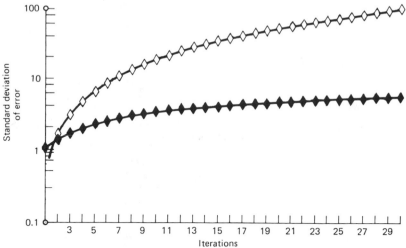

Figure 10.7. Growth of the integrated random walk ($\Diamond$) for the Position–Velocity model versus the random walk ($\blacklozenge$) for the Position model.

In many vehicular applications, such as an aircraft or an automobile in near-constant velocity motion, the random perturbations to the intended straight-line path are greater in the horizontal plane than in the vertical. We can take advantage of this in the locally level reference frame chosen by selecting a value for S_p that is lesser for the altitude channel (states x_5 and x_6 here) than for the other two horizontal ones. The variance associated with a random-walk process, such as those of the velocity and the range drift states in the model given above, grows linearly with time. However, for the integrated random-walk, which is used for position in the PV model, the variance grows in proportion with the cubic function of time, as indicated by Eq. (10.3.3). Figure 10.7 compares the rates of growth of the standard deviation between the integrated random-walk and the random-walk processes for the case where $S_p = 1$. Although this information may seem trivial where the data update rate is high, usually once a second for a GPS receiver, it is nevertheless useful to optimize a filter in certain situations where the data are either interrupted or their quality may be poor for brief periods of time.

Not surprisingly then, the PV model also becomes inadequate for cases where the near-constant velocity assumption does not hold, that is, in the presence of severe accelerations. To accommodate acceleration in the process model, it is appropriate to add another degree of freedom for each position state. Although we can easily add one more integrator to obtain a Position–Velocity–Acceleration (PVA) model, a stationary process, such as the Gauss–Markov process, is perhaps more appropriate than the nonstationary random walk for acceleration (Fig. 10.8). This is in accordance with real-life physical situations, where vehicular acceleration is usually brief and seldom sustained. The state vector for this PVA model then becomes 11-dimensional with the addition of three more acceleration states. Derivation of the corre-

Figure 10.8. Position – Velocity – Acceleration model for high (accelerative) dynamics observer.

Table 10.1 Correlated Errors Affecting Pseudorange: Approximate Statistical Parameters for a Stationary Observer (10, 11, 12)

Error Component	Standard Deviation[a]	Time Constant	Determining Factors
Satellite position	3, 3 m	$\frac{1}{2}$ h	GPS Control Segment
Iono refraction with correction	1, 5 m	$\frac{1}{2}$ h	Solar flare activity; local time of day
Tropo refraction with correction	2, 2 m	1 h	Local atmospheric conditions
Multipath	1, 5 m	10 min	Low elevations ($< 15°$); interfering signal paths
Selective availability	0, 30 m	3 min	GPS Control Segment

[a]The two numbers given are for the PPS (authorized) and SPS (unauthorized) operational modes, respectively.

sponding state transition and the process noise covariance matrices will be left as part of an exercise in Problem 10.2. It should be pointed out here that, although acceleration is being accounted for in some stochastic form, the exercise is an approximation at best. There is no exact way to fit deterministic accelerations into any of the random process dynamics suitable for a Kalman filter model. With the PVA model, the Kalman filter will do a better job of estimation than the PV model, but it may still be inadequate by other measures by optimality. To address deterministic dynamics, the best form of modeling will lie in using, if available, direct measurements of acceleration, or even higher-order dynamics. Further discussion of this topic will be presented in Section 10.6.

Generally, the models above are adequate for the Kalman filtering of GPS measurements. However, greater complexity can be added to the model by accounting for some of the correlated errors alluded to earlier in the chapter and listed in Table 10.1 with their approximate statistical characteristics.

EXAMPLE 10.2

One way to handle the error components given in Table 10.1 is to lump them together into a single process. A good way to model the lumped variable is with a scalar Gauss–Markov process due to its stationary property as well as its simplicity (one state variable). Assigning one state to each satellite adds as many variables to the state vector as there are satellites. Any correlation in the error states between satellites can be considered negligible. From Table 10.1 it is obvious that the dominant component is selective availability if it is operated at its nominal levels. Hence, a good choice of parameters to use for the Gauss–Markov process that is designed to approximate the lumped error state would be a variance of $(30 \text{ m})^2$ and a time constant of 3 min.

$$x_{k+1} = \phi x_k + w_k$$

where $\phi = e^{-\Delta t/180} = 0.994$ $(\Delta t = 1 \text{ sec})$

$$E\{x_{k+1}^2\} = E\{(\phi x_k + w_k)^2\}$$

$$= \phi^2 E\{x_k^2\} + E\{w_k^2\}$$

$$E\{w_k^2\} = E\{x^2\}(1 - \phi^2) = 9.945 \text{ m}^2$$

where $E\{x^2\} = E\{x_k^2\} = (30 \text{ m})^2$ for all k.   ∎

Measurement Linearization

The measurement situation for GPS is clearly nonlinear from Eq. (10.1.1). Linearization of a measurement of this form has already been covered in Section 9.1 and will not be reiterated here. We will simply evaluate the partial derivatives necessary to obtain the linearized equations about an approximate observer location $\mathbf{x}_0 = [x_0, y_0, z_0]^T$. This nominal point of linearization $\mathbf{x}_0$ is sometimes based on an estimate of the true observer location $\mathbf{x}$ although, in general, its choice may be arbitrary.

$$\frac{\partial \psi_i}{\partial x} = \frac{-(X_i - x_0)}{\sqrt{(X_i - x_0)^2 + (Y_i - y_0)^2 + (Z_i - z_0)^2}}$$

$$\frac{\partial \psi_i}{\partial y} = \frac{-(Y_i - y_0)}{\sqrt{(X_i - x_0)^2 + (Y_i - y_0)^2 + (Z_i - z_0)^2}}$$

$$\frac{\partial \psi_i}{\partial z} = \frac{-(Z_i - z_0)}{\sqrt{(X_i - x_0)^2 + (Y_i - y_0)^2 + (Z_i - z_0)^2}} \qquad (10.3.7)$$

for $i = 1, \ldots, 4$.

From a geometrical perspective, the partial derivative vector for each satellite i, $[(\partial \psi_i/\partial x), (\partial \psi_i/\partial y), (\partial \psi_i/\partial z)]^T$ as given in Eq. (10.3.7), is actually the unit direction vector pointing from the satellite to the observer, the direction being specified by the negative sign in the equation. In classical navigation geometry, the components of this unit vector are often called *direction cosines*. The resulting measurement vector equation for pseudorange as the observable is then given by (without noise):

$$
\begin{bmatrix} \psi_1 \\ \psi_2 \\ \psi_3 \\ \psi_4 \end{bmatrix} - \begin{bmatrix} \hat{\psi}_1(\mathbf{x}_0) \\ \hat{\psi}_2(\mathbf{x}_0) \\ \hat{\psi}_3(\mathbf{x}_0) \\ \hat{\psi}_4(\mathbf{x}_0) \end{bmatrix} = \begin{bmatrix} \dfrac{\partial \psi_1}{\partial x} & \dfrac{\partial \psi_1}{\partial y} & \dfrac{\partial \psi_1}{\partial z} & 1 \\[2ex] \dfrac{\partial \psi_2}{\partial x} & \dfrac{\partial \psi_2}{\partial y} & \dfrac{\partial \psi_2}{\partial z} & 1 \\[2ex] \dfrac{\partial \psi_3}{\partial x} & \dfrac{\partial \psi_3}{\partial y} & \dfrac{\partial \psi_3}{\partial z} & 1 \\[2ex] \dfrac{\partial \psi_4}{\partial x} & \dfrac{\partial \psi_4}{\partial y} & \dfrac{\partial \psi_4}{\partial z} & 1 \end{bmatrix} \begin{bmatrix} \Delta x \\ \Delta y \\ \Delta z \\ c\,\Delta t \end{bmatrix}
$$

where

$$\psi_i = \text{measured pseudorange}$$
$$\mathbf{x}_0 = \text{nominal point of linearization based on } [x_0, y_0, z_0]^T \text{ and predicted receiver time}$$
$$\hat{\psi}_i(\mathbf{x}_0) = \text{predicted pseudorange based on } \mathbf{x}_0$$
$$[\Delta x, \Delta y, \Delta z]^T = \text{difference vector between true location } \mathbf{x} \text{ and } \mathbf{x}_0$$
$$c\,\Delta t = \text{range equivalent of the receiver timing error}$$

To formulate the measurement equation for the Kalman filter, we start with Eq. (10.2.1) where the measured pseudorange is given by

$$\rho = \psi + \beta_\rho + v_\rho$$

On linearizing this equation, we get

$$\rho - \hat{\rho}(\mathbf{x}_0) = \frac{\partial \psi}{\partial \mathbf{x}} \Delta \mathbf{x} + \beta_\rho + v_\rho \tag{10.3.8}$$

From the above equation, the residual term on the left-hand side becomes the measurement for the filter. From the right-hand side, $\partial \psi/\partial \mathbf{x}$ becomes the filter parameter $\mathbf{H}$, and $\Delta \mathbf{x}$ is the state vector. As pointed out previously, the time-correlated term β_ρ is usually ignored in the model because of the difficulty in estimating it.

There are two choices available when it comes to implementing a linearization for the Kalman filter. We shall refer the reader to earlier discussions in Chapter 9 for details on the *linearized* and the *extended* Kalman filters.

To complete the description of the eight-state Kalman filter model of Example 10.1, the four-satellite measurement equation is given by

$$
\begin{bmatrix} z_1 \\ z_2 \\ z_3 \\ z_4 \end{bmatrix} = \begin{bmatrix} h_x^{(1)} & 0 & h_y^{(1)} & 0 & h_z^{(1)} & 0 & 1 & 0 \\ h_x^{(2)} & 0 & h_y^{(2)} & 0 & h_z^{(2)} & 0 & 1 & 0 \\ h_x^{(3)} & 0 & h_y^{(3)} & 0 & h_z^{(3)} & 0 & 1 & 0 \\ h_x^{(4)} & 0 & h_y^{(4)} & 0 & h_z^{(4)} & 0 & 1 & 0 \end{bmatrix} \begin{bmatrix} x_1 \\ x_2 \\ x_3 \\ x_4 \\ x_5 \\ x_6 \\ x_7 \\ x_8 \end{bmatrix} + \begin{bmatrix} v_{\rho 1} \\ v_{\rho 2} \\ v_{\rho 3} \\ v_{\rho 4} \end{bmatrix} \quad (10.3.9)
$$

where

z_i = measurement residuals—measured minus predicted pseudorange

$v_{\rho i}$ = pseudorange measurement noise sequence

$$[h_x^{(i)}, h_y^{(i)}, h_z^{(i)}] = \left[\frac{\partial \psi_i}{\partial x}, \frac{\partial \psi_i}{\partial y}, \frac{\partial \psi_i}{\partial z} \right] \text{ for } i = 1, \ldots, 4$$

To summarize, Eq. (10.3.9) can be written in a vector form as

$$
\mathbf{z}_\rho = \boldsymbol{\rho} - \hat{\boldsymbol{\rho}}(\mathbf{x}_0) = \mathbf{H}_\rho \mathbf{x} + \mathbf{v}_\rho \quad (10.3.10)
$$

The corresponding covariance matrix for the measurement noise sequence $\mathbf{v}_\rho$ for C/A-code pseudorange data is

$$
\mathbf{R}_\rho = \begin{bmatrix} r_\rho & 0 & 0 & 0 \\ 0 & r_\rho & 0 & 0 \\ 0 & 0 & r_\rho & 0 \\ 0 & 0 & 0 & r_\rho \end{bmatrix} \quad (10.3.11)
$$

Equation (10.3.11) assumes uncorrelated measurement errors among satellites. Based on this assumption, the measurements may be sequentially processed as scalar measurement data (see Section 6.3). Although it was noted in Section 10.2 that the variance of the pseudorange measurement white noise is rather small, in actual practice a slightly larger variance is often chosen for r_ρ. A conservative choice of $(15 \text{ m})^2$, for example in the absence of Selective Availability, will do well to help offset effects of other correlated measurement noise unaccounted for in the PV or the PVA models given here.

10.4
CRYSTAL CLOCK MODELING

The two-state clock model that was embedded in the GPS Kalman filter in Section 10.3 makes good sense intuitively. It simply says that we expect both the oscillator frequency and phase to random walk over reasonable spans of time. We now wish to look at how the Q parameters of the state model may be determined from the more conventional Allan variance parameters that are often used to describe clock drift. We begin by writing the expressions for the variances and covariances for the general two-state model shown in Fig. 10.9.

Following the state assignments used in Eq. (10.3.1), the clock states given here are x_7 and x_8. Let the elapsed time since initiating the white noise inputs be Δt. Then, using the methods given in Chapter 3, we have

$$E\left[x_7^2(\Delta t)\right] = \int_0^{\Delta t}\int_0^{\Delta t} 1 \cdot 1 \cdot S_f \cdot \delta(u - v)\, du\, dv$$

$$+ \int_0^{\Delta t}\int_0^{\Delta t} u \cdot v \cdot S_g \cdot \delta(u - v)\, du\, dv$$

$$= S_f\, \Delta t + \frac{S_g\, \Delta t^3}{3} \tag{10.4.1}$$

$$E\left[x_8^2(\Delta t)\right] = \int_0^{\Delta t}\int_0^{\Delta t} 1 \cdot 1 \cdot S_g \cdot \delta(u - v)\, du\, dv = S_g\, \Delta t \tag{10.4.2}$$

$$E\left[x_7(\Delta t)x_8(\Delta t)\right] = \int_0^{\Delta t}\int_0^{\Delta t} 1 \cdot v \cdot S_g \cdot \delta(u - v)\, du\, dv = \frac{S_g\, \Delta t^2}{2} \tag{10.4.3}$$

Now let us concentrate on the variance of state x_7. In particular, let us form the rms value of x_7 time-averaged over the elapsed time Δt.

$$\text{Average rms } x_7 = \left\{\sqrt{S_f\, \Delta t + \frac{S_g\, \Delta t^3}{3}}\right\}(\Delta t)^{-1}$$

$$= \sqrt{\frac{S_f}{\Delta t} + \frac{S_g \Delta t}{3}} \tag{10.4.4}$$

Figure 10.9. General two-static model describing clock errors. The independent white noise inputs u_7 and u_8 have spectral amplitudes of S_f and S_g.

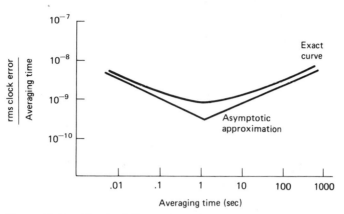

Figure 10.10.   Sketch of Allan variance for two-state clock model.

Next, it is helpful to sketch the response given by Eq. (10.4.4) by using logarithmic scales for both the ordinate and abscissa. This is shown in Fig. 10.10. The asymptotic approximation is shown as a V-shaped plot, and the exact response is shown as a smooth curve. Plots like those shown in Fig. 10.10 are known as Allan variance plots, and they are a convenient way of presenting clock stability information. All one has to do to find the amount of rms drift that occurs during a specified period Δt is to read off the appropriate ordinate and multiply by Δt.

Drift characteristics of real crystal clocks have been studied extensively over the past few decades (13, 14). Figure 10.11 shows an asymptotic Allan variance plot for a typical crystal clock with good (but not exceptional) stability (15). Note that the plot for the real-life crystal oscillator shows a flat portion at the bottom of the "V," whereas the plot for our two-state model exhibits no such flat spot. This flat portion of the experimental plot is sometimes called the

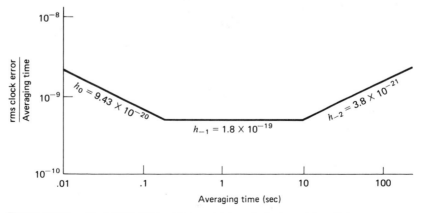

Figure 10.11.   Asymptotic Allan variance plot for a typical crystal oscillator.

flicker floor and is due to the flicker noise, which was omitted in the two-state model. There was good reason for this omission. Flicker noise gives rise to a term in the variance expression that is of the order of Δt^2, and it is impossible to model this term exactly with a finite-order state model (16). Thus we have a modeling dilemma. One approximate solution to the problem is to simply evaluate the theoretical V of the two-state model so as to obtain a better match in the flicker floor region.* This leads to a compromise model that exhibits somewhat higher drift than the true experimental values for both small and large averaging times. However, it yields a better match in the flicker floor region than would be obtained if one were to ignore the flicker component completely. The amount of elevation of the V that is required to give a good compromise will depend on the width of the flicker floor for the particular oscillator under consideration; hence, this calls for a certain amount of engineering judgment.

We now return to the original problem of determining the white noise spectral amplitudes S_f and S_g of our two-state model. The Allan variance plot of Fig. 10.11 will be used to illustrate the procedure. The key Allan variance parameters that describe clock drift for averaging times a few decades on either side of 1 sec are h_0, h_{-1}, and h_{-2} (15). We begin by ignoring the flicker term completely. We then compare terms of similar order in Δt in Eq. (10.4.1) and the q_{11} term in Eq. (60) of Ref. (15), which is repeated in the footnote for convenience. This leads to the correspondence

$$S_f \sim \frac{h_0}{2}$$

$$S_g \sim 2\pi^2 h_{-2}$$

Now, if we were to choose S_f and S_g according to these relationships, our two-state model would match the sloping portions of the actual experimental plots perfectly, but we would have considerable mismatch in the flicker floor

---

* The approximate solution given here is simpler and less rigorous than the one given in Ref. (15). A word of caution is in order, though, about using the model given in Ref. (15). There are mistakes in the Q matrix given by Eq. (60) in the reference. The correct expressions for the q_{12} and q_{22} terms are

$$q_{12} = q_{21} = h_{-1}\,\Delta t + \pi^2 h_{-2}\,\Delta t^2$$

$$q_{22} = \frac{h_0}{2\,\Delta t} + 4h_{-1} + \tfrac{8}{3}\pi^2 h_{-2}\,\Delta t$$

The q_{11} term is correct in the reference and is repeated here for convenience:

$$q_{11} = \frac{h_0}{2}\,\Delta t + 2h_{-1}\,\Delta t^2 + \tfrac{2}{3}\pi^2 h_{-2}\,\Delta t^3$$

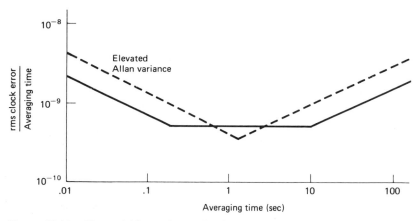

Figure 10.12. Elevated Allan variance plot shown as dashed plot.

region. Therefore, for the clock of Fig. 10.11, it would seem reasonable to elevate the V of the two-state model by a factor of about two to get a better all-around compromise. This is shown as the dashed V plot in Fig. 10.12. Thus, in our final design we must let S_f and S_g be a factor of four larger than the values obtained with the flicker noise neglected. The final design values for S_f and S_g are

$$S_f = 2h_0 \qquad (10.4.5)$$

$$S_g = 8\pi^2 h_{-2} \qquad (10.4.6)$$

Once we have set the values of S_f and S_g, we can compute the q_{11}, q_{12}, and q_{22} terms of our clock model from Eqs. (10.4.1) through (10.4.3), given the step size Δt of the Kalman filter.

When one considers the rounding effect in the true variance plots (in contrast to the asymptotic plots), it can be verified that the compromise choice of S_f and S_g given by Eqs. (10.4.5) and (10.4.6) yields a good match in the flicker floor region. By coincidence, in this example, this region also corresponds to an averaging time of about 1 sec, which is a typical Kalman filter update interval for a GPS receiver. Thus, if our receiver clock had the Allan variance characteristic shown in Fig. 10.11, we would have achieved a good match in the region of special interest. Some mismatch outside this region is acceptable in this application, because the system is completely observable with four satellites in view, and the system is not highly sensitive to the choice of Q parameters. Furthermore, erring on the side of degraded performance is probably better than being overly optimistic about clock stability. Thus we have a safe clock model for this example.

Before closing, it should be mentioned that precision clocks can have widely diverse Allan variance characteristics, even within the family of crystal

Table 10.2   Typical Allan Variance Parameters for Various Timing Standards

Timing Standard	h_0	h_{-1}	h_{-2}
Crystal	$2(10^{-19})$	$7(10^{-21})$	$2(10^{-20})$
Ovenized Crystal	$8(10^{-20})$	$2(10^{-21})$	$4(10^{-23})$
Rubidium	$2(10^{-20})$	$7(10^{-24})$	$4(10^{-29})$

clocks. Thus, one must treat each case individually and work out an approximate model that fits the application at hand. Table 10.2 gives typical values of h_0, h_{-1}, and h_{-2} for various types of timing standards widely used in GPS receivers.

10.5
EFFECTS OF POOR SATELLITE GEOMETRY

One of the more prominent characteristics of GPS is the nongeostationary nature of its satellite orbits. This has the effect of making the measurement (**H**) matrix time-varying. As a consequence, the observability of the system also varies with time. The instantaneous measure of the degree of observability commonly used, called the Geometric Dilution of Precision* (GDOP), is literally a "snapshot" rms error for the position and time solution caused by one meter rms of error in the measurements.

For a given set of satellites, there may be times when the GDOP condition becomes singular. A typical plot for such an instance can be seen in Fig. 10.13, a condition sometimes figuratively called the "GDOP chimney." GDOP chimneys are generally avoidable by changing satellite combinations if enough satellites are available to do so. However, we now consider the case where a GDOP chimney is unavoidable, and look at its effect on a Kalman filter solution.

Within the vicinity of the GDOP chimney, the high GDOP condition causes a deterioration in position solution accuracy. As an example, the error for an instantaneous (unfiltered) solution when the GDOP is 50, assuming a measurement of accuracy of 10 m, would then be an intolerable 500 m (17). Fortunately, some relief can be obtained with filtering by relying partially on data from outside the chimney window that are at lower GDOP levels. In this regard, the Kalman filter can be considered an *information manager* that weights each piece of data it uses according to the observability conditions at the time of acquisition. In other words, the information collected during the

---

* For the measurement equation $\mathbf{z} = \mathbf{Hx} + \mathbf{v}$, GDOP $= \sqrt{\text{trace}(\mathbf{H}^T\mathbf{H})^{-1}}$ (see Ref. 17).

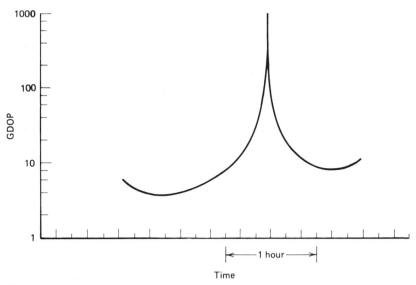

Figure 10.13.   A typical incidence of a "GDOP chimney."

GDOP chimney is given proportionately less weight by the filter than the information collected before the onset of the poor geometric condition.

At the other end of the balance, if the GDOP is constant throughout, the accuracy of the information collected, with regard to the random process it serves to estimate, also degrades with time. When the two opposing effects of GDOP and the aging of data are put together, the compromise results in situations depicted by the plots of solution error shown in Fig. 10.14. In this comparison, the three cases shown differ in the relative spectral amplitudes of their process noise. This reflects the dependence of the maximal peak of position error (near the point of GDOP singularity) on the anticipated state dynamics that are modeled in the Kalman filter. The more stable (i.e., more predictable) the observer's vehicle and clock dynamics are, the lesser the effect of the GDOP chimney on position accuracy will be felt. On the other hand, under high dynamical stress, the random process with a larger process noise spectral amplitude will cause the filter to "forget" past information faster, and it has to depend on more recently collected data, thereby reacting with a response that follows more closely to the actual GDOP chimney. Note that the latter is really a limiting condition when the process noise gets infinitely large, and the filter becomes "memoryless" from one iteration to the next.

This property of the Kalman filter has been somewhat forgotten in GPS applications. It is easily overlooked when the observability conditions relating to satellite geometry are favorable. However, where we must deal with situations that use the satellite resources to the fullest, particularly problems that require redundant data from more than four satellites, favorable satellite geometries are not always assured. This is where optimal estimation with Kalman filtering becomes indispensable.

Figure 10.14. Comparison of the choice of process noise spectral amplitudes on the peak error during a GDOP chimney encounter.

10.6
GPS WITH INTERNAL AIDING

The primary mode of positioning with GPS is achieved by means of ranging on the pseudorandom code. But, although the individual pseudorange measurement is relatively noisy, improved position accuracy can be realized by using some kind of averaging process to reduce the variance of the estimation noise. The longer the averaging time, the smaller the average variance becomes. Unfortunately, the price one pays for a long averaging time is a sluggish response to high-frequency changes in the state dynamics, such as those encountered in a highly maneuverable jet fighter. Such a sluggish response also introduces errors. To avoid the shortcomings that are associated with the noisiness of the pseudorange measurements, supplementary measurements are sometimes used.

Delta Range (Velocity) Aiding

The most common supplementary measurement used in almost all standard GPS receivers is the delta range, which is an approximate measure of velocity obtained from the Doppler frequency. To formulate the delta range measurement equation for the Kalman filter, we start with Eq. (10.2.2) from Section

10.2, where the measured carrier phase is given by

$$\phi = \psi + N_\phi + \beta_\phi + v_\phi$$

Delta range is formed by the finite difference between two carrier phase measurements made over a small time interval:

$$\delta \triangleq \phi - \phi^-$$
$$= (\psi - \psi^-) + (\beta_\phi - \beta_\phi^-) + (v_\phi - v_\phi^-)$$

N_ϕ is the same in both ϕ and ϕ^- and hence cancels out. When the above equation is linearized, we get*

$$\delta - \hat{\delta}(\mathbf{x}_0) = \frac{\partial \psi}{\partial \mathbf{x}}(\Delta \mathbf{x} - \Delta \mathbf{x}^-) + (\beta_\phi - \beta_\phi^-) + (v_\phi - v_\phi^-)$$

If the delta range is properly normalized by the time interval, we get

$$\delta' - \hat{\delta}'(\mathbf{x}_0) \approx \frac{\partial \psi}{\partial \mathbf{x}} \Delta \dot{\mathbf{x}} + \dot{\beta}_\phi + v_\delta$$

where $\delta' = \delta/\Delta t$, $\hat{\delta}'(\mathbf{x}_0) = \hat{\delta}(\mathbf{x}_0)/\Delta t$, $\Delta \dot{\mathbf{x}}$ is the time derivative of position, and $\dot{\beta}_\phi$ is the time derivative of the time-correlated error. As for the pseudorange, the residual term on the left-hand side of the above equation becomes the measurement for the filter. On the right-hand side, $\partial \psi/\partial \mathbf{x}$ becomes the filter parameter **H**, and $\Delta \dot{\mathbf{x}}$ forms the velocity portion of the state vector. The measurement model for the delta range is given in Eq. (10.6.1). The term $\dot{\beta}_\phi$ is negligible and is thus ignored.

$$\begin{bmatrix} \delta_1 \\ \delta_2 \\ \delta_3 \\ \delta_4 \end{bmatrix} - \begin{bmatrix} \hat{\delta}_1(\mathbf{x}_0) \\ \hat{\delta}_2(\mathbf{x}_0) \\ \hat{\delta}_3(\mathbf{x}_0) \\ \hat{\delta}_4(\mathbf{x}_0) \end{bmatrix} = \begin{bmatrix} 0 & h_x^{(1)} & 0 & h_y^{(1)} & 0 & h_z^{(1)} & 0 & 1 \\ 0 & h_x^{(2)} & 0 & h_y^{(2)} & 0 & h_z^{(2)} & 0 & 1 \\ 0 & h_x^{(3)} & 0 & h_y^{(3)} & 0 & h_z^{(3)} & 0 & 1 \\ 0 & h_x^{(4)} & 0 & h_y^{(4)} & 0 & h_z^{(4)} & 0 & 1 \end{bmatrix} \begin{bmatrix} x_1 \\ x_2 \\ x_3 \\ x_4 \\ x_5 \\ x_6 \\ x_7 \\ x_8 \end{bmatrix} + \begin{bmatrix} v_{\delta 1} \\ v_{\delta 2} \\ v_{\delta 3} \\ v_{\delta 4} \end{bmatrix}$$

$$(10.6.1)$$

---

* The term $\hat{\delta}(\mathbf{x}_0)$ is usually approximated by using a finite difference, such as

$$\hat{\delta}(\mathbf{x}_0) \approx \frac{\hat{\rho}(\mathbf{x}_0)|_t - \hat{\rho}(\mathbf{x}_0)|_{t-\Delta}}{\Delta}$$

The $\hat{\delta}(\mathbf{x}_0)$ approximation, over the time interval in question, takes into account the changing satellite position relative to observer motion made in accordance with the state estimation dynamics.

where

$$\delta_i = \text{measured delta range (in units of velocity)}$$
$$\hat{\delta}_i(\mathbf{x}_0) = \text{predicted delta range based on the nominal}$$
$$\text{point of linearization } \mathbf{x}_0$$
$$\upsilon_{\delta i} = \text{delta range measurement noise sequence}$$
$$[h_x^{(i)}, h_y^{(i)}, h_z^{(i)}] = \left[\frac{\partial \psi_i}{\partial x}, \frac{\partial \psi_i}{\partial y}, \frac{\partial \psi_i}{\partial z}\right] \text{ for } i = 1, \dots, 4$$
$$[x_1, x_2, \dots, x_8]^T = \text{state vector defined in Eq. (10.3.1)}$$

Here also, just as with Eq. (10.3.8), Eq. (10.6.1) can be written in a vector form as

$$\mathbf{z}_\delta = \boldsymbol{\delta} - \hat{\boldsymbol{\delta}}(\mathbf{x}_0) = \mathbf{H}_\delta \mathbf{x} + \mathbf{v}_\delta \qquad (10.6.2)$$

The corresponding covariance matrix for the measurement noise sequence $\mathbf{v}_\delta$ in Eq. (10.6.2) is

$$\mathbf{R}_\delta = \begin{bmatrix} r_\delta & 0 & 0 & 0 \\ 0 & r_\delta & 0 & 0 \\ 0 & 0 & r_\delta & 0 \\ 0 & 0 & 0 & r_\delta \end{bmatrix}$$

In addition to velocity, the delta range measurements also contribute some position information to the Kalman filter. The velocity is regarded as constant over the entire sampling interval between measurements. This is understandably a better approximation in low dynamics than in high dynamics. In the case of the latter, significant velocity changes can occur within the sampling interval. However, since the Kalman filter needs to assume constant velocity over the entire duration, the true velocity profile cannot properly be accounted for in high dynamics using the model given by Eq. (10.6.1).

Carrier Aiding

To provide the Kalman filter with a more accurate picture of the velocity profile, another form of measurement can be used. Instead of integrating the Doppler frequency for a short interval for each measurement sample as was the case for the delta range, a continuously running integration of the Doppler can be carried out (see Section 10.2). The resulting samples of continuous carrier phase are all referred to an arbitrary but common starting point. We start out with the carrier phase measurement model of Eq. (10.2.2) and then form the difference between the carrier phase measurements at time t_k and the reference time t_0. [By itself, Eq. (10.2.2) is not very useful because of the gross error uncertainty in N_ϕ.]

$$\phi = \psi + N_\phi + \beta_\phi + \upsilon_\phi$$
$$\phi_k - \phi_0 = (\psi_k - \psi_0) + (\beta_{\phi k} - \beta_{\phi 0}) + (\upsilon_{\phi k} - \upsilon_{\phi 0}) \qquad (10.6.3)$$

Linearizing the above equation, we get

$$\left[\phi_k - \hat{\phi}_k(\mathbf{x}'_k) \right] - \left[\phi_0 - \hat{\phi}_0(\mathbf{x}'_0) \right]$$

$$= \left\{ \frac{\partial \psi_k}{\partial \mathbf{x}} \Delta \mathbf{x}_k - \frac{\partial \psi_0}{\partial \mathbf{x}} \Delta \mathbf{x}_0 \right\} + \left(\beta_{\phi k} - \beta_{\phi 0} - \upsilon_{\phi 0} \right) + \upsilon_{\phi k}$$

$$= \left\{ \frac{\partial \psi_k}{\partial \mathbf{x}} \Delta \mathbf{x}_k - \frac{\partial \psi_0}{\partial \mathbf{x}} \Delta \mathbf{x}_0 \right\} + \xi_k + \upsilon_{\phi k}$$

where we define a lumped error term $\xi_k \triangleq \beta_{\phi k} - \beta_{\phi 0} - \upsilon_{\phi 0}$.

Rewriting, the measurement model for the continuous carrier phase for each satellite is given by Eq. (10.6.4).

$$\left[\phi_k - \hat{\phi}_k(\mathbf{x}'_k) \right] - \left[\phi_0 - \hat{\phi}_0(\mathbf{x}'_0) \right]$$

$$= \mathbf{h}_k(\mathbf{x}'_k) \Delta \mathbf{x}_k(\mathbf{x}'_k) - \mathbf{h}_0(\mathbf{x}'_0) \Delta \mathbf{x}_0(\mathbf{x}'_0) + \xi_k + \upsilon_{\phi k} \qquad (10.6.4)$$

where

$$\phi_k = \text{measured phase at } t_k$$
$$\hat{\phi}_k(\mathbf{x}'_k) = \text{predicted phase linearized about } \mathbf{x}'_k \text{ at } t_k$$
$$\mathbf{x}'_k = \text{nominal point of linearization about } \mathbf{x}_k$$
$$\mathbf{h}_k = \left[\frac{\partial \psi}{\partial x}, \frac{\partial \psi}{\partial y}, \frac{\partial \psi}{\partial z}, 1 \right]$$
$$\Delta \mathbf{x}_k(\mathbf{x}'_k) = \text{state estimate error vector } (\mathbf{x}_k - \mathbf{x}'_k) = [\Delta x \; \Delta y \; \Delta z \; c \Delta t]^T$$
$$\xi_k = \text{lumped time-correlated error state}$$
$$\upsilon_{\phi k} = \text{carrier phase measurement noise}$$

The lumped time-correlated error state ξ_k is a random-walk–type process used to accommodate the long-term instability of a position solution obtained from the carrier phase data. This nonstationarity is due to small frequency-type errors in the Doppler that, when integrated, results in an unbounded error in position over time (8).

We then combine the measurement model for the continuous carrier phase of Eq. (10.6.4) with the model for pseudorange of Eq. (10.3.8). Although it is outside our scope of examination, it should be pointed out that a correlation structure does exist between the time-correlated terms β_ρ, associated with pseudorange, and ξ, associated with carrier phase. For our purpose here, we shall ignore β_ρ as was done in Eq. (10.3.10). The measurement equations for this integrated Kalman filter are described by

$$z_{\rho k} \triangleq \rho_k - \hat{\rho}(\mathbf{x}'_k) = \mathbf{h}_k(\mathbf{x}'_k)\Delta\mathbf{x}_k(\mathbf{x}'_k) + v_{\rho k}$$

$$z_{\phi k} \triangleq \left[\phi_k - \hat{\phi}_k(\mathbf{x}'_k) \right] - \left[\phi_0 - \hat{\phi}_0(\mathbf{x}'_0) \right]$$

$$= \mathbf{h}_k(\mathbf{x}'_k)\Delta\mathbf{x}_k(\mathbf{x}'_k) - \mathbf{h}_0(\mathbf{x}'_0)\Delta\mathbf{x}_0(\mathbf{x}'_0) + \xi_k + v_{\phi k}$$

When the above equations are rewritten in matrix form, we then have for each satellite

$$\begin{bmatrix} z_{\rho k} \\ z_{\phi k} \end{bmatrix} = \begin{bmatrix} \mathbf{h}_k(\mathbf{x}'_k) & \mathbf{0} & 0 \\ \mathbf{h}_k(\mathbf{x}'_k) & -\mathbf{h}_0(\mathbf{x}'_0) & 1 \end{bmatrix} \begin{bmatrix} \Delta\mathbf{x}_k(\mathbf{x}'_k) \\ \Delta\mathbf{x}_0(\mathbf{x}'_0) \\ \xi_k \end{bmatrix} + \begin{bmatrix} v_{\rho k} \\ v_{\phi k} \end{bmatrix} \quad (10.6.5)$$

The results of each processing step from the integrated Kalman filter will include the state estimates for $\Delta\hat{\mathbf{x}}_k(\mathbf{x}'_k)$ and $\Delta\hat{\mathbf{x}}_0(\mathbf{x}'_0)$. When these error quantities are added to their respective nominal points of linearization, the total solutions can be reconstructed as follows:

$$\hat{\mathbf{x}}_k = \mathbf{x}'_k + \Delta\hat{\mathbf{x}}_k(\mathbf{x}'_k)$$

$$\hat{\mathbf{x}}_0 = \mathbf{x}'_0 + \Delta\hat{\mathbf{x}}_0(\mathbf{x}'_0)$$

To avoid the complexity of formally implementing the integrated filter of Eq. (10.6.5), a simpler alternative exists in using a complementary filter approach (8). The Kalman filter scheme shown in Figure 10.15 carries the same philosophy of data integration as the complementary filter introduced in Chapter 4. The continuous carrier phase data are used to obtain a reference trajectory based in large part on the dynamical motion of the observer. The reference trajectory therefore paints a very faithful picture of that motion. In the complementary scheme of Figure 10.15, this reference information is then subtracted out of the pseudorange measurement data with the residual fed into

Figure 10.15.   Complementary Kalman filter combining continuous carrier phase and pseudorange data from GPS signal measurements.

a Kalman filter. With the dynamical part of the information totally extracted from the pseudorange, the residual data then appears to the Kalman filter as if the whole measurement situation were a stationary one. The Kalman filter can then afford to lengthen its averaging time constant to significantly reduce the large noise variance associated with the pseudorange measurements. The filter output when recombined with the reference trajectory data provides a final position solution that is highly responsive to any kind of dynamical motion and has small estimation error. This is the complementary facet of this form of data integration.

Instead of starting out with Eq. (10.6.3), let us modify this equation to form a difference between carrier phase measurements at t_k and t_{k-1}:

$$\phi_k - \phi_{k-1} = (\psi_k - \psi_{k-1}) + \xi_k' + v_{\phi k} \tag{10.6.6}$$

where $\xi_k' = \beta_{\phi k} - \beta_{\phi k-1} - v_{\phi k-1}$. Then linearizing Eq. (10.6.6) about the nominal point $\mathbf{x}_{k-1}$:

$$\left[\phi_k - \hat{\phi}_k(\mathbf{x}_{k-1})\right] - \left[\phi_{k-1} - \hat{\phi}_{k-1}(\mathbf{x}_{k-1})\right]$$

$$= \left\{\frac{\partial \psi_k}{\partial \mathbf{x}}(\mathbf{x}_k - \mathbf{x}_{k-1}) - \frac{\partial \psi_{k-1}}{\partial \mathbf{x}}(\mathbf{x}_{k-1} - \mathbf{x}_{k-1})\right\} + \xi_k' + v_{\phi k}$$

$$= \frac{\partial \psi_k}{\partial \mathbf{x}}(\mathbf{x}_k - \mathbf{x}_{k-1}) + \xi_k' + v_{\phi k}$$

The pseudorange model given in Eq. (10.2.1) is reiterated here for time t_k and then linearized about the nominal point $\mathbf{x}_{k-1}$:

$$\rho_k = \psi_k + \beta_{\rho k} + v_{\rho k}$$

$$\rho_k - \hat{\rho}(\mathbf{x}_{k-1}) = \frac{\partial \psi_k}{\partial \mathbf{x}}(\mathbf{x}_k - \mathbf{x}_{k-1}) + \beta_{\rho k} + v_{\rho k}$$

Of course, the exact value of $\mathbf{x}_{k-1}$ is unknown, although $\hat{\mathbf{x}}_{k-1}$ is a close enough substitute for it. Any error that arises from this approximation is small and can be absorbed by $\beta_{\rho k}$.

When the carrier-phase–derived reference trajectory is subtracted from the pseudorange measurement (see Figure 10.15), information pertaining to the dynamical motion (described below by $\mathbf{x}_k - \mathbf{x}_{k-1}$) is completely removed:

$$\left[\rho_k - \hat{\rho}(\mathbf{x}_{k-1})\right] - \left\{\left[\phi_k - \hat{\phi}_k(\mathbf{x}_{k-1})\right] - \left[\phi_{k-1} - \hat{\phi}_{k-1}(\mathbf{x}_{k-1})\right]\right\}$$

$$= \frac{\partial \psi_k}{\partial \mathbf{x}}(\mathbf{x}_k - \mathbf{x}_{k-1}) + \beta_{\rho k} + \upsilon_{\rho k} - \frac{\partial \psi_k}{\partial \mathbf{x}}(\mathbf{x}_k - \mathbf{x}_{k-1}) - \xi'_k - \upsilon_{\phi k}$$

$$= \beta_{\rho k} - \xi'_k + \upsilon_{\rho k} - \upsilon_{\phi k}$$

$$\approx s_k + \upsilon_{\rho k}$$

where $s_k \triangleq \beta_{\rho k} - \xi'_k$, becomes a state of the Kalman filter model representing one satellite, and we also treat the carrier phase measurement noise $\upsilon_{\phi k}$ as negligible compared with the pseudorange measurement noise $\upsilon_{\rho k}$. As a minimum, a four-state Kalman filter model is needed to represent four satellites.

To summarize, the Kalman filter model description for multiple satellites is as follows:

Process model:

$$\mathbf{x}_{k+1} = \mathbf{x}_k + \mathbf{w}_k$$

Measurement model:

$$\left[\rho_k - \hat{\rho}(\mathbf{x}_{k-1})\right] - \left\{\left[\phi_k - \hat{\phi}_k(\mathbf{x}_{k-1})\right] - \left[\phi_{k-1} - \hat{\phi}_{k-1}(\mathbf{x}_{k-1})\right]\right\} = \mathbf{x}_k + \mathbf{v}_{\rho k}$$

The carrier-derived position is highly responsive to any dynamical motion, but its low-frequency error grows over time without bound. As such, the feedback mode of an extended Kalman filter arrangement is necessary to provide stability to this scheme. This complementary filter is also very similar in idea to the aided inertial navigation system block diagram of Fig. 6.2 in a feedback arrangement, with the continuous carrier phase playing the role of the inertial system and the pseudorange coming from the "Aiding Sources" block.

Before moving on, we should point out that one of the main difficulties usually associated with the use of the continuous carrier phase has been the issue of cycle slips. Recall that the continuous carrier phase is a piece of data that is referred to a known initial point, from where incremental changes in phase are integrated so that the exact number of whole cycles accumulated, although unobservable at any instant in time, is properly accounted for. If the signal is weak or, worse, altogether lost, the integration process becomes prone to error, which affects most critically the integer portion of the carrier phase count. This pathological situation is called cycle slip. There are various methods to detect cycle slips, and we refer the interested reader to the suggested references for detailed information (8, 18).

10.7
GPS WITH EXTERNAL INERTIAL AIDING

Although the signal-plus-noise character of the continuous carrier phase is in many ways similar to inertial measurements, their operational contributions differ significantly. The carrier phase is part of the GPS signal, and as long as a receiver is tracking the GPS signal, the carrier phase is a free piece of data that can be used to improve the accuracy of the navigation solution computed by the receiver. However, if the signal is interrupted, both code and carrier data are lost. This is where an autonomous inertial measurement unit is beneficial for a whole different reason. There are applications where signal obscuration occurs fairly frequently and the application demands continuity in navigation despite interrupted GPS availability. A typical obscuration example is that of a car going through a tunnel; another might be fuselage shading of the antenna when an aircraft makes a sharp banking maneuver. If, as with these examples, the interruptions are brief, inertial or any type of aiding data for that matter can enable the navigation system to coast along until GPS signal tracking can be reestablished.

As with continuous carrier phase data, inertial measurements when doubly integrated give position information with a very good high-frequency response to dynamical motion. This inertially derived position information can then be used to generate a reference trajectory much the same way as the continuous carrier phase was utilized previously. This is the so-called *complementary filter* approach that was used before. However, inertial measurements contain additional information beyond that of spatial position and velocity; platform attitude is also available. In the following example we show how this is taken into account.

EXAMPLE 10.3

We begin by forming a nine-dimensional reference trajectory from the inertial measurements that consist of three each of position, velocity, and attitude states. In the complementary arrangement, the Kalman filter's role is to estimate the deviation of the reference from the truth trajectory. This is then the established state space. Hence, it is important to note that the state model for the Kalman filter here consists of inertial *error* quantities, rather than "total" dynamical quantities, as is true of the stand-alone GPS case. Correspondingly, the random process model should correspond to the random character of the errors in the inertial sensor.

For the inertial error model, the chosen navigation reference frame will follow the locally level convention established back in Section 10.1 where x points East, y points North, and z is the altitude above the reference WGS-84 ellipsoid. Without going into the details of their derivation, we simply work from the undamped inertial error equations for a slow-moving vehicle (19). The error models are represented in Figs. 10.16, 10.17, and 10.18, which show separate channels for East, North, and altitude. These can be regarded, in part, as

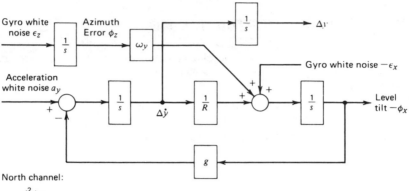

North channel:

$$\frac{d^2\Delta y}{dt^2} = a_y + g\,\phi_x$$

$$-\frac{d\phi_x}{dt} = \frac{1}{R}\frac{d\Delta y}{dt} + \omega_y\phi_z - \epsilon_x$$

Azimuth error:

$$\frac{d\phi_z}{dt} = \epsilon_z$$

Figure 10.16. North channel error model. (ω_y = Platform angular rate about y axis.)

generic models that are representative of a typical North-oriented locally level platform INS. Let the state variables be organized in the following order for the purpose of convenience in partitioning:

$$x_1 = \text{East position error}$$

$$x_2 = \text{East velocity error}$$

$$x_3 = \text{tilt about North } (y) \text{ axis}$$

$$x_4 = \text{North position error}$$

$$x_5 = \text{North velocity error}$$

$$x_6 = \text{tilt about East } (x) \text{ axis}$$

$$x_7 = \text{altitude error}$$

$$x_8 = \text{altitude rate error}$$

$$x_9 = \text{azimuth attitude error}$$

$$x_{10} = \text{range bias state}$$

$$x_{11} = \text{range drift state}$$

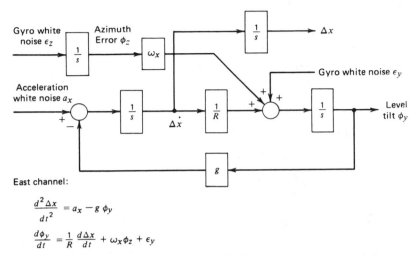

East channel:

$$\frac{d^2 \Delta x}{dt^2} = a_x - g\, \phi_y$$

$$\frac{d\phi_y}{dt} = \frac{1}{R}\frac{d\Delta x}{dt} + \omega_x \phi_z + \epsilon_y$$

Figure 10.17 East channel error model. (ω_x = Platform angular rate about x axis.)

Altitude channel:

$$\frac{d^2 \Delta z}{dt^2} - 2\frac{g}{R}\Delta z = a_z$$

Figure 10.18. Altitude channel error model.

The state transition matrix approximated with $e^{-\mathbf{A}\,\Delta t} \approx \mathbf{I} + \mathbf{A}\,\Delta t$ is given by

$$
\boldsymbol{\phi} \approx
\begin{bmatrix}
1 & \Delta t & 0 & 0 & 0 & 0 & 0 & 0 & 0 & 0 & 0 \\
0 & 1 & -g\,\Delta t & 0 & 0 & 0 & 0 & 0 & 0 & 0 & 0 \\
0 & \dfrac{\Delta t}{R} & 1 & 0 & 0 & 0 & 0 & 0 & \omega_x\,\Delta t & 0 & 0 \\
0 & 0 & 0 & 1 & \Delta t & 0 & 0 & 0 & 0 & 0 & 0 \\
0 & 0 & 0 & 0 & 1 & -g\,\Delta t & 0 & 0 & 0 & 0 & 0 \\
0 & 0 & 0 & 0 & \dfrac{\Delta t}{R} & 1 & 0 & 0 & \omega_y\,\Delta t & 0 & 0 \\
0 & 0 & 0 & 0 & 0 & 0 & 1 & \Delta t & 0 & 0 & 0 \\
0 & 0 & 0 & 0 & 0 & 0 & \dfrac{-2g\,\Delta t}{R} & 1 & 0 & 0 & 0 \\
0 & 0 & 0 & 0 & 0 & 0 & 0 & 0 & 1 & 0 & 0 \\
0 & 0 & 0 & 0 & 0 & 0 & 0 & 0 & 0 & 1 & \Delta t \\
0 & 0 & 0 & 0 & 0 & 0 & 0 & 0 & 0 & 0 & 1
\end{bmatrix}
$$

The measurement vector associated with the Kalman filter consists only of GPS pseudorange measurements. The measurement matrix $\mathbf{H}$, as a result, comprises the usual components of the unit direction vectors for the requisite four satellites needed for the model:

$$
\mathbf{H} = \begin{bmatrix}
h_x^{(1)} & 0 & 0 & h_y^{(1)} & 0 & 0 & h_z^{(1)} & 0 & 0 & 1 & 0 \\
h_x^{(2)} & 0 & 0 & h_y^{(2)} & 0 & 0 & h_z^{(2)} & 0 & 0 & 1 & 0 \\
h_x^{(3)} & 0 & 0 & h_y^{(3)} & 0 & 0 & h_z^{(3)} & 0 & 0 & 1 & 0 \\
h_x^{(4)} & 0 & 0 & h_y^{(4)} & 0 & 0 & h_z^{(4)} & 0 & 0 & 1 & 0
\end{bmatrix}
$$

Since the contribution of measurement noise from the inertial unit is comparatively minuscule, the measurement noise covariance matrix $\mathbf{R}$ should account mostly for the GPS pseudorange noise. Use of the $\mathbf{R}$ matrix from Eq. (10.3.11) is quite suitable here. ∎

The model given in the example above is a simple scheme to provide aiding for a GPS navigation system with inertial data. For tutorial purposes, the example is quite adequate in illustrating the fundamentals of the system integration. Clearly, the complexity of the model, especially for the inertial part, can be raised significantly to improve the estimation accuracy. It should be pointed out that the vertical channel (altitude) used for the example model here is unstable when it is required to operate in the absence of GPS. However, without any other source of data available, the double integration of vertical acceleration is about the best we can do to obtain altitude. Unfortunately, this process leads to an altitude error that builds up exponentially. This arrangement is probably quite acceptable if the loss of GPS is brief, that is, on the order of a few minutes. In instances where the inertial portion has to operate by itself for extended durations of time, the addition of another sensor such as the barometric altimeter is necessary to bound the error growth. The barometric altitude might then be combined together with the inertial data in another complementary coupling such as the one suggested in Problem 4.14.

Before we conclude this section, we must mention an often encountered problem in GPS/INS integration. It is not always that the designer or system integrator commands the luxury of having to solve problems without design constraints. When one is dealing with avionics, for instance, the modularity of the equipment might necessitate the treatment of a GPS receiver as an instrument sensor that outputs position and velocity data. To use these data to aid an Inertial Navigation System (INS), a system integrator is faced with using the output of some Kalman filter in the GPS receiver to feed into another integrated Kalman filter, which blends the data from all sensors to calibrate the INS. This situation is commonly known as the *cascaded Kalman filters* problem where the output of the first filter, corrupted by time-correlated estimation error, is fed into a second filter, which "thinks" the noise corrupting the data is white. This problem has been studied somewhat extensively, sometimes under the heading of embedded Kalman filters, chiefly because of its potential usefulness in large modular systems (20, 21). We will not pursue this further here, though.

10.8
DIFFERENTIAL CARRIER PHASE POSITIONING

Thus far, we have been concerned with only the more conventional applications of GPS, where code measurements provide the primary source of positioning information. However, some innovative work that began in the field of terrestrial surveying at the close of the 1970s has led to the blossoming of a very exciting and productive application area during the past decade (22, 23). Terrestrial surveying requires precisions an order of magnitude or more better than those designed for conventional GPS positioning. Interferometric techniques developed in the past for radiating sources such as quasars were adapted for GPS by using the pure tone of the carrier signal only and, in turn, reaping the benefits of its accuracy. The method is a differential one where the solution obtained is only for the position of one observer with respect to another reference observer. The errors that corrupt the observations of two observers that are relatively close together are strongly correlated and can be removed in the differential GPS solution. Today, static surveying using the GPS carrier phase observable has matured beyond its experimental stages and is widely used out in the field.

EXAMPLE 10.4

In static surveying where the object is to determine the baseline vector between two points, the observable consists of the difference between carrier phase measurements made simultaneously at the two ends of the baseline. The one-dimensional situation depicted in Fig. 10.19 shows the relationship between the total phase delay and the baseline separation through the known geometry.

When the receivers make their initial measurements, the total phase delay is, of course, unobservable. What is measurable as the phase delay, however, is known to be in error by an integral number of cycles, a quantity called the *initial integer ambiguity*. Thus, this one-dimensional measurement equation contains two variables to be solved for: (1) the baseline separation, and (2) the initial integer ambiguity. The Kalman filter is certainly suited for handling estimation problems of such a nature.

In the GPS situation, where only the position-related quantities are of interest, we can eliminate the range (or clock)-related biases by carrying out a second difference across satellites of the single difference measurements that are originally formed across the two stations (see Eqs. 10.7.1 and 10.8.2). It is

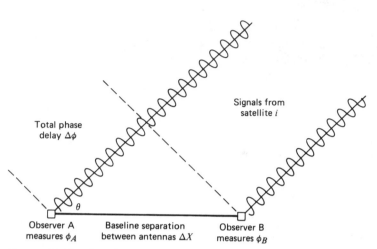

Figure 10.19. One-dimensional phase difference measurement model. Baseline separation must be small enough to maintain linearity assumptions.

assumed that the set of phase measurements used to form the double differences are made simultaneously in order to do this (24).

Single difference:

$$\Delta\phi^{(i)} = \phi_A^{(i)} - \phi_B^{(i)} \tag{10.8.1}$$

Double difference:

$$\nabla\Delta\phi^{(i)} = \Delta\phi^{(i)} - \Delta\phi^{(i+1)} \tag{10.8.2}$$

for $i = 1, 2\dots$, where ϕ_A is the phase measured by Observer A and ϕ_B is the phase measured by Observer B from satellite i.

The example used here is based on a model by Brown and Hwang [see Ref. (25)] where the initial integer ambiguities, although integer in nature, are estimated as continuous random variables.

$$
\begin{bmatrix} \nabla\Delta\phi^{(1)} \\ \nabla\Delta\phi^{(2)} \\ \nabla\Delta\phi^{(3)} \end{bmatrix}_{k+1}
=
\begin{bmatrix}
\nabla h_x^{(1)} & \nabla h_y^{(1)} & \nabla h_z^{(1)} & 1 & 0 & 0 \\
\nabla h_x^{(2)} & \nabla h_y^{(2)} & \nabla h_z^{(2)} & 0 & 1 & 0 \\
\nabla h_x^{(3)} & \nabla h_y^{(3)} & \nabla h_z^{(3)} & 0 & 0 & 1
\end{bmatrix}
\begin{bmatrix} \Delta x \\ \Delta y \\ \Delta z \\ \nabla N^{(1)} \\ \nabla N^{(2)} \\ \nabla N^{(3)} \end{bmatrix}_k
+
\begin{bmatrix} \nabla v^{(1)} \\ \nabla v^{(2)} \\ \nabla v^{(3)} \end{bmatrix}_k
$$

$$\tag{10.8.3}$$

where

$$\nabla\Delta\phi^{(1)}, \nabla\Delta\phi^{(2)}, \nabla\Delta\phi^{(3)} = \text{double difference measurement set}$$

$$\nabla h_x^{(i)}, \nabla h_y^{(i)}, \nabla h_z^{(i)} = i\text{th unit direction vector differenced across satel-}$$
lites

$$\Delta x, \Delta y, \Delta z = \text{position vector}$$

$$\nabla N^{(1)}, \nabla N^{(2)}, \nabla N^{(3)} = \text{initial integer ambiguity vector}$$

$$\nabla v^{(1)}, \nabla v^{(2)}, \nabla v^{(3)} = \text{measurement noise vector}$$

The process model is trivial, since all elements of the state vector are random biases. Therefore, the **Q** matrix is zero. Here, we have an example of where it is safe to add no process noise only because this problem is confined to establishing the fixed values of the state vector and does not run indefinitely.

The measurement noise covariance matrix depends on the combination used for the double difference. For the combinations shown in Eq. (10.8.2), the resulting double differences formed are no longer uncorrelated, as can be assumed for the single differences of Eq. (10.8.1). For $i = 1, 2, 3$:

$$E\, \nabla\Delta\phi^{(i)}\, \nabla\Delta\phi^{(i+1)} = E\left[\Delta\phi^{(i)} - \Delta\phi^{(i+1)}\right]\left[\Delta\phi^{(i+1)} - \Delta\phi^{(i+2)}\right]$$

$$= -E\, \Delta\phi^{(i+1)}\, \Delta\phi^{(i+1)}$$

$$= -E\left[\Delta\phi^{(i)}\right]^2$$

since

$$E\, \Delta\phi^{(i)}\, \Delta\phi^{(j)} = \begin{cases} r_\phi, & i = j \\ 0, & i \neq j \end{cases}$$

Therefore, the covariance matrix for the differenced measurement noise vector $[\nabla v^{(1)}, \nabla v^{(2)}, \nabla v^{(3)}]^T$ is given by

$$\mathbf{R}_{\nabla\Delta\phi} = \begin{bmatrix} 2r_\phi & -r_\phi & 0 \\ -r_\phi & 2r_\phi & -r_\phi \\ 0 & -r_\phi & 2r_\phi \end{bmatrix} \qquad \blacksquare$$

In practice, r_ϕ is in the range of 1 to 5 mm for a stationary receiver as is the case here with static surveying. Here again, as with r_ρ for a code-ranging measurement model, the choice of a value for r_ϕ may also depend greatly on time-correlated errors that are unaccounted for in this model. These time-correlated errors, which may originate from all the usual GPS error sources (see

Table 10.1), arise in this situation only when the differential assumption, made when posing this problem, starts to break down.

It should be noted here that the simple formulation of Example 10.4 can be carried further and the accuracy can be refined by taking advantage of the certainty that the initial integer ambiguity states are, in fact, integers. In Refs. (25) and (26), the Magill Adaptive scheme (see Section 9.3) was used where a parallel bank of Kalman filters, each assuming an integer value for a hypothesis, processes the same double difference phase measurement sequence. Although cumbersome at first glance, the parallel filter viewpoint serves as an excellent conceptual starting point, whereupon many simplifications are available to make the problem a tractable one indeed, especially in a computational sense! This statistical scheme provides a convenient use of Kalman filtering to solve the multiple-hypothesis-testing problem formulated in an optimal way.

10.9
OTHER AREAS OF APPLICATIONS

The applications discussed in this chapter by no means cover the entire myriad of GPS problems that are or can be solved by Kalman filtering techniques. Attention here has mostly been devoted to the primary user issues of navigation, positioning, and timing. Even so, the coverage has not been comprehensive in the interest of trying to keep the chapter to a tutorial pace. For those who are interested in delving further into other applications of Kalman filtering in the area of GPS, several references are hereby suggested.

The effects of Selective Availability (SA) and its modeling are often foremost concerns for the nonauthorized user of the system (12, 27). Kinematic positioning takes the carrier phase methods of static surveying one step further by introducing motion to the problem (28). The result appears to be, in addition to more efficiently run survey operations, a promising extension of the high accuracies associated with carrier phase surveying to differential positioning (29). GPS has also found itself forming synergistic collaborations with other types of instrumentation (30, 31, 32). In the most natural of combinations, GPS has found an inertial navigation field in the midst of a technological surge in miniaturization activity sparked by the advent of fiberoptic gyros (33). The GPS/inertial combination is especially attractive because the inertial system makes it possible to "coast" through short periods of poor satellite geometry or brief satellite obscurations.

Other GPS areas outside of user equipment where Kalman filtering is being used include orbit determination, satellite oscillator characterization, and ionospheric model parameter estimation. All these involve the postprocessing of data collected by the network of ground monitoring stations of the GPS Control Segment. Smoothing techniques are used to adjust the Kalman estimator states modeled in order to provide the most accurate predictive information transmitted by the satellites (34, 35). Another form of monitoring that is in only the formative stages of development is the checking of GPS transmission integrity. This is an area where Kalman filtering could play a central role in the area of failure detection (36). Also in its formative stages is autonomous navigation, a proposed scheme to enable the GPS satellites to navigate off each other in the event of a breakdown in the principal elements of the Ground Control segment (37).

It would not be an overstatement to say that the usefulness of GPS has, by this time, far surpassed the purpose originally envisioned by its early designers. Despite more than 10 years of extensive development, it does not appear that the well of ideas for GPS applications is about to dry up. This optimism ensures that applied Kalman filtering will play a continued role in the wealth of GPS applications already in development as well as those that are yet untapped.

PROBLEMS

10.1. Measurement models for processing GPS pseudorange measurements generally do not take into account the bias errors that result from ionospheric delays. Instead, deterministic corrections for these biases can be made to the measurements prior to filtering. These predictive corrections are derived from a global model that takes into account variabilities such as satellite elevation, local time of day, and geographic latitude. It has been suggested, however, that because of their crudeness in accuracy, the corrections actually do little to improve positioning accuracy and need not be used (38). Without any deterministic corrections, a major portion of the ionospheric delay common to all measurements will be absorbed into the time error state, thus affecting time but not position.

Assume that the ionospheric delay (D) has a Gaussian probability density with mean m and variance s^2. Also assume that the deterministic model used to predict the ionospheric delay (d) does so with a density function identical to that for D. Without going into a detailed derivation, suppose that the variance of any one of the three position states x can be written in the form

$$\text{Var } x = G_1^2 \, \text{Var} \, (D_1 - D_2) + G_2^2 \, \text{Var} \, (D_2 - D_3) + G_3^2 \, \text{Var} \, (D_3 - D_4)$$

where G_1, G_2, G_3 are scale factors dependent on satellite geometry.

(a) Show that to obtain an improvement in the accuracy of a position state x with the use of the deterministic correction, the correlation coefficient between the delay D and its corresponding correction d must be greater than 0.5.

(b) From experimental data, it has been found that the deterministic correction reduces the root mean square (rms) of the iono delay by 60 percent of its uncorrected value [i.e., $E(D - d)^2 = 0.4 ED^2$] (10). Find the range of values for m in terms of the variance s^2 for which the use of the deterministic correction is justified.

(c) Suppose that the same experimental data taken in 1985 suggests the distribution of the iono delay can be approximated by a Gaussian probability density function with a mean of 12 m and a variance of $(5\text{ m})^2$. Based on these data, does the deterministic correction to the iono delay in fact clearly help improve position accuracy?

10.2.   Derive the process noise matrix $\mathbf{Q}$ for the Position–Velocity–Acceleration (PVA) dynamical model whose transfer function block diagram is shown in Figure 10.8 in terms of the integration interval Δt. Choose the three states required for a complete description to be position, velocity, and acceleration.

10.3.   Unlike most real-life positioning problems, the positioning of a train is unique in being essentially a one-dimensional measurement situation—its one degree of freedom being along the track to which it is constrained. To take advantage of this reduction in the model dimensionality, the exact trajectory of the railroad track must be known and linearized for the approximate vicinity of the train's location.

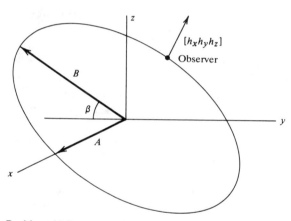

Problem 10.3

(a) Begin by assuming the standard four-variable GPS measurement model:

$$\tilde{z} = \rho - \hat{\rho} = \begin{bmatrix} h_x(t) h_y(t) h_z(t) 1 \end{bmatrix} \begin{bmatrix} \Delta x \\ \Delta y \\ \Delta z \\ c\,\Delta t \end{bmatrix}$$

Let the rail track (see figure) be described by the set of parametric equations:

$$x = A \cos \psi$$

$$y = B \sin \psi \cos \beta$$

$$z = B \sin \psi \sin \beta$$

where ψ is the parametric variable and β is a fixed angle of inclination
of the elliptical track whose semimajor and semiminor axes are A and
B. Using a linearized approximation of the above set of parametric
equations, rewrite the measurement model such that the state variables
comprise just ψ and $c\,\Delta t$. What is the new linear measurement connection vector **h** written in terms of $h_x(t), h_y(t), h_z(t)$?
(b) What is the minimum number of satellites required to solve the
positioning problem of the train with given track trajectory information
as is the case here?
(c) Formulate a similar measurement model that allows for dynamical
motion [start out with the Position–Velocity model of Eq. (10.3.9)] by
also assuming the same track described above.
(d) What is the random process model (specify ϕ and **Q** matrices) for the
modified state vector that corresponds to that given in Eq. (10.3.6) for
the Position–Velocity model.

10.4.* Sometimes, a simple *moving average* process is used to filter GPS data.
A direct solution of the position is obtained from the linearized measurement
equations (Eqs. 10.3.7) for each set of measurements made. This unfiltered
solution constitutes one data point. When a fixed number N of consecutive data
points are collected, an average is computed. Then when a new data point is
available, it displaces the oldest data point in the collection. With this, the
averaging time window slides over to absorb new information and "forgets" old
information after N samples, thereby providing the averaging needed to get
solution accuracy while concurrently adapting to changes in the solution.

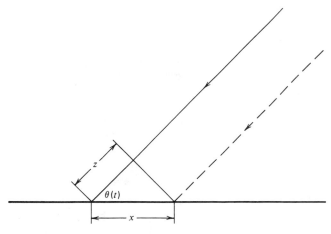

Problem 10.4

The main philosophical difference between this crude estimator and the
Kalman filter lies in the weighting of the data samples. The simple average
draws equally weighted information from among all the samples in the window

(see Section 5.1). To compare the moving average and the Kalman filter, a one-dimensional model can be simulated.

Process:

$$x_{k+1} = x_k + w_k; \qquad Q = E[w_k^2] = 100 \text{ m}^2$$

Measurement:

$$z_k = [-\cos \theta(t_k)] \cdot x_k + v_k; \qquad R = E[v_k^2] = (15 \text{ m})^2$$

Let

$$\theta(t_k) = 75° + \frac{t_k}{43082} \cdot 360°; \qquad t_k = 10k, \quad k = 0, 1, \ldots, 300$$

(a) Generate a sample state trajectory and the corresponding measurement data for the process given above. Then use the measurement data in the Kalman filter to obtain a time profile of the position estimates. The error profile can then be obtained by taking the difference between the position estimates and the truth-state trajectory generated.

(b) Use the same measurement data generated in part (a) to assess the moving-average process as well. To obtain the unfiltered solution, the Kalman filter can be used by setting the value of Q to be very high, say 10^{10}. The unfiltered solution is then filtered with the moving average process

$$\hat{x}_k = \frac{1}{N} \sum_{i=k-N}^{k} X_k; \qquad k = N, N+1, N+2, \ldots$$

Use $N = 10$. Here, again, as in part (a), generate the error profile by taking the difference between the moving-average solution and the truth-state trajectory.

(c) Compare the results obtained for the error profiles using the Kalman filter and the moving-average filter. Explain the differences.

It should be pointed out that the software provided can be used to obtain part (a) and the unfiltered solution of part (b) by using the Monte Carlo simulation mode of operation. The function to perform the moving-average process, however, is not provided in the software and must be made independently.

10.5.* An elementary five-state Kalman filter model for a GPS receiver is discussed in Section 10.3. Even though this model is used only rarely in practice because of its limitations, it is still useful as a means of assessing the effect of varying key filter parameters. We will now pursue this further.

We begin by defining the basic five-state model as follows:

$x_1 = \Delta x$

$x_2 = \Delta y$ The position perturbations from the nominal (x, y, z)

$x_3 = \Delta z$

$x_4 = c\,\Delta t$ Range equivalent of clock bias

$x_5 = c\,\Delta\dot{t}$ Range equivalent of clock drift

The x_1, x_2, x_3 states will each be modeled as independent random-walk processes as shown in Figure 10.4. The range (clock) states x_4, x_5 will be modeled as shown in Figure 10.9 with appropriate scaling to account for units of range and range rate. First, write out the general expressions for the state transition matrix $\boldsymbol{\phi}_k$ and the process noise matrix $\mathbf{Q}_k$. Let the step size be denoted as Δt.

We will assume that the measurements consist of four pseudoranges at each update point (no Dopplers) and that the model has been linearized. Therefore, the measurement model will be of the form

$$\mathbf{z}_k = \mathbf{H}_k \mathbf{x}_k + \mathbf{v}_k$$

where $\mathbf{z}_k$ is 4×1, $\mathbf{H}_k$ is 4×5, $\mathbf{x}_k$ is 5×1, and $\mathbf{v}_k$ is 4×1. We will make the usual assumption that the measurement errors are white and have zero cross-correlation. Therefore, $\mathbf{R}_k$ is diagonal. The motion of the GPS satellites will be assumed to be slow relative to the time span of interest here, so we will regard $\mathbf{H}_k$ to be a constant matrix as follows:

$$\mathbf{H}_k = \begin{bmatrix} -0.3523 & -0.0495 & 0.9346 & 1.0 & 0.0 \\ 0.6199 & 0.7406 & 0.2593 & 1.0 & 0.0 \\ -0.9506 & -0.2553 & 0.1764 & 1.0 & 0.0 \\ 0.9613 & 0.2129 & 0.1747 & 1.0 & 0.0 \end{bmatrix}$$

[The specified $\mathbf{H}_k$ matrix would provide a snapshot solution GDOP of about 3.0, which is typical for GPS.] The model is now complete except for numerical values.

   (a) Let each of the white noise spectral amplitudes that drive the random-walk position states be

$$S_p = 1.0 \ \frac{(\text{m/sec})^2}{\text{rad/sec}}$$

(This allows for a modest amount of receiver motion.) Also, let the clock model spectral amplitudes be

$$S_f = 0.4\,(10^{-18})\text{ sec}$$

$$S_g = 1.58\,(10^{-18})\text{ sec}^{-1}$$

Equations (10.4.1), (10.4.2), and (10.4.3) can then be used to find the clock **Q** parameters. [Note that the results of these calculations must be rescaled by the square of the velocity of light (c^2) to yield the correct units.] Also let

$$\text{Step size } \Delta t = 1.0 \text{ sec}$$

$$\text{Measurement error variance} = 225 \text{ m}^2$$

Finally, the filter error covariance matrix is to be initialized as a diagonal matrix as follows:

$$\mathbf{P}_0^- = \begin{bmatrix} (100\text{ m})^2 & 0 & 0 & 0 & 0 \\ 0 & (100\text{ m})^2 & 0 & 0 & 0 \\ 0 & 0 & (100\text{ m})^2 & 0 & 0 \\ 0 & 0 & 0 & (300\text{ m})^2 & 0 \\ 0 & 0 & 0 & 0 & (30\text{ m/sec})^2 \end{bmatrix}$$

Specification of the numerical values is now complete.

It is desirable to have the receiver filter settle to steady-state condition (or nearly so) in about one minute. To test this, cycle the filter error covariance equations through 60 steps and note the rate of convergence. The filter parameters have been assumed to be constant, so the filter's characteristic roots in the z-plane can also be calculated (see Section 7.6). Calculate the characteristic roots and see if they are consistent with the results of the covariance analysis.

(b) Now increase the elements of the $\mathbf{Q}_k$ matrix by a factor of 10. Repeat the experiment of part (a) and note how this change affects both the error-covariance convergence and the characteristic roots. Comparative plots of key elements of the **P** matrices will be helpful here.

(c) Use the same inflated values of $\mathbf{Q}_k$ as in part (b), and now increase the terms of the $\mathbf{R}_k$ matrix by a factor of 10. Perform an analysis similar to that of part (b) and note the effect of increasing the terms of the $\mathbf{R}_k$ matrix.

Before leaving this problem, it should be mentioned that the 5-state filter is overly simple because it does not model velocity. It was chosen here for tutorial purposes, and it should be recognized that the 8-state Position–Velocity model or the 11-state Position–Velocity–Acceleration model are more accurate in the presence of modest or high dynamics.

10.6.* In addition to relative positioning, the continuous-carrier-phase measurement can also be used for absolute *point* positioning as well. A similar type of measurement principle was originally used with the Navy's Transit satellite system (39). Point positioning refers to the determination of a location that is fixed or stationary.

Starting with the general measurement equation for continuous carrier phase given by Eq. (10.6.4), we recognize that in this special case the position states in $\Delta \mathbf{x}_k(\mathbf{x}'_k)$ are constant and the same as those in $\Delta \mathbf{x}_0(\mathbf{x}'_0)$. On the other hand, as before, the timing error state in $\Delta \mathbf{x}_k(\mathbf{x}'_k)$ is dynamic. That equation can be rewritten as follows (recall that this is for a single satellite):

$$
z_k - z_0 = \begin{bmatrix} h_{xk} h_{yk} h_{zk} 1 \; 0 \end{bmatrix} \begin{bmatrix} \Delta x \\ \Delta y \\ \Delta z \\ c\,\Delta t_k \\ c\,\Delta \dot{t}_k \end{bmatrix} - \begin{bmatrix} h_{x0} h_{y0} h_{z0} 1 \; 0 \end{bmatrix} \begin{bmatrix} \Delta x \\ \Delta y \\ \Delta z \\ c\,\Delta t_0 \\ c\,\Delta \dot{t}_0 \end{bmatrix} + \xi_k + v_{\phi k}
$$

On recombining terms, we get

$$
z_k - z_0 = \begin{bmatrix} (h_{xk} - h_{x0})(h_{yk} - h_{y0})(h_{zk} - h_{z0}) 1 \; 0 \end{bmatrix} \begin{bmatrix} \Delta x \\ \Delta y \\ \Delta z \\ c(\Delta t_k - \Delta t_0) \\ c(\Delta \dot{t}_k - \Delta \dot{t}_0) \end{bmatrix} + \xi_k + v_{\phi k}
$$

The measurement model above can be treated as a five-state model by absorbing ξ_k into the position states $\Delta x, \Delta y, \Delta z$, which are otherwise random constants. In absorbing ξ_k, the position states should then be modeled as random-walk processes. The states $c(\Delta t_k - \Delta t_0)$ and $c(\Delta \dot{t}_k - \Delta \dot{t}_0)$ take on the same character as the usual range (clock) states (see Problem 10.5) except that the initial value of the $c(\Delta t_k - \Delta t_0)$ state at t_0 is known to be zero with a probability of one (error variance of zero).

(a) Set up and run a Kalman filter simulation for the model formulated above. For the random walk states Δx, Δy, and Δz, use a spectral amplitude of $S_p = (0.1 \text{ m})^2$ for its process noise. For the timing states, use the model given in Problem 10.5 and a spectral amplitude of $S_f = 0.4(10^{-18})$ sec and $S_g = 1.58(10^{-18})$ sec^{-1}. Let the measurement noise $\mathbf{R} = (0.005)^2 \mathbf{I}$ (**I** is a 4×4 identity matrix). Use the idealized geometry given for a typical GPS example in Appendix B (choose Satellites 1, 2, 3 and 4) to obtain the unit direction vectors for the **H** matrix, and plot the variances of the Δx, Δy, and Δz states over an interval of 300 sec.

(b) Make a second run using the same parameters except with only one satellite (use Satellite 1). Graphically compare in general terms the convergence of the variance of the position states Δx, Δy, and Δz for both cases.

10.7.* The Geometric Dilution of Precision (commonly called GDOP) introduced in Section 10.5 is a measure of the position solution accuracy as a function of satellite geometry. It is the "snapshot" rms error of the combined position and time solution caused by one meter rms of error in the measurements. The term *snapshot* implies that the solution is entirely dependent on the four satellite measurements made at a specific instance in time. In other words, the solution is unfiltered.

The Kalman filter software can be made to set up the required conditions above to derive the necessary terms to obtain the GDOP. Start out by using a four-state model (three position states and a range state) and run the software in the Covariance Analysis mode of operation. For the time-varying $\mathbf{H}$ matrix, use the idealized satellite equations described in Appendix B for Satellites 1, 2, 3, and 5. Choose proper constant values for the $\boldsymbol{\phi}$, $\mathbf{Q}$, $\mathbf{R}$, and $\mathbf{P}_0$ matrices. The square-root of the trace of the updated $\mathbf{P}$ matrix is defined as the GDOP. What is the GDOP value for this set of satellites at $t = 1000$ sec?

REFERENCES CITED IN CHAPTER 10

1. B. W. Parkinson and S. W. Gilbert, "NAVSTAR: Global Positioning System—Ten Years Later," *Proceedings of the IEEE*, 71:10, 1177–1186 (October 1983).
2. *Global Positioning System*, Vols. I, II, and III, Washington, D.C.: The Institute of Navigation, 1980–1986.
3. *Department of Defense World Geodetic System 1984—Its Definition and Relationships with Local Geodetic Systems*, Defense Mapping Agency Technical Report 8350.2, September 1987.
4. J. C. Rambo, "Receiver Processing Software Design of the Rockwell International DoD Standard GPS Receivers," *Proceedings of the 2nd International Technical Meeting of the Satellite Division of the Institute of Navigation*, Colorado Springs, Colo., September 27–29, 1989, pp. 217–225.
5. R. M. Kalafus, J. Vilcans, and N. Knable, "Differential Operation of NAVSTAR GPS," *Navigation, The Jour. of the Inst. of Navigation*, 30:3, 187–204 (Fall 1983).
6. E. G. Blackwell, "Overview of Differential GPS Methods," *Navigation, The Jour. of the Inst. of Navigation*, 32:2, 114–125 (Summer 1985).
7. D. J. Henson, E. A. Collier, and K. R. Schneider, "Geodetic Applications of the Texas Instruments TI4100 GPS Navigator," *Proceedings of the 1st International Symposium on Precise Positioning with the Global Positioning System*, Vol. I, Rockville, Md., April 15–19, 1985, pp. 191–200.

8. P. Y. C. Hwang and R. G. Brown, "GPS Navigation: Combining Pseudo-range with Continuous Carrier Phase Using a Kalman Filter," *Navigation, The Jour. of the Inst. of Navigation*, *37*:2 (Summer 1990).

9. J. M. Ashjaee, "GPS Doppler Processing for Precise Positioning in Dynamic Applications," *Navigation, The Jour. of the Inst. of Navigation*, *32*:4, 370–385 (Winter 1985–86).

10. R. I. Abbot, Y. Bock, C. C. Counselman, R. W. King, S. A. Gourevitch, and B. J. Rosen, "Interferometric Determination of GPS Satellite Orbits," *Proceedings of the 1st International Symposium on Precise Positioning with the Global Positioning*, Vol. I, Rockville, Md., April 15–19, 1985, pp. 63–72.

11. W. A. Feess and S. G. Stephens, "Evaluation of GPS Ionospheric Time Delay Algorithm for Single-Frequency Users," *Proceedings of the IEEE Position Location and Navigation Symposium (PLANS '86)*, Las Vegas, Nev., November 4–7, 1986, pp. 206–213.

12. G. T. Kremer, R. M. Kalafus, P. V. W. Loomis, and J. C. Reynolds, "The Effect of Selective Availability on Differential GPS Corrections," *Proceedings of the 2nd International Technical Meeting of the Satellite Division of the Institute of Navigation*, Colorado Springs, Colo., September 27–29, 1989, pp. 307–312.

13. B. E. Blair (Ed.), *Time and Frequency: Theory and Fundamentals*, NBS Monograph 140 (May 1974).

14. J. A. Barnes, "Models for the Interpretation of Frequency Stability Measurements," *NBS Technical Note 683* (August 1976).

15. A. J. Van Dierendonck, J. B. McGraw, and R. G. Brown, "Relationship Between Allan Variances and Kalman Filter Parameters," *Proceedings of the 16th Annual Precise Time and Time Interval (PTTI) Applications and Planning Meeting*, NASA Goddard Space Flight Center, November 27–29, 1984, pp. 273–293.

16. R. G. Brown, "Kalman Filter Modeling," *Proceedings of the 16th Annual Precise Time and Time Interval (PTTI) Applications and Planning Meeting*, NASA Goddard Space Flight Center, November 27–29, 1984, pp. 261–272.

17. R. J. Milliken and C. J. Zoller, "Principle of Operation of NAVSTAR and System Characteristics," *Global Positioning System*, Vol. I, Washington, D.C.: Institute of Navigation, 1980.

18. J. Westrop, M. E. Napier, and V. Ashkenazi, "Cycle Slips on the Move: Detection and Elimination," *Proceedings of the 2nd International Technical Meeting of the Satellite Division of the Institute of Navigation*, Colorado Springs, Colo., September 27–29, 1989, pp. 31–34.

19. G. R. Pitman (Ed.), *Inertial Guidance*, New York: Wiley, 1962.

20. R. B. Dayton and J. T. Nielson, "A Flight Test Comparison of Two GPS/INS Integration Approaches," *Proceedings of the 2nd International Technical Meeting of the Satellite Division of the Institute of Navigation*, Colorado Springs, Colo., September 27–29, 1989, pp. 269–273.

21. N. A. Carlson, "Federated Filter for Fault-Tolerant Integrated Navigation

Systems," *Proceedings of the IEEE Position Location and Navigation Symposium* (*PLANS '88*), Orlando, Fla., November 29–December 1, 1988, pp. 110–119.

22. C. C. Counselman and S. A. Gourevitch, "Miniature Interferometer Terminals for Earth Surveying: Ambiguity and Multipath with Global Positioning System," *IEEE Transactions on Geoscience and Remote Sensing*, *GE-19*:4, 244–252 (1981).

23. J. D. Bossler, C. C. Goad, and P. L. Bender, "Using the Global Positioning System (GPS) for Geodetic Positioning," *Bulletin Geodesique*, *54*:4, 553–563 (1980).

24. B. W. Remondi, "Modeling the GPS Carrier Phase for Geodetic Applications," *Proceedings of the 1st International Symposium on Precise Positioning with the Global Positioning System*, Vol. I, Rockville, Md., April 15–19, 1985, pp. 325–336.

25. R. G. Brown and P. Y. C. Hwang, "A Kalman Filter Approach to Precision GPS Geodesy," *Navigation, The Jour. of the Inst. of Navigation*, *30*:4, 338–349 (Winter 1983–84).

26. P. Y. C. Hwang and R. G. Brown, "GPS Geodesy: Experimental Results Using the Kalman Filter Approach," *Proceedings of the IEEE Electronics and Aerospace Systems Conference* (*EASCON '85*), Washington, D.C., October 28–30, 1985, pp. 9–18.

27. P. W. McBurney, "A Robust Approach to Reliable Real-Time Kalman Filtering," *Proceedings of the IEEE Position Location And Navigation Symposium* (*PLANS '90*), Las Vegas, Nev., March 21–23, 1990, pp. 549–556.

28. B. W. Remondi, "Performing Centimeter-Level Surveys in Seconds with GPS Carrier Phase: Initial Results," *Navigation, The Jour. of the Inst. of Navigation*, *32*:4, 386–400 (Winter 1985–86).

29. P. Y. C. Hwang, "Kinematic GPS: Resolving Integer Ambiguities on the Fly," *Proceedings of the IEEE Position Location and Navigation Symposium* (*PLANS '90*), Las Vegas, Nev., March 21–23, 1990, pp. 579–586.

30. L. Chin, "Feasibility Study of Using GPS to Calibrate an Instrumentation Radar," *Navigation, The Jour. of the Inst. of Navigation*, *32*:1, 57–67 (Spring 1985).

31. P. Braisted, R. Eschenbach, and A. Tiwari, "Combining LORAN and GPS —The Best of Both Worlds," *Global Positioning System*, Vol. III, Washington, D.C.: Institute of Navigation, 1986.

32. P. Axelrad and B. W. Parkinson, "Closed Loop Navigation and Guidance for Gravity Probe B Orbit Insertion," *Navigation, The Jour. of the Inst. of Navigation*, *36*:1, 45–61 (Spring 1989).

33. A. Mathews, "Utilization of Fiber-Optic Gyros in Inertial Measurement Unit," *Proceedings of the IEEE Position Location and Navigation Symposium* (*PLANS '90*), Las Vegas, Nev., March 21–23, 1990, pp. 147–160.

34. S. S. Russell and J. H. Schaibly, "Control Segment and User Performance," *Global Positioning System*, Vol. I, Washington, D.C.: Institute of Navigation, 1980.

35. E. R. Swift, "NSWC's GPS Orbit/Clock Determination System," *Proceedings of the 1st International Symposium on Precise Positioning with the Global Positioning System*, Vol. I, Rockville, Md., April 15–19, 1985, pp. 51–62.
36. R. G. Brown and P. Y. C. Hwang, "GPS Integrity: Failure Detection by Autonomous Means Within the Cockpit," *Navigation, The Jour. of the Inst. of Navigation*, *33*:4, 335–353 (Fall 1986).
37. M. P. Ananda, H. Bernstein, W. A. Feess, and T. C. Paugstat, "Global Positioning System (GPS) Autonomous User System," *Navigation, The Jour. of the Inst. of Navigation*, *35*:2, 197–216 (Summer 1988).
38. P. S. Jorgenson, "An Assessment of Ionospheric Effects on the GPS User," *Navigation, The Jour. of the Inst. of Navigation*, *36*:2, 195–204 (Summer 1989).

Appendix A

Laplace and Fourier Transforms

Elementary treatments of Laplace and Fourier transforms usually gloss over matters of convergence and formal inversion of these transforms by the inversion integral. There are problems in signal analysis where these matters are important, though (e.g., Wiener filter theory), and thus we will embellish on these ideas here. It is not the intent here to teach linear transform theory from the beginning. We assume that the reader has the usual manipulative skills in Laplace and Fourier transforms that one would normally get in an undergraduate electrical engineering program. [For example, see Mayhan (1) or Kamen (2) for Fourier transforms and Dorf (3) or D'Azzo and Houpis (4) for Laplace transforms.] Here the emphasis will be to place Laplace and Fourier transforms in perspective relative to each other, and to discuss, in particular, formal inversion of these transforms by the inversion integral. We begin with the one-sided Laplace transform.

A.1
THE ONE-SIDED LAPLACE TRANSFORM

Electrical engineers usually first encounter Laplace transforms in circuit analysis, and then again in linear control theory. In both cases the central problem is one of finding the system response to an input initiated at $t = 0$. Since the time history of the system prior to $t = 0$ is summarized in the form of the initial conditions, the ordinary one-sided Laplace transform serves us quite well. Recall that it is defined as

$$F(s) = \int_{0+}^{\infty} f(t)e^{-st}\, dt \qquad (A.1)$$

Table A.1   Common One-Sided Laplace Transform Pairs[a]

Name	Pictorial Description	Laplace Transform
Unit impulse (Area is to right of origin)	$\delta(t)$ ... 0 ... t	1
Unit step	1 ... $f(t) = 1,\quad t > 0$ ... 0 ... t	$\frac{1}{s}$
Unit ramp	$f(t) = t,\quad t \geq 0$ ... 0 ... t	$\frac{1}{s^2}$
[b] General power of t	$f(t) = t^n,\quad t \geq 0$ ... 0 ... t	$\frac{n!}{s^{n+1}}$
Damped exponential	1 ... $f(t) = e^{-at},\quad t > 0$ ... 0 ... t	$\frac{1}{s+a}$
Sine wave	$f(t) = \sin bt,\quad t \geq 0$ ... 1 ... 0 ... t	$\frac{b}{s^2 + b^2}$
Cosine wave	$f(t) = \cos bt,\quad t > 0$ ... 1 ... 0 ... t	$\frac{s}{s^2 + b^2}$

Table A.1 *(Continued)*

Name	Pictorial Description	Laplace Transform

Damped sine wave — $f(t) = e^{-at} \sin bt, \quad t \geq 0$ — $\dfrac{b}{(s+a)^2 + b^2}$

Damped cosine wave — $f(t) = e^{-at} \cos bt, \quad t > 0$ — $\dfrac{s+a}{(s+a)^2 + b^2}$

Delayed positive-time function — $f(t-T), \quad T > 0$ — $F(s)e^{-Ts}$

---

[a] Time functions having a discontinuity at $t = 0$ are intentionally left undefined at the origin in this table. The missing value does not affect the direct transform. See Exercise A.1 at the end of this Appendix for a discussion of the appropriate choice of $f(0)$ to assure compatibility with the inversion integral.

[b] When n is not an integer, $n!$ must be interpreted as the gamma function $\Gamma(n + 1)$.

The defining integral is, of course, insensitive to $f(t)$ for negative t; but, for reasons that will become apparent shortly, we *arbitrarily* set $f(t) = 0$ for $t < 0$ in one-sided transform theory. The integral of Eq. (A.1) has powerful convergence properties because of the e^{-st} term. We know it will always converge somewhere in the right-half s plane, provided that we consider only inputs (and responses) that increase no faster than at some fixed exponential rate. This is usually the case in circuits and control problems, and hence the actual region of convergence is of little concern. A common region of convergence is tacitly assumed to exist somewhere, and we simply adopt a table look-up viewpoint for getting back and forth between the time and complex s domains. For reference purposes a brief list of common transform pairs is given in Table A.1. Note again that we have intentionally defined all time functions in the table to be zero for $t < 0$. We will have occasion later to refer to such functions as *positive-time* type functions. It is also worth mentioning that the impulse function of one-sided transform theory is considered to have all its area to the right of the origin in the limiting process. Thus it is a positive-time function. (The word *function* is abused a bit in describing an impulse, but this is common usage, so it will be continued here.)

A.2
THE FOURIER TRANSFORM

The Fourier transform is used widely in communications theory where we often wish to consider signals that are nontrivial for both positive and negative time. Thus a two-sided transform is appropriate. Recall that the Fourier transform of $f(t)$ is defined as

$$F(j\omega) = \int_{-\infty}^{\infty} f(t)e^{-j\omega t}\,dt \qquad (A.2)$$

We know, through the evolution of the Fourier transform from the Fourier series, that $F(j\omega)$ has the physical significance of signal spectrum. The parameter ω in Eq. (A.2) is $(2\pi) \times$ (frequency in hertz), and in elementary signal analysis we usually consider ω to be real. This leads to obvious convergence problems with the defining integral, Eq. (A.2), and is usually circumvented simply by restricting the class of time functions being considered to those for which convergence exists for real ω. The two exceptions to this are constant (*d-c*) and harmonic (sinusoidal) signals. These are usually admitted by going through a limiting process that leads to Dirac delta functions in the ω domain. Even though the class of time functions allowed is somewhat restrictive, the Fourier transform is still very useful because many physical signals just happen to fit into this class (e.g., pulses and finite-energy signals). If we take convergence for granted, we can form a table of transform pairs, just as we did with Laplace transforms, and Table A.2 gives a brief list of common Fourier transform pairs.

For those who are more accustomed to one-sided Laplace transforms than Fourier transforms, there are formulas for getting from one to the other. These are especially useful when the time functions have either even or odd symmetry. Let $f(t)$ be a time function for which the Fourier transform exists, and let

$$\mathscr{F}[f(t)] = \text{Fourier transform of } f(t)$$

$$F(s) = \text{One-sided Laplace transform of } f(t)$$

Then, if $f(t)$ is even,

$$\mathscr{F}[f(t)] = F(s)|_{s=j\omega} + F(s)|_{s=-j\omega} \qquad (A.3)$$

Table A.2 Common Fourier Transform Pairs

Name	Pictorial Description	Fourier Transform
Damped exponential		$\dfrac{2a}{\omega^2 + a^2}$
Rectangular pulse		$T\,\dfrac{\sin(\omega T/2)}{(\omega T/2)}$
Triangular pulse		$\dfrac{T}{2}\left[\dfrac{\sin(\omega T/4)}{(\omega T/4)}\right]^2$
Gaussian pulse		$\dfrac{\sqrt{\pi}}{a}\,e^{-(\omega^2/4a^2)}$
Symmetric impulse		1
Sinc function (sinc $2Wt$)		$F(j\omega) = \begin{cases} \dfrac{1}{2W}, & \lvert\omega\rvert < 2\pi W \\ 0, & \lvert\omega\rvert > 2\pi W \end{cases}$

Table A.2 *(Continued)*

Name	Pictorial Description	Fourier Transform
Damped sine wave	$e^{-a\|t\|}\sin bt$	$\dfrac{ja}{a^2 + (\omega + b)^2}$ $-\dfrac{ja}{a^2 + (\omega - b)^2}$
Damped cosine wave	$e^{-a\|t\|}\cos bt$	$\dfrac{a}{a^2 + (\omega + b)^2}$ $+\dfrac{a}{a^2 + (\omega - b)^2}$
Constant	1	$2\pi\,\delta(\omega)$
Sine wave	$\sin bt$	$j\pi\,\delta(\omega + b) - j\pi\,\delta(\omega - b)$
Cosine wave	$\cos bt$	$\pi\,\delta(\omega + b) + \pi\,\delta(\omega - b)$

or if $f(t)$ is odd,

$$\mathscr{F}[f(t)] = F(s)\big|_{s=j\omega} - F(s)\big|_{s=-j\omega} \tag{A.4}$$

These formulas follow directly from the defining integrals of the two transforms.

A.3
TWO-SIDED LAPLACE TRANSFORM

The usual Fourier transform with ω real serves us well for much of signal analysis, but there are occasions where we wish to consider functions for which this transform does not exist. In such cases the so-called two-sided Laplace transform is sometimes useful. It is defined as

$$F(s) = \int_{-\infty}^{\infty} f(t)e^{-st}\, dt \tag{A.5}$$

It can be seen to be identical in form to the Fourier transform except for notation, that is, $j\omega$ is replaced with s. Innocent as this change may appear at first glance, it is important, because we now wish to let s be complex. Furthermore, we do not wish to place any restrictions on the type of time functions being considered, other than to say the integral of Eq. (A.5) must converge *somewhere* in the s plane. Now the region of convergence becomes important, as an example will illustrate.

EXAMPLE A.1

Consider the two time functions shown in Fig. A.1. Their respective two-sided Laplace transforms may be found from Eq. (A.5).
For signal $f_1(t)$ of Fig. A.1a,

$$F_1(s) = \int_{-\infty}^{0} e^{at}e^{-st}\, dt = \frac{-1}{s-a}, \qquad \text{for } \mathrm{Re}\,[s] < a$$

For signal $f_2(t)$ of Fig. A.1b,

$$F_2(s) = \int_{0}^{\infty} e^{at}e^{-st}\, dt = \frac{1}{s-a}, \qquad \text{for } \mathrm{Re}\,[s] > a$$

Here we have the uncomfortable situation of two different time functions having the same transform except for sign! There is a saving feature though—their regions of convergence are nonoverlapping. Hence, if we add the qualification of the convergence region to the functional form in the s domain, we then restore the one-to-one transform-pair relationship that we must have in any transform algebra. In this case the entry in our table of transform pairs might

Figure A.1 Time functions for Example A.1. (*a*) Negative-time function. (*b*) Positive-time function.

Table A.3   Two-Sided Laplace Transforms

Time Function	Two-sided Laplace Transform	Region of Convergence
	$\dfrac{2a}{-s^2 + a^2}$	
	$\dfrac{1}{-s + a}$	
	$\dfrac{1}{s - a}$	
	$\dfrac{1}{s}$	
	$\dfrac{1}{-s}$	

Table A.3   *(Continued)*

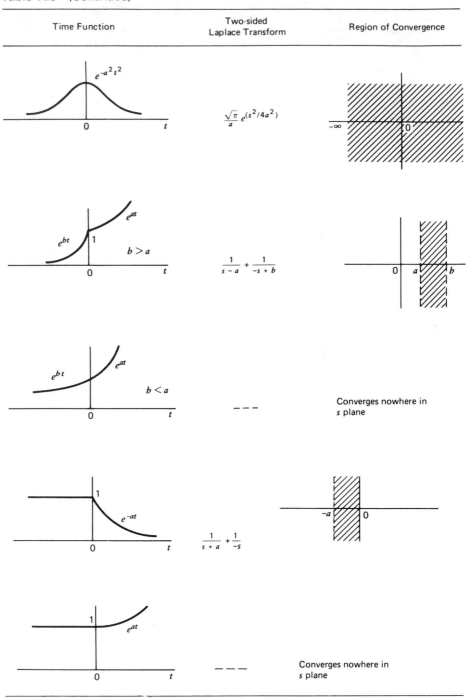

Time Function	Two-sided Laplace Transform	Region of Convergence
$e^{-a^2 t^2}$	$\dfrac{\sqrt{\pi}}{a} e^{(s^2/4a^2)}$	$-\infty$ to ∞
e^{bt}, e^{at}, $b > a$	$\dfrac{1}{s-a} + \dfrac{1}{-s+b}$	a, b
e^{bt}, e^{at}, $b < a$	– – –	Converges nowhere in s plane
e^{-at}	$\dfrac{1}{s+a} + \dfrac{1}{-s}$	$-a$, 0
e^{at}	– – –	Converges nowhere in s plane

appear as follows:

Time Function		Transform
$f(t) = \begin{cases} e^{at}, & t < 0 \\ 0, & t > 0 \end{cases}$		$\dfrac{-1}{s-a}$ *and* region of convergence is $\mathrm{Re}\,[s] < a$
$f(t) = \begin{cases} 0, & t < 0 \\ e^{at}, & t > 0 \end{cases}$		$\dfrac{1}{s-a}$ *and* region of convergence is $\mathrm{Re}\,[s] > a$

Tacking on the region of convergence in our table of transform pairs is a bit cumbersome, but it must be done, or at least tacitly implied somehow.

Table A.3 shows a few common time functions, their two-sided Laplace transforms, and their respective regions of convergence. The list is by no means complete. It is simply intended to show the variety of situations that can occur. Note that a number of the entries have regions of convergence that do not include the $j\omega$ axis, and therefore usual Fourier transforms will not exist for these time functions. Also, two of the time functions shown do not have two-sided Laplace transforms, because the defining integral does not converge anywhere in the s plane. ∎

The Inversion Integral

To compile and work with two-sided transform tables such as Table A.2 would be awkward, to say the least. The addition of the specification of convergence region to the functional form is like adding an extra dimension to the transform. Thus, rather than use cumbersome two-sided transform tables for inversion, it is more convenient to use one-sided transform tables with the appropriate interpretation as to positive- and negative-time parts of the time function. This is where the formal rules of inversion are helpful. Beginning with the Fourier transform, we have the inversion formula

$$f(t) = \frac{1}{2\pi} \int_{-\infty}^{\infty} F(j\omega) e^{j\omega t}\, d\omega \qquad (A.6)$$

Recall that this came from a limiting situation of the complex Fourier series where the expansion interval was allowed to approach infinity. The corresponding inversion integral for the Laplace transform is

$$f(t) = \frac{1}{2\pi j} \int_{c-j\infty}^{c+j\infty} F(s) e^{st}\, ds \qquad (A.7)$$

This formula applies to both the one- and two-sided transforms. Note that the integration is along a vertical line in the s plane, and the constant c is chosen such that *the path of integration lies within the strip of convergence of F(s).* Knowledge of the convergence region is the key to getting the correct inverse transform (i.e., time function) in the inversion process. An example will illustrate this.

EXAMPLE A.2

Consider the transform pair shown as the seventh entry in Table A.3. The transform and the strip of convergence are

$$F(s) = \frac{1}{s-a} + \frac{1}{-s+b}; \qquad a < \text{Re}[s] < b$$

The theory says to choose the path of integration between the poles at a and b as shown in Fig. A.2. One could evaluate the integral by treating the complex function as a sum of real and imaginary parts, and then use real variable calculus for each part. This, however, is working the problem the hard way (see Exercise A.1). An easier way is by means of residue theory (5). It can be shown (6) that the addition of an infinite semicircular path to the left as shown in Fig. A.3 contributes nothing to the integral of Eq. (A.7) *for positive t.* (This can be seen intuitively by noting that the real part of s has a large negative value along this path.) Thus, the positive-time part may be evaluated by the method of residues as follows:

$$f(t) = \frac{1}{2\pi j} \int_{c-j\infty}^{c+j\infty} F(s)e^{st}\, dt$$

$$= \frac{1}{2\pi j} \cdot 2\pi j \cdot \left[\text{sum of residues of } F(s)e^{st} \text{ within contour } \Gamma\right]$$

$$= \left[\left(\frac{1}{s-a} + \frac{1}{-s+b}\right)(e^{st})(s-a)\right]_{s=a}$$

$$= e^{at} \qquad (\text{for } t > 0 \text{ only})$$

Similarly, it should be apparent that if we close the contour to the right, we get the negative-time portion of $f(t)$. In this example it is

$$f(t) = -\frac{1}{2\pi j} \cdot 2\pi j \cdot \left[\begin{array}{l}\text{sum of residues of } F(s)e^{st} \text{ within} \\ \text{clockwise contour closed to the right}\end{array}\right]$$

$$= \left[-\left(\frac{1}{s-a} + \frac{1}{-s+b}\right)(e^{st})(s-b)\right]_{s=b}$$

$$= e^{bt} \qquad (\text{for } t < 0 \text{ only}) \qquad \blacksquare$$

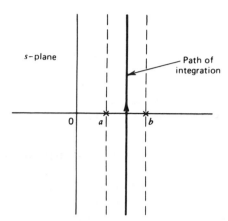

Figure A.2   Convergence strip and path of integration for Example A.2.

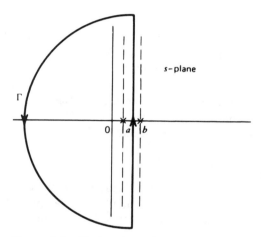

Figure A.3   Closing the path to left to get the positive-time part of $f(t)$.

   The procedure illustrated in Example A.2 can be simplified even further by simply noting that poles of $F(s)$ to the left of the vertical path of integration contribute to the positive-time part of $f(t)$, and those to the right yield the negative-time part. Thus, if $F(s)$ is rational, a simpler inversion algorithm may be stated:

1.   First think of writing $F(s)$ in terms of a partial fraction expansion and grouping terms together in the form

$$F(s) = \begin{bmatrix} \text{Terms with poles to left} \\ \text{of path of integration} \end{bmatrix} + \begin{bmatrix} \text{Terms with poles to right} \\ \text{of path of integration} \end{bmatrix}$$

$$(A.8)$$

Note that $F(s)$ is to be resolved into an *additive* combination, and thus this is not the same as spectral factorization.

2. Then the inverse transform of $F(s)$ is obtained by evaluating the positive- and negative-time parts separately. The first term of Eq. (A.8) yields the positive-time part, and the second gives the negative-time part. Each may be evaluated using one-sided transform methods. [See Exercise A.1 for a discussion of the evaluation of $f(t)$ at $t = 0$.]

Note that if there are no poles to the right of the path of integration (e.g., one-sided Laplace transforms), the inversion integral automatically yields zero for the negative-time part. This is why we must consider $f(t)$ to be zero for negative time in one-sided transform theory, even though the defining integral is insensitive to the negative-time portion of the function. Also note that if the strip of convergence includes the imaginary axis, the two-sided Laplace transform degenerates to the Fourier transform with a simple change in notation (i.e., $j\omega$ is replaced with s). Thus resolving the transform into positive- and negative-time parts is also useful in the evaluation of inverse Fourier transforms.

Our presentation of Laplace and Fourier transforms has been largely an overview, and it was intended to place these transforms in proper perspective relative to each other. Many details have thus been omitted. Also, there are other linear transforms that are of interest in signal analysis, and these have not even been mentioned. Thus, much of linear transform theory has been left unsaid here. For further reading, see Bracewell (7). This is an especially readable book on the subject and it includes many applications.

Exercise A.1

It was stated earlier that the inversion integral automatically yields zero for $t < 0$ in one-sided transform theory. This statement becomes more convincing after actually performing the integration for a sample problem. Consider the one-sided transform pair:

$$f(t) = \begin{cases} e^{-2t}, & t > 0 \\ 0, & t < 0 \end{cases} \tag{A.9}$$

$$F(s) = \frac{1}{s + 2}, \qquad \mathrm{Re}\,[s] > -2 \tag{A.10}$$

Obviously the "strip" of convergence is the entire semi-infinite s plane to the right of $\mathrm{Re}\,[s] = -2$. Thus it includes the $j\omega$ axis. A legitimate inversion integral would then be

$$\text{Inverse of } F(s) = \frac{1}{2\pi j} \int_{-j\infty}^{j\infty} \frac{1}{j\omega + 2} e^{j\omega t}\, d(j\omega) \tag{A.11}$$

(a) Rewrite Eq. (A.11) as the sum of real and imaginary parts, and then integrate each part separately using ordinary real calculus methods. The end result should be just $f(t)$ as stated in the transform-pair statement. [*Hint*: Take advantage of even and odd symmetry wherever possible. Also, you may find it necessary to use tables of integrals in evaluating the resulting real integrals. The tables given in Dwight (8) are adequate for this problem.]

(b) In Eq. (A.9) the value of $f(t)$ was intentionally left undefined at the point of discontinuity, that is, at $t = 0$. Find the value of $f(t)$ at $t = 0$ as dictated by the inversion integral, Eq. (A.11). Does this seem reasonable in terms of Fourier *series* theory? (*Hint*: Evaluate the Fourier series for a square wave at the point of discontinuity and compare the result with that obtained from the inversion integral.)

REFERENCES

1. R. J. Mayhan, *Discrete-Time and Continuous-Time Linear Systems*, Reading, MA: Addison-Wesley, 1984.
2. E. Kamen, *Introduction to Signals and Systems*, New York: Macmillan, 1987.
3. R. C. Dorf, *Modern Control Systems*, 5th ed., Reading, Mass.: Addison-Wesley, 1989.
4. J. J. D'Azzo and C. H. Houpis, *Linear Control System Analysis and Design*, 3rd ed., New York: McGraw-Hill, 1988.
5. R. V. Churchill, J. W. Brown, and R. F. Verhey, *Complex Variables and Applications*, 3rd ed., New York: McGraw-Hill, 1976.
6. S. Goldman, *Transformation Calculus and Electrical Transients*, Englewood Cliffs, N.J.: Prentice-Hall, 1949.
7. R. N. Bracewell, *The Fourier Transform and Its Applications*, 2nd ed., New York: McGraw-Hill, 1978.
8. H. B. Dwight, *Tables of Integrals and Other Mathematical Data*, 4th ed., New York: Macmillan, 1961.

Appendix B

Typical Navigation Satellite Geometry

The exact description of a satellite orbital trajectory can be very complex. A perfectly elliptical orbit can be fully described by six parameters: size and flatness of the ellipse, orientation of the elliptical plane, orientation of the ellipse in its plane, and position of the satellite in the elliptical orbit. These are known as *Keplerian parameters*. To account for external forces other than the ideal satellite–earth interaction, many perturbation correction parameters are included with the standard set of six, listed above, for the purpose of describing the GPS satellite orbits to meet its accuracy requirements. For computer simulation purposes, though, it is seldom important to place the satellites and move them with the kind of detailed precision that must be accounted for in the real physical situation. Although these finer perturbations from the ideal elliptical trajectories may vary up to several hundreds of meters, in relationship to the large satellite–observer distances, they hardly affect the angular relationships that dictate the satellite geometries. It is the latter and its rate of change in time that determine the time-varying observability of the measurement situation, an aspect of GPS that is often the foremost issue addressed by computer simulations.

To aid the reader in running computer simulations of GPS scenarios, we shall derive here equations that analytically describe typical satellite orbits. For simplicity, these orbits are assumed to be perfectly circular with a radius R (26 560 km) and inclined 55 degrees to the equatorial plane. As seen in Figure B.1, the parameter Ω, called the *right ascension*, is the longitude where the orbital plane intersects the equatorial plane as the orbit crosses over from the southern to the northern hemisphere. Although Ω is very nearly constant in inertial space, it changes with respect to an Earth-fixed frame of reference because of the rotation of the Earth. A second parameter θ represents the location of the satellite as the angular phase in the circular orbit using the right ascension as the reference point. For the assumption of a circular orbit, θ changes at a

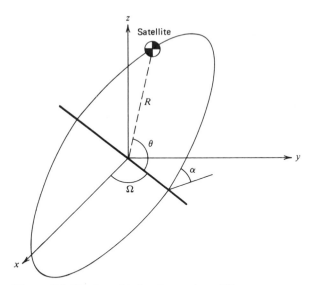

Figure B.1 Circular orbital trajectory ($\alpha = 55°$).

constant rate, which gives rise to a period of approximately 43 082 sec (slightly less than one half a day). Hence, the Earth-Centered Earth-Fixed coordinates for the satellite position in the orbital model of Figure B.1 is given by

$$x = R[\cos \theta \cos \Omega - \sin \theta \sin \Omega \cos 55°]$$

$$y = R[\cos \theta \sin \Omega + \sin \theta \cos \Omega \cos 55°]$$

$$z = R \sin \theta \sin 55°$$

where

$$\times$$

$$\theta = \theta_0 + (t - t_0) \cdot \frac{360}{43\,082} \text{ deg}$$

$$\Omega = \Omega_0 - (t - t_0) \cdot \frac{360}{86\,164} \text{ deg}$$

$$R = 26\,560\,000 \text{ m}$$

By way of concatenating a series of translational and rotational transformations, a direct linear transformation from the Earth-Centered Earth-Fixed coordinate frame to the locally level coordinate frame shown in Figure 10.1 can be obtained as

$$\begin{bmatrix} x' \\ y' \\ z' \end{bmatrix} = \begin{bmatrix} -\sin \theta_u & \cos \theta_u & 0 & x_u \sin \theta_u - y_u \cos \theta_u \\ -\sin \phi_u \cos \theta_u & -\sin \phi_u \sin \theta_u & \cos \phi_u & \begin{matrix} x_u \sin \phi_u \cos \theta_u \\ +y_u \sin \phi_u \sin \theta_u \\ -z_u \cos \phi_u \end{matrix} \\ \cos \phi_u \cos \theta_u & \cos \phi_u \sin \theta_u & \sin \phi_u & \begin{matrix} -x_u \cos \phi_u \cos \theta_u \\ -y_u \cos \phi_u \sin \theta_u \\ -z_u \sin \phi_u \end{matrix} \end{bmatrix} \begin{bmatrix} x \\ y \\ z \\ 1 \end{bmatrix}$$

where $[x_u, y_u, z_u]$ are the observer's coordinates, and ϕ_u and θ_u are the local reference latitude and longitude respectively. One way to use the locally level coordinate frame and yet avoid the complications of the transformation is to adopt as the observer's local reference, the "zero-zero" location ($\phi_u = 0°$; $\theta_u = 0°$) where we simply have $x' = y$, $y' = z$, and $z' = x - r$ (r is the Earth radius − 6 380 km).

Based on a proposed "Optimal" 21-satellite GPS constellation, five satellite trajectories are selected, their parameters tabulated accordingly.

Satellite ID	$\Omega_0(°)$	$\theta_0(°)^a$
1	326	68
2	26	340
3	146	198
4	86	271
5	206	90

[a] θ_0 is the value of θ at the reference time t_0. Ω_0 is the value of Ω at the reference time t_0.

The satellite positions in the locally level coordinate frame chosen above can be written as functions of time as follows:

$$x'(t) = R[\cos\theta(t)\sin\Omega(t) + \sin\theta(t)\cos\Omega(t)\cos 55°]$$

$$y'(t) = R\sin\theta(t)\sin 55°$$

$$z'(t) = R[\cos\theta(t)\cos\Omega(t) - \sin\theta(t)\sin\Omega(t)\cos 55°] - r$$

The components of the unit direction vector can be formulated as functions of time t in the following manner:

$$h_x = \frac{-x'(t)}{\sqrt{x'^2(t) + y'^2(t) + z'^2(t)}}$$

$$h_y = \frac{-y'(t)}{\sqrt{x'^2(t) + y'^2(t) + z'^2(t)}}$$

$$h_z = \frac{-z'(t)}{\sqrt{x'^2(t) + y'^2(t) + z'^2(t)}}$$

The recommended satellite visibility window runs from $t = 0$ to $t = 3600$, a total of 60 min in duration. The combination set $(1, 2, 3, 4)$ provides good GDOP conditions throughout whereas the other combination $(1, 2, 3, 5)$ goes through a GDOP chimney over the same time interval.

REFERENCE CITED IN APPENDIX B

1. G. B. Green, P. D. Massatt, and N. W. Rhodus, "The GPS 21 Primary Satellite Constellation," Navigation, *The Journal of the Institute of Navigation, 36:*1, 9–24 (Spring 1989).

Appendix C

Kalman Filter Software

The main thrust of this book has been about Kalman filtering. In engineering applications it is always the discrete Kalman filter that is implemented—hence, its special importance. It is, of course, a recursive numerical procedure, and therefore many of the exercises that we want to use to reinforce the theory simply cannot be done with paper-and-pencil methods with a reasonable amount of effort. Thus, a diskette containing a package of Kalman filter and related programs accompanies this book. The purpose of this software is to assist the student (and instructor) in working problems that would otherwise have to be avoided because of the drudgery of hand computations. A limited number of these problems have been included as "starred" problems at the end of the chapters of this book. We also encourage students and instructors alike to invent their own exercises. Once you become comfortable with the software package, you will find that all sorts of new problem-solving vistas will open up to you. All of a sudden, what was drudgery will become fun!

Before describing the capabilities of the included software package, we wish to emphasize that its programs were written primarily for instructional purposes. No extraordinary attempt was made to make the software especially efficient, in terms of either speed, program, or memory capacity. Quite to the contrary, "user friendliness" was the prime consideration in the software development. All programs in the software package are menu oriented, and the user is prompted at each step as to what needs to be done to get to the next step. Thus, the user will find that even on a personal computer with modest speed, most assigned tasks will be executed expediently enough for tutorial purposes.

C.1
OVERVIEW OF THE SOFTWARE
CAPABILITIES

Covariance Analysis for Filtering and Smoothing

In covariance analysis, the random input (measurement) and output (estimate) processes are not generated. Rather, the program simply cycles through the

474

error covariance equations recursively beginning with the initial covariance. To do this, of course, the user must specify the four key parameters ϕ_k, Q_k, H_k, and R_k, plus the initial P_0^-. The end result that the user would want to derive from this processing is a sequence of error covariance matrices. These data contain information about the estimation accuracy under optimal conditions. There are separate functions available for filtering and smoothing. The required parameters are identical for both, except that in the latter, the smoothing interval must be specified in addition.

Suboptimal Analysis

It is not difficult to imagine a sequence of filter gains that are suboptimal. They could be the result of a mismodeled situation, or the suboptimal gain sequence might be specified completely arbitrarily with no connection to any model. Regardless, once the gain sequence is specified, the structure of the filter dictates that the estimate update and projection be done in accordance with

$$\hat{x}_k = \hat{x}_k^- + K_k(z_k - K_k\hat{x}_k^-) \tag{C.1}$$

$$\hat{x}_{k+1}^- = \phi_k\hat{x}_k \tag{C.2}$$

The theory allows us to compute the error covariance associated with this operation for *any* arbitrary gain K_k (with the proper dimensions, of course). The general recursive equations for the error covariance (optimal or suboptimal) are

$$P_k = (I - K_kH_k)P_k^-(I - K_kH_k)^T + K_kR_kK_k^T \tag{C.3}$$

$$P_{k+1}^- = \phi_kP_k^-\phi_k^T + Q_k \tag{C.4}$$

The suboptimal analysis program simply implements these equations. Obviously, the user must specify the true ϕ_k, Q_k, H_k, and R_k, which define the structure of the filter. This is sometimes referred to as the "truth" model. The user must also specify the gain sequence K_k. If this comes from an "untruth" model, then this system must be run first to generate the appropriate number of computational steps, the gains must be saved in a data file, and then these suboptimal gains must be used in the suboptimal component program.

Monte Carlo Kalman Filter and Smoother Simulation

In a Monte Carlo filter simulation, a sample realization of the input (measurement) process must first be generated by using the random process generation function described below. Separate Monte Carlo processing functions are available for filtering and smoothing. Therefore, in addition to parameters specified above for the covariance analysis, the user must specify the measurement sequence $\mathbf{z}$ as well as an initial state estimate $\hat{\mathbf{x}}_0^-$. The filter or smoother then yields a corresponding output (estimate) process that constitutes simply one sample realization for the specified number of computational steps. This can, of course, be repeated over and over again with different noisy measurement sequences made from the random process generation. In doing so, the user obtains an ensemble of experiments that may be analyzed statistically. This makes a vivid demonstration that the error covariance sequence associated with the experiment does, in fact, yield meaningful information about the estimation accuracy.

Random Process Generation

In Monte Carlo simulation work we often have occasion to generate a discrete random sequence in state-space form, that is,

$$\mathbf{x}_{k+1} = \boldsymbol{\phi}_k \mathbf{x}_k + \mathbf{w}_k \tag{C.5}$$

$$\mathbf{z}_k = \mathbf{H}_k \mathbf{x}_k + \mathbf{v}_k \tag{C.6}$$

where $\mathbf{w}_k$ and $\mathbf{v}_k$ are Gaussian white sequences. This form allows the generation of any stationary Gaussian process with a given rational spectral function by using the modeling methods of Chapter 5. Clearly, if the user wishes to generate a "noiseless" version of the process, the variance of the parameter $\mathbf{v}_k$ can simply be set to zero. This program requires the user to specify the usual $\boldsymbol{\phi}_k, \mathbf{Q}_k, \mathbf{H}_k$, and $\mathbf{R}_k$ parameters, and also the initial conditions on $\mathbf{x}_k$. The end results are files for both $\mathbf{x}_k$ and $\mathbf{z}_k$, which can then be used in subsequent programs.

Steady-State Characteristic Roots Computation

It was mentioned in Chapter 7 that a Kalman filter behaves much the same as any other constant-parameter digital filter when (and if) the filter gain reaches a steady-state condition. The user needs to specify only the usual $\boldsymbol{\phi}_k, \mathbf{Q}_k, \mathbf{H}_k$, and $\mathbf{R}_k$ parameters, and the program computes the steady-state gain. The end result of this program is the set of characteristic roots of the filter in the z-plane.

State Transition Matrix (STM) and Q Matrix Computations

Calculation of the STM and Q matrices in higher-order systems is an onerous task with paper-and-pencil methods. This program relieves the user of this task. We begin with the continuous model for the system state vector $\mathbf{x}$:

$$\dot{\mathbf{x}} = \mathbf{F}\mathbf{x} + \mathbf{G}\mathbf{u}$$

where $\mathbf{F}$ and $\mathbf{G}$ are constant matrices, and $\mathbf{u}$ is the input vector whose elements are independent unity white noises. The user specifies $\mathbf{F}$ and $\mathbf{G}$ in the model and the step size Δt, and the program does the rest. The end products are data files for both $\boldsymbol{\phi}_k$ and $\mathbf{Q}_k$ that can be used in subsequence programs.

Discrete Fourier Transform (DFT), Inverse DFT, and Periodogram

In signal analysis, it is important to be able to go back and forth between the time and frequency domains. Thus, direct and indirect Fourier transform programs have been included in the software package. Note that the straightforward Discrete Fourier Transform (DFT) is implemented in this software (in contrast to the FFT), so the number of data points is not restricted to an integer power of 2. The DFT takes a little longer to execute than the FFT, but from a tutorial viewpoint it is better because it is less restrictive. In the included programs, the data stream may be of any reasonable length and may be either real or complex in both the time and frequency domains. It is worth noting that the DFT program transforms an N-point time function into an N-point complex function in the frequency domain. If the original time function is real, this of course leads to redundancy in the frequency domain, with points above the foldover frequency being complex conjugates of the corresponding lower-frequency points. We all know that this is excess baggage for the special case of real-valued time-domain functions, but it does no harm to do so and is correct mathematically.

The periodogram of a time signal is also of special interest in signal analysis. Recall from Chapter 2 that the periodogram is proportional to the square of the Fourier transform and that it has a close statistical connection to the autocorrelation function. Therefore, in the DFT program, an additional output is provided that is proportional to the square of the magnitude of the Fourier transform, that is, the periodogram.

Autocorrelation Function

The autocorrelation function is a basic descriptor of a stationary random process. There is, of course, no way that one can determine the true autocorrelation function from just one sample realization of a process. First, there is the matter of the sample length; and second there is the problem of ergodicity. However, if the data span is long relative to the pertinent correlation times, and if we assume that the sample realization at hand is representative of all possible samples (i.e., assume ergodicity), then we can compute something that approximates the true autocorrelation function. This is done simply by multiplying the time function by itself shifted τ units in time, and then averaging. The autocorrelation function program included in this software package does just that. The approximation so obtained does have limitations, though, and these should be recognized, especially when the number of shifts (the τ parameter) is appreciable relative to the data span (see Chapter 2).

Data Graphing

Who was it that said, "A picture is worth a thousand words"? How true this is, especially in analysis work. For this reason, data plotting capability has been built into this software package in two ways. First, in procedures where it is appropriate, the user is provided with a prompt such as "Press ENTER to view the data" after a procedure is executed. This provides a quick look at the data in bar graph form without formal plotting. Second, a more formal data graphing capability is provided in a separate program whereby several data files may be plotted together on the same X-Y or line (unspecified independent variable) graphs.

Note on Learning to Use the Software

"A User's Guide" giving detailed instructions on how to use the accompanying software follows in the next section. After installation of the programs on your computer, we suggest that you first become familiar with the various features provided in this software package by working simple examples of your own creation. For example, to check out the DFT procedure, you might try creating a data file for a simple 16-point unit impulse in the time domain. Then take the DFT of this function (which we know to have a constant magnitude); and, finally, form the inverse of the DFT to see if you get the unit impulse back again (within roundoff error, of course). In this way, you not only become familiar with the various procedures, but you gain a degree of confidence that the programs are doing the job and giving the correct answers.

You are now ready to begin using the accompanying software package. Be forewarned that a modest amount of practice is necessary before you will begin to feel completely comfortable with the software. This is true with any new software package, and it is to be expected. We trust that the "learning curve" will be brief. Once you get past the initial familiarization phase, we think you will find that *Kalman filtering can be fun*!

C.2
USER'S GUIDE FOR KALM 'N SMOOTH (VERSION 1.2)

Hardware Requirements

The programs provided will run on an IBM PC, PS/2, XT, or AT (or a 100 percent compatible computer) with 640 KBytes of internal memory. Most of the standard display adapters and printers are supported. Color displays are recommended. DOS Version 3.2 or later should be used.

Installing Programs

The programs can be executed from the floppy disk provided or from a hard disk. If it is desired to have the programs from your original diskette transferred to another diskette or installed onto a hard disk, simply copy all the programs over with any file-copying utility. In DOS, the file-copying utility COPY is suitable for such a task. *Important:* All programs and files furnished on the original diskette must be located on the same directory path. If installing on a hard disk, a (program's) subdirectory should be set up on the hard disk for the purpose of containing the furnished programs and files. For efficient management, it is recommended that a separate (work) subdirectory be set up on the hard disk for new data files created in the process of using these programs. Starting from this work subdirectory, the needed program can be executed by properly specifying the proper path and the appropriate program name. The following example shows the execution of the program KALM, which resides in the programs subdirectory named PROGS, while all data files specified during execution will by default be stored or read from the work subdirectory named KALMDATA:

```
C:\KALMDATA>\PROGS\KALM
```

Programs and Files

On the 720-KByte 3.5-inch diskette furnished with the book, there are a total of 13 files and a subdirectory of files intended for a demonstration. Of those in the main directory, four are executable programs that work together to provide Kalman filter–related functions: KALM.EXE, PROCESOR.EXE, EIGNPROC.EXE, STMQPROC.EXE. A fifth executable program, N-LIES.EXE, contains simple utilities intended to provide the user some additional analysis and graphing capabilities. In addition to these programs, the diskette also contains several display drivers for some of the more common graphics standards available today for IBM-compatible PC systems. These driver files are identified by the extension .BGI:

CGA.BGI	For IBM CGA or MCGA
EGAVGA.BGI	For IBM EGA or VGA
HERC.BGI	For Hercules Monochrome
ATT.BGI	For AT&T 6300 (400 line)
PC3270.BGI	For IBM 3270 PC
IBM8514.BGI	For IBM-8514

The capability for printing onscreen-generated graphics is also supported. The two files associated with this capability are named DEFFONT.GLF and PRINTER.CFG. PRINTER.CFG is an ASCII text file that contains user-definable parameters pertaining to the desirable printer and printing configurations. Prior to doing any printing with these programs, PRINTER.CFG must be properly set up. See the section on Printer Configurations for instructions to do this, using the Data Editor program function provided with the N-LIES program.

The collection of files in the subdirectory GPSDEMO of the supplied diskette is meant to be used for a self-guided demonstration of the software operation. Instructions for this demonstration can be obtained by printing out the ASCII file READ.ME.

KALM

The main program KALM.EXE handles input data files and runs in conjunction with several processing programs to provide Kalman filtering and other utility functions. They comprise the following:

PROCESOR.EXE   Handles all the Kalman filtering functions including generation of random processes.

EIGNPROC.EXE   Computes eigenvalues and eigenvectors of the state space model and also the steady-state covariance of a time-invariant state model.

STMQPROC.EXE   Computes the state transition matrix and the process noise covariance matrix for a discrete-time model from a continuous-time equivalent.

There is a limit to the size of the Kalman filter that can be modeled with the KALM 'N SMOOTH programs. The state dimensionality must be no larger than 20. The limit on the dimensionality of the measurement space is also 20. Therefore, the largest H matrix allowable is 20×20.

The two principal data elements utilized here to manage the flow of information are the Parameter and Data files. They are described in detail in the discussion that follows.

Parameter File   By convention, the full set of Kalman parameters that may be used consists of the following, although in certain cases, some are not necessary:

1. State Transition Matrix (Φ)
2. Process Noise Covariance Matrix (Q)
3. Measurement Matrix (H)
4. Measurement Noise Covariance Matrix (R)
5. Initial State (x_0)
6. Initial Error Covariance (P_0)
7. Measurement (z)
8. State Output (x)

In addition to the names of the Data Files for the Kalman parameters above, a Parameter File also records several other items, including the mode of data entry for the respective Kalman parameter as well as the dimensions of the state and measurement spaces. Although these parameters are important for the data processing that follows, they are primarily intended for housekeeping functions by the programs that use them and should really be transparent to the user.

Data File   A Data File is an ASCII text file that allows the user to input information for a particular Kalman parameter. There are three standard modes of data entry into the processing Kalman filter: (1) constant, (2) file, and (3) equation (see Mode of Data Entry). In each Data File, the mode of data entry must be specified, along with the dimensions of the matrix representing the parameter. Both are located in the very first line of each Data File. The dimensions of the matrix must be delimited by brackets and separated (number of rows versus columns) by a colon. If the parameter represented is a vector instead, the dimension may be represented as a single number without use of the colon. The data that follow are read in sequence across the columns of the matrix and down the rows. *Comments*, which are ignored by the data processing, can be added to each line by placing them to the right of a colon. Equations used in the equation mode of data entry must be in the form of function of the cycle time, t or T (see Examples 3 and 4 below). The following are examples of various formats that can be used for a Data File:

1. A Data *File* with 2×3 matrices to be read at each iteration, with comments:

   [2:3] File
   3.2   1.1   4.8
   2.0   8.2   5.2E-2:0
   3.1   1.1   5.0
   2.5   8.4   5.2E-2:1
   3.0   1.1   5.1
   3.0   8.4   5.2E-2:2
       .
       .

2. A Data File with 2×3 *constant* matrix:

   [2:3] Constant
   3.2   1.1   4.8
   2.0   8.2   5.2E-2

3. A Data File with 2×2 *equations* matrix using semicolon separators:

   [2:2] Equation
   k1 = 2*t + 3;   k2 = cos (2*pi*T)

$1$1 = k1*k2; $1$2 = k1 + 4

$2$1 = t; $2$2 = 3

4. A Data File with 2 × 2 *equations* matrix with separate lines:

[2:2] Equation

k1 = 2*T + 3

k2 = cos (2*pi*T)

$1$1 = k1*k2

$1$2 = k1 + 4

$2$1 = T

$2$2 = 3

5. A Data File with 3 × 3 *symmetric* matrix:

[3:3] Constant

3.2

2.0   8.2

7.9   2.5   0.5

6. A Data File with 4-tuple *constant* vector:

[3:1] Constant

3.2

2.0

7.9

7. A Data File with 4-tuple *constant* vector:

[3] Constant

3.2

2.0

7.9

Starting Up

To execute the main program, type KALM followed by a carriage return. If executing from a directory not containing these programs, a directory path must be placed in front of the program name KALM. The first screen that appears is the cover screen. The cover screen sequence can be interrupted by hitting any key. Following that, a series of menus will appear. The commands that are solicited from the user will, for the most part, be obvious. Where one command is to be selected from a number of alternatives, the highlighted character describing that particular command should be entered from the keyboard in order to select it.

Choosing Operating Modes

After the brief appearance of the cover screen, which can be interrupted by hitting any key, the "Functions Menu" is presented for the user to select an operating mode and specify a parameter file name (see below). From this menu, either a program function can be selected by entering a number **1, 2, 3, 4, 5, 6, F1, F2,** or **F3** according to the menu below, or a parameter file name may be specified. If a parameter file containing names of the data files needed was created in a previous run and is needed for the current run, simply type the name of the parameter file from the keyboard. When the full file name has been entered, complete the entry with a carriage return. To aid the user with a list of available files, directory assistance should be consulted (see **On-Line Directory Help.**) A program function can be selected without first specifying a parameter file name. A selection can also be made with the name of a file that does not yet exist; a new parameter file will be created with that name. If the parameter file chosen contains a valid operating mode, it will be displayed on the screen (see below).

```
SELECT PROGRAM FUNCTION:

    1.  Kalman Filtering        ⎫
                                 ⎬  Covariance Analysis
    2.  Kalman Smoothing         ⎪
                                 ⎭
    3.  Suboptimal Analysis

    4.  Kalman Filtering        ⎫
                                 ⎬  Monte Carlo Processing
    5.  Kalman Smoothing         ⎪
                                 ⎭
    6.  Random Process Generation

   F1.  STM/Q Computations      ⎫
                                 ⎬  Miscellaneous Utilities
   F2.  Steady-State Computations⎪
                                 ⎭
   F3.  N-LIES Analysis Program

NAME OF PARAMETER FILE:              MODE TYPE:

                                     xxxxxxxxxx
_____

[ESC] To Exit Program                [ENTER] To Choose Selected
                                     Mode Type
```

With a specified operating mode, a carriage return [ENTER] will confirm the choice of the operating mode. Regardless, any one of the other operating modes can still be chosen to override the one shown.

Parameter Menu

The data entry menu that pops up on the selection of a program function consists of several boxes, each allocated for a Kalman filter parameter that is needed for the chosen program function. The boxes in Areas A, B, C, D, E and F (see picture below) are reserved for the entry of data file names corresponding to parameters required for the chosen program function. As such, the number of boxes used for each program function will vary accordingly. Each box will contain the name of the parameter it represents.

Each box can be chosen by using the cursor or arrow keys ↓, ↑, ←, → to move the highlighted box to the appropriate location. When a box is highlighted, a data file name can be entered from the keyboard. This entry from the keyboard is completed by a carriage return. To obtain a reference to available file names, see **On-Line Directory Help**. A carriage return made via the keyboard but not entered by the user as part of a file name entry will invoke a screen-oriented data editor (see next section for Data Editor). If the data file to be edited corresponds to a parameter that is a covariance matrix, a reminder will flash momentarily in the box, accompanied by a beeping tone, indicating that the matrix (if the parameter is multidimensional) is *symmetric* (see Mode of Data Entry for special symmetric matrix format).

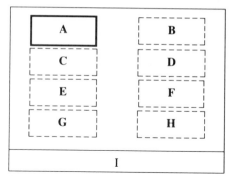

Data File manager screen.

To quit the program from the top-level menu, the ESC key must be used. To proceed from the Data File Manager after all data file names have been entered, Ctrl-Enter should be used. The validity of the data files specified is verified, and the matrix dimensions of the Kalman parameters are checked for consistency. If any data file name entry is new or changed, the user will be asked if the parameter file needs to be saved or updated. This prompt is made in Area I at the bottom of the screen (see picture). If everything is in order, the chosen processing program is then executed. To exit the Data File Manager and return to the top-level menu, simply use the ESC key.

On-Line Directory Help

On-line directory help is available at two areas of file name input—the Functions Menu and the Parameters Menu. In either case, the on-line directory help is invoked by typing a string label that includes the wild card character *, e.g. *.* or *.PRM. The directory list shows up in a window at the bottom of the screen. The highlighted file name can be changed by using the arrow keys. After the correct one is selected, it is selected by hitting the [ENTER] key.

Directory Help In The Functions Menu. The list of files first shown when invoked belongs to the current directory from which the program KALM was executed. If the file being sought resides in another directory, the current directory must be changed. To do this, simply select the entry located at the end of the list marked ''\..''. this changes the current directory to the parent directory (one level closer to the root directory). For example, if C:\PROGS is originally the current directory, choosing \.. changes it to C:\.

Directory Help In the Parameters Menu. The list of files first shown when invoked belongs to the current directory. If the file being sought resides in another directory, the directory list can be changed. To do this, select the entry located at the end of the current list marked ''\..''. Unlike the Functions Menu, this does not change the current directory but merely the directory list that is shown. If the file ultimately selected is from the current directory, only the file name will be annunciated (e.g. TRIAL.STM). However, if it is selected from a directory other than the current, the full path name for the file is also annunciated with its name (e.g. C:\PROGS\TRIAL.STM).

Data Editor

When the data editor is invoked, an entirely new screen will show up. The data editor will create a new file with the specified name if that data file does not already exist. Otherwise, the specified file is brought up on-screen to be edited. The menu of commands for the data editor is given below. Of these, the most notable command is "Save and exit," which is accomplished by either Ctrl-K X or F2. To abandon a file, or to quickly exit an untouched one, use Ctrl-K Q.

Function	Keystrokes
Character left	Left arrow or Ctrl-S
Character right	Right arrow or Ctrl-D
Word left	Ctrl-Left arrow or Ctrl-A
Word right	Ctrl-Right arrow or Ctrl-F
Line up	Up arrow or Ctrl-E
Line down	Down arrow or Ctrl-X
Scroll up	Ctrl-W
Scroll down	Ctrl-Z
Page up	PgUp or Ctrl-R
Page down	PgDn or Ctrl-C
Beginning of file	Ctrl-PgUp or Ctrl-Q R
End of file	Ctrl-PgDn or Ctrl-Q C
Beginning of line	Home or Ctrl-Q S
End of line	End or Ctrl-Q D

Function	Keystrokes
Top of screen	Ctrl-Home or Ctrl-Q E
Bottom of screen	Ctrl-End or Ctrl-Q X
Go to line	Ctrl-J L
Go to column	Ctrl-J C
Top of block	Ctrl-Q B
Bottom of block	Ctrl-Q K
Jump to marker 0..9	Ctrl-Q 0..Ctrl-Q 9
Set marker 0..9	Ctrl-K 0..Ctrl-K 9
Previous cursor position	Ctrl-Q P
New line	Enter or Ctrl-M
Insert line	Ctrl-N
Insert control character	Ctrl-P
Tab	Tab or Ctrl-I
Delete current character	Del or Ctrl-G
Delete character left	Backspace or Ctrl-H
Delete word	Ctrl-T
Delete to end of line	Ctrl-Q Y
Delete line	Ctrl-Y
Find pattern	Ctrl-Q F
Find and replace	Ctrl-Q A
Find next	Ctrl-L
Abandon file	Ctrl-K Q
Save and continue edit	Ctrl-K S
Save and exit to DOS	Ctrl-K X or F2
Save to file	Ctrl-K N
Add window	Ctrl-O A or Shift-F3
Next window	Ctrl-O N or F6
Previous window	Ctrl-O P or Shift-F6
Resize current window	Ctrl-O S
Begin block	Ctrl-K B or F7
End block	Ctrl-K K or F8
Copy block	Ctrl-K C
Move block	Ctrl-K V
Delete block	Ctrl-K Y
Hide block	Ctrl-K H
Mark current word as block	Ctrl-K T
Read block from file	Ctrl-K R

Data Files

At the top-level menu, the program function is chosen to specify one of six available operating modes. In the listings below, the parameters tagged with a {#} are output in nature. For these, a file name will need to be specified even though a data file associated with that name may not actually yet exist. Certain parameters tagged with a {*} are considered optional in their specification; if specified, the corresponding data files must be valid and correct. Those parameters that are tagged with both symbols {#,*} are considered output optional. If a data file name is specified, here again, the data file with that name need not actually yet exist.

1. *Covariance Analysis (for Kalman Filtering and Smoothing).* The difference between this mode and Monte Carlo Processing mode described before is that no measurements are assimilated here and, hence, no states are estimated. The Kalman filter computes only the error covariance matrix and parameters associated with it. The error covariance (P) is the output optional parameter. The data parameters requested in this mode are

   (a) State Transition Matrix (Φ)
   (b) Process Noise Covariance (Q)
   (c) Measurement Connection (H)
   (d) Measurement Noise Covariance (R)
   (e) Error Covariance (P) {#,*}
   (f) Initial Error Covariance (P_0)

2. *Suboptimal Analysis.* A covariance-type analysis is implied here. The data entry configuration is the same as that for the covariance analysis mode except for the addition of the suboptimal gain parameter. Note that if a data file of filter gains generated by a previous Kalman filter run is used, it must first be modified at the very first line of the data file with the proper specification of the data entry mode and dimensionality of the gain matrix. This can be accomplished by using the data file editor utility provided in the N-LIES package (discussed later). The error covariance (P) is the output optional parameter. The data parameters requested in this mode are

   (a) State Transition Matrix (Φ)
   (b) Process Noise Covariance (Q)
   (c) Measurement Connection (H)
   (d) Measurement Noise Covariance (R)
   (e) Error Covariance (P) {#,*}
   (f) Initial Error Covariance (P_0)
   (g) Suboptimal Gain (K)

3. *Monte Carlo Processing (for Kalman Filtering and Smoothing).* In this program function, the Kalman filter processes noisy measurement data to simulate a sample realization that may be part of an ensemble of runs

designed to determine the statistics of the estimation process. The state estimate (x) is an optional output parameter, whereas all the others are required input parameters. The data parameters requested in this mode are

(a) State Transition Matrix (Φ)

(b) Process Noise Covariance (Q)

(c) Measurement Connection (H)

(d) Measurement Noise Covariance (R)

(e) Initial State Estimate (x_0)

(f) Initial Error Covariance (P_0)

(g) Measurement Vector(s) (z)

(h) State Estimate (x) {#,*}

4. ***Random Process Generation.*** This is a non-Kalman filter mode but uses the same model structure as that of a Kalman filter. It generates Monte Carlo data according to the specified state model. The measurement generated (z) is a required output parameter, whereas the state generated (x) is an optional output parameter. The initial state (x_0) and the initial covariance (P_0) are optional input parameters. The initial state and initial covariance parameters define the mean and covariance of the Gaussian distribution on which the randomized initial state vector is based. If either of the two parameters is left unspecified, it is assumed to be zero. Clearly, if P_0 is defaulted to zero, the resulting state is deterministic and set exactly to the value specified by x_0. And if both are unspecified, the initial state vector is zero. The data parameters requested for in this mode are

(a) State Transition Matrix (Φ)

(b) Process Noise Covariance (Q)

(c) Measurement Connection (H)

(d) Measurement Noise Covariance (R)

(e) Initial State (x_0) {*}

(f) Initial Covariance (P_0) {*}

(g) Measurement Generated (z) {#}

(h) State Process Generated (x) {#,*}

5. ***Steady-State Computations.*** The characteristic roots of the specified time-invariant state model is computed here. Another product of this program function is the computation of the steady-state error covariance matrix. The parameters requested to specify the state model are

(a) State Transition Matrix (Φ)

(b) Process Noise Covariance (Q)

(c) Measurement Connection (H)

(d) Measurement Noise Covariance (R)

6. ***STM / Q Computations.*** The state transition matrix and the process noise covariance matrix are computed here from a continuous-time state model. The finite-time increment needed for the conversion from a continuous to a discrete-time model is also requested. The input parameters are

(a) State Coefficient Matrix (F)
(b) Noise Coefficient Matrix (G)

Mode of Data Entry

For each of the program functions described above, the data parameters listed must be specified as a data file. The data entry can be accomplished in one of three ways:

1. ***Constant.*** The data for the parameter are read once at the start of the run from the specified data file and then assumed constant for the remainder of the run. The first line specifies the dimension of the matrix and the "Constant" mode of data entry. It should be in the format "[xx:yy] Constant" where the matrix has xx rows and yy columns. In the subsequent lines, any two elements of data on the same line must be separated by a space. Comments placed to the right of a colon for a particular line are ignored. If the parameter is a covariance-type matrix, a reminder message shows up before the editor is invoked that the matrix to be entered is symmetric. For symmetric matrices, it is important that only the *lower triangular* portion of the matrix be entered. Otherwise, if not followed, the sequence of data read by the program will be incorrect.

2. ***File.*** The data for the parameter are read from the specified data file once every iteration of the processing run. The first line specifies the dimension of the matrix and the "File" mode of data entry. It should be in the format "[xx:yy] File" where the matrix has xx rows and yy columns. In the subsequent lines, as before, any two elements of data on the same line must be separated by a space. Comments placed to the right of a colon for a particular line are ignored. For symmetric matrices, it is important that only the *lower triangular* portion of the matrix be entered.

3. ***Equation.*** The data for the parameter are evaluated once every iteration of the processing run from a set of equations that is read once at the start from a file. The algebraic description can be broken up into several lines, each being no more than 80 characters long. Each equation line must be assigned ($=$) to a variable name. Variable names beginning with the character "$" are reserved for final assignments to the appropriate matrix component. For instance, the equation line

$$\$2\$3 = \cos(6*\mathrm{pi})$$

assigns the value of $\cos 6\pi$ to the component in the 2nd row and 3rd column of that matrix. To accommodate long equations or equations with repetitive terms, intermediate assignments to arbitrarily defined variable names are also allowed. For example,

$$\mathrm{temp1} = 6*\mathrm{pi}$$

$$\$2\$3 = \cos(\mathrm{temp1})$$

Variable names may mix alpha with numeric characters but must begin with an alpha as the first character. Equation lines are separated by being placed on different lines or, if on the same line, by a semicolon (;). The standard arithmetic rules of priority in operation as well as the requirement of matched pairs of parentheses will apply. For a function that carries an argument, the argument must be isolated within parentheses, for example, sqrt ($3*t + 1$). Values preceded by a sign must also be isolated with parentheses unless the sign is the first character of an equation. The following are some syntax examples.

Correct	Incorrect
A3 = 5*(- t)	A3 = 5* - t
$1$3 = 14 + (- 36*t^(- 2))	$1$3 = 14 + -36*t^- 2
x_out = -cos(3*t)	x - out = -cos(3*t)

The mathematical functions available for the Equation mode of data entry and their corresponding symbols are listed as follows (uppercase characters are allowable):

Function	Symbol
addition	+
subtraction	-
multiplication	*
division	/
exponential	^
modulus	mod

Function	Symbol
square	sqr
square root	sqrt
absolute value	abs
integer	int
sine (radians)	sin
cosine (radians)	cos
tangent (radians)	tan
arcsine	asin
arccosine	acos
arctangent	atan
natural logarithm	ln
natural exponential	exp
base-10 logarithm	log
π	pi

Data Processing Screen

A new screen appears when the data processing program is invoked. In all, there are five general areas where output information is displayed. Of the five areas, three can be reconfigured by the user, whereas the other two cannot (see figure). The latter consist of a partitioned area (A) at the top of the screen depicting cycle/time information, and an "information bar" (E) at the very bottom of the screen. The remaining three (B, C, and D) show updated information on the Kalman filter state and covariance. The current information window is highlighted and the cursor keys, ← , → , ↓, can be used to access the areas B, C, and D, respectively. The five areas are described below in detail.

A. Header Area   The information displayed in the Header Area panel of this screen includes the following:

1. *Running step#*. The iteration step number that starts at zero and is incremented by one at the end of each iteration.
2. *Display step#*. The iteration step at which the information being displayed in Areas B, C, and D *and* to any open output files was last updated.
3. *Display refresh rate*. The display refresh rate at which the information displayed in Areas B, C, and D is updated, in terms of iteration step (also determines the rate at which data are written to the output files).
4. *Interrupt at step#*. The target iteration step to interrupt the processing on. If it is not shown in the Header Area display, then no target iteration has been set. This parameter can be set or changed from Control Area E. When the processing is interrupted at the specified

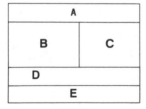

Screen partitions.

step, a prompt will appear in Area D:

** Run Interrupted **

Continue? [Y/N]

If "Y"(es) is chosen, the program will continue but pauses immediately after, to allow the user the choice of resetting the next Iteration Step to interrupt on. If "N"(o) is chosen instead, the program reverts back to the Data File Manager menu.

B. Filter State, Variance, and Gain Information Area   In this Information Area, there are four columns of information:

1. The index of the state variable in the state vector.
2. *A priori* (prior) or *a posteriori* (updated) or *smoothed* states if applicable (applies to Monte Carlo processing only).
3. *A priori* (prior) or *a posteriori* (updated) or *smoothed* variances if applicable.
4. Filter gains.

A menu bar at the bottom of the window allows the user to configure the display in this area (if the measurement situation is nonscalar, then there are filter gain vectors for each of the components in the measurement vector):

P —*a priori* (prior) information

U—*a posteriori* (updated) information

S —smoothed information (for smoothing functions)

<—decrements the index of the measurement vector

>—increments the index of the measurement vector

C. Miscellaneous Information Area   The information in this area is displayed in two columns, the first of which is the index of the matrix or vector data that is shown in the second. The data shown can be selected from the following:

Error Covariance—off-diagonal terms of the error covariance matrix

Residuals—measurement residuals vector

Residual Variance—variance of the residuals vector

Residual Covariance—covariances of the residuals vector

D. Control Panel   There are four changeable parameters to configure the display window or the Kalman filter. For each, there are several suboptions to choose from. All available options to change are listed (suboptions are indented):

    Display Parameters

        Refresh Rate

        Interrupt Point

    Filter Parameters

        Φ (enter "**f**" or "**F**") for the state transition matrix

        w for the process noise covariance matrix

        H for the measurement matrix

        v for the measurement noise covariance matrix

    Output Files List

        1. a priori est.

        2. updated est.

        3. kalman gain

        4. a priori error cov.

        5. updated error cov.

        6. measurement residual

        7. residual covariance

    Single Step Toggle—turns on the single-step mode

E. Status Bar   When the processing is in pause, the message **[ESC] to Resume** is displayed at the bottom left-hand corner of the screen. When the processing is in progress, the message **[F1] to Interrupt** will appear instead in place of the previous one. The SINGLE STEP message appears on the bottom right-hand corner of the screen when the single-step mode is invoked. When the processing is paused after the first iteration, the message $\leftarrow \downarrow \rightarrow$ **To Change Panels** appears at the center of the Status Bar to indicate that Area B or Area C can be accessed to change what is displayed, or Area D accessed to change parameter settings.

    If there are more than 11 elements of the matrix or vector displayed in either Area B or Area C, then only 11 will be shown. To see the rest, a message appears at the bottom of the screen instructing the user to tap the SPACE BAR in order to scroll through the rest of the elements.

Data Processing (Utility Functions)

Steady-State Computations   In this mode, the following menu appears. The OUTPUT options do not show up at the onset until the appropriate computations have been carried out, whereupon the corresponding output data are then available.

```
COMPUTE

    Error Covariance (Steady State)

    Kalman Gain (Steady State)

    Roots of the Kalman Filter
```

```
OUTPUT
    A. Computed Error Covariance
    B. Computed Kalman Gain
    C. Computed Kalman Filter Roots
QUIT
```

STM / Q Computations   In this mode, the time increment between successive iterations must be specified with the following prompt:

TIME INCREMENT (in seconds)?

Then the following menu appears. Here also, the SAVE options do not show up at the onset until the appropriate computations have been carried out, whereupon the corresponding output data are then available.

```
COMPUTE
    State Transition Matrix
    Process Noise Covariance
SAVE
    A. State Transition Matrix
    B. Process Noise Covariance Matrix
QUIT
```

N-LIES

A program that can be used for analysis and graphing is also included with the software package under the executable file name N-LIES. This program provides a Discrete Fourier Transform (or its corresponding inverse) and an Autocorrelation Function to process data that is input via a text data file. The results can be displayed on the screen. An additional function to graph or edit the data file itself is also included.

Starting Up

To directly execute the program, type N-LIES followed by a carriage return. If executing from a directory not containing the program, a directory path must precede the program name N-LIES. The first screen that appears is the cover screen. If N-LIES is executed from KALM, no cover screen will appear. Following that, a series of menus will appear. The commands that are solicited from the user will, for the most part, be obvious. Where one command is to be selected from a number of alternatives, the highlighted character describing that particular command should be entered from the keyboard in order to select it.

Choosing Program Function

After a brief appearance of the cover screen, a top-level menu is presented for the user to specify a data file name and to select a program function:

```
SELECT PROGRAM FUNCTION:

     1.   Discrete Fourier Transform
     2.   Inverse Discrete Fourier Transform
     3.   Autocorrelation Function
     4.   Graphing Data
     5.   Edit Data File
     6.   Data Extraction


NAME OF INPUT DATA FILE:
_____


[ESC] To Exit Program
```

In this menu, either a program function can be selected by entering number **1, 2, 3, 4, 5,** or **6** or a data file name may be typed from the keyboard (see **On-Line Directory Help** for file listing). When the full file name has been entered, complete the entry with a carriage return. A program function cannot be selected without first specifying a data file name. If the selection is done with the name of a data file that does not exist, the name will be erased from the screen to indicate to the user that it is not a valid choice.

Initializing the Data File. If any of the functions other than #5 is chosen above with a proper file name, a sample of the data file will be displayed in an upper window of the screen while the following series of prompts will show up in the lower window requesting the appropriate user response via the keyboard:

Number Of Lines To Skip (Hit [ENTER] if none)?
Number Of Lines To Read (Hit [ENTER] for all)?
Number Of Lines To Periodically Jump (Hit [ENTER] if none)?
Hit [ENTER] to read data or [R] to read square root of data

This allows the data file to be "sampled" in the sense of extracting lines of data at a specified frequency. Consider, for example, the following symmetric 2x2 matrix:

```
1.432 :[1][1] 0
3.212 :[1][2] 0
0.231 :[2][2] 0
1.400 :[1][1] 1
2.988 :[1][2] 1
0.100 :[2][2] 1
1.402 :[1][1] 2
2.554 :[1][2] 2
0.002 :[2][2] 2
    •
    •
```

For instance, in order to read only the 2,2-term of the covariance matrix, every third line of the above example data file should be sampled starting with the third line. The appropriate response in this case would be to "skip" 2 lines, "read" all lines, and "periodically jump" 2 lines. The last option allows the user to read the data in as a square root function of the original data in the file. This provides a means to plot the variance data as a graph of standard deviation.

1. *Discrete Fourier Transform.* When the Discrete Fourier Transform program function is chosen and after the data file is initialized, the following series of prompts will appear:

   There are xx columns of data

   ```
   Enter column # for data samples [real]-
   Enter column # for data samples [imaginary]-
   <Hit [ENTER] to ignore>

   Total number of samples to evaluate [xxx]?
   Output File Name (Hit [ENTER] if none)-

   CHOOSE ONE
   [ENTER] to view Discrete Fourier Transform
   [CTRL-ENTER] to output graph to printer
   [ESC] to exit from program
   ```

   The input data can be complex numbers. If so, these numbers must be generated in the rectangular form with the real and imaginary components placed in separate columns.

2. *Inverse Discrete Fourier Transform.* The Inverse Discrete Fourier Transform program function prompts a series of questions that are identical to that for the Discrete Fourier Transform given in the foregoing discussion.

3. *Autocorrelation Function.* When the Autocorrelation Function program function is chosen and after the data file is initialized, the following series of prompts will appear:

   There are xx columns of data

   ```
   Enter column # for data samples-
   Enter column # for Time Mark -
   <Hit [ENTER] for Auto Time Mark>

   Total number of samples to evaluate [xxx]?
   Output File Name (Hit [ENTER] if none)-
   Total Number of Lags to be computed (<xx)-
   ```

```
CHOOSE ONE
[ENTER] to view Autocorrelation Function
[CTRL-ENTER] to output graph to printer
[ESC] to exit from program
```

If the Time Mark is not associated with any data column, the indexing for the Time Mark is automatically generated.

4. *Graphing Data.* On choosing this program function and after the data file is initialized, the following menu for the graphing program function appears:

```
CHOOSE ONE—File #1
    Time Mark (auto-forward)
    Plot A (column *)
    Plot B (column *)
    Plot C (column *)
    X-axis (Auto Scaling)
    Y-axis (Auto Scaling)
    Format (Lines)
    Labels
    [ENTER] to view graph
    [CTRL-ENTER] to print
    [↓] to set up File #2
    [ESC] to exit program
```

This program function allows the user to generate several overlaying plots of the X-Y type. Up to four separate data files may be opened for data input. On-line directory help is available. The first file is already specified at the top-level menu. Then, using [↓] allows the user to open a second file. At this point when in the File #2 menu, [↑] brings back the preceding File #1 menu. To open a third file, the menu for File #2 must be showing (use [↓] to get from the menu of File #1 to that of File #2). Then, [↓] automatically opens the new file with a request of the file name. This can be repeated for a fourth data file. Each file is allotted one independent variable (Time Mark), and three dependent variables (A, B, and C). Therefore, up to three plots can be specified for each data file. Note here that if a data file contains more than three sets of data to be plotted, it can be respecified as a second data file with the same name to obtain space for three more plots. To generate a plot, the column in the current data file to be plotted must be properly identified, where at least one of the dependent variables must be associated with a data column. If the Time Mark is not associated with any data column, it is assumed to be automatically generated (auto-forward). In other words, an incrementing index is automatically associated with each data point in the associated data column. To select an automatically decrementing index, select a value of -1 for the Time Mark (auto-reverse). This is useful for viewing output files generated by smoothing functions. Other graphing functions that are available on the menu are the scaling (manual or automatic) of the axes, pasting labels on the graph, and the format of the graph, whether lines only, symbols only, or both lines and symbols.

5. *Edit Data File.* The input data file can be edited using a screen-oriented editor by choosing this program function. The file specified can be either an existing or a new one. The menu of commands is given in a table in the Data Editor section. This editor can also be used to install the appropriate printer and printing configurations in the file PRINTER.CFG. See the section Printer Configurations for instructions on how to do so.

6. *Data Extraction.* When this program function is chosen and the data file is initialized, the following prompt will appear:

Output File Name:

The "extracted" file specified by the responses to the prompts when the data file was first read in will be stored under the above output file name.

Printer Configurations

The ASCII text file PRINTER.CFG contains the user-defined parameters for setting up the desirable printer and printing configurations. It consists of only three lines, an example of which is given as follows:

EpsonFX
Medium Resolution
Landscape

The entries may be made up of a mix of upper- and lowercase characters. The first parameter specifies the type of printer used. The options available for this parameter are listed as follows:

1. Epson9
2. EpsonMX
3. EpsonRX
4. EpsonFX
5. EpsonLX
6. EpsonLQ
7. IBMPro
8. IBMQuiet
9. NEC24
10. Toshiba24
11. HPLaserJet
12. HPDeskJet

The second parameter specifies the resolution of the printing mode. The options available for this parameter are:

1. Low resolution
2. Medium resolution
3. High resolution

The third parameter specifies the orientation of the printed output. The options available for this parameter are:

1. Landscape
2. Portrait

Index